Communications
in Computer and Information Science　　　2058

Rationale

The CCIS series is devoted to the publication of proceedings of computer science conferences. Its aim is to efficiently disseminate original research results in informatics in printed and electronic form. While the focus is on publication of peer-reviewed full papers presenting mature work, inclusion of reviewed short papers reporting on work in progress is welcome, too. Besides globally relevant meetings with internationally representative program committees guaranteeing a strict peer-reviewing and paper selection process, conferences run by societies or of high regional or national relevance are also considered for publication.

Topics

The topical scope of CCIS spans the entire spectrum of informatics ranging from foundational topics in the theory of computing to information and communications science and technology and a broad variety of interdisciplinary application fields.

Information for Volume Editors and Authors

Publication in CCIS is free of charge. No royalties are paid, however, we offer registered conference participants temporary free access to the online version of the conference proceedings on SpringerLink (http://link.springer.com) by means of an http referrer from the conference website and/or a number of complimentary printed copies, as specified in the official acceptance email of the event.

CCIS proceedings can be published in time for distribution at conferences or as post-proceedings, and delivered in the form of printed books and/or electronically as USBs and/or e-content licenses for accessing proceedings at SpringerLink. Furthermore, CCIS proceedings are included in the CCIS electronic book series hosted in the SpringerLink digital library at http://link.springer.com/bookseries/7899. Conferences publishing in CCIS are allowed to use Online Conference Service (OCS) for managing the whole proceedings lifecycle (from submission and reviewing to preparing for publication) free of charge.

Publication process

The language of publication is exclusively English. Authors publishing in CCIS have to sign the Springer CCIS copyright transfer form, however, they are free to use their material published in CCIS for substantially changed, more elaborate subsequent publications elsewhere. For the preparation of the camera-ready papers/files, authors have to strictly adhere to the Springer CCIS Authors' Instructions and are strongly encouraged to use the CCIS LaTeX style files or templates.

Abstracting/Indexing

CCIS is abstracted/indexed in DBLP, Google Scholar, EI-Compendex, Mathematical Reviews, SCImago, Scopus. CCIS volumes are also submitted for the inclusion in ISI Proceedings.

How to start

To start the evaluation of your proposal for inclusion in the CCIS series, please send an e-mail to ccis@springer.com.

Hai Jin · Yi Pan · Jianfeng Lu
Editors

Artificial Intelligence and Machine Learning

First International Artificial Intelligence Conference, IAIC 2023
Nanjing, China, November 25–27, 2023
Revised Selected Papers, Part I

 Springer

Editors
Hai Jin
Huazhong University of Science
and Technology
Wuhan, Hubei, China

Yi Pan
Chinese Academy of Science
Shenzhen, China

Jianfeng Lu (iD)
Nanjing University of Science
and Technology
Nanjing, China

ISSN 1865-0929   ISSN 1865-0937 (electronic)
Communications in Computer and Information Science
ISBN 978-981-97-1276-2   ISBN 978-981-97-1277-9 (eBook)
https://doi.org/10.1007/978-981-97-1277-9

This Springer imprint is published by the registered company Springer Nature Singapore Pte Ltd.
The registered company address is: 152 Beach Road, #21-01/04 Gateway East, Singapore 189721, Singapore

Paper in this product is recyclable.

Preface

These conference proceedings are a collection of the papers accepted by IAIC 2023 – the 2023 International Artificial Intelligence Conference, held on November 25–27, 2023 in Nanjing, China.

The conference was organized by Nanjing University of Science & Technology, and Tech Science Press. IAIC 2023 aimed to provide a platform for the exchange of ideas and the discussion of recent developments in artificial intelligence. The conference showcased a diverse range of topics, including machine learning, natural language processing, computer vision, robotics, and ethical considerations in AI, among others.

The reviewing process for IAIC 2023 was meticulous and thorough. We received an impressive number of qualified submissions, reflecting the growing interest and engagement in the field of artificial intelligence. The number of the final accepted papers for publication is 86. The high standard set for acceptance resulted in a competitive selection, with a commendable acceptance rate that attests to the caliber of the contributions presented at the conference.

We extend our gratitude to the authors for their outstanding contributions and dedication, as well as to the reviewers for ensuring the selection of high-quality papers, which made these conference proceedings possible.

We also would like to thank the organizers and sponsors whose generous support made IAIC 2023 possible. Their commitment to advancing the field of artificial intelligence is commendable, and we acknowledge their contributions with sincere appreciation. The logos of our esteemed sponsors can be found on the following pages.

We hope this volume serves as a valuable resource for researchers, academics, and practitioners, contributing to the ongoing dialogue that propels the field forward.

December 2023   IAIC 2023 Organizing Committee

Organization

General Chairs

Hai Jin   Huazhong University of Science and Technology, China

Yi Pan   Shenzhen Institute of Advanced Technology, Chinese Academy of Sciences, China

Jianfeng Lu   Nanjing University of Science and Technology, China

Technical Program Chairs

Yingtao Jiang   University of Nevada Las Vegas, USA

Q. M. Jonathan Wu   University of Windsor, Canada

Technical Program Committee Members

Yudong Zhang   University of Leicester, UK

Shuwen Chen   Jiangsu Second Normal University, China

Xiaoyan Zhao   Nanjing Institute of Technology, China

Wentao Li   Southwest University, China

Chao Zhang   Shanxi University, China

Huiyan Zhang   Chongqing Technology and Business University, China

Tao Zhan   Southwest University, China

Muhammad Attique Khan   HITEC University, Pakistan

Tallha Akram   COMSATS University Islamabad, Pakistan

Zhewei Liang   Mayo Clinic, USA

Yi Ding   University of Electronic Science and Technology of China, China

Xianhua Niu   Xihua University, China

Yingjie Zhou   Sichuan University, China

Dajiang Chen   University of Electronic Science and Technology of China, China

Fang Liu   Hunan University, China

Zhiping Cai	National University of Defense Technology, China
Zongshuai Zhang	Chinese Academy of Sciences, China
Daniel Xiapu Luo	Hong Kong Polytechnic University, China
Jieren Cheng	Hainan University, China
Xinwang Liu	National University of Defense Technology, China
Qiang Liu	National University of Defense Technology, China
Xiangyang (Alex X.) Liu	Michigan State University, USA
Wei Fang	Nanjing University of Information Science and Technology, China
Victor S. Sheng	Texas Tech University, USA
Jinwei Wang	Nanjing University of Information Science and Technology, China
Leiming Yan	Nanjing University of Information Science and Technology, China
Jian Su	Nanjing University of Information Science and Technology, China
Zheng-guo Sheng	University of Sussex, UK
Si-guang Chen	Nanjing University of Posts and Telecommunications, China
Yanchao Zhao	Nanjing University of Aeronautics and Astronautics, China
Hao Han	Nanjing University of Aeronautics and Astronautics, China
Hao Wang	Ratidar Technologies LLC, China

Publication Chair

Zhihua Xia	Jinan University, China

Publicity Chairs

Lei Chen	Shandong University, China
Yuan Tian	Nanjing Institute of Technology, China

Organization Committee Members

Laith Abualigah	Al Al-Bayt University, Jordan
Muhammad Azeem Akbar	LUT University, Finland
Farman Ali	Sejong University, South Korea
Shuwen Chen	Jiangsu Second Normal University, China
Chien-Ming Chen	Nanjing University of Information Science and Technology, China
Dajiang Chen	University of Electronic Science and Technology of China, China
Ting Chen	University of Electronic Science and Technology of China, China
Ke Feng	National University of Singapore, Singapore
Honghao Gao	Shanghai University, China
Xiaozhi Gao	University of Eastern Finland, Finland
Ke Gu	Changsha University of Science and Technology, China
Mohammad Kamrul Hasan	Universiti Kebangsaan Malaysia, Malaysia
Celestine Iwendi	University of Bolton, UK
Heming Jia	Sanming University, China
Deming Lei	Wuhan University of Technology, China
Peng Li	Nanjing University of Aeronautics and Astronautics, China
Huchang Liao	Sichuan University, China
Mingwei Lin	Fujian Normal University, China
Anfeng Liu	Central South University, China
Xiaodong Liu	Edinburgh Napier University, UK
Niancheng Long	Shanghai Jiao Tong University, China
Jeng-Shyang Pan	Shandong University of Science and Technology, China
Danilo Pelusi	University of Teramo, Italy
Kewei Sha	University of Houston, USA
Shigen Shen	Huzhou University, China
Xiangbo Shu	Nanjing University of Science and Technology, China
Adam Slowik	Koszalin University of Technology, Poland
Jin Wang	Changsha University of Science and Technology, China
Kun Wang	Fudan University, China
Changyan Yi	Nanjing University of Aeronautics and Astronautics, China
Yudong Zhang	University of Leicester, UK
Chengwen Zhong	Northwestern Polytechnic University, China

Junlong Zhou   Nanjing University of Science and Technology,
                                China
Xiaobo Zhou   Tianjin University, China
Fa Zhu   Nanjing Forestry University, China

Contents – Part I

Contents – Part II

Contents – Part III

Machine Printed Page Number Anomaly Detection Method Based on Multi-scale Self Attention Encoding Decoding

Xiangchao Shao$^{(\boxtimes)}$, Xueli Xiao, and Yingxiong Leng

Dongguan Power Supply Bureau of Guangdong Power Grid Corp., Dongguan 523000, Guangdong, China
354277894@qq.com

Abstract. Paging is a particularly important step in the process of organizing archive files, and page numbers often appear missing, blurry, and other abnormal phenomena during the printing process. Effective detection of machine printed page numbers is of great practical significance for improving the efficiency of archive organization work; A machine printed page anomaly detection method based on multi-scale self attention encoding decoding is proposed to address the issue of poor performance of current general object detection methods and anomaly detection methods such as automatic encoders when migrating to the field of machine printed page anomaly detection; On the basis of the automatic encoder model, a deep network is adopted to replace the backbone network of the native automatic encoder encoding module, improving the model's feature extraction ability; A multi-layer network with self attention module acting on encoding and decoding modules was proposed to mine the information features of key areas of image targets, improve the image reconstruction ability of the decoding module, and enable the model to generate high-quality reconstructed images. By calculating the distance between the reconstructed image and the original image, anomaly detection of machine printed page numbers was achieved; Through experiments, it has been proven that the detection accuracy of this method reaches 98.7%, which better meets the needs of practical engineering.

Keywords: archive organization · page number · abnormal detection · self attention · multi scale · encoding · decoding · Third Keyword

1 Introduction

With the continuous increase of China's population and the rapid economic development, the total amount of various types of information is showing an explosive growth trend. As one of the important carriers of information, texts are widely involved in daily life and work, especially in the field of file sorting. How to effectively save and process text information has become an urgent need. Currently, technologies like computer vision can preliminarily achieve automatic detection and recognition of archival text information [1–3]. However, there is a large recognition error for some non-standardized contents

H. Jin et al. (Eds.): IAIC 2023, CCIS 2058, pp. 1–15, 2024.
https://doi.org/10.1007/978-981-97-1277-9_1

of archives. For example, when using general computer vision methods to detect and recognize page numbers in archives, if the typed page numbers belong to standardized text, the recognition accuracy is high and the error is small [4]. But when machine-typed page numbers have missing, tilted, blurred, double images and other abnormalities, they belong to non-standardized text [5], and the recognition results show larger errors, failing to meet engineering needs.

Currently, for such non-standardized text, it requires manual verification and correction. However, the traditional method of relying on manual page-by-page checks is too costly and inefficient. In addition, the frequency of non-standardized machine-typed page numbers is low, the sample size is small, the distribution of non-standard samples and standard samples is severely imbalanced, and shows a long-tailed distribution feature [6]. At the same time, the categories of non-standard samples are diverse, and there is no universal abnormal definition rule, resulting in traditional image classification algorithms [7–10] such as statistical models, matrix decomposition, feature pyramids, and template matching, failing to design stable and robust models to implement detection of non-standardized machine-typed page numbers.

Currently, object detection algorithms and anomaly detection algorithms are widely used in various fields, and the detection effects are continuously improving, greatly promoting the intelligence and unmanned progress in various fields. Current deep learning-based object detection algorithms mainly fall into three categories: First, one-stage detection algorithms represented by the YOLO (You Only Look Once) series [11–14] and the SSD (Single Shot MultiBox Detector) [15]; second, two-stage detection algorithms represented by R-CNN (Region-CNN) [16]; third, methods combining the advantages of one-stage and two-stage algorithms such as RetinaNet [17] and emerging methods based on Transformer structures like DETR (DEtection TRansformer) [18]. However, due to the specificity of the content and rules of machine-typed page number anomaly detection, as well as the limitations of training samples, these methods, similar to image classification algorithms, are difficult to apply cross-domain in the current page number anomaly detection scene. In comparison, current image anomaly detection methods based on autoencoders [19, 20] and generative adversarial networks [21] perform better in this field and are widely used in hyperspectral remote sensing images [22] and X-ray images [23].

In 2015, Emrecan et al. [24] applied the autoencoder model structure to the field of image anomaly detection. They encoded and decoded the HIS (Hue-Intensity-Saturation) values of images and then calculated the reconstruction error to achieve anomaly detection for hyperspectral images. Subsequently, many improved autoencoder-based image anomaly detection algorithms were proposed. For instance, method [25] used a stacked encoder structure to distinguish between image background and anomaly areas, merging spatial and spectral features of both areas to produce better reconstruction images. Method [26] sparsified the autoencoder and trained encoders on partitioned input images, achieving better results. The principle of these methods is basically the same as the original autoencoder, training pixel by pixel, largely losing the spatial relationship between pixels, which is a crucial factor affecting detection accuracy.

Methods [27, 28] use a semi-supervised approach, leveraging the powerful generative ability of generative adversarial networks to reconstruct image backgrounds. They then compared it with error images to achieve image anomaly detection. However, using generative adversarial networks for image anomaly detection is still in its infancy, with unstable model training and precision in the anomaly detection field that still needs improvement.

In response to the above situation, to effectively achieve machine-typed page number anomaly detection, this paper proposes a multi-scale self-attention encoding-decoding method for machine-typed page number anomaly detection. Based on the autoencoder model, combining deep networks and introducing multi-scale self-attention modules, the model's feature extraction and image reconstruction capabilities are enhanced, showing better detection performance in the task of machine-typed page number anomaly detection. The main contributions of this article are:

1. Analyzing the anomaly categories of machine-typed page numbers for practical application, and constructing a machine-typed page number detection dataset based on the frequency of actual anomalies to promote more applicable model training for real scenarios.
2. Proposing a self-attention module that can effectively mine information features from key areas of image targets, maximizing the use of original image information.
3. Introducing a machine-typed page number anomaly detection method based on multi-scale self-attention encoding-decoding. By combining deep network structures, cross-entropy loss, and localization loss functions, and applying self-attention modules to multiple network layers, it supplements the feature information of reconstructed images, producing higher-quality reconstructed images. Thus, making it easier for the model to distinguish between abnormal and normal images, achieving good performance in machine-typed page number detection tasks.

2   Overall Design for Anomaly Detection

The machine-printed page number anomaly detection system is constructed from two parts: data acquisition and anomaly detection. The data acquisition part converts paper text data into image data, mainly realized with the help of scanning platforms. The main components of the anomaly detection part are the image preprocessing algorithm, anomaly detection algorithm, and detection application platform. The overall framework of the detection system is shown in Fig. 1.

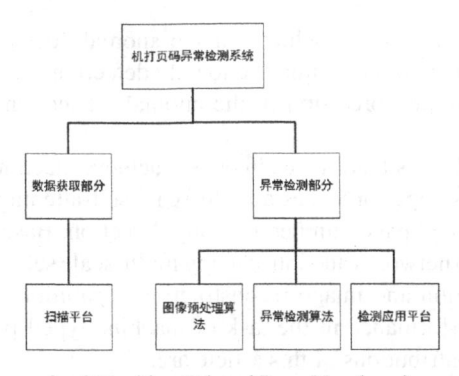

Fig. 1. Overall Framework of Machine-Printed Page Number Anomaly Detection System.

2.1 Data Acquisition

The data acquisition part mainly consists of a scanning platform, the purpose of which is to convert data forms. It is responsible for converting paper documents into image data and storing them, and it is also the input for the subsequent anomaly detection part. As shown in Fig. 2, the scanning platform used in this paper is the Ricoh MP C2503SP digital multifunction machine. Its relevant important parameters are as follows: color is in full color, scanning speed is 54ppm, output formats are TIFF and JPEG, scanning resolution is 600 dpi, and the hard drive capacity is 200 G.

Fig. 2. Scanning Platform.

2.2 Anomaly Detection Section

The image processing algorithm in the system anomaly detection part aims to acquire the page number area from the images outputted by the scanning platform. There are two methods for this:

1. The location of the machine-printed page number can be determined using prior knowledge. Since the page number of the file appears in a fixed area at the bottom of the image, this prior knowledge can be used to define corresponding rules to obtain the candidate area of the page number.

2. Use current general object detection technology to preliminarily detect the candidate area of the page number. Preliminarily determining the page number candidate area can reduce the computational load of the subsequent anomaly detection algorithm and interference from the complex background of the original image's non-page number area.

   The page number candidate area of the image is then input into the machine-printed page number anomaly detection algorithm based on multi-scale self-attention encoding-decoding proposed in this paper to perform anomaly detection, thus obtaining the final detection results.

   The detection application platform is written in C# language, embedding the image preprocessing algorithm and anomaly detection algorithm model, and includes a result display interface. After performing anomaly detection on the scanned image, the results will be displayed on the interface, and the anomalous images will be automatically saved in the database.

3 Anomaly Detection Algorithm for Machine-Printed Page Numbers

3.1 Introduction and Analysis of the Original Autoencoder Structure

The autoencoder is a typical model of self-supervised learning, mainly composed of an Encoder block and a Decoder block. The specific structure of the autoencoder is shown in Fig. 3.

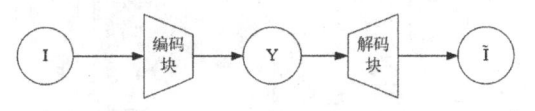

Fig. 3. Autoencoder Structure.

The encoding block of the autoencoder is used to effectively encode the input data I into hidden state data Y. Y has a smaller data dimension compared to I. Essentially, this is a nonlinear dimension reduction or feature extraction method. The decoding block is for feature reconstruction, which maps the hidden state data Y back to the same size as $\tilde{I}$. Essentially, this is a mapping function. For example, in the field of computer vision, the encoding block typically uses the downsampling method of convolutional neural networks to encode three-dimensional image data into feature vectors; while the decoding block typically uses the upsampling method of convolutional neural networks to restore the feature vector to the initial dimension of the original image.

During the training process of the autoencoder, the reconstruction error function is continuously optimized, aiming to make the reconstructed data as similar as possible to the input of the encoding block. The objective function of the autoencoder is shown in Eq. 1.

$$MinimizeLoss = d\left(I, \tilde{I}\right) \tag{1}$$

Where d is the distance function between the reconstructed output $\tilde{I}$ from the decoder and the input I to the encoder, typically using the MSE (Mean Squared Error) function.

The training process of the autoencoder does not require labeled input data, making it easy to collect and widely applicable to anomaly detection scenarios with strong generalization capabilities.

The quality of the autoencoder's generated results hinges on the encoding outcome of the intermediate hidden layer. However, during the feature extraction process of the encoder network, it considers the importance of all image pixels to be the same. In actual visual perception, certain parts of an image (such as the image's foreground targets and edge information) are more significant, leading the autoencoder to fail in extracting information features from the key areas of the image target. This inability to fully leverage the original image information limits the quality of the reconstructed image and affects its efficacy in anomaly detection.

3.2   Improved Multi-scale Self-attention Encoding-Decoding Anomaly Detection Model

The multi-scale self-attention encoder-decoder model proposed in this paper is detailed in Fig. 4.

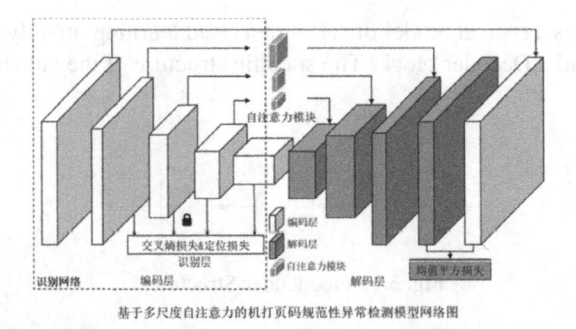

基于多尺度自注意力的机打页码规范性异常检测模型网络图

Fig. 4. Machine-Printed Page Number Anomaly Detection Model Structure based on Multi-scale Self-attention Encoding-Decoding.

Compared to the autoencoder, the main improvements of the multi-scale self-attention encoder-decoder model proposed in this article are: Firstly, backbone networks such as VGG, GoogLeNet, and ResNet replace the neural network structure used by the native autoencoder, combined with cross-entropy loss and localization loss, in order to achieve better feature representation and better adaptability in the field of computer vision. Secondly, a self-attention module is introduced. During the encoding process, the self-attention module further extracts self-attention features of different scales from the encoding module, digging deep into the key information characteristics of the image target region. Then, the multi-scale self-attention features extracted by the encoding module are mapped to the corresponding network layer of the decoding module to supplement the feature information of the reconstructed image, producing a higher quality reconstructed image, thereby facilitating the model's ability to distinguish between abnormal and normal images.

Encoding Module. The encoding module adopts the backbone network structures of VGG, GoogLeNet, ResNet, etc., which have been pre-trained on the VOC dataset. At the same time, dimensional features from three scales with larger sizes in the backbone network are extracted for use in the self-attention feature module.

Self-Attention Module. The self-attention module takes the feature maps from three stages of the encoder's main network as input, with its basic structure shown in Fig. 5.

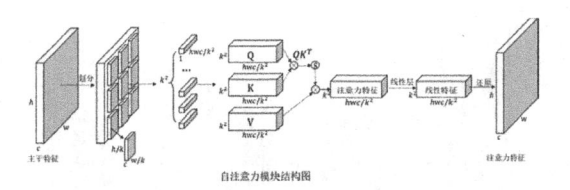

Fig. 5. Structure of the Self-attention Module.

The self-attention module first divides the feature map into k^2 segments in the spatial dimension. Each segment is then integrated in terms of its feature region, resulting in k^2 features each of size hwc/k^2. The specific expression is shown in Eq. 2.

$$F = split(F_{in}) \tag{2}$$

Wherein F_{in} represents the input feature, with a dimension $R^{d \in (c,h,w)}$, split indicates the division, F denotes the output, and $R^{d \in (K^2, hwc/k^2)}$ is its feature dimension.

Then, using the coefficient matrix operation, we obtain the three layers of features, Q, K, and V. The specific expression is shown in Eq. 3.

$$Q, K, V = \max trix(T, F) \tag{3}$$

Wherein T is a learnable matrix with a dimension of $R^{d \in (3K^2, k^2)}$, and the feature dimensions of Q, K, and V are $R^{d \in (K^2, hwc/k^2)}$;

Multiply Q with the transpose of K to obtain feature QK^T with a feature dimension of $R^{d \in (K^2, k^2)}$; then apply *soft* max and coefficient factor, and finally perform matrix multiplication with V to get a feature dimension of $R^{d \in (K^2, hwc/k^2)}$. After that, combining with two linear layers to retrieve the attention feature. The specific expression is shown in formula 4. The obtained output feature is then restored to obtain an attention feature map with the same dimension as the input feature, and its operation is opposite to the partition operation.

$$F_{out} = linear\left(norm\left(soft \max\left(\frac{QK^T}{\sqrt{d}}\right)V\right)\right) \tag{4}$$

Wherein d is the coefficient factor, *soft* max represents the soft attention calculation, *norm* denotes normalization, *linear* signifies two linear layers, and F_{out} indicates the output feature with a dimension of $R^{d\in(K^2,hwc/k^2)}$.

Decoding Module. The structure of the decoding layer is opposite to that of the encoding layer, enlarging the feature layer step by step. Bilinear interpolation is used here for upsampling. After upsampling, the features are combined with two 3 x 3 convolutions, then element-wise multiplied with the attention features to get the upsampled features. The specific structure is shown in Fig. 6.

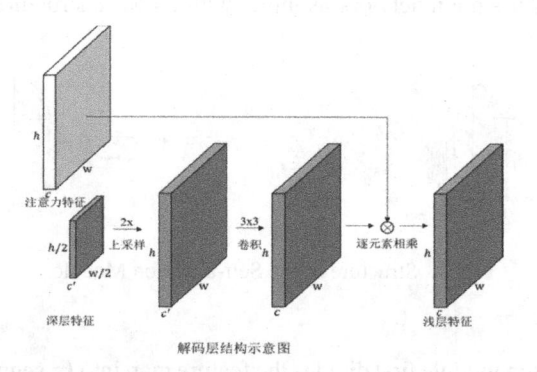

Fig. 6. Structure Schematic of the Decoding Layer.

Loss Function. To better adapt the encoding module for the field of image anomaly detection, the features dimensions from 4 blocks of scales in the main network are extracted as input. A 3x3 convolution is adopted for multi-scale feature extraction, allowing for further target feature acquisition in detection tasks. By combining cross-entropy loss and Smooth L1 localization loss, preliminary localization and classification of the target are achieved. Its specific expression is shown as Eq. (5).

$$L(x, c, p, g) = \frac{1}{N}(L_{cls}(x, c) + \lambda L_{loc}(x, p, g)) \tag{5}$$

Where x represents the input, c represents the actual category, p represents the predicted frame, and g represents the actual label frame. λ denotes the influence factor, which is usually set to 1.0. N represents the selected matched target, L represents the total loss, L_{cls} represents the classification loss, and L_{loc} represents the localization error.

3.3 Anomaly Detection Layer

The main purpose of this layer is to calculate the distance between the predicted features outputted from the decoding layer and the original image. The calculation method uses Mean Squared Error (MSE) to determine the distance between them. The aim is to obtain the average value of the squared differences between the predicted features and the actual values at corresponding positions in the original image feature. By optimizing to make this distance converge close to 0, it implies that the original image and the prediction are close, and the model has learned the general features of the normal sample dataset. Its specific expression is shown in Eq. (6).

$$MSE = \frac{1}{n} \sum_{i=1}^{n} (I - \widetilde{I})^2 \tag{6}$$

Where MSE represents the Mean Squared Error, n represents the number of predicted values,I represents the original image, and $\widetilde{I}$ represents the reconstructed image.

During the training phase of the anomaly detection model, the range interval of the model's normal samples will be recorded as the area for threshold setting. In the inference phase, the anomaly detection model has good extraction and representation capabilities for positive samples, and the difference between its output features and the original image should be within the threshold. When faced with an abnormal sample, if the calculated distance by the model exceeds the threshold, it indicates an anomaly.

4 Experimental Results and Analysis

4.1 Dataset and Anomaly Introduction

This study, set within the context of real-world archival reviews, collected images scanned from non-confidential machine-printed files in the review process. A total of 6,000 images were collected. Among these, 4,800 images of standard machine-printed page number archive files were used as the training set for the detection model. Meanwhile, 1,200 images of non-standard machine-printed page number files were used as the test set for the anomaly detection model. The types of non-standard machine-printed page numbers in the test set include: missing page numbers, skewness, blurriness, and double images. The number of images for each type is evenly distributed, with 300 images each. To more intuitively showcase the differences between types of non-standard machine-printed page numbers in the dataset, this study cropped the page number area from the archive images. A comparison is shown in Fig. 7.

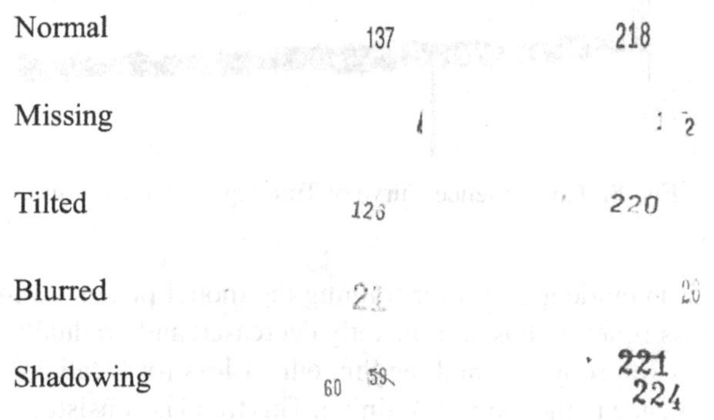

Normal   137   218

Missing

Tilted   12ᵒ   220

Blurred

Shadowing   60   59   221 / 224

Fig. 7. Comparison of Types of Machine-Printed Page Numbers.

4.2   Experimental Strategy and Environment

The model training in this study is divided into two stages. First, the training set from this study's dataset is fed into the encoding module, which has been pre-trained on a machine-printed number dataset. This is done to individually train the encoding module, enabling it to possess the capability for page number recognition and detection. Subsequently, parameters in the main network of the encoding module are frozen, and the entire anomaly detection model is then trained. During this training phase, the parameters of the encoding module are not adjusted. The goal of this training phase is to converge the Mean Squared Error (MSE) value between the reconstructed image of the decoding module and the input image of the anomaly detection model.

The experiments for this study's model were conducted on devices running the Ubuntu 14.04 operating system, with the PyTorch deep learning framework, an Intel Core i7-9700F CPU, 64 GB RAM, and an NVIDIA GeForce RTX 2080Ti GPU with 16 GB of video memory.

During the training process, Adam was chosen as the optimizer, with the beta1 parameter adjusted to 0.5 and beta2 adjusted to 0.999. The batch size was set to the maximum value that the video memory could accommodate. The initial learning rate was set to 0.001, and the number of epochs was 120.

4.3   Training Loss Function Convergence Curve

The convergence graph of the MSE loss function for the native autoencoder and the multi-scale self-attention encoder-decoder model proposed in this paper is shown in Fig. 8.

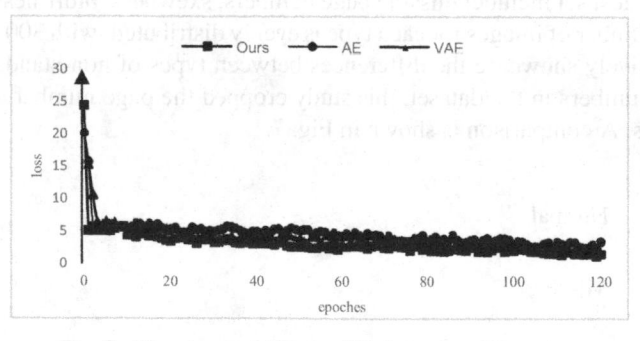

Fig. 8. Convergence Curve of Training Loss Function.

From Fig. 8, it is evident that after training the model presented in this article for 120 epochs, the loss function has significantly decreased and gradually converged. The addition of the cross-entropy loss and the Smooth L1 loss for localization did not cause instability or divergence in the model's training. This trend is consistent with the original autoencoder model (AE) and variational autoencoder (VAE).

4.4 Ablation Study

To verify the effectiveness of each part of the model proposed in this paper, the following ablation studies were designed: Firstly, VGG, GoogLeNet, and Resnet were respectively used as the backbone networks for the encoding module in combination with the self-attention module for experimentation. Subsequently, experiments were conducted by progressively removing the multi-scale self-attention module. The results of the ablation studies are presented in Table 1, where the evaluation metric of accuracy is the ratio of correctly predicted samples to the total samples in the test set.

Table 1. Ablation study results.

Method	Accuracy (%)
Encoder with VGG	78.4
Encoder with VGG + Self-attention Module	84.6
Encoder with VGG + Self-attention Module + Multi-scale	86.5
Encoder with GoogLeNet	80.3
Encoder with GoogLeNet + Self-attention Module	84.4
Encoder with GoogLeNet + Self-attention Module + Multi-scale	86.2
Encoder with Resnet	86.3
Encoder with Resnet + Self-attention Module	96.6
Encoder with Resnet + Self-attention Module + Multi-scale	98.7

From Table 1, it can be observed that without the addition of the self-attention module, the encoder's backbone network using the Resnet structure has an accuracy rate that is 7.9% and 6% higher than that of the VGG and GoogLeNet structures, respectively. This is due to the Resnet structure capturing richer shallow and deep image features, which is an advantage of the residual connections. Meanwhile, with the addition of the self-attention module, the encoder's backbone network using VGG, GoogLeNet, and Resnet structures respectively have shown a significant improvement in accuracy, all exceeding an increase of 4%. Particularly, the Resnet structure with the self-attention module added showed the most notable enhancement, with an increase of 10.3%. This is because the Resnet network structure possesses a broader feature representation space, facilitating gradient backflow to adjust the shallow information feature space, enabling the self-attention module to exhibit superior performance. Therefore, in the subsequent comparative experiments of this paper, the Resnet structure is used for the encoder's backbone network.

Additionally, it's worth noting that this paper also performed ablation on multi-scale levels. The specific approach was to apply the self-attention module to a single layer or multiple layers of the encoder and decoder, with the application to multiple layers corresponding to multi-scale. Applying the self-attention module to multiple layers, compared to a single layer, has led to an improvement of about 2% for each. This is because the multi-scale captures information from different receptive fields, enriching more precise detail features. Simultaneously, it integrates the high-level abstract semantic information of the target into the feature map, making it more suitable for detecting obscured targets, small targets, or blurred or overlapping targets. The results of the ablation study fully validate the effectiveness of each component of the method proposed in this paper.

4.5 Comparative Experiments

To fully verify the effectiveness of the model proposed in this paper for detecting anomalies in machine-printed page numbers, we first compared it with the widely used SSD and YOLO series of object detection algorithms on the test dataset. It is worth noting that, in this paper, the comparison target detection method treats abnormal samples as targets. At the same time, the training set and test set of this study were merged and re-divided, then re-trained and tested, and finally, only the detection accuracy was compared. Additionally, this paper also compared with the autoencoder and its variant, the variational autoencoder. The comparison results are shown in Table 2.

Table 2. Comparative experiment results.

Model	Accuracy (%)
SSD	78.1
Yolo	73.2
Yolov3	76.4
AE (Autoencoder)	86.3
VAE (Variational Autoencoder)	89.5
Model in this paper	98.7

From Table 2, it can be observed that the widely-used object detection algorithms perform poorly in detecting anomalies in machine-printed page numbers. The reason is primarily that the difference between anomalous and normal samples of machine-printed page numbers is subtle, making their variations less conspicuous. This leads to a higher false detection rate for these object detection algorithms. Moreover, the uneven distribution of positive and negative samples is also a factor. The number of positive samples significantly outweighs negative ones, resulting in unstable model training, a tendency to overfit, and consequently, poor overall performance.

Autoencoders, compared to deep object detection algorithms, demonstrate superior performance. This is because the essence of autoencoders lies in focusing on the differences between images, achieving relatively better results. The Variational Autoencoder (VAE), when compared to the standard autoencoder, showed a 3.2% improvement in accuracy. The reason being, VAEs encode the input data in latent space following a specific distribution, exhibiting the ability to generate higher-quality new samples.

The method proposed in this paper, based on multi-scale self-attention encoding-decoding for machine-printed page number anomaly detection, can effectively mine and utilize the information features of the key regions of the image target. This supplements the feature information of the reconstructed image, making it easier for the model to distinguish the differences between abnormal and normal images, thereby enhancing anomaly detection precision. The method of this paper improved the detection accuracy by 12.4% compared to the autoencoder model, and by 9.2% compared to the VAE.

In summary, the multi-scale self-attention encoding-decoding method for machine-printed page number anomaly detection proposed in this paper has achieved excellent results when compared with deep object detection algorithms, autoencoders, and many other models. This validates the practical effectiveness of the method presented in this paper for actual page number anomaly detection.

4.6 Experimental Results Presentation

To more intuitively demonstrate the advantages of the model proposed in this paper, we compared the actual detection results of several models on the four types of abnormal page number data in the test set. Due to page constraints, the example data in this paper only shows the page number area, as shown in Fig. 9.

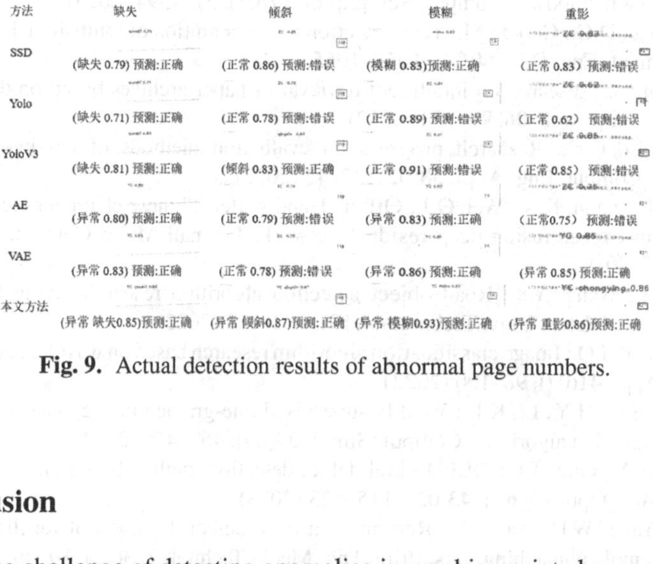

方法	缺失	倾斜	模糊	重影
SSD	(缺失 0.79) 预测:正确	(正常 0.86) 预测:错误	(模糊 0.83)预测:正确	(正常 0.83) 预测:错误
Yolo	(缺失 0.71) 预测:正确	(正常 0.78) 预测:错误	(正常 0.89) 预测:错误	(正常 0.62) 预测:错误
YoloV3	(缺失 0.81) 预测:正确	(倾斜 0.83) 预测:正确	(正常 0.91) 预测:错误	(正常 0.85) 预测:错误
AE	(异常 0.80) 预测:正确	(正常 0.79) 预测:错误	(异常 0.83) 预测:正确	(正常0.75) 预测:错误
VAE	(异常 0.83) 预测:正确	(正常 0.78) 预测:错误	(异常 0.86) 预测:正确	(异常 0.85) 预测:正确
本文方法	(异常 缺失0.85)预测:正确	(异常 倾斜0.87)预测:正确	(异常 模糊0.93)预测:正确	(异常 重影0.86)预测:正确

Fig. 9. Actual detection results of abnormal page numbers.

5 Conclusion

To address the challenge of detecting anomalies in machine-printed page numbers and further enhance the detection efficacy, this paper introduces a method based on multi-scale self-attention encoding-decoding for machine-printed page number anomaly detection. Building on the foundation of the autoencoder model, we adopted a deep neural network to replace the main network of the original autoencoder's encoding module. This combined with cross-entropy loss and localization loss. Moreover, a self-attention module was introduced. During the encoding process, the self-attention module further extracts self-attention features of different scales from the encoding module, aiming to mine the information features of key regions in the image. Subsequently, the multi-scale self-attention features extracted by the encoding module are mapped to the corresponding network layer of the decoding module, supplementing the feature information of the reconstructed image and producing a higher-quality reconstructed image. By calculating the distance between the reconstructed image and the original image, effective anomaly detection of machine-printed page numbers is achieved.

Experimental results demonstrate that the multi-scale self-attention encoding-decoding method for machine-printed page number detection proposed in this paper outperforms the current mainstream object detection models and other anomaly detection models, such as autoencoders. It effectively meets the real-world needs in areas like archival review. Future research will further categorize the anomalies, analyze the frequency of various anomalies, adjust the dataset distribution, optimize the model structure, and enhance the detection efficacy of the anomaly detection model. Additionally, further exploration will be undertaken to apply the model introduced in this paper to anomaly detection in other domains.

References

1. Zhou, Y., Wei, Q.B., Liao, J.W.: Text detection and end-to-end recognition in natural scenes: deep learning methods. J. Comput. Sci. Explor. **17**(03), 577–594 (2023)
2. Hou, Y., Gao, D.G., Gao, H.M.: Text detection and recognition of multi-font Tibetan print in Ujain. Comput. Eng. Des. **44**(04), 1058–1065 (2023)
3. Yi, L., Zou, B.: Research on intelligent retrieval of paper archives based on deep learning. Mechatron. Ship Arch. **06**, 99–103 (2022)
4. Xiao, X., Li, C.C.: Research progress on evaluation methods of handwritten Chinese characters. Comput. Eng. Appl. **58**(02), 27–42 (2022)
5. Zhang, X.L., Zhou, K.X., Wei, Q.J.: Offline Handwritten Chinese character recognition with multi-channel cross fusion deep residual network. J. Small Micro Comput. Syst. **40**(10), 2232–2235 (2019)
6. Wang, Z.H., Wang, Y.S.: Road object detection algorithm research facing long-tail data distribution. J. Wuhan Univ. Technol. **44**(10), 102–108 (2022)
7. Che, D.Q., Lü, J.Q.: Image classification algorithm research based on wavelet pooling. Autom. Technol. Appl. **41**(07), 98–100 (2022)
8. Cui, X.N., Sun, H.Y., Li, K.L.: Weakly supervised fine-grained image classification method based on Bayesian algorithm. Comput. Simul. **39**(09), 467–470 (2022)
9. Liu, L., Ye, Y., Guo, T.L.: OLED pixel defect detection method based on extended feature pyramid. Acta Optica Sinica **43**(02), 115–123 (2023)
10. Li, B.Y., Yang, W.H., Yang, L.: Research on print defect detection of fertilizer packaging based on template matching. Res. Print. Dig. Media Technol. **223**(02), 39–49 (2023)
11. Redmon, J., Divvala, S., Girshick, R., et al.: You only look once: unified, real-time object detection. In: Proceedings of the IEEE Conference on Computer Vision and Pattern Recognition, pp. 779–788 (2016)
12. Farhadi, A., Redmon, J.: Yolov3: an incremental improvement. In: Computer Vision and Pattern Recognition, vol. 1804, pp. 1–6. Springer, Heidelberg (2018)
13. Bochkovskiy, A., Wang, C.Y., Liao, H.Y.M.: Yolov4: optimal speed and accuracy of object detection. arXiv preprint arXiv:2004.10934 (2020)
14. Zhu, X., Lyu, S., Wang, X., et al.: TPH-YOLOv5: improved YOLOv5 based on transformer prediction head for object detection on drone-captured scenarios. In: Proceedings of the IEEE/CVF International Conference on Computer Vision, pp. 2778–2788 (2021)

15. Liu, W., Anguelov, D., Erhan, D., et al.: SSD: single shot multibox detector. In: Computer Vision–ECCV 2016: 14th European Conference, Part I 14, pp. 21–37. Springer International Publishing, Amsterdam (2016)
16. Girshick, R., Donahue, J., Darrell, T., et al.: Region-based convolutional networks for accurate object detection and segmentation. IEEE Trans. Pattern Anal. Mach. Intell. **38**(1), 142–158 (2015)
17. Lin, T.Y., Goyal, P., Girshick, R., et al.: Focal loss for dense object detection. In: Proceedings of the IEEE International Conference on Computer Vision, pp. 2980–2988 (2017)
18. Carion, N., Massa, F., Synnaeve, G., et al.: End-to-end object detection with transformers. In: Computer Vision–ECCV 2020: 16th European Conference, pp. 213–229 (2020)
19. Hinton, G.E., Osindero, S., Teh, Y.W.: A fast learning algorithm for deep belief nets. Neural Comput. **18**(7), 1527–1554 (2006)
20. Kingma, D.P., Welling, M.: Auto-encoding variational bayes. arXiv preprint arXiv:1312.6114 (2013)
21. Creswell, A., White, T., Dumoulin, V., et al.: Generative adversarial networks: an overview. IEEE Signal Process. Mag. **35**(1), 53–65 (2018)
22. Pu, L.: Research on deep learning classification model of hyperspectral images. J. Surv. Mapp. **52**(01), 172 (2023)
23. Zhou, Y.: GAN-based X-ray image multimodal fusion. Laser J. **42**(09), 139–143 (2021)
24. Bati, E., Çalışkan, A., Koz, A., et al.: Hyperspectral anomaly detection method based on auto-encoder. In: Image and Signal Processing for Remote Sensing XXI, vol. 9643, pp. 220–226. SPIE (2015)
25. Zhao, C., Zhang, L.: Spectral-spatial stacked autoencoders based on low-rank and sparse matrix decomposition for hyperspectral anomaly detection. Infrared Phys. Technol. **92**, 166–176 (2018)
26. Yang, Y., Zhang, J., Song, S., et al.: Hyperspectral anomaly detection via dictionary construction-based low-rank representation and adaptive weighting. Rem. Sens. **11**(2), 192 (2019)
27. Arisoy, S., Nasrabadi, N.M., Kayabol, K.: GAN-based hyperspectral anomaly detection. In: 2020 28th European Signal Processing Conference (EUSIPCO), pp. 1891–1895. IEEE (2020)
28. Zhong, J., Xie, W., Li, Y., et al.: Characterization of background-anomaly separability with generative adversarial network for hyperspectral anomaly detection. IEEE Trans. Geosci. Remote Sens.Geosci. Remote Sens. **59**(7), 6017–6028 (2020)

A Novel MEC Framework for Extractive Summarization Using Semantic Role Graph and Semantic Matching

Tiantian Li[1(✉)], Bingchuan He[1], Huangfei Cheng[2], and Bin Cao[1]

[1] Zhejiang University of Technology, Hangzhou 310023, China
ttli89@zjut.edu.cn
[2] Peoples Government of Shiliang Town, Kecheng District, Quzhou 324015, China

Abstract. With the progressive development of internet technologies, people are exposed to huge amount of information in various formats every day, and need to deal with them anytime and anywhere through different terminal devices. Among which, text information usually costs too much time for length and redundancy. Hence, providing an efficient text summarization approach is quite necessary. However, such approaches usually depend on artificial intelligence techniques requiring greater computing power that the terminal devices cannot afford. Therefore, in this paper we propose a Mobile Edge Computing framework for text summarization, in which the summarizing requests submission and results display are done on terminal devices, while the computationally intensive summarization work is offloaded to mobile edge servers deployed with summarization model. The efficiency and accuracy of the results obtained based on this framework depend largely on the summarization model used. Hence, we also propose an extractive summarization model using semantic role graph (SRG) and semantic matching model (SMM). First, construct a SRG based on the semantic roles obtained by semantic role labeling; then generate candidate summaries by traversing the SRG and choose the best one using a SMM based on Bidirectional Encoder Representation Transformers. SRG guarantees conciseness and accuracy of the abstracts at syntactic level though reducing the extraction granularity to semantic roles. SMM further guarantees accuracy at semantic level based on the prior knowledge learned through pre-training. In a word, the proposed framework together with the extractive summarization model using SRG and SMM can realize an efficient and accurate text summarization approach. Experiments on both in-house and public datasets validate this conclusion through comparison with existing methods.

Keywords: Mobile Edge Computing · Text Summarization · Semantic Role Graph · Semantic Matching · BERT

1 Introduction

Nowadays, information in varieties of formats increases exponentially with the continuous development of Internet technology. People are exposed to huge amount of information either actively or passively every day. They usually need to handle such

H. Jin et al. (Eds.): IAIC 2023, CCIS 2058, pp. 16–31, 2024.
https://doi.org/10.1007/978-981-97-1277-9_2

information real-timely anytime and anywhere though different terminal devices. As one of the most common and typical formats, text information usually costs a lot of time to get the main points due to the length of one document and the redundancy existing in multiple documents on the same topic. Therefore, an efficient and automatic text summarization approach is quite necessary.

In recent years, with the development of artificial intelligence (AI) techniques, automatic text summarization based on deep learning has emerged as the mainstream approach. However, such approaches usually have quite high requirements for computing power, especially GPU resources, which cannot be afforded by the terminal devices like mobile phones, laptops, personal computers, etc. To address this problem, we propose a novel framework for text summarization based on the newly emerging promising paradigm: Mobile Edge Computing (MEC) [1]. In this framework, terminal devices are only responsible for posting summarization requests and displaying the generated abstracts, while mobile edge servers (MES) deployed with summarization model are responsible for the offloaded computationally intensive summarization work.

As to the summarization model adopted in this framework, we propose an extractive summarization model using semantic role graph (SRG) and semantic matching model (SMM) to further ensure efficiency and accuracy. In specific, first construct a SRG based on the semantic roles obtained by semantic role labeling (SRL) technique [2]; then generate candidate summaries by traversing the SRG and choose the best one using a SMM based on Bidirectional Encoder Representation Transformers (BERT) [3]. SRG guarantees conciseness and accuracy of the generated summary at syntactic level with a finer extraction granularity, while SMM further guarantees accuracy at semantic level based on the prior knowledge learned through pre-training.

In a word, the proposed MEC framework together with the extractive summarization model using SRG and SMM can provide an efficient and accurate text summarization approach. Since the computationally intensive training and updating of the summarization model are both completed on MES, users can quickly and accurately extract key information from massive texts. With this framework, people can better cope with massive text information real-timely anytime and anywhere, without being limited by the computing power and places of the terminal devices.

1.1 Motivation

Nowadays, people have great demands on mobile applications to deal with huge amounts of data real-timely, while the terminal devices cannot afford the required computing and power resources. To promote the performance of mobile applications, especially those depending on AI techniques such as the widely used text summarization [4–7] focused here, the terminal devices' computations should be offloaded to external more powerful devices to ease their computation burden and energy consumption. To achieve this goal, the classic cloud computing paradigm [8] can be employed, but the fixed location feature of its data center cannot satisfy the real-time communication demand of processing data anytime and anywhere. For this reason, we decide to adopt another more appropriate paradigm, mobile edge computing (MEC) [1, 9–12], as the framework to power the text summarization application.

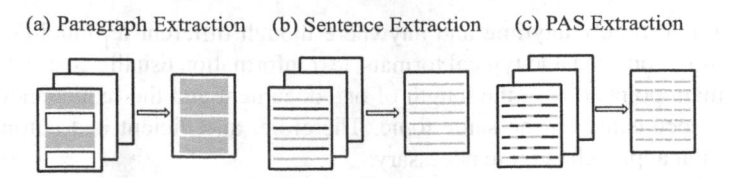

Fig. 1. Extractive MDSs with different granularities.

Table 1. Comparison of text summarization with different extraction granularities.

Original Text: Newspapers reported wednesday that three top libyan officials have been tried and jailed in the Lockerbie case, but libyan dissidents said the reports appeared to be a political ploy by libya's leader, col.moammar Gadhafi. Libyan leader moammar Gadhafi said the suspects in the lockerbie bombing are "very happy" to be tried in the netherlands, and he hoped the trial would lead to a better relationship with the U.S
Secretary-general kofi annan said wednesday that he may travel to libya next week in hopes of closing a deal to try two libyan suspects in the pan am lockerbie bombing

Paragraph Extraction: Newspapers reported wednesday that three top libyan officials have been tried and jailed in the Lockerbie case, but libyan dissidents said the reports appeared to be a political ploy by libya's leader, col.moammar Gadhafi. Libyan leader moammar Gadhafi said the suspects in the lockerbie bombing are "very happy" to be tried in the netherlands, and he hoped the trial would lead to a better relationship with the U.S

Sentence Extraction: Newspapers reported wednesday that three top libyan officials have been tried and jailed in the lockerbie case, but libyan dissidents said the reports appeared to be a political ploy by libya's leader, col.moammar Gadhafi

PAS Extraction: Three top Libyan officials jailed in Lockerbie case, the trial would lead to a better relationship with the U.S

Gold Summary: Three top Libyan officials reported jailed in Lockerbie case. Gadhafi sees better U.S.-Libyan relations after trial of Lockerbie suspects

Beside efficiency, accuracy is another indicator that text summarization cares. For automatic text summarization, single-document summarization is a special case of multi-document summarization (MDS) [4], so we focus on MDS directly. Currently, MDS can be completed in the following two paradigms: extractive MDS and abstractive MDS. Extractive MDS assembles summaries exclusively from spans taken from the source documents, while abstractive MDS may generate novel words and phrases not featured in the source documents [13]. Considering that abstractive MDS cannot guarantee baseline levels of grammaticality and accuracy, we choose extractive MDS in this paper.

Further, existing MDS approaches usually take sentence or paragraph as extraction granularity. Such granularities are not fined enough to guarantee the summarization conciseness and accuracy [14, 15]. There will still exist intra-sentence redundancy in the extracted sentences, and sentences filtered may still contain key information. In view of this, we innovatively propose to use SRG to represent the text information with semantic arguments extracted from multiple documents by SRL. SRL uses predicate-argument structure (PAS) to express "who did what to whom", thus filtering redundant information in the original sentence while preserving the core semantics with a more fine-grained extraction. Extractive MDS methods with the three different granularities mentioned above are shown in Fig. 1. The advantages of our proposed smaller extraction granularity (PAS Extraction) can be exhibited intuitively in the example shown in Table 1: The summary generated by PAS is the most concise and closest to the gold summary.

In addition to the syntactic-level assurance measures, semantic-level measures should also be taken to further guarantee the summarization accuracy. To this end, we use a BERT-based SMM to learn the knowledge that can choose the best one from the candidate summaries generated by traversing the SRG.

1.2 Contribution

The main contributions of this paper are as follows:

- Propose a MEC framework for extractive summarization, allowing users (1) to efficiently deal with massive text information anytime and anywhere without being limited by the computing power and places of the terminal devices; (2) to use this service without interruption even when updating the summarization model;
- Propose a summarization model which innovatively (1) uses semantic role graph to represent the text information, making the summarization much more concise and accurate by reducing the extraction granularity to semantic roles; (2) uses semantic matching to further guarantee accuracy based on the prior knowledge learned through pre-training;
- The proposed extractive summarization approach outperforms the counterpart methods by a large margin on both public and in-house datasets measured by ROUGE scores.

2 Preliminary

In this section, we briefly introduce the mechanism of semantic role labeling (SRL) technique [3], whose various implementations like [16] have been widely employed in many natural language processing (NLP) applications. SRL aims to extract all predicate-argument structures (PASs) from a sentence, thus determining essentially "who did what to whom," "when" and "where". The predicate in a sentence typically indicates a particular action, while its syntactic arguments are associated with the participants in this action. For example, for sentence "*Mike lost the book*", *loss* is the action, while *Mike* is the agent and *the book* is the patient.

Semantic roles are defined as the relationships between syntactic arguments and predicates [17]. As shown in Table 2, arguments of predicates are indicated as different labels. ARG0 and ARG1 separately indicate *agent* and *patient*, while ARG2 indicates *instrument*, *benefactive*, or *attribute*. Labels prefixed with AGRM represent *modifier*, such as ARGM-LOC (Location), ARGM-TMP (Temporal), and ARGM-ADV (Adverbial). For the example above, *Mike* will be labeled as AGR0, and *the book* labeled as AGR1.

Table 2. Common semantic roles and corresponding labels.

Semantic Roles	Labels
agent	ARG0
patient	ARG1
instrument, benefactive, attribute	ARG2
modifier (Location, Temporal, Adverbial)	ARGM (AGRM-LOC, AGRM-TEM, AGRM-ADV)

3 MEC Framework for Extractive Summarization

The proposed MEC framework for extractive summarization is illustrated in Fig. 2. It consists of three main parts. The first part "Internet" usually plays the role of text sources providers. The second part "User Terminal" is only responsible for sending summarization requests and displaying the abstracts generated. Hence, it can be any ordinary lightweight device without strong computing capability and abundant energy, such as mobile phones, laptops, and personal computers. The third part "Mobile Edge Server (MES)" handles the offloaded summarization work, thus requiring powerful computability like GPUs with Compute Unified Device Architecture (CUDA).

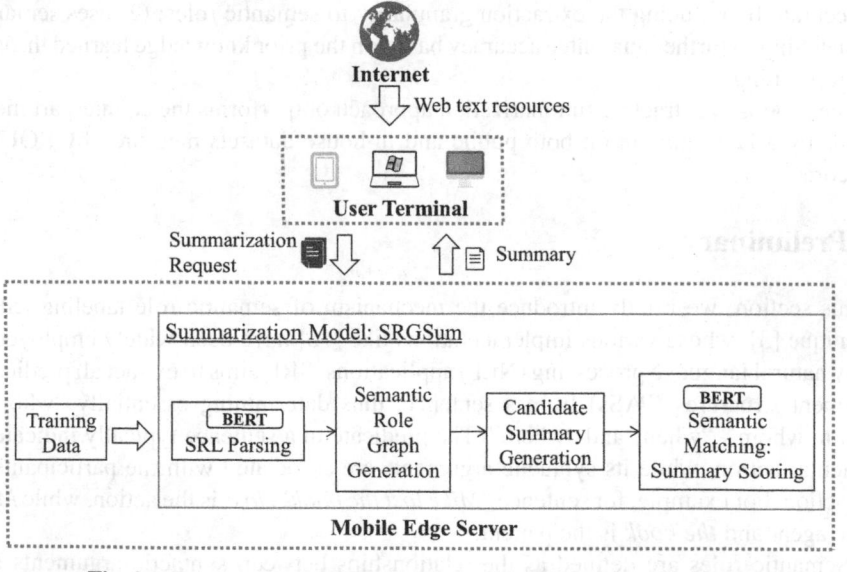

Fig. 2. The proposed MEC framework for extractive summarization.

The working mechanism of this framework is as follows. First, based on the prepared training data of MDS, we train the proposed summarization model denoted as SRGSum on MES equipped with GPU. Once the model training is done, users can submit summarization requests through various terminal devices to MES. Text sources can be either local or remote from the internet. Moreover, data features may change over time, meaning that the summarization model needs retraining to ensure accuracy. Hence, to provide the summarization service without interruption, we equip MES with at least two GPUs for service providing and model training respectively.

The training mechanism of the critical summarization model SRGSum is as follows. First, parse each sentence using a BERT-based semantic role labeling model [18]. Second, based on the parsing results, construct a semantic role graph (SRG) for each document by heuristic rules, then merge them to eliminate redundant information. We also design a rating method based on TextRank [19] to measure the importance of each node in SRG. Third, generate candidate summaries by traversing the SRG under constraints such as summary length and grammar rules. The candidate summaries are then sorted by the sum of node scores, and only the top-k ones are remained. Finally, represent the source documents and candidate summaries as vectors using BERT, and calculate their similarity as the summary score. Candidate summary having the highest score will be output as the summarization result.

3.1 SRL Parsing

For the multiple documents to be summarized, we parse each document separately. Since SRL parsing is based on sentences, we first split each document into sentences by period. Then, different from the traditional SRL model, we leverage a pre-trained model BERT[1] to implement semantic role labeling. Such implementation does not need to rely on lexical or syntactic features, thus reducing the parsing difficulty.

SRL parsing consists of the following two steps:

(1) **Predicate Identification.** First feed each sentence into the BERT encoder to get the contextual representation, and then use a one-hidden-layer Multilayer Perceptron (MLP) to predict the predicate.
(2) **Argument Identification..** Detect the argument spans or syntactic argument heads and assign them the corresponding semantic role labels. On the basis of the network structure used in step 1), a one-layer Bidirectional Long Short-Term Memory (BiLSTM) is added to obtain the hidden states before MLP.

For each sentence, one or more different Predicate-Argument Structures (PASs) may be produced as the parsing results. Table 3 shows the five possible PASs obtained for the sample sentence. This means that semantic roles in a sentence are not unique, but all SRL parsing results should be considered when constructing SRG in the next step.

---

[1] BERT [3] is a bidirectional encoder representation model based on Transformers and has been widely used in many NLP tasks as the pre-trained model due to its outstanding performance.

Table 3. Five possible PASs obtained for the sample sentence.

Original Text: Newspapers reported wednesday that three top libyan officials have been tried and jailed in the Lockerbie case, but libyan dissidents said the reports appeared to be a political ploy by libya's leader, col.moammar Gadhafi

PAS1: [ARG0: Newspapers] [V: reported] [ARGM-TMP: wednesday] [ARG1: that three top libyan officials have been tried and jailed in the Lockerbie case], but libyan dissidents said the reports appeared to be a political ploy by libya's leader, col.moammar Gadhafi

PAS2: Newspapers reported wednesday that [ARG1: three top libyan officials] have been [V: tried] and jailed [ARGM-LOC: in the Lockerbie case], but libyan dissidents said the reports appeared to be a political ploy by libya's leader, col.moammar Gadhafi

PAS3: Newspapers reported wednesday that [ARG1: three top libyan officials] have been tried and [V: jailed] [ARGM-LOC: in the Lockerbie case], but libyan dissidents said the reports appeared to be a political ploy by libya's leader, col.moammar Gadhafi

PAS4: Newspapers reported wednesday that three top libyan officials have been tried and jailed in the Lockerbie case, but [ARG0: libyan dissidents] [V: said] [ARG1: the reports appeared to be a political ploy by libya's leader, col.moammar Gadhafi]

PAS5: Newspapers reported wednesday that three top libyan officials have been tried and jailed in the Lockerbie case, but libyan dissidents said [ARG1: the reports] [V: appeared] [ARG2: to be a political ploy by libya's leader, col.moammar Gadhafi]

3.2 Semantic Role Graph Generation

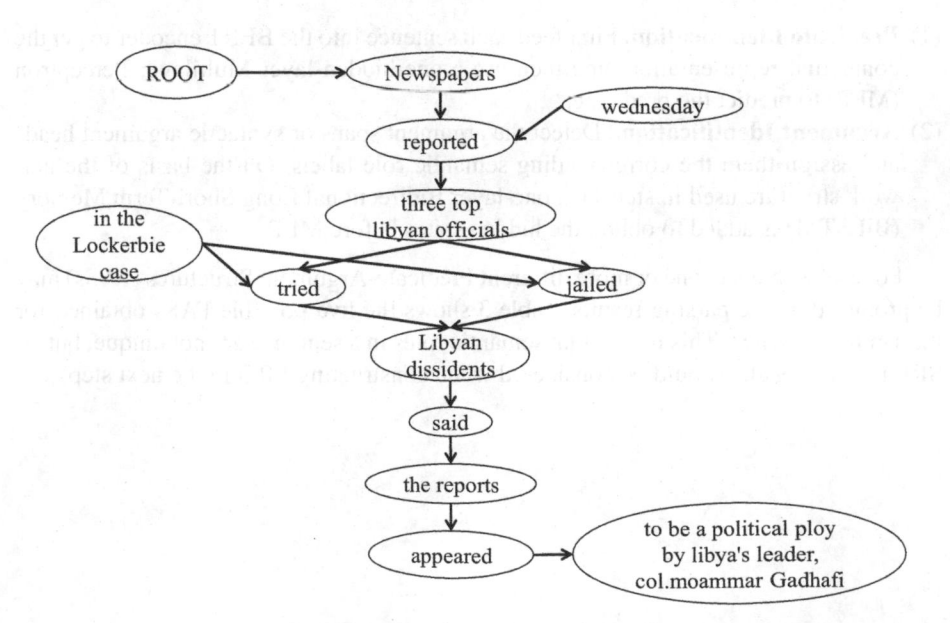

Fig. 3. A sample of semantic role graph.

Based on the SRL parsing result, we first construct a SRG for each document, then merge them to obtain a multi-document SRG without redundancy. Single SRG is constructed by the following two rules:

1. **Nodes** in SRG correspond to predicates and semantic roles that do not further contain other predicates. Besides, a virtual node ROOT is added as the initial node for each document.
2. **Edges** in SRG are all directed. AGR0 points to predicate, predicate points to AGR1, AGR1 points to ARG2, and ARGM points to predicate.

In Table 3, PAS2 is contained in PAS1 as ARG1 and PAS5 is contained in PAS4 as ARG1. In fact, when the semantic role of one PAS contains another PAS, it means that this semantic role has not yet been fully parsed. Rule (1) is made to ensure that all nodes in SRG can be no longer further parsed. Rule (2) is made to specify the connection direction of semantic roles in the same sentence. Semantic roles of different sentences are connected in the same order as appeared in the original text. In specific, the last semantic role of the previous sentence points to the first semantic role of the next sentence. In addition, each node has a start flag and an end flag, which are used to determine whether it is the beginning or the end of a complete sentence. Figure 3 shows the SRG generated for the example in Table 3.

After generating SRGs for every document, we merge them into one graph G. The merging method is as follows. Let p denote the traversal pointer to the current node of graph G, then:

1. Initialize p pointing to ROOT;
2. Add all nodes and edges of the first SRG to G;
3. Traverse the second SRG from ROOT: If the current node g does not exist in G, add g to G, and let g point to the node pointed by p, and p point to g; otherwise, let p point to the location of g in G;
4. Repeat step (3) until all nodes of the second SRG are traversed;
5. Repeat steps (3) and (4) until all single-document SRGs have been traversed.

The multi-document semantic role graph G obtained in this way contains all information of the source documents, and has no redundant information at the same time.

3.3 Candidate Summary Generation

By traversing the semantic role graph G, we generate candidate summaries. The pruning rules used for traversal are as follows:

1. The start node should contain a start flag, and the end node should contain an end flag;
2. The length of the generated summary obtained by traversing cannot exceed that of the gold summary;
3. It cannot form a loop.

For example, for the original text corresponding to the SRG shown in Fig. 3, if we assume the length of gold summary is 18, then the candidate summary generated should

Table 4. Candidate summaries for the SRG shown in Fig. 3.

Original Text: Newspapers reported wednesday that three top libyan officials have been tried and jailed in the Lockerbie case, but libyan dissidents said the reports appeared to be a political ploy by libya's leader, col.moammar Gadhafi
Candidate 1: Newspapers reported wednesday three top libyan officials tried in the Lockerbie case
Candidate 2: Newspapers reported wednesday three top libyan officials jailed in the Lockerbie case
Candidate 3: three top libyan officials tried in the Lockerbie case
Candidate 4: three top libyan officials jailed in the Lockerbie case
Candidate 5: libyan dissidents said the reports appeared to be a political ploy by libya's leader, col.moammar Gadhafi

contain no more than 18 words. Candidate summaries generated by traversing graph G are shown in Table 4.

Even pruned by the rules mentioned above, it may still generate many candidate summaries with similar semantics, which is not good for training in the next summary scoring step. To address this problem, we use a keyword scoring method based on TextRank to select the top-k candidate summaries. TextRank is essentially a way of deciding the importance of a vertex within a graph based on global information recursively drawn from the entire graph. The specific steps are as follows.

First, for each sentence in a document, perform word segmentation and part-of-speech tagging, filter out stop words, and keep only words with specified parts of speech like nouns, verbs, and adjectives.

Second, construct a directed graph $G = (V, E)$, in which vertices V are words passing the syntactic filter, and edges E are produced between those lexical units co-occurring within a window of words. The score of each vertex is initialized as 1, and then is updated as follows:

$$S(V_i) = (1 - d) + d * \sum_{j \in In(V_i)} \frac{1}{|Out(V_j)|} S(V_j) \tag{1}$$

where d is the damping factor ranged in $[0, 1]$, $In(V_i)$ is the predecessor set of vertex V_i, $Out(V_j)$ is the successor set of vertex V_j, and $S(V_i)$ indicates the importance of vertex V_i based on the relationship with its predecessors in G. By Eq. (1), the score of each vertex is iteratively calculated until convergence.

Third, build a dictionary D, whose key is the word corresponding to each vertex and value is its corresponding score. Then, calculate the keyword scores for each candidate summary based on D, and add them up to get the summary score. Finally, sort candidate summaries by summary score in descending order and output the top-k candidate summaries to the next summary scoring step.

3.4 Semantic Matching: Summary Scoring

Among the top-k candidate summaries, the one having the highest semantic similarity with the source text should be chosen as the final summary. To achieve this goal, we adopt a BERT-based semantic matching model. As to the weight updating strategy in BERT1, we adopt the same margin-based triplet loss as work [15]. The basic idea is to let the gold summary have the highest matching score, i.e., the higher the matching score, the better. Further, the widely acknowledged metric Recall-Oriented Understudy for Gisting Evaluation (ROUGE) F1 [20] is used to evaluate the quality of a candidate summary. Among which, ROUGE-1 and ROUGE-2 are used to assess the informativeness through unigram and bigram overlap separately, while ROUGE-L is used to assess the fluency through the longest common subsequence. Average of the three scores is taken as the candidate summary's ground truth.

Once the last step of semantic matching model (SMM) training is done, the whole summarization model (SRGSum) placed on the mobile edge server can be put to use. For the source documents submitted by terminal users, SRGSum first extracts the semantic roles through SRL parsing, based on which generate the semantic role graph and output the top-k candidate summaries. Finally, it derives the semantically meaningful embeddings in source text and each candidate summary leveraging the trained SMM, then calculates the cosine similarity of their embeddings as the matching score, and sends back the candidate summary with the highest score to the terminal users.

4 Experiments

4.1 Datasets and Evaluation Metrics

The proposed extractive summarization approach is validated on the following two datasets:

- **Telecom:** an in-house dataset from Zhejiang Telecom, which includes 512 topics and each topic contains 20 short customer complaint handling reports.
- **LCSTS:**[21] a public dataset named Large-scale Chinese Short Text Summarization, which is constructed from the Chinese microblogging website Sina Weibo. This corpus consists of over 2 million real Chinese short texts along with short summaries provided by the original writers.

We evaluate the summarization quality using the widely acknowledged metric ROUGE F1 [22], including ROUGE-1, ROUGE-2, and ROUGE-L. ROUGE-1 and ROUGE-2 assess the informativeness through unigram and bigram overlap separately, while ROUGE-L assesses the fluency through the longest common subsequence.

4.2 Baselines

To the best of our knowledge, existing text summarization approaches usually aim at different scenarios with this work from aspects like language type or summarization category. But the models they based on can be extracted for comparison. Here, we compare our summarization model SRMSum against a broad spectrum of strong baselines shown as follows:

- **LEAD::** a commonly used baseline method for text summarization, which simple concatenates the first sentence of each document as the final abstract.
- **TextRank** [19]: an unsupervised graph-based ranking model, which regards sentences in the text as nodes, uses the similarity between sentences as the edge weights, and calculates the importance scores of sentences according to the voting mechanism.
- **RNN**[22]: a model widely used in natural language processing, speech recognition, machine translation and automatic dialog response.
- **BertSum**[23]: a model applying BERT to extractive summarization for the first time, which improves extractive summarization through taking full advantage of the prior knowledge generated by BERT pre-training.

4.3 Experimental Setup

All models are trained on two GPUs (NVIDIA 3090) for 5 epochs. Batch size is set to 16, weight decay is set to 0.001, learning rate is set to 0.0001, margin is set to 0.01, and the number of warmup steps is set as 2000. Base version of BERT is used. The maximum co-occurrence window in keywords extraction is set to 6, and the damping factor d is set to 0.85. The top 20 candidate summaries are retained.

4.4 Results

Accuracy Analysis. Table 5 shows the evaluation results of accuracy by ROUGE F1 metrics on Telecom and LCSTS. ROUGE-1 (R-1) and ROUGE-2 (R-2) separately assess the informativeness through unigram and bigram overlap, while ROUGE-L (R-L) assesses the fluency through the longest common subsequence.

Table 5. Evaluation results on Telecom and LCSTS.

Model	Telecom			LCSTS		
	R-1	R-2	R-L	R-1	R-2	R-L
LEAD	45.90	30.10	43.30	27.85	16.13	25.07
TextRank	46.10	30.68	42.01	26.76	15.88	23.14
RNN	48.36	32.94	46.21	29.90	17.40	27.20
BertSum	50.30	35.83	46.93	32.32	21.70	28.30
SRGSum (The proposed)	**53.57**	**37.08**	**51.53**	**35.57**	**23.03**	**31.36**

From this table, we can see that our model SRGSum outperforms all counterpart models by a large margin on both datasets. For Telecom, it improves 7.67%, 6.98%, 8.23% compared with LEAD, and 3.27%, 1.25%, 4.60% compared with BertSum, on ROUGE-1, ROUGE-2, and ROUGE-L, respectively. For LCSTS, it improves 7.72%, 6.90%, 6.29% compared with LEAD, and 3.25%, 1.33%, 3.06% compared with BertSum, on ROUGE-1, ROUGE-2, and ROUGE-L, respectively.

Further, the following conclusions can be drawn based on the experimental results: 1) Supervised models (RNN, BertSum, SRGSum) perform better than unsupervised models (LEAD, TextRank); 2) Models with finer extraction granularity (SRGSum) result in higher accuracy.

Efficiency Analysis. For the proposed MEC framework for extractive summarization, time efficiency depends on network transmission efficiency and summary generation efficiency. Since the former one is affected by outside factors and costs not much time, we only focus on the latter one here. Figure 4 shows the time efficiency scalability with the number of documents under multi-document dataset Telecom. Since the text of each topic provided by telecom contains only 20 documents, we choose the number of 1, 5, 10, 15, 20 for the experiment. For each case, we run dozens of tests and take their average value as the running time.

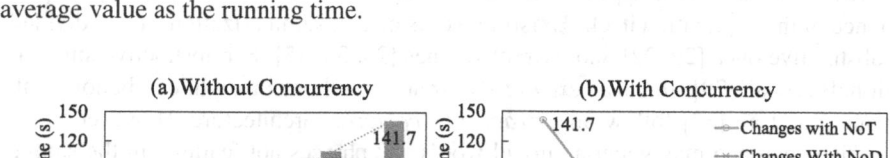
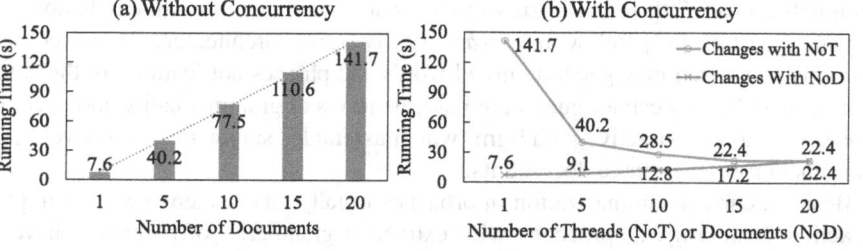

Fig. 4. Time efficiency scalability of the proposed approach.

Figure 4(a) shows the running time of our proposed model with SRG generation without concurrency. We can see that it has a near-linear relationship with the number of documents. On average, it costs 7–8 s to summarize one document. When the number increases to 20, the time cost is 141.7 s. Through analysis, we find that a considerable amount of time is spent on the SRG generation. So, we can easily improve its efficiency by increasing concurrency. Figure 4(b) shows the corresponding results, where the line with circle marks represents the changing trend of running time with the number of threads while the line with cross marks represents the achievable extremum efficiency under different number of documents. We can see that the increasing speed of the time cost is largely decreased to a level sufficient for daily use of terminal users.

Actually, for most optimization problems, satisfying all optimization objectives is usually impossible, especially for those contradictory ones. The extractive summarization approach proposed in this paper achieves quite high accuracy improvement while guaranteeing a sufficient time efficiency for daily use. It works quite well for summarization scenario of many short documents.

Based on the above experimental above, we can conclude that the innovatively proposed MEC framework for extractive summarization as a whole is capable of providing an efficient and accurate summarization service by offloading the time-consuming work to MES and using SRG and SMM to construct the summarization model.

5 Related Work

Mobile edge computing (MEC) [1] is a promising paradigm to promote the performance of real-time applications. Mobile edge server (MES) in this paradigm supports this goal through providing low-latency, high-availability, low-cost computing resources. The performance of offloaded tasks will be largely enhanced by MEC for time- and energy-limited terminal devices [9, 10]. Numerous research works have greatly facilitated the application of MEC [24–28]. Considering the many advantages of this paradigm, we choose it as the basic framework for our text summarization approach, so as to improve the efficiency through offloading the time-consuming summarization work to the mobile edge server.

However, for a specific application, the user experience is also influenced by the performance of the application itself. Existing works on text summarization can be divided into abstractive ones [29–32] and extractive ones [23, 33–35]. For abstractive summarization, Bae et al. [29] adopts an *extract-then-rewrite* architecture, while Lebanoff et al. [31] and Xu et al. [32] follow an *extract-then-compress* architecture. However, since abstractive approach may generate novel words and phrases not featured in the source documents [13], they cannot guarantee baseline levels of grammaticality and accuracy. Hence, we choose extractive paradigm, which assembles summaries exclusively from spans taken from the source documents.

Besides, current summarization approaches usually take sentence or paragraph as extraction granularity. Approaches with extraction granularity of paragraph have the greatest redundancy followed by sentence extraction [14, 15]. To reduce redundancy, traditional non-neural methods are proposed. Carbonell et al. [33] propose the Maximal Marginal Relevance (MMR) criterion for multi-document construction; Erkan et al. introduce a stochastic graph-based method for computing the relative importance of textual units [34]. The above two methods take sentence as the extraction granularity and score each sentence for summarization. In addition to the methods by sentence scoring, Radev et al. [35] generates summary for multiple documents using cluster centroids produced by a topic detection and tracking system. What's more, neural networks have also been used for extractive summarization in recent years. Yang Liu [23] first applies BERT to extractive summarization, making full use of the prior knowledge generated by BERT pre-training to select sentences. However, such approaches with extraction granularity of sentence are still not fined enough to guarantee the summarization conciseness and accuracy. There will still exist intra-sentence redundancy in the extracted sentences, and sentences filtered may still contain key information.

Based on the mentioned summarization works above, to further improve the performance of text summarization, we propose a novel extractive summarization model using SRG and SMM, which 1) has a finer extraction granularity of semantic role, 2) has the lowest redundancy, 3) has the highest semantic similarity by leveraging the prior knowledge based on BERT. Overall, our model can be viewed as an *extract-deduplicate-match* architecture, which extracts semantic roles from the original documents first, then generates a semantic role graph without redundant semantic role nodes, and finally selects the summary with the highest semantic matching score based on the prior knowledge learned through BERT pre-training.

Certainly, there also exists several text summarization methods with semantic role as granularity like research works [36–39]. However, research works [36, 37] aims at English documents which have significant grammatical differences between Chinses documents; research [38, 39] focuses on abstractive multi-document summarization which cannot always guarantee accuracy since it may generate novel words and phrases not featured in the source documents. Hence, it is quite necessary to propose a new extractive summarization approach based on semantic role specifically for multiple Chinese documents.

6 Conclusion

This paper proposes a novel MEC framework for extractive summarization using semantic role graph and semantic matching. To improve the efficiency of summarization service, this framework leverages the low-latency, high-availability, low-cost features of mobile edge servers to undertake the offloaded summarization work. To guarantee the accuracy of summarization, this framework embeds a summarization model with finer extraction granularity. This model improves the accuracy from both syntactic and semantic perspectives by semantic role graph and semantic matching techniques separately.

Experiments on both in-house and public datasets show that the innovatively proposed MEC framework for extractive summarization is capable of providing an accurate and efficient enough summarization service to mobile users for daily use. In future work, we will continue to explore methods to further improve accuracy while reducing time consumption.

Funding Statement. This work was funded by the Natural Science Foundation of Zhejiang Province, China, under Grant No. (LQ21F020019), and Key Research Project of Zhejiang Province, China, under Grant No. (2022C01145).

References

1. Mao, Y., You, C., Zhang, J., Huang, K., Letaief, K.B.: A survey on mobile edge computing: the communication perspective. IEEE Commun. Surv. Tutor. **19**(4), 2322–2358 (2017)
2. He, L., Lee, K., Lewis, M., Zettlemoyer, L.: Deep semantic role labeling: What works and what's next. In: Barzilay, R., Kan, M. Y. (eds.) ACL 2017, vol. 1, pp. 473–483. ACL, Vancouver (2017)
3. Devlin, J., Chang, M.W., Lee, K., Toutanova, K.: BERT: pre-training of deep bidirectional transformers for language understanding. In: Burstein J., Doran C., Solorio T. (eds.) NAACL-HLT 2019, vol. 1, pp. 4171–4186. ACL, Minneapolis (2019)
4. Fabbri, A.R., Li, I., She, T., Li, S., Radev, D.: Multi-news: a large-scale multi-document summarization dataset and abstractive hierarchical model. In: ACL 2019, vol. 1, pp. 1074–1084. ACL, Florence (2019)
5. Liu, P.J., Saleh, M., Pot, E., Goodrich, B., Sepassi, R., et al.: Generating Wikipedia by summarizing long sequences. CoRR abs/1801.10198, pp. 1–18 (2018)
6. Gerani, S., Mehdad, Y., Carenini, G., Ng, R.T., Nejat, B.: Abstractive summarization of product reviews using discourse structure. In: Moschitti, A., Pang, B., Daelemans, W. (eds.) EMNLP 2014, pp. 1602–1613. ACL, Doha (2014)

7. Yasunaga, M., Kasai, J., Zhang, R., Fabbri, A.R., Li, I., et al.: Scisummnet: a large annotated corpus and content-impact models for scientific paper summarization with citation networks. In: AAAI 2019, pp. 7386–7393. AAAI Press, Hawaii (2019)
8. Sandhu, A.K.: Big data with cloud computing: discussions and challenges. Big Data Min. Anal. **5**(1), 32–40 (2022)
9. Xu, X.L., Jiang, Q.T., Zhang, P.M., Cao, X.F., Khosravi, M.R., et al.: Game theory for distributed IoV task offloading with fuzzy neural network in edge computing. IEEE Trans. Fuzzy Syst. **30**(11), 4593–4604 (2022)
10. Nath, S., Wu, J.: Deep reinforcement learning for dynamic computation offloading and resource allocation in cache-assisted mobile edge computing systems. Intell. Converg. Netw. **1**(2), 181–198 (2020)
11. Yan, C., Zhang, Y.K., Zhong, W.Y., Zhang, C., Xin, B.: A truncated SVD-based ARIMA model for multiple QoS prediction in mobile edge computing. Tsinghua Sci. Technol. **27**(2), 315–324 (2022)
12. Zhang, Y., Zhang, H., Cosmas, J., Jawad, N., Zarakovitis, C.C.: Internet of radio and light: 5G building network radio and edge architecture. Intell. Converg. Netw. **1**(1), 37–57 (2020)
13. See, A., Liu, P.J., Manning, C.D.: Get to the point: summarization with pointer-generator networks. In: Barzilay, R., Kan, M.Y. (eds.) ACL 2017, vol. 1, pp. 1073–1083. ACL, Vancouver (2017)
14. Liu, Y., Lapata, M.: Hierarchical transformers for multi-document summarization. In: Korhonen, A., Traum, D.R., Marquez, L. (eds.) ACL 2019, vol. 1, pp. 5070–5081. ACL, Florence (2019)
15. Zhong, M., Liu, P., Chen, Y., Wang, D., Qiu, X., et al.: Extractive summarization as text matching. In: Jurafsky, D., Chai, J., Schluter, N., Tetreault, J.R. (eds.) ACL 2020, vol.1, pp. 6197–6208. ACL, Seattle (2020)
16. Tan, Z.X., Wang, M.X., Xie, J., Chen, Y.D., Shi, X.D.: Deep semantic role labeling with self-attention. In: McIlraith, S.A., Weinberger, K.Q. (eds.) AAAI 2018, pp. 4929–4936. AAAI Press, New Orleans, USA (2018)
17. Aksoy, C., Bugdayci, A., Gur, T., Uysal, I., Can, F.: Semantic argument frequency-based multi-document summarization. In: ISCIS 2009, pp. 460–464. IEEE, Guzelyurt (2009)
18. Shi, P., Lin, J.: Simple bert models for relation extraction and semantic role labeling. CoRR abs/1904.05255, pp. 1–6 (2019)
19. Mihalcea, R., Tarau, P.: Textrank: bringing order into text. In: EMNLP 2004, pp. 404–411. ACL, Barcelona (2004)
20. Lin, C.Y.: Rouge: a package for automatic evaluation of summaries. In: ACL 2004, pp. 74–81. ACL, Barcelona (2004)
21. Hu, B.T., Cnen, Q.C., Zhu, F.Z.: LCSTS: a large scale chinese short text summarization dataset. In: Marquez, L., Callison-Burch, C., Su, J., Pighin, D., Marton, Y. (eds.) EMNLP 2015, pp. 1967–1972. ACL, Trivandrum (2015)
22. Zaremba, W., Sutskever, I., Vinyals, O.: Recurrent neural network regularization. CoRR abs/1409.2329, pp. 1–8 (2014)
23. Liu, Y.: Fine-tune BERT for extractive summarization. CoRR abs/1903.10318, pp. 1–6 (2019)
24. Tong, Z., Ye, F., Yan, M., Liu, H., Basodi, S.: A survey on algorithms for intelligent computing and smart city applications. Big Data Min. Anal. **4**(3), 155–172 (2021)
25. Xu, X.L., Li, H.Y., Xu, W.J., Liu, Z.J., Yao, L., et al.: Artificial intelligence for edge service optimization in internet of vehicles: a survey. Tsinghua Sci. Technol. **27**(2), 270–287 (2022)
26. Dong, J., Wu, W., Gao, Y., Wang, X., Si, P.: Deep reinforcement learning based worker selection for distributed machine learning enhanced edge intelligence in internet of vehicles. Intell. Converg. Netw. **1**(3), 234–242 (2020)

27. Xu, X.L., Tian, H., Zhang, X.Y., Qi, L.Y., He, Q., et al.: DisCOV: distributed COVID-19 detection on X-Ray images with edge-cloud collaboration. IEEE Trans. Serv. Comput. **15**(3), 1206–1219 (2022)
28. Yang, Y.H., Yang, X., Heidari, M., Srivastava, G., Khosravi, M.R., et al.: ASTREAM: data-stream-driven scalable anomaly detection with accuracy guarantee in IIoT environment. IEEE Trans. Netw. Sci. Eng. (2022). https://doi.org/10.1109/TNSE.2022.3157730
29. Bae, S., Kim, T., Kim, J., Lee, S.G.: Summary level training of sentence rewriting for abstractive summarization. CoRR abs/1909.0875, pp. 1–11 (2019)
30. Zhang, H., Cai, J., Xu, J., Wang, J.: Pretraining-based natural language generation for text summarization. CoRR abs/1902.09243, pp. 1–9 (2019)
31. Lebanoff, L., Song, K.Q., Dernoncourt, F., Kim, D.S., Kim, S., et al.: Scoring sentence singletons and pairs for abstractive summarization. In: ACL 2019, vol. 1, pp. 2175–2189. ACL, Florence (2019)
32. Xu, J., Durrett, G.: Neural extractive text summarization with syntactic compression. CoRR abs/1902.00863, pp. 1–12 (2019)
33. Carbonell, J., Goldstein, J.: The use of MMR, diversity-based reranking for reordering documents and producing summaries. In: ACM SIGIR 1998, pp. 335–336. ACM, Melbourne, Vic. (1998)
34. Erkan, G., Radev, D.R.: Lexrank: graph-based lexical centrality as salience in text summarization. J. Artif. Intell. Res. **22**(1), 457–479 (2004)
35. Radev, D.R., Jing, H., Styś, M., Tam, D.: Centroid-based summarization of multiple documents. Inf. Process. Manage. **40**(6), 919–938 (2004)
36. Yan, S., Wan, X.J.: SRRank: leveraging semantic roles for extractive multi-document summarization. IEEE-ACM Trans. Audio Speech Lang. Process. **22**(12), 2048–2058 (2014)
37. Mohamed, M.A., Oussalah, M.: SRL-ESA-TextSum: a text summarization approach based on semantic role labeling and explicit semantic analysis. Inf. Process. Manage. **56**(4), 1356–1372 (2019)
38. Su, M.H., Wu, C.H., Cheng, H.T.: A two-stage transformer-based approach for variable-length abstractive summarization. IEEE-ACM Trans. Audio Speech Lang. Process. **28**(1), 2061–2072 (2020)
39. Li, W., Zhuge, H.: Abstractive multi-document summarization based on semantic link network. IEEE Trans. Knowl. Data Eng. **33**(1), 43–54 (2021)

A Paper Citation Link Prediction Method Using Graph Attention Network

Zhixuan Zou, Yiwen Sun, Weiguo Li[✉], Yiqi Li, and Yintong Wang

School of Artificial Intelligence, Nanjing Xiaozhuang University, Nanjing, Jiangsu, China
wglee460@hotmail.com

Abstract. Link prediction utilizes existing data to prediction the relational ties between any pair of nodes within a relationship network, and then holds the potential to recover or anticipate absent relationship data. Nevertheless, the existing methods encounter the problems of performance degradation, insufficient utilization of attribute information of nodes or edges caused by the increase of network size, which brings extremely adverse effects to the citation prediction of papers. We proposed a novel approach to paper citation prediction using graph attention network (GAT), which uses a dual attention mechanism, amalgamating both the structural and content-related facets of network nodes, to adeptly gauge the influence exerted by neighboring nodes upon the central node. Additionally, it adeptly integrates attribute data associated with network nodes and edges, optimizing the graph attention network for the specific task of link prediction. The experimental results shown that our method can achieve a significant improvement in the prediction accuracy of the paper citation link on the Cora dataset.

Keywords: Link Prediction · Graph Attention Network · Paper Citation Prediction

1 Introduction

As a key task in graph structure data, link prediction aims to infer the possible future connection relationship between nodes in the graph, which is widely used in many fields, such as social networks, bioinformatics, citation networks, recommendation systems, etc.. Academic citation network is a complex network composed of academic papers as citation relationships between nodes and edges. Its link prediction is to predict new citation relationships that may occur in the future based on existing network structure, node or edge attribute information. It has important value in discovering new academic trends, recommending related papers, and evaluating the influence of papers.

Traditional link prediction methods can be broadly categorized into two principal approaches, statistical features and matrix decomposition. Among them, the statistical feature method is used to construct the topological structure network, such as the number of shared neighbors, Jaccard coefficient, etc., so as to calculate the probability of connection between two nodes. Traditional link prediction method of matrix decomposition predicts the missing edges by decomposing the adjacency matrix or similarity matrix of

© The Author(s), under exclusive license to Springer Nature Singapore Pte Ltd. 2024
H. Jin et al. (Eds.): IAIC 2023, CCIS 2058, pp. 32–41, 2024.
https://doi.org/10.1007/978-981-97-1277-9_3

the graph. Previous studies have shown that these two methods are overly dependent on the topology of the network, while ignoring the attribute information of nodes or edges in the network. The link prediction method based on deep learning uses the embedded representation of learning nodes to predict the link, which can well consider the attribute information of nodes and edges in the network. However, when dealing with large-scale complex networks, the efficiency and performance of deep learning methods are seriously degraded.

In order to solve the above problems, we proposed a paper citation link prediction method using graph attention network. The specific works are as follows: (1) Theoretically sort out how to consider the network topology and the attribute information of nodes or edges in link prediction. (2) Introduce the modeling method and emphasize how to use graph attention network technology to deal with link prediction tasks. (3) The experimental analysis on the Cora dataset shown that our method has improved the prediction accuracy of the paper citation link.

2 Related Work

Citation frequency is the most representative, simplest, most standard and most objective index to measure academic influence, which runs through the whole process of scientific research activities. This section will introduce the current paper citation prediction algorithms from the perspective of the use of features, including three types of methods which are method based on fixed-length feature vectors, method based on variable-length citation timing features, and method based on citation network features.

Method based on fixed-length feature vectors. This kind of method usually represents each paper as a fixed-length feature vector, and then uses traditional machine learning algorithms, such as support vector machine, random forest, etc., for classification or regression prediction. It has the advantages of fast calculation speed and easy implementation, but it cannot make full use of the rich information in the paper, so there are some limitations in the prediction accuracy. Geng et al. [1] tried seven algorithms: Naive Bayes, Logistic Regression, Support Vector Machine, Gradient Boosting Decision Tree, XGBoost, AdaBoost and Random Forest, and found that XGBoost and Random Forest can achieve the best prediction results.

Method based on variable-length citation timing features. This kind of method usually represents each paper as an indefinite length sequence, where each element represents the number of times the paper is cited at a certain time point or other relevant information, and then uses deep learning models such as recurrent neural network and convolutional neural network to model the sequence. These methods can make full use of time series information and adapt to different length sequences, but they cannot make full use of other rich information in the paper, such as title, abstract, author, etc.. Jiang et al. [2] predicted the cited time series of newly published papers in the dynamic citation network, and proposed the HINT framework encoding the historical information before the publication of the paper, the cited time series of the new paper is predicted from the publication date, which can be used to solve the cold start problem. Su et al. [3] proposed a method that integrates multi-dimensional features for predicting and analyzing the citation frequency of academic papers. This method regards the citation frequency of papers as a continuous value, and can simultaneously use fixed-length features, citation network features and citation timing features, and fuse and predict the extracted features through a multi-head attention mechanism.

Method based on citation network features. Such methods usually represent each paper as a node, and the citation relationship between the papers as an edge, and then use the graph neural network to model the nodes and edges, and use the classification or regression model to predict. Zhang et al. [4] proposed a paper citation prediction based on multi-task learning framework. GCN and Transformer are used to deal with the network topology and text features of the paper, respectively. At the same time, the paper's own attributes, author attributes, journal attributes and other characteristics are considered, and knowledge transfer between different tasks is realized by sharing parameters and gating mechanism. Zhu et al. [5] proposed a paper citation prediction method based on heterogeneous feature fusion, which uses multiple GAT layers to extract the topological features, content features and meta-information features of the paper respectively, and uses a fusion layer to integrate these features, so as to realize heterogeneous feature fusion and weight distribution between different types of sequences. A path-aware graph neural network model is proposed by Yang et al. [6], which can use the interaction structure and attribute information between nodes for link prediction.

3   Paper Citation Link Prediction Method Using Graph Attention Network

3.1   Paper Citation Link Prediction Framework

Graph Attention Network (GAT) is a graph neural network based on self-attention mechanism. Its basic idea is to give different weights when aggregating neighbor node information, and the weights are learned by the attention mechanism to adaptively weigh the influence of neighbor nodes on the central node [7, 8]. The framework of the paper citation link prediction method of the graph attention network is shown in Fig. 1. The specific processes are as follows.

(1) Database Preparation. The initial step entails the compilation of a comprehensive paper database. Subsequently, pertinent information pertaining to nodes and edges is extracted from this repository to serve as the algorithm's input.

(2) Node Pre-representation. In this phase, each node is imbued with the representation of a paper, while edges signify the citation relationships. The paper's attributes encompass attributes such as title, abstract, authorship details, and publication timestamps. Attribute information for each node is systematically extracted and encoded, yielding a fixed-length feature vector that serves as the initial node representation.

(3) Graph Attention Network Construction. Employing the Graph Attention Network, nodes and edges are meticulously modeled to yield an evolved graph neural network. Notably, GAT is instrumental in dynamically adjusting the weightage of neighboring nodes' influence on the focal node, thereby engendering updates to the feature representations of individual nodes.

(4) Subgraph Extraction and Feature Fusion. Subsequent to the construction of the GAT-based graph, multiple subgraphs are extracted, and feature fusion operations are conducted on each of these subgraphs. The rationale behind this operation is to leverage subgraphs as intermediate outputs, thereby mitigating computational complexity and enhancing computational efficiency. The self-attention network is deployed to compute attention weights between nodes, thereby facilitating updates to node feature representations.

(5) Graph Neural Network Generation and Training. A novel graph neural network generator is invoked to formulate an augmented graph neural network. Training data, alongside an appropriate loss function, drive the adaptation of parameters within the generator and the graph neural network, thereby optimizing their capacity to accurately capture the underlying data distribution.

(6) Paper Citation Prediction. The trained graph neural network is subsequently employed to predict linkages between any two nodes (representing papers). This comprehensive approach yields predictions for paper citation relationships, thereby contributing to the advancement of citation prediction methodologies within academic contexts.

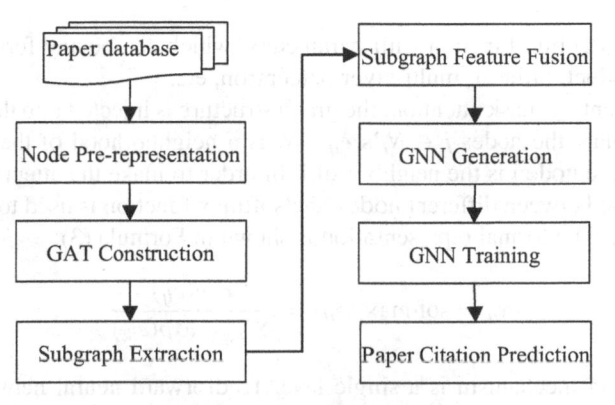

Fig. 1. The framework of Paper Citation link Prediction.

3.2 Paper Citation Link Prediction Method

In the graph attention network, the single-head attention mechanism converts the input into an output through an attention weight function [7]. The multi-head attention mechanism runs multiple attention operations in parallel, and the model can capture multiple different types of information at the same time. Each 'head' has its own parameters. Therefore, each 'head' can learn different features of the input in different representation spaces, independently apply multiple sets of parameters for feature transformation, and finally perform splicing or averaging, thereby enhancing the expression ability and generalization ability of the model.

Construct a single graph attention layer, and its input is a set of node features $\mathbf{h} = \{\vec{h}_1, \vec{h}_2, ..., \vec{h}_N\}$, $\vec{h}_i \in \mathbb{R}^F$, N is the number of nodes, and F is the number of features of each node. This layer generates a new set of node features (potentially different cardinalities $F\prime$), the outputs are $\mathbf{h} = \{\vec{h\prime}_1, \vec{h\prime}_2, ..., \vec{h\prime}_N\}$, $\vec{h\prime}_i \in \mathbb{R}^{F\prime}$.

Node Representation Learning: Given a graph $G = (V, E)$, V is the set of nodes and E is the set of edges. Each node $v \in V$ has a d-dimensional input feature vector. Firstly, these input feature vectors are transformed through a graph convolution layer to obtain a new node representation. The formal representation is shown in Formula (1):

$$h'_v = Wh_v \tag{1}$$

where W is a $d\prime \times d$ weight matrix, $\vec{h\prime}_v$ is a new feature vector of node v, and the dimension is $d\prime$.

Attention Weight Calculation: The attention mechanism is introduced to calculate the influence weight of adjacent nodes on the current node. For each pair of nodes (i, j), their representations are first spliced or other types of combinations, and then their attention coefficients are calculated through a shared attention network. The formal representation is shown in Formula (2):

$$e_{ij} = a(W\vec{h_i}, W\vec{h_j}) \tag{2}$$

where a is an attention function with parameters, which can be any form of function, such as dot product, bilinear, multi-layer perceptron, etc.

By implementing mask attention, the graph structure is injected into the mechanism, and only calculate the nodes $j \in N_i$'s e_{ij}, N_i is a neighborhood of the node i in the graph, that is, the node j is the neighbor of i. In order to make the attention coefficient easy to compare between different nodes, the softmax function is used to normalize all the choices of j. The formal representation is shown in Formula (3):

$$\alpha_{ij} = \text{softmax}_j(e_{ij}) = \frac{exp(e_{ij})}{\sum_{k \in N_i} exp(e_{ik})} \tag{3}$$

The attention mechanism is a single-layer feedforward neural network, which is parameterizationed by weight vector $\vec{a} \in \mathbb{R}^{2F\prime}$, and the LeakyReLU nonlinear function is applied. The formal representation is as follows:

$$\alpha_{ij} = \frac{exp(\text{LeakyReLU}(\vec{a}^T[W\vec{h_i}||W\vec{h_j}]))}{\sum_{k \in N_i} exp(\text{LeakyReLU}(\vec{a}^T[W\vec{h_i}||W\vec{h_k}]))} \tag{4}$$

where | | denotes the concat operation.

Feature Combination: After calculating all the attention coefficients, the features of the adjacent nodes of each node v are weighted and summed. The weight is the attention coefficient, and a new representation of the node v is obtained. This step takes into account the topological structure and node characteristics of the graph. The formal representation is shown in Formula (5):

$$\vec{h'_i} = \sigma(\sum_{j \in N_i} \alpha_{ij} W\vec{h_j}) \tag{5}$$

where σ is a nonlinear activation function, such as ReLU, tanh, etc. α_{ij} is the normalized attention weight, which is calculated by the softmax function.

Link prediction: after obtaining a new representation of all nodes, link prediction can be performed. For each pair of nodes (i, j), the existence of edges between them is predicted by comparing their representations, such as calculating the dot product. By predicting all nodes, the prediction results of the citation network can be obtained. The Formula is expresses as:

$$h_i'' = Concat(Head_1(h_i'), Head_2(h_i'), ..., Head_k(h_i')) \tag{6}$$

where h_i' is the update feature of node i in the previous step, $Head_n$ denotes n attention heads, Concat represents the connection of multiple vectors into a large vector. k is the number of attention heads set in this paper.

3.3 Algorithm Implementation

In the paper citation prediction, the dual attention mechanism of node content and structure is the focus of this paper. The content attention mechanism calculates the similarity between nodes based on the content of the paper, such as title, abstract and other features. The structural attention mechanism is based on the citation relationship of the paper, calculates the structural similarity between the nodes, and captures the citation pattern of the paper. The paper citation prediction algorithm of graph attention network is implemented as follows:

Step 1: Initialize node features: The initial node features (Eq. 1) come from the academic citation data. Each node represents a paper, and the characteristics of each node correspond to the content of the paper. An adjacency matrix is also constructed to describe the structure of the citation network.

Step 2: Initialize the attention head: Multiple attention heads are assigned to each layer of GAT to capture the influence of multiple fields or topics that a paper may involve. Each attention head corresponds to a specific weight matrix $W \in \mathbb{R}^{F' \times F}$, which will be learned and adjusted during the training process.

Step 3: Node feature transformation and calculation of attention weight: Each attention head uses content attention and structural attention to calculate the attention coefficient of its neighbor nodes (Eq. 2). Firstly, the linear transformation of the node features is carried out, and then the dot product of the transformed node features is calculated. The dot product values are transformed into weights by functions softmax (Eq. 3). Then the mapped feature vectors are spliced together, and the spliced vectors are mapped to real numbers through a feedforward neural network, and the attention value is obtained through the LeakyReLU activation function (Eq. 4). In order to make the attention coefficient comparable and sum to 1, the attention values of all neighbor nodes of each node are normalized (Eq. 5).

Step 4: Combination of multi-head results: The above steps are executed in parallel under the multi-head attention mechanism, and the node features obtained by each attention head are combined to generate the final node feature representation. A common way to combine the results of each head is to connect the results of each head, because each head may capture the influence of different topics, and the connection can retain this information(Eq. 6).

4 Experiment and Result Analysis

4.1 Experimental Data Sets

As shown in Table 1, Cora dataset includes 2708 scientific papers as nodes and 5429 citation relations as edges. There are seven categories in total, which are neural network, reinforcement learning, rule learning, probability method, genetic algorithm, theoretical research and case correlation. The features of each paper are obtained by the bag-of-words model. The dimension is 1433. Each dimension represents a word, 1 indicates that the word has appeared in this article, and 0 indicates that it has not appeared. This part corresponds to the input of the ' content ' attention mechanism. This part corresponds to the input of the ' content ' attention mechanism. The information of the adjacency matrix corresponds to the input of the ' structure ' attention mechanism. The goal of the data set is to predict the domain category of each node based on the characteristics and reference relationships of the nodes.

Table 1. The description information of Cora dataset.

Characteristics	Description
Data type	Citation network data sets, including scientific papers and citation relationships
Number of nodes	2708
Number of edges	5429
Node category	7 kinds
Node characteristics	The 1433-dimensional word vector indicates whether there are corresponding words in the paper

4.2 Experimental Parameter Settings

In terms of parameter setting, the number of layers of GAT, the number of hidden units in each layer, the number of attention heads, learning rate, training times and other factors are mainly considered. The details are shown in Table 2 as below.

In the experiment, the data set is divided into training set, validation set and test set, accounting for 80%, 10% and 10% respectively. The training set and verification set will be used to train and adjust the model, and finally the test set will be tested to evaluate the performance of the model.

4.3 Evaluating Indicators

In the experimental results, the accuracy of link prediction is mainly concerned. In this paper, the accuracy Precision and the area under the ROC curve (AUC) are used as evaluation indicators [9].

Table 2. The model parameter settings.

Parameter name	Parameter value	Parameter declaration
Number of layers of GAT	2	The GAT model consists of two graph attention layers
Number of attention heads	8	The number of multi-head attention mechanisms used in each GAT layer
The number of hidden layer nodes per layer	8	The output feature dimension of each GAT layer
Learning rate	0.005	Control the speed of model learning, the larger the value, the faster the learning speed, but too large may lead to failure to converge
Optimizer	Adam	Update network weight to reduce the training error
L2 regularization coefficient	0.005	The coefficient of the L2 regularization term, used to prevent overfitting
Training cycle	1000	The number of training of the model
Early stop cycle	20	If the accuracy of the validation set is not significantly improved after 20 training cycles, the training is stopped

Accuracy Precision: refers to calculating the proportion of truly positive samples in all samples that are predicted to be positive. The calculation formula of accuracy is:

$$Precision = \frac{TP}{TP + FP} \tag{7}$$

where TP is the real number of cases, namely the target of correct prediction, and FP is the false positive number of cases, namely the target of false prediction.

AUC: This index takes into account the classification ability of the classifier for both positive and negative cases. In the case of unbalanced samples, it can still make a reasonable evaluation of the classifier. It also represents the area under the ROC curve, and the value of this area will not be greater than 1. Because the ROC curve is generally above the straight line y = x, the value range of AUC is between 0.5 and 1. The closer the AUC is to 1.0, the higher the authenticity of the detection method is. When the value is equal to 0.5, the authenticity is the lowest and has no application value.

4.4 Experimental Results

In order to verify the effectiveness of the proposed method, the shared neighbor number method and the Jaccard coefficient method are selected as the benchmark methods. Both of these methods are based on statistical characteristics. According to the topology of the network to calculate the probability of connection between two nodes, but the attribute information of nodes or edges is ignored. The experimental results are shown in Table 3. It can be seen from the table that the GAT model is superior to the shared neighbor number method and the Jaccard coefficient method in terms of accuracy and AUC value, indicating that the GAT model can make better use of the information in the citation network and improve the accuracy of link prediction.Meanwhile, the GAT model has the best performance when considering the node content and structure at the same time, indicating that the combination of node content and structure information can enhance the expression ability and generalization ability of the model.

Table 3. Comparison of experimental results.

Model	Precision	AUC
GAT (considering both node content and structure)	0.932	0.952
GAT (only consider the node content)	0.895	0.912
GAT (only considering the node structure)	0.910	0.928
Shared neighbor number method	0.720	0.780
Jaccard coefficient method	0.650	0.730

Figure 2 is the convergence curve of the GAT model and the shared neighbor number method and the Jaccard method on the link prediction task. As shown in Fig. 2, the convergence speed of the GAT model is faster than the shared neighbor number method and the Jaccard coefficient method, indicating that the GAT model can learn the features in the citation network faster. At the same time, the convergence curves of the GAT model are relatively smooth, indicating that the GAT model can fit the data distribution more stably and avoid over-fitting or under-fitting.

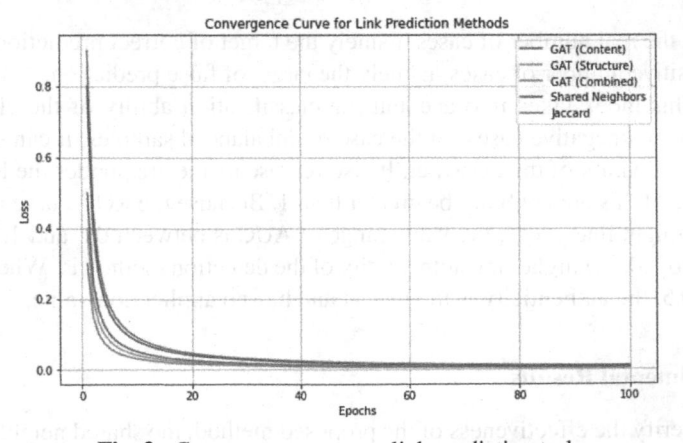

Fig. 2. Convergence curve on link prediction task.

5   Conclusion

This paper proposed a paper citation network link prediction method based on the dual attention mechanism of node content and structure. GAT can adaptively adjust the influence weight of each adjacent node on the current node. It does not need to pre-define the meta-path or walk strategy, nor does it need to perform matrix decomposition or fill missing values, and has better scalability and adaptability. This method can use both node content and structure information at the same time, and the expression ability and generalization ability of the model are improved to avoid the performance degradation caused by the lack of certain information. In the future work, we should also consider the quality and influence of the paper, distinguish the papers in different fields, and analyze the papers in different time periods, so as to further optimize the method.

References

1. Geng, Q., Jing, R., Jin, J., Luo, Q.: Citation prediction and influencing factors analysis on academic papers. Libr. Inf. Serv. **62**(14), 29–40 (2018)
2. Jiang, S., Koch, B., Sun, Y.: HINTS: citation time series prediction for new publications via dynamic heterogeneous information network embedding. In: Proceedings of the Web Conference 2021, New York, NY, USA, pp. 3158–3167 (2021)
3. Su, Z.: Prediction and analysis of citation counts for academic papers based on multi-dimensional features. Libr. Sci. Res. Work **4**, 49–55 (2023)
4. Zhang, D., Mao, Y., Zhang, S., Cheng, Y., Shi, C.: Predicting paper citations via multi-task learning. J. Minnan Normal Univ. (Natural Science) **35**(3), 46–53 (2022)
5. Zhu, D., Huang, X.: Paper citation prediction method based on heterogeneous feature fusion. J. Data Acquis. Process. **37**(5), 1134–1144 (2022)
6. Yang, S., et al.: Inductive link prediction with interactive structure learning on attributed graph. In: Oliver, N., Pérez-Cruz, F., Kramer, S., Read, J., Lozano, J.A. (eds.) Machine Learning and Knowledge Discovery in Databases. Research Track. ECML PKDD 2021. Lecture Notes in Computer Science(), vol 12976. Springer, Cham (2021). https://doi.org/10.1007/978-3-030-86520-7_24
7. Velickovic, P., Cucurull, G., Casanova, A., Romero, A.: Graph Attention Networks. In: International Conference on Learning Representations, arXiv:1710.10903v3, pp. 1–12 (2018)
8. Wang, J., Kong, L., Huang, Z., Xiao, J.: Survey of graph neural network. Comput. Eng. **47**(4), 1–12 (2021)
9. Zhang, C., Zhu, L., Yu, L.: Review of attention mechanism in convolutional neural networks. Comput. Eng. Appl. **57**(20), 64–72 (2021)

Research on Emotional Analysis of Tibetan Short Text Based on Fusion Sentiment Lexicon

Wan Ma Dao Ji[1]([✉]), An Jian Cai Rang[1,2,3], Jan Yang Cuo[1], Gong Bao Jia[1],
De Ji Cuo[1], Zan La Gong[1], and Yang Qin[1]

[1] School of Computer Science, Qinghai University for Minzu, Xining 810007, China
3312139609@qq.com
[2] State Key Laboratory of Intelligent Information Processing and Application of Tibetan
Language, Jointly Constructed By the Ministry of Provincial Affairs and the Ministry of
Education of China, Xining, China
[3] Qinghai Key Laboratory of Tibetan Information Processing and Machine Translation, Item
No. 2021-Z-001, Xining, China

Abstract. Short-text sentiment analysis plays an integral role in predicting sentiment polarity. The current Tibetan short-text sentiment analysis model applies deep neural networks to learn some local grammatical structure information but ignores known sentiment lexicon information in modeling. Aiming at the above problems, this paper proposes a method for sentiment analysis of Tibetan short-text that incorporates sentiment lexicon information. The method firstly uses the word features extracted by Word2vec and Glove to fuse with the sentiment dictionary information, and inputs them into CNN & BiLSTM networks to extract text context features; secondly, adjusts the weights of the sentiment features through the self-attention mechanism; and finally, uses Softmax to infer the sentiment categories. The experimental results show that the accuracy of this method can reach 93.18% on the self-built sentiment analysis corpus. In addition, compared with the methods of extracting syllable features and word features, the accuracy of this paper's method can be improved by 4.92 and 3.43% respectively, which proves the effectiveness of the fused sentiment lexicon.

Keywords: Short Tibetan Text · Emotional Analysis · CNN · BiLSTM · Self-Attention · Emotional Lexicon

1 Introduction

Sentiment analysis is an important research in the field of Tibetan information processing. Tibetan text sentiment analysis can be regarded as a text classification task to some extent. Sentiment analysis of Tibetan text refers to analyzing, processing, summarizing, and judging the emotional tendency of subjective text with emotional color. Short-text sentiment analysis mainly focuses on subjective short-texts such as WeChat public numbers and comments on major Tibetan websites for sentiment classification, in which the sentiment tendency mainly includes positive sentiment and negative sentiment and so on. With the continuous development of network information technology, the demand

H. Jin et al. (Eds.): IAIC 2023, CCIS 2058, pp. 42–57, 2024.
https://doi.org/10.1007/978-981-97-1277-9_4

for public opinion analysis, news event detection, commodity recommendation, and academic research has gradually increased. However, analysing positive or negative views is a task that requires in-depth analysis and a deep understanding of the context of the text, making use of common sense and domain knowledge as well as linguistic knowledge. Even for humans, the interpretation of opinions can be biased due to the different experiences and positions of each individual. Therefore, this paper attempts to automate the sentiment classification of microblog comments through deep learning models and aims to achieve accurate and stable model results.

The current short-text sentiment classification methods can be roughly divided into three categories: the first category is the sentiment classification method based on sentiment dictionaries, which realizes the division of sentiment polarity under different granularity according to the sentiment polarity of the sentiment words provided by different sentiment dictionaries; the second category is the sentiment classification method based on traditional machine learning, which is a kind of learning method that trains a model by given data and predicts the results through the model method; the third category is the deep learning based sentiment classification method, which is carried out using neural networks, and the method can be further categorized into the sentiment classification method of single neural network, the sentiment classification method of hybrid (combined, fused) neural network, the sentiment classification method that introduces the attention mechanism, and the sentiment classification using pre-trained models. Du Hui et al. [1] used machine learning methods such as the word2vec word vector technique to study the sentiment text commented by users on microblogs as a text categorization problem; some scholars have also used sentiment lexicon combined with machine learning and deep learning algorithm models to conduct text categorization research [2–4]; Zhao Zhiyang et al. [5] combined the attention model of graphical convolutional neural networks and word sentiment sequence features to complete the Chinese sentiment analysis. Ding Meirong et al. [6] constructed a pre-training model based on the BiLSTM neural network for experiments, and at the same time, with the basic lexicon as the main body, constructed an extended sentiment lexicon applicable to hotel reviews to realize the sentiment analysis of hotel reviews. He Xu et al. [7] firstly use the BERT pre-training model to dynamically represent the evaluation object and its context respectively, and then extract the semantic information of the two through the BiLSTM network, and then input the interactive attention module to get the interaction relationship between the two and the extracted semantic information to reorganize the final representation of the evaluation object and its context, and then complete the Chinese sentiment analysis. The above study used pre-trained models and traditional recurrent neural networks to classify the sentiment of text, but the model selection was based on natural language processing models and not on models originally used in image processing, such as text convolution.

In terms of sentiment classification of Tibetan text, Yan et al. [2] realized a sentence sentiment classification of Tibetan text based on a sentiment dictionary, which constructed a dictionary including base words, negatives, double negatives, adverbs of degree, and transitions manually, combined polarity words and modifiers to form a polarity phrase as the basic unit of polarity computation, and investigated the effect of transitions on the sentiment polarity of a sentence to propose a method of sentence sentiment analysis based on polarity lexicon of Tibetan text sentence sentiment analysis method. The sentiment lexicon in this paper includes middle school and elementary school level sentiment lexicons respectively, and its size is slightly better than the size of the sentiment lexicon proposed by the authors. Anbo et al. [8] proposed a Tibetan text categorization method based on a pre-trained language model, whose pre-trained language model can significantly improve the performance of Tibetan text categorization (the F_1 value is improved by 9.3% on average), which verifies the value of the pre-trained language model in the task of Tibetan text categorization. Meng Xianghe et al. [9] proposed a research method to complete the sentiment classification of Tibetan commented text using deep learning models such as CNN, BiLSTM, and Multi-HeadedSelf-Attention mechanism with Tibetan syllables and Tibetan words as the basic textual representation at the same time. The above studies have improved the accuracy of sentiment analysis of Tibetan text based on the existing techniques and methods respectively, but there are also limitations in terms of corpus and lexicon sizes, and the impact of the effectiveness of the word vector model on sentiment classification is not mentioned. In this paper, we address these issues by using word2vec and Glove models for text vectorisation, and expanding the corpus and the sentiment lexicon in order to achieve the best results in the experiments.

Self-AttentionMechanism is proposed by Vaswani et al. [10], which is applied to the field of machine translation and gets good results. Wu Xiaohua et al. [11] represent short-text in word vectorization, use the BiLSTM network to extract text contextual relationship features, and dynamically adjust feature weights through the self-attention mechanism, and Softmax classifier to get the sentiment categories; Yao [12] captures semantic features in both directions of the text through the BiLSTM network combined with the Attention mechanism to complete the sentiment analysis of the short-text; Guan Pengcheng [13] uses BiLSTM network to capture the semantic features in both directions of the text, and thus completes the sentiment analysis of the short-text; Guan Pengcheng [14] uses the Attention mechanism to capture the semantic features in both directions of the text. Text sentiment analysis; Pengfei Guan et al. [13], proposed an attention-enhanced bidirectional LSTM model. The model uses the attention mechanism to learn the weight distribution of each word on the sentiment tendency of the sentence directly from the word vector, to learn the words that can enhance the classification effect, and uses bidirectional LSTM to learn the semantic information of the text, and improves the classification effect through parallel fusion. The above methods have achieved remarkable results in the tasks of Chinese text classification or Chinese text sentiment analysis, so it is worth a try in Tibetan text sentiment analysis. In this paper, these deep neural network models are better applied to Tibetan text sentiment classification by combining the characteristics of Tibetan short-text.

In summary, the CNN model can capture representative local text features, the BiLSTM model can extract contextual semantic knowledge, and the Attention mechanism can highlight the importance of words and strengthen the learning and generalization ability of the network model. Based on the above research, this paper proposes a short-text categorization model that integrates sentiment lexicon and CNN & BiLSTM, and the whole model plays the advantages of CNN, BiLSTM, and Attention as much as possible to improve the accuracy of Tibetan short-text categorization, and considers the complementarity and coordination of short-text categorization from multiple perspectives.

This paper focuses on the construction of emotion lexicon in subject knowledge domains, aiming at better focusing on emotion characteristics to optimise the emotion classification method and improve the emotion classification effect of Tibetan short texts. In terms of research ideas, firstly, we construct a corpus of Tibetan short texts containing multiple domains and a corpus of emotion dictionaries of subject knowledge domains. Secondly, we set up three groups of experiments and each group of experiments adopts three models respectively, so as to better reflect that our proposed model is feasible.

2 Pre-Processing

2.1 Text Slicing

Unlike the form of English words divided by spaces, Tibetan and Chinese texts have no separator existing between words. To a certain extent, Tibetan text segmentation is the basic work of other Tibetan text information processing research such as text categorization. At present, there are three main types of Tibetan text segmentation methods, namely, rule-based segmentation, statistical-based Tibetan text segmentation, and segmentation methods based on deep learning models. In this paper, the Tibetan text in the experiment is adopts the Tibetan text particle method proposed in the literature [14], which not only retains the advantages of the traditional statistical learning particle method but also adds the neural network technology to strengthen the particle effect. The experimental results show that this particle method is better than the rule-based and statistical-based particle methods in its experimental data.

2.2 Pre-Training

Tibetan were conducted using Glove and tools introduced by Stanford University in 2014 for the work of Tibetan text feature vectorization. Glove was developed by Pennington et al. [15], and its core idea is to use global statistical informationto learn the relationship between words. The corpus used for text feature vectorization in this method is more than 30,000 pieces of comment texts of WeChat public numbers and other platforms, etc. All text corpus is subjected to lexical operation and processed into the text format required by the Glove tool, and finally, the feature vectorization representation is performed by using the Glove model to represent each text feature as a 50-dimensional vector.

3 A Sentiment Analysis Model for Short Tibetan Texts Incorporating Sentiment Lexicon

The structure of the Tibetan short-text classification model incorporating sentiment lexicon is shown in Fig. 1. The whole model is mainly composed of 3 parts: embedding layer, coding layer, and inferred fault.

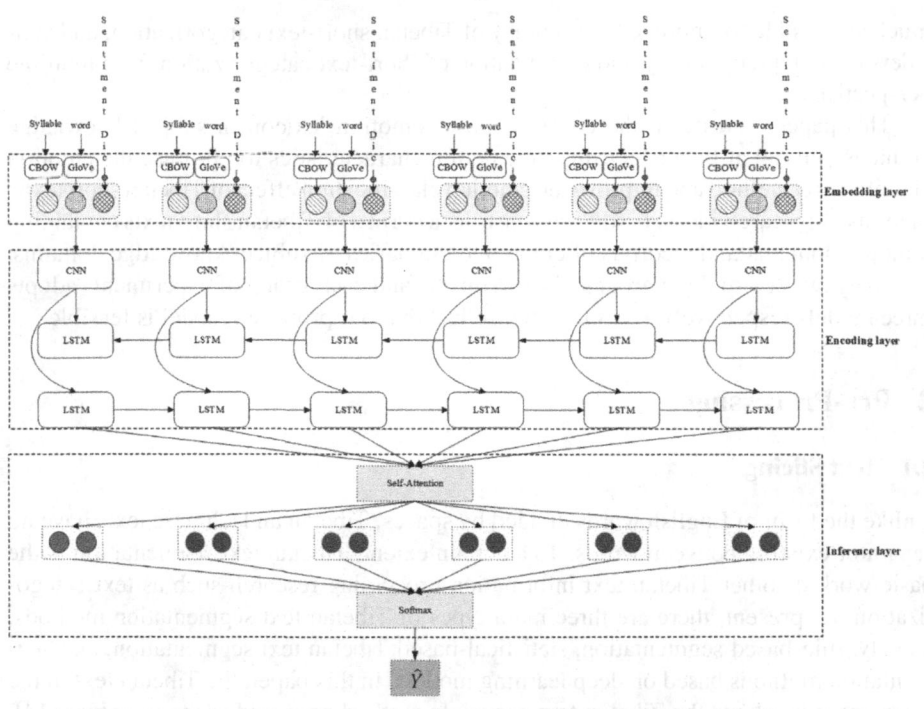

Fig. 1. The model is divided into three parts, which are embedding layer, coding layer and inferring layer. In the first embedding layer, syllable embedding is vectorised using CBOW, word embedding is vectorised using Glove, and lexicon embedding is vectorised using One-hot. The second coding layer uses one layer of CNN and two layers of Bi-LSTM.The third inferencing layer uses Self-Attention and Softmax for inferring sentiment categories and outputting results.

3.1 Embedding Layer

The text classification model incorporating the sentiment dictionary provides semantic understanding from the multi-feature semantic level, which in turn improves the accuracy of text classification. For the text classification model incorporating a sentiment dictionary, it is first necessary to model the semantic features of the preprocessed text data to be classified, including syllable feature embedding, word feature embedding, and sentiment dictionary information embedding, to construct three different input representations; then the processed feature representations are input into CNN & BiLSTM for a local semantic feature and contextual semantic feature acquisition, respectively, to enhance the feature learning ability of the model with a self-attention mechanism to enhance the feature learning ability of the model; finally, the three different semantic features are fused to enrich the feature information and input into the Softmax layer for classification and prediction to get the desired category labels $\hat{Y}$ The model is then used to predict the desired category labels.

3.1.1 Syllable Feature Embedding

In the syllable feature embedding stage, Word2Vec is used for syllable feature modeling firstly, the text to be classified is sliced by syllable by using the word-splitting tool, and then the word vectors are trained with each sliced syllable (length ≥ 1) as a constituent unit.Word2Vec [16] includes the CBOW model and the Skip-Gram model, considering that the model computational computation of the Skip-Gram Considering that the model of Skip-Gram has higher computational complexity and larger computational volume, to save time and obtain higher computational accuracy, this paper uses the CBOW model to realize word vector training.CBOW model includes an input layer, mapping layer, and output layer. In the training process, the text to be classified as S is segmented to obtain the sentence $s = (w_1, w_2, w_3, \cdots, w_n)$, where n denotes the length of the sentence, and w_i denotes the ith syllable, and w_i the length of the contained string is l.

In the training process, the CBOW model encodes the n words in the context of the word, then uses the sliding window to calculate the weights, and finally obtains the vector representation of the current word. The model takes into account the contextual information of the word in the word vector representation, so the final word vector represents a certain semantic information. In the text word vector training, each syllable representation is fixed as a 50-dimension vector e_i^w the computation process is shown in Eq. (1).T_1 as an input to the model, the T_1 is constructed as shown in Eq. (2), the E_1 for the CBOW vectorization process.

$$e_i^w = E_1(w_i) \tag{1}$$

$$0 \cdot T_1 = [e_i^w] \tag{2}$$

3.1.2 Word Feature Embedding

Word feature embedding, i.e., phrase embedding sequence. Considering the sparse text features and other problems in the word feature embedding stage, the global semantic information is modeled as word vector features using Glove, which firstly cuts the text to be classified using the word-splitting tool, and then trains the word vectors with each word cut as a constituent unit.

During training the Glove model optimizes the representation of word vectors by minimizing an objective function. This objective function aims to make the word vectors capture the semantic relationships and similarities between words. Eventually, the final word vector representation for each word is obtained by iteratively optimizing the training process. In the text word vector training, each word representation is fixed as a vector of 50 dimensions, and the computational process is shown in Eq. (3), where J is the objective function, $h(x_{ij})$ is the weight function, and u_j and v_i are the words u_j and v_i are the word vector representations, and b_i and c_j denote the deviation terms, and $\log x_{ij}$ is the logarithmic value of the number of co-occurrences. The construction is shown in Eq. (4), where X_{ij} is the element of the i-th row and j-th column of the co-occurrence matrix w_i, w_j denotes the number of co-occurrences in the corpus.

$$J = \sum_{i \in v} \sum_{j \in v} h(x_{ij}) \left(u_j^T v_i + b_i + c_j - \log x_{ij} \right)^2 \tag{3}$$

$$X_{ij} = \text{cooccurrence}(w_i, w_j) \tag{4}$$

3.1.3 Emotional Lexicon Information Embedding

The experiment-related emotion dictionary is mainly divided into positive keywords and negative emotion words, which is further divided into secondary school emotion dictionary and elementary school emotion dictionary. The emotion dictionary is mainly based on the Tibetan language textbooks of secondary schools and elementary schools to collect and supplement some emotion keywords that are used in daily life but are not in the textbooks, and the dictionary covers a total of 24,460 emotion words. The scoring of emotion intensity uses the emotion labeling of two readers and the author. For example, if the word ༼beautiful༽ is labeled as positive (1) by two people, then it is a positive emotion word, and if only one person indicates that it is positive, then it is a negative emotion word (0). The exact size and samples of the sentiment lexicon are shown in Table 1.

Table 1. Emotional Lexicon Size and Example

Main Class	Subcategory	Size	Example
positive	Primary schools	13508	ཡག (Beautiful) བཟང (Good)
	Middle school		ཞིམ་པོ (Delicious) དྲིན་ཅན (Benefactor)
Negative	Primary schools	10952	མ་རབས (Naughty) སྐྱོ་བ (Sad)
	Middle school		སྐྲག་པ (Fear) ཡིད་མི་ཆེས་པ (Disbelief)

Next, the feature words and weights of the sentiment lexicon are converted into the form of word vectors with 50 dimensions, where the weights of the word vectors corresponding to the sentiment words are additive, and the intrinsic semantic knowledge and correlation information between the words are further characterized by the word vectors, which are finally passed as an input matrix into the Self-Attention and BiLSTM-CNN models to achieve sentiment classification.

3.2 Encoding Layer

Since the traditional feature extraction is done manually, it requires a high level of professionalism from the researcher, and at the same time, since the interactive texts are often acronyms and networked buzzwords, it will result in a time-consuming and laborious feature extraction process. To overcome the disadvantages of manual extraction, this study adopts the method of convolutional neural network to extract the local key features of interactive text. CNN consists of a convolutional layer, pooling layer, and fully-connected layer, and the syllable features, lexical features, and sentiment lexicon vectors, which are obtained from the pre-training of CBOW and Glove as well as the

One-hot coding, are used as inputs to the CNN network. $w_i \epsilon R^d$ as inputs to the CNN network, where d denotes the word vector dimension. The CNN builds multiple convolutional layers based on convolutional kernels of different sizes, which are used to extract the local feature information of syllable features, phrase features, and sentiment dictionary text. The convolution calculation formula:

$$c_i = f(\omega w_{i:i+h-1}) + b \tag{5}$$

where $w_{i:i+h-1}$ denotes the number of words taken per convolution operation; $b \epsilon R$ denotes the bias term; f denotes a nonlinear activation function such as the tanh function or the corrected linear unit (ReLU) function. Currently, most of the convolutional neural networks use the ReLU function because it enables the convolutional neural network to achieve better results. After the convolutional layer feature matrix is used as an input to the pooling layer, the pooling layer is used to compress the amount of data and parameters, screen the effective features, and reduce the total number of features; while screening the features, the pooling layer ensures the invariance and robustness of the data features. Here the maximum pooling layer is used to reduce the dimensionality, and the merged vector after the pooling layer is used as the final output of the CNN network layer.

After the CNN neural network layer inputs the local feature vector data into the BiLSTM layer, this paper adopts the LSTM model proposed by Hochreiter et al. [16] as a reference to construct the BiLSTM model for extracting contextual semantic features in the input data. The LSTM model is a kind of special recurrent neural model for solving the long-term dependency problem of the RNN, gradient vanishing, and gradient exploding problems. The LSTM model incorporates an adaptive gating mechanism to ensure that the LSTM neural units can preserve the previous state and memorize the current input neural units for feature extraction, to obtain the long-term dependencies in the text, and to achieve a more ideal understanding of the text semantics from the perspective of the text as a whole.

The gate structure in the memory unit of the LSTM model consists of 3 parts: input gate, forgetting gate and output gate. The gate structure is used to memorize and update the information of the memory neural unit. Among them, at the current time t, the equations for updating the state of each gating mechanism and unit are as follows:

$$i_t = \sigma(w_i \cdot [h_{t-1}, x_t] + b_i)$$

$$f_t = \sigma(w_f \cdot [h_{t-1}, x_t] + b_f)$$

$$o_t = \sigma(w_o \cdot [h_{t-1}, x_t] + b_o)$$

$$g_t = \sigma(w_c \cdot [h_{t-1}, x_t] + b_c) \tag{6}$$

$$c_t = f_t \cdot c_{t-1} + i_t \cdot g_t) \tag{7}$$

After the completion of each iteration, the output information of the current neural network, which is determined by the output gate o_t is determined, which is calculated as shown in Eq. (8):

$$h_t = o_t \cdot \tanh(c_t) \tag{8}$$

where x denotes the input interactive text word vector, t denotes the current moment and t − 1 denotes the previous moment, h denotes the cell output of the LSTM, the h_{t-1} denotes the generative state of the previous layer, c denotes the memory cell value, and σ denotes the Sigmoid function. w_i, w_f, w_o, b_i, b_f, b_o denote the input gate i_t, forget gate f_t and output gate o_t and the weights and biases of the output gates. Equation (6) represents the update of the memory cell, which serves to forget part of the information and update the information in the current input information that needs to be updated to the current memory cell. Equation (7) represents the calculation of the current output, which is finally determined by output gate o_t which determines the output information.

The current memory cell value is obtained from the memory cell value of the previous layer and the information generated by the current cell. However, the standard LSTM network ignores the following semantic feature information when processing the time series task. In this paper, based on the LSTM model, a two-layer LSTM model is used as the forward propagation channel and the backward propagation channel of the BiLSTM layer in the CNN-BiLSTM-Attention model, respectively. The forward propagation channel is used to extract the above semantic features of the text sequence, and the backward propagation channel is used to extract the below semantic features of the text sequence, and finally the semantic features extracted by the two channels are combined to obtain the contextual semantic features.

3.3 Inference Layer

To retrieve the relevance of sentence s semantic information to feature information, attention is calculated using Self-Attention, which is shown in Eqs. (9) to (11).

$$q^i = W^q a^i \tag{9}$$

$$k^i = W^k a^i \tag{10}$$

$$v^i = W^v a^i \tag{11}$$

The final classification representation is obtained $v^i \epsilon R^{2d}$ after that a fully connected layer mapping is used to get the output x, which is predicted to get the labeled values by softmax classification $\hat{Y}$ as shown in Eq. (12).

$$\hat{Y} = \operatorname{argmax}_y[\operatorname{softmax}(W^{(s)}Q + b^{(s)})] \tag{12}$$

where $y \epsilon R^m$ is the estimated probability of each class obtained from softmax. The convolutional neural network objective cost function is the cross-entropy loss function as shown in Eq. (13).

$$\text{Loss} = \sum\nolimits_{i=1}^{m} s_i \log(y_i) + \lambda \|\theta\|^2 \tag{13}$$

where $s \epsilon R^m$ is the true label of the document, represented by coding with One-hot; m is the number of target classifications, and λ denotes the number of L2 regularization hyperparameters, and θ denotes the parameters for model training.

4 Experimental

4.1 Experimental Data and Parameters

4.1.1 Experimental Data

In the research work of Tibetan text sentiment classification, the experimental corpus has three problems: (1) there is no public and authoritative experimental corpus; (2) the existing Internet platform has relatively little Tibetan text review data; (3) the existing short-text data has multiple problems such as typos. Aiming at the above problems, in this paper, on the text dataset of more than 30,000 texts obtained through crawler technology, taking into account the diversity of text corpus sources, text semantic fluency, and text sentiment classification accuracy and other factors, a total of more than 10,000 short-texts in Tibetan were screened out. The manual translation method is used to clean and correct the text of the corpus, and then form a corpus of more than 10,000 short Tibetan texts used in the experiments of this paper, of which 5,021 are positive comment texts and 5,005 are negative comment texts. The domains covered by the comment texts collected in the experiment are electronic products, hotel accommodation, books, dairy products, home appliances, fruits and vegetables, toiletries, nomadic knowledge, and primary and secondary school Tibetan textbook knowledge. The average sentence length and data

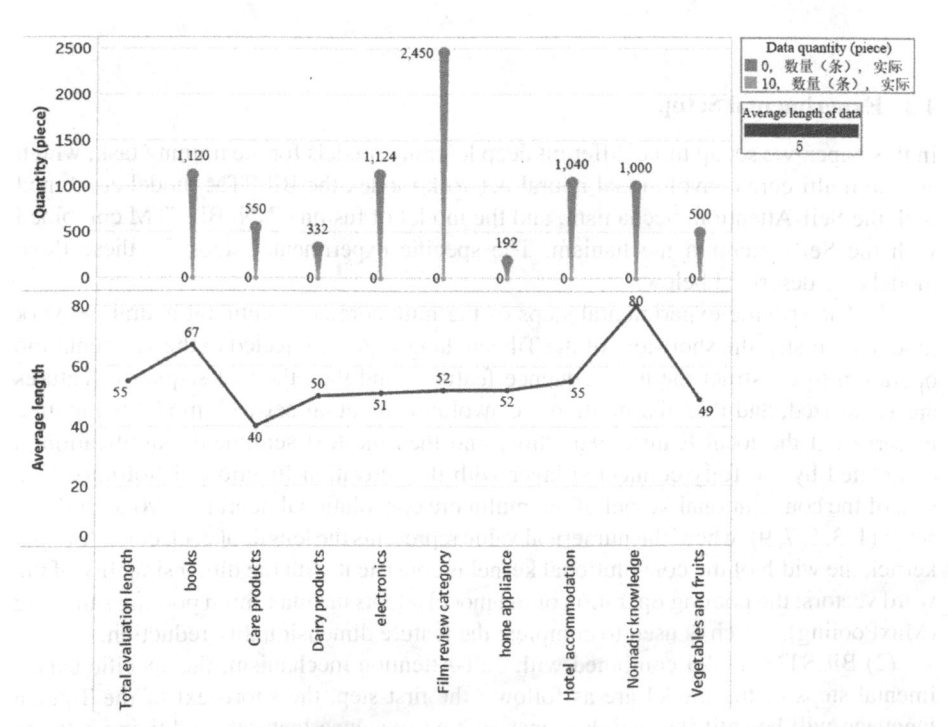

Fig. 2. At the bottom of Fig. 1, the nine domains contained in the text data are shown horizontally, and the Quantity section shows the amount of text data in each of the nine domains. The Averge length section shows the average length of the text data for each domain, and the average length of the overall text data was calculated.

proportion of each domain are shown in Fig. 2. The training set, test set, and validation set of the experiment are sliced according to the ratio of 8/1/1, where the number of samples in the training set is 6016, the number of samples in the test set is 2005, and the number of samples in the validation set is 2005.

4.1.2 Experimental Parameters

The main model parameters related to this experiment are learning rate (R), neuron dropout (Drop_out), and the number of model hidden layer units (Hidden_size). The specific parameter settings of the relevant model are shown in Table 2.

Table 2. Parameter Settings Related to the Experiment

Model	Parameters	Parameter value
CNN&BiLSTM	R	0.0001
CNN&BiLSTM	Drop_out	0.2
CNN	Hidden_size	512
BiLSTM	Hidden_size	256

4.2 Experimental Setup

In this paper, we set up three different deep learning models for the training task, which are the multi-core convolutional neural network model, the BiLSTM model combined with the Self-Attention mechanism, and the model of fusion CNN, BiLSTM combined with the Self-Attention mechanism. The specific experimental steps for these three models are described below:

(1) The specific experimental steps of the multicore convolutional neural network model are firstly, the short-text of the Tibetan language is subjected to the segmentation operation to construct the text sequence features, and then the text sequence features are vectorized, and then the multicore convolutional neural network model is inputted to carry out the local feature extraction, and then the text sentiment classification is completed by the fully-connected layer with the activation function of Softmax. The size of the convolutional kernel of the multicore convolutional neural network model is set to $(1, 3, 5, 7, 9)$, where the numerical value represents the length of each convolutional kernel, the width of the convolutional kernel is consistent with the dimensionality of the word vectors; the pooling operation of the model adopts the maximum pooling sampling (MaxPooling), which is used to complete the feature dimensionality reduction.

(2) BiLSTM model combined with Self-Attention mechanism, the specific experimental steps of the model are as follows: the first step, the short-text of the Tibetan language will be split into words, construct text sequence features, and then vectorize the text sequence features; the second step, the vectorized text will be firstly obtained the sequence features of the text by BiLSTM, and then further obtain the internal information of the sequence by Self-Attention mechanism. Attention mechanism to further

obtain the internal information of the sequence, followed by pooling and dimensionality reduction of the two feature matrices, this experiment uses AveragePooling pooling operation, and finally, the two feature vectors will be spliced; in the third step, the spliced feature vectors will be used to complete the sentiment classification of the text through the fully connected layer with the activation function of Softmax.

(3) Fusion of CNN, BILSTM combined with Self-Attention mechanism, the specific experimental steps of the model are as follows: firstly, the short-text of Tibetan language is subjected to the segmentation operation to construct the text sequence features of syllables, words, and dictionaries, and then vectorized representation of the text sequence features; secondly, the multicore convolutional neural network model is adopted to capture the local features of the text, and BiLSTM model combined with Self-Attention mechanism to capture the sequence features of the context of the text and sequence internal information; finally, the spliced feature vectors are spliced; the spliced feature vectors are then spliced with the internal information of the text. Secondly, a multi-core convolutional neural network model is used to capture the local features of the text and a BiLSTM model combined with the Self-Attention mechanism is used to capture the contextual sequence features of the text as well as the internal information of the sequence, and then the acquired features are spliced together; finally, the spliced feature vectors are used to complete the sentiment classification of the text via the fully-connected layer with the activation function of Softmax.

To ensure that the experimental results are comparable, this paper sets up three groups of experiments to carry out the study, namely, experiments based on lexical features, experiments based on syllabic features, and experiments incorporating sentiment dictionaries.

The first set of experiments is sentiment text classification based on lexical features, which uses Tibetan lexical items as the basic text representation features. The experimental parameter of the lexicon size (Vocabulary_size) is 4531, that is, the number of kinds of Tibetan words contained in the corpus text; the experimental parameter of the Tibetan text sequence length (max_sen_len) is set to 85, which is taken because the interval of text sequence lengths from 10 to 85 is derived from the lexical and statistical analyses of the corpus text, and then through the three kinds of models training, it is determined that the model is the optimal effect when the text sequence length is 85.

The second group of experiments is the syllable-based sentiment text classification experiments; this group of experiments uses Tibetan syllables as the basic text representation features; the experimental parameter of the lexicon library size (Vocabulary_size) is 1657, which is the number of kinds of Tibetan syllables contained in the corpus text, the experimental parameter of the Tibetan text sequence length (max_sen_len) is set to 50, which is taken as the value because of the analysis on the experimental parameter Tibetan text sequence length (max_sen_len) is set to 50, which is taken because the text of the corpus is divided into words and statistically analyzed, which results in the range of text sequence lengths from 10 to 50, and then through the training of the three models, it is determined that the model can get the optimal effect when the text sequence length is 50.

The second group of experiments is the emotion text classification experiments based on the fusion of emotion lexicon; this group of experiments fuses the Tibetan emotion words and uses the lexical items as the basic text representation features; the size of the lexicon library (Vocabulary_size) in the experimental parameters is 6010, that is, the number of types of Tibetan syllables contained in the corpus text; the length of the text sequences (max_sen_len) in the experimental parameters is set at 85, this value is taken because after the text of the corpus is divided into words and statistically analyzed, it is concluded that the length of the text sequence is in the range of 10 to 85, and then through the training of the three models, it is determined that the model can get the optimal effect when the length of the text sequence is 85.

4.3 Analysis of Experimental Results

In this paper, when evaluating the sentiment classification algorithm of Tibetan text, we adopt Accuracy as the evaluation index, which evaluates the classification accuracy of the classification algorithm as a whole, and the experimental results are shown in Table 3. In the "Experimental Groups" column of the table, "Group 1" corresponds to the experiments of emotion text classification based on word features; "Group 2" corresponds to the experiments of emotion text classification based on syllable features; and "Group 3" corresponds to the experiments of emotion text classification based on fusion of emotions. The "Experimental Model" column in the table gives the model used in each group of experiments, in which "CNN" is a multicore convolutional neural network model; "BiLSTM + Att" is a BiLSTM model combined with a Sentiment classification model; and "BiLSTM + Att" is a BiLSTM model combined with a Sentiment classification model. BiLSTM + Att" is a BiLSTM model combined with a Self-Attention mechanism; "CNN&BiLSTM + Att" is a fusion of a multi-core convolutional neural network model and a BiLSTM model combined with a Self-Attention mechanism. The values given in the "Accuracy" column in Table 3 are the optimal text categorization accuracies obtained by each group of experiments corresponding to the relevant models on the validation set after the continuous adjustment of hyperparameters. From the classification accuracy results obtained during the experiments in this paper, the following three conclusions can be drawn.

First, in this paper's sentiment categorization experiments on short Tibetan texts, the word-based model is overall better than the syllable-based model. Analyzing, it is due to the following reasons: because Tibetan text cut by syllable will make the semantic sparse and individual syllables can't form words with semantic information, so the accuracy rate of word cut is better than that of syllable cut, so the word-based model under the deep neural network model is relatively able to obtain more correct and effective upper semantic feature information; in the word-based model, the feature sparsity phenomenon caused by a large number of words will cause the model overfitting. In the word-based model, the feature sparsity phenomenon caused by a large number of words will result in the overfitting of the model. However, to address this problem, this paper uses the Glove word vector technique, which is more suitable for the Tibetan language, to train word vectors, so that the overfitting problem can be effectively solved to a certain extent. Secondly, from the classification accuracy of the experiment, it can be seen that the "CNN&BiLSTM + Att" model is slightly better than the "BiLSTM + Att" and "CNN The main reason is that the "CNN&BiLSTM + Att" model integrates the advantages of the "BiLSTM + Att" and "CNN" models. The reason is that the "CNN&BiLSTM + Att"

Table 3. Experimental Results

Experimental grouping	Experimental model	% Accuracy
Group I (words)	CNN	88.65
	BiLSTM + Att	87.12
	CNN&BiLSTM + Att	89.75
Group II (sylls)	CNN	84.21
	BiLSTM + Att	86.37
	CNN&BiLSTM + Att	88.26
Group III (dic&words&sylls)	CNN	**90.1**
	BiLSTM + Att	**92.32**
	CNN&BiLSTM + Att	**93.18**

model integrates the advantages of "BiLSTM + Att" and "CNN", which makes the fused model able to obtain more comprehensive text feature information, and thus improves the accuracy rate of sentiment classification of Tibetan short-text. Third, according to the experimental results, it can be concluded that the sentiment text classification model fused with a sentiment dictionary is superior to the model based on syllable features and the model based on word features. In the third set of experiments, the "CNN&BiLSTM + Att" model achieves the best results, with an accuracy of 93.18% on the test set. The main reason why the text categorization method incorporating sentiment lexicon achieves good classification accuracy is that it increases the total number of text sequence features in the short-text corpus to a certain extent, which better solves the problem of sparse text features in the commentaries; secondly, the "CNN&BiLSTM + Att" model is proposed, which firstly uses the multicore model to classify the text sequence in the short-text corpus, and secondly, the "CNN&BiLSTM + Att" model is proposed. Secondly, the "CNN&BiLSTM + Att" model is proposed, which firstly obtains the local features of the text through the multi-core convolutional neural network model, and then adopts the BiLSTM model combined with the Self-Attention mechanism to obtain the contextual sequence features of the text and the internal features between the sequences, and then completes the task of classifying the textual sentiment after the feature splicing. The model proposed in this paper obtains more textual semantic features of Tibetan short-text, and then it can achieve better text sentiment classification results.

5   Conclusion

In this paper, we propose a method of fusing sentiment lexicon to analyze text sentiment, which uses a convolutional neural network model, bi-directional long and short-term memory network, and attention mechanism, respectively, which makes it possible to further improve the accuracy of sentiment analysis. In this paper, the sentiment lexicon is used as a feature and fused with traditional syllable-based and word-based features to fuse multiple features to achieve sentiment analysis.

In this paper, the effectiveness of the proposed model is verified and experiments are conducted on the constructed dataset with better results. The experiments show that the classification effect of CNN & BiLSTM + Self-Attention is improved compared with the existing classification models.CNN & BiLSTM + Self-Attention can extract the local features of the text as well as the contextual features of the text efficiently and improve the ability of sentiment analysis.CNN & BiLSTM + Self-Attention have great flexibility in the choice of word vector generation mode and can be adjusted to meet different task requirements.

In the future, we will continue to discuss the effectiveness and efficiency of CNN & BiLSTM + Self-Attention. For example, in this paper's experiment positive and negative sample pairs were randomly selected, and other methods such as choosing samples with stronger emotional tendencies may achieve better results. The next research direction is to carry out relevant experimental studies on other deep neural network models such as Multilayer Attention Networks (Hierarchical Attention Networks) and Graph Convolutional Networks (Graph Convolutional Networks) with the method of fusing the emotion lexicon proposed in this paper as the basic object of representation of Tibetan text.

References

1. Du, H., Xu, X., Wu, D., Liu, Y., Yu, Z., Cheng, X.: Microblog sentiment classification based on sentiment word vectors. J. Chin. Inf. **31**(03), 170–176 (2017)
2. Xiaodong, Y., Tao, H.: Sentiment classification of Tibetan text sentences based on sentiment dictionary. J. Chin. Inf. **32**(02), 75–80 (2018)
3. Haijie, X.: Research on sentiment analysis based on deep learning. Northeast Petroleum University (2022). https://doi.org/10.26995/d.cnki.gdqsc.2022.000496
4. Zhao, Y., Qin, B., Shi, Q., Liu, T.: Construction of large-scale sentiment dictionary and its application in sentiment classification J. Chin. Inf. **31**(02), 187–193 (2017)
5. Zhao, Z., Shao, X., Lin, X.: An attention model combining graph convolutional neural networks for aspectual sentiment analysis. J. Chin. Inf. **36**(07), 154–163 (2022)
6. Meirong, D., Weisen, F., Rongxiang, H., et al.: Sentiment analysis of hotel reviews based on a pre-trained model and base dictionary extension. Comput. Syst. Appl. **31**(11), 296–308 (2022). https://doi.org/10.15888/j.cnki.csa.008779
7. He, X., Yang, L., Huang, Y.: Aspect-level sentiment analysis model based on BER T-Bi-IAN. J. Henan Eng. College (Natural Science) **35**(01), 65–70 (2023). https://doi.org/10.16203/j.cnki.41-1397/n.2023.01.009
8. Bo, A., Congjun, L.: Tibetan text categorization based on pre-trained language models. J. Chin. Inf. **36**(12), 85–93 (2022)
9. Meng, X., Yu, H.: A study on sentiment classification of Tibetan text by integrating syllable and word features. J. Chin. Inf. **37**(02), 80–86 (2023)
10. Vaswani, A., Shazeer, N., Parmar, N., et al.: Attention Is All You Need. arXiv:1706.03762 (2017). https://doi.org/10.48550/arXiv.1706.03762
11. Wu, X., Chen, L., Wei, T., Fan, T.: Sentiment analysis of Chinese short text based on Self-Attention and Bi-LSTM. J. Chin. Inf. **33**(06), 100–107 (2019)
12. Xianglu, Y.: Attention-based BiLSTM Neural Networks for Sentiment Classification of Short Texts (2017)
13. Guan, P., Li, B., Lu, X., Zhou, J.: Bidirectional LSTM sentiment analysis for attention enhancement. J. Chin. Inf. **33**(02), 105–111 (2019)

14. Fangyu, C.: Research on BiLSTM ___ CRF Tibetan Segmentation Technique by Fusing Attention. Qinghai Normal University (2023). https://doi.org/10.27778/d.cnki.gqhzy.2023. 000199
15. Pennington, S.R., Manning, C.G.: Global vectors for word representation. In: Conference on Empirical Methods in Natural Language Processing (2014). https://doi.org/10.3115/v1/D14-1162
16. Hochreiter, S., Schmidhuber, J.: Long short-term memory. Neural Comput. 9(8), 1735–1780 (1997)

A Review of Solving Non-IID Data in Federated Learning: Current Status and Future Directions

Wenhai Lu[1,4], Jieren Cheng[1,4(✉)], Xiulai Li[2,4], and Ji He[3]

[1] School of Computer Science and Technology, Hainan University,
Haikou 570228, China
[2] School of Cyberspace Security Academy (Cryptography Academy),
Hainan University, Haikou 570228, China
[3] School of Engineering and Information Science, University of Wollongong,
Wollongong 2500, Australia
[4] Hainan Blockchain Technology Engineering Research Center, Haikou 570228, China
cjr22@163.com

Abstract. Federated learning (FL), as a machine learning framework, has garnered substantial attention from researchers in recent years. FL makes it possible to train a global model through coordination by a central server while ensuring the privacy of data on individual edge devices. However, the data on edge devices that participate in FL training are not independently and identically distributed (IID), resulting in challenges related to heterogeneity data. In this paper, we introduce the challenges generated by non-IID data to FL and provide a detailed classification of non-IID data. Then, we summarize the existing solutions to non-IID data in FL from the perspectives of data and process. To the best of our knowledge, despite the considerable efforts achieved by many researchers in solving the non-IID problem, some issues remain unsolved. This paper provides researchers with the latest findings and analyzes the potential future directions for solving non-IID in FL.

Keywords: Federated learning · Non-IID data · Heterogeneous data

1 Introduction

In the digital era, the use of edge devices has become widespread. These devices generate vast amounts of data at extraordinary rates. The traditional method of training machine learning models requires the centralized aggregation of data generated by these edge devices. However, this practice raises concerns about user privacy and data security. At the same time, to protect the privacy of individual data, the General Data Protection Regulation [38] was introduced, making centralized data collection less desirable [36,39]. Instead, researchers have shifted their focus to methods that enable data to remain on edge devices while facilitating model aggregation. Federated Learning (FL), initially proposed by McMahan [28], allows model training to take place without centralizing the

H. Jin et al. (Eds.): IAIC 2023, CCIS 2058, pp. 58–72, 2024.
https://doi.org/10.1007/978-981-97-1277-9_5

raw data. This paradigm plays a pivotal role in addressing concerns related to data privacy and enabling distributed learning.

Federated Learning has become a pivotal method for addressing contemporary challenges in the fields of data science and machine learning. There are many applications of FL, such as safeguarding data privacy among various tiers of hospitals [16], object detection [42], and wireless communications [30]. However, challenges persist, particularly in the context of the widespread adoption of edge devices. The substantial variability in data distribution across individual devices can compromise the robustness of the final results. Furthermore, challenges arise in practical FL applications due to factors like data heterogeneity and the accompanying communication and computational overhead between edge devices and central servers.

This paper focuses on addressing the issue of data heterogeneity in FL, which arises from non independently and identically distributed (IID). Many scholars in the field have operated under the assumption of IID data, an idealized data distribution. However, in the real world, data generated by various edge devices inherently exhibit non-IID characteristics [22]. Numerous scholars have dedicated substantial efforts to tackle the challenge of non-IID data in FL. For instance, McMahan [27] highlighted non-IID data as a primary challenge. As more researchers have entered the FL field, they have introduced new improvements [12]. These advancements have propelled FL to make groundbreaking progress in handling non-IID data and achieving enhanced performance while ensuring privacy protection, improving communication efficiency, and reducing communication costs [10, 11, 52].

Table 1 lists the top 10 review articles related to FL, ranked by citation count. Our focus is on summarizing the principal contributions and research findings of these articles, with a particular emphasis on the only [56] article that delves into the topic of non-IID data in FL. We contend that this article's coverage of existing solutions for addressing non-IID data in FL is not comprehensive. Furthermore, research pertaining to data heterogeneity is gaining momentum, and such review articles hold timeliness. Therefore, the primary objective of this paper is to provide a systematic review and summary of the latest literature on addressing non-IID data.

The primary contributions of this paper are outlined below:

- This paper analyzes and summarizes the challenges posed by non-IID data in FL, while also categorizing non-IID data and providing examples for illustration.
- This paper provides a comprehensive and in-depth summary of current research on solving non IID data problems in federated learning.
- This paper presents future trends for addressing non-IID data issues in the field of FL, to promote further research in related domains.

This paper is structured in Sect. 2, a brief introduction to the FL framework and non-IID data is provided, along with the issue posed by non-IID data. Section 3 analyzes and summarizes the latest research status from the perspectives of data and process. Section 4 proposes possible future research directions. Finally, in Sect. 5, the paper concludes with a summary of the research conducted.

Table 1. Research directions in Federated Learning

Ref	Major contributions
[41]	An introduction to statistical challenges and privacy concerns in FL, along with an exhaustive review of research in healthcare field
[17]	An introduction to the application in industrial engineering to guide for the future landing application
[45]	An in-depth and comprehensive review article focusing on the techniques and challenges of privacy preserving FL
[12]	An introducing FL as a solution to address security and privacy issues in traditional data collection while driving urban intelligence using IoT devices
[3]	An introduction was provided regarding the diverse practical uses of FL, as well as the prospective future directions within the market for FL applications
[24]	A systematic review from a software engineering perspective and identify future trends
[2]	A systematic review on FL in the context of EHR data for healthcare applications
[50]	A brief review of the state-of-the-art FL techniques and detailedly discuss the improvement of FL
[44]	A comprehensive treatment of implementing FL techniques for wireless communications
[56]	An introducing the concept of non-IID data, this paper proceeds to conduct a comprehensive review of current research devoted to resolving these concerns

2 Background

2.1 Overview of Federated Learning

Federated learning is a machine learning approach, the core idea of which is that model training no longer requires data to be stored centrally on a single server. Instead, it allows for model training without sharing raw data. See Fig. 1 the training process for FL can be divided into four steps [28]: initialization, where the training process starts with the initialization of a central model, which is usually a neural network or other machine learning model; local training, where the central model is sent to the local devices of each participant, and each participant trains the central model using its own local dataset. This step is performed locally without sending the raw data to the central server; model update, after local training is completed, each participant sends an update of the model to the server; and aggregation, after the server receives the model updates from participants, it performs the aggregation of the model parameters. This process typically involves averaging or weighted averaging of model updates from different participants to update the central model [26,31]. The above steps are iterated several times, with each iteration improving the performance of the central model. This process is repeated until the model converges or reaches a predetermined number of training rounds.

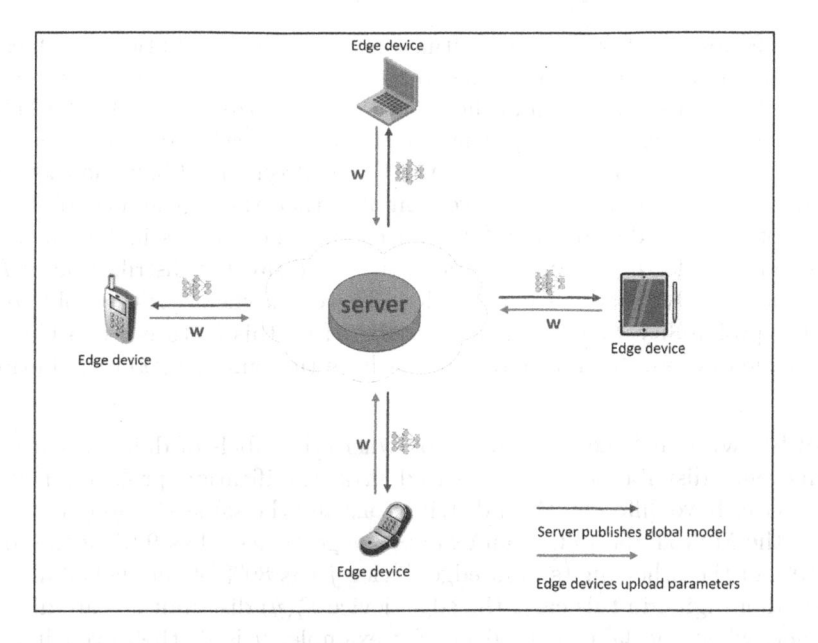

Fig. 1. Federated Learning Training Overview

2.2   Overview of Non-IID Data

In FL, the composition of training data on each edge device is notably influenced by the specific usage patterns of local devices, giving rise to the potential for substantial dissimilarities in data distribution among devices. This phenomenon, recognized as Non-IID [27], can lead to significant challenges such as model divergence. Data samples may originate from diverse data sources, resulting in significant disparities in features, attributes, and data distributions. Such heterogeneity renders it challenging to describe data samples with a single probability distribution that encompasses them all. In non-IID data, samples from different categories or labels may exhibit uneven distributions. Additionally, the data distribution may evolve over time, location, or other factors, thereby introducing challenges to model performance. The existence of non-IID data presents a formidable challenge to machine learning and data analysis as it contravenes the assumptions underpinning many conventional process [18,40].

We assume that the data distribution of one of the edge devices i is P_i and the data distribution of the other edge device j is P_j. In the supervised task, x and y are used to denote the features and labels of the samples, respectively. $P_i(y|x)$ denotes the probability distribution of obtaining the outcome of y given that the features are x. In this paper, we use the non-IID data of the edge devices in FL to represent the probability of obtaining the result of y. Similar to [15,18,25], we discuss categories of non-IID data in FL are subdivided into five categories:

Feature Skew. Feature skew is a situation where the distribution of values of a feature in a dataset shows unevenness or asymmetry. Different edge devices have different $P_i(x)$ distributions and the same $P(y|x)$ distribution. Take the MNIST dataset [6] as an example, different people can be regarded as different edge devices, the first person i prefers to write the bold version of 3, while the second person j prefers to write the slim version of 3, then the probability of the bold version of 3 in the distribution $P_i(x)$ on the first person i is higher, while the opposite is true for the second person j, that is to say, the distribution of $P_i(x)$ is not the same for different edge devices. However, when x is the bold version of 3, the probability that the label y predicted by this feature x is 3 is similar across edge devices, different edge devices have the same $P(y|x)$ distribution.

Label Skew. Label skew is a situation where the labels of different categories are unevenly distributed or unbalanced in a classification problem. Different edge devices have different $P_i(y)$ distributions and the same $P(x|y)$ distribution. Taking the MNIST dataset as an example, edge device i has 90% of the digit 3 and 10% of the other digits, and edge device j has 80% of the digit 6 and 20% of the other digits, in this case, the edge device $P_i(y)$ distributions are different. However, when the label y is given, for example, y is 6, then even if it is a different edge device, the corresponding feature probability is the shape of 7 so that $P(x|y)$ distribution is the same.

Quantity Skew or Unbalancedness. Quantity skew or unbalancedness is a situation where the number of samples from different categories or labels in a dataset is unevenly distributed. This means that some categories have far more samples than others, resulting in an unbalanced dataset. For example, there are 100 data samples in edge device i and 20,000 data samples in edge device j.

Same Label and Different Features. Same label and different features mean that in a data set, different samples have the same label or category, but their characteristics can be different. Different edge devices $P(x|y)$ distribution, $P_i(y)$ can be distributed the same. It can be understood that the same cat has the same label, but the hair type is different, such as a hairless cat and a haired cat, that is, the corresponding labels of different edge device samples are different. Although the breeds of these two cats are different, they both belong to the cat category label.

Same Features and Different Label. Same features and different labels mean that in a dataset, different samples have the same features or attributes, but they are assigned to different labels or categories. Different edge devices have different $P(y|x)$ distributions and the same $P_i(x)$ distribution. For example, different individuals treat the same news story differently.

2.3   Challenges Posed by Non-IID Data in FL

Figure 2 depicts a statistical representation of the challenges encountered in FL [24]. We observed that the challenge of data heterogeneity ranked second only to communication efficiency. In real-world scenarios, data is generated on individual edge devices, diverging from the assumption of IID data often made in most studies. In the FL framework, data is decentralized on each edge device for training, which results in a decline in the accuracy of FL due to the presence of non-IID data. particularly on datasets exhibiting a high degree of skewness, where this effect is more pronounced. Non-IID data results in numerous issues with FL model performance, including reduced accuracy, delayed model communication, slow model convergence, and more [19,33,54].

In addition, the local edge devices are uploaded to the central server for model aggregation after training, and the resulting aggregated model may not increase but decrease in performance locally under non-IID data and is inferior to the local model before aggregation [9,32,48].In the next section, we will provide a comprehensive overview of the current approaches and methods used to address non-IID data in the context of FL.

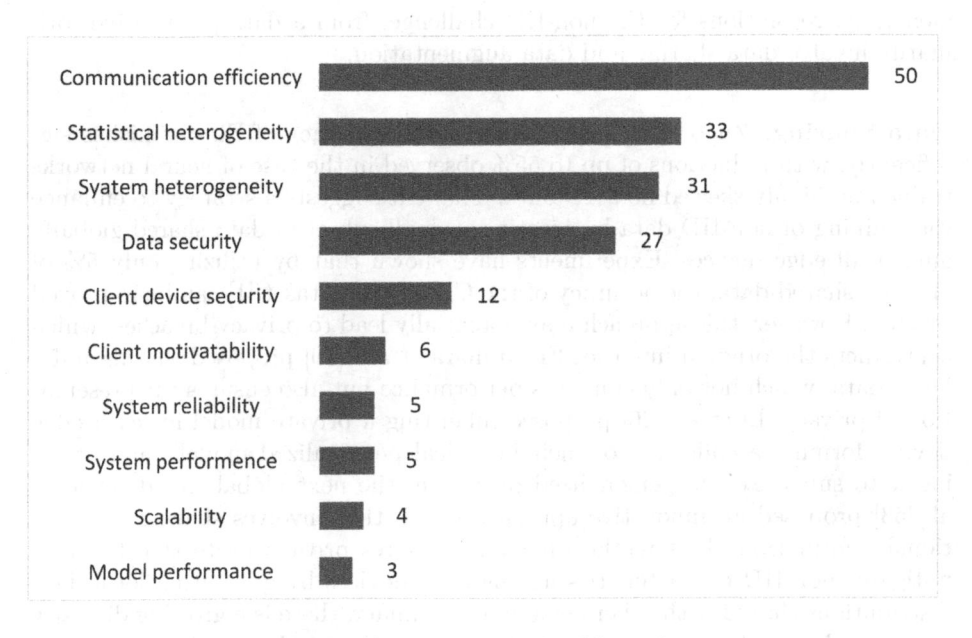

Fig. 2. Challenges in Federal Learning

3   Main Solutions to the Problems of Non-IID Data in FL

As previously stated, researchers have increasingly focused on addressing the non-IID data issue in FL. Hence, we undertake an extensive review of the most recent research papers, systematically categorizing and summarizing them.

Figure 3 provides a illustrating the various approaches employed to tackle this challenge. This paper offers a comprehensive summary of existing methodologies for mitigating the non-IID data in FL from the perspectives of data and process. In this section, we will delve into an in-depth discussion of these research methodologies.

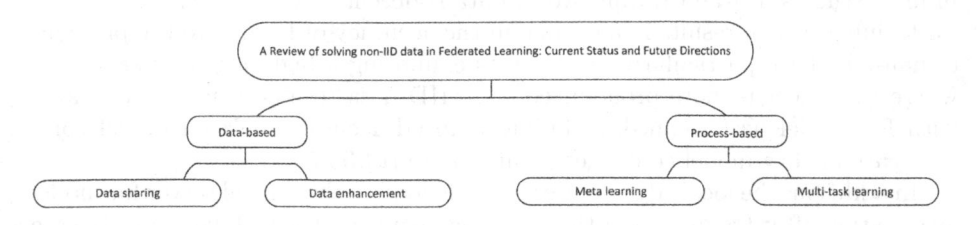

Fig. 3. Illustration of the Solution Architecture

3.1   Data-Based

Generally, resolutions for the non-IID challenge, from a data perspective, primarily involve data sharing and data augmentation.

Data Sharing. Zhao et al. [54] show that the accuracy of FL diminishes significantly, with reductions of up to 55% observed in the case of neural networks trained on highly skewed non-IID data. The text suggests a strategy to enhance the training of non-IID data by creating a small subset of data shared globally among all edge devices. Experiments have shown that by utilizing only 5% of globally shared data, the accuracy of the CIFAR-10 dataset [1] can be improved by 30%. However, this approach may potentially lead to privacy breaches, which contradicts the original intent of FL. Itahara et al. [10] proposed sharing unlabeled data, which not only enhances performance but also ensures the preservation of privacy. Li et al. [20] proposed inheriting a private model for each edge device, forming a collection of their historical personalized models, and utilizing it to supervise the personalized process in the next global round. Zhao et al. [53] proposed an innovative approach to FL that involves sharing distributional information. This method not only ensures privacy protection but also mitigates non-IID characteristics among edge devices by generating local data distributions shared with other devices. In summary, there is a growing diversity of approaches to data sharing. Data sharing can address the non-IID issue at its core, but it invariably involves compromising privacy. Striking the right balance is a key consideration in all of these studies.

Data Augmentation. Data augmentation [37] entails applying various transformations and manipulations to the raw data, thereby generating a more diverse set of data samples. This serves as a means to address the issue of non-IID data within the context of FL. Using a generator to produce data constitutes an effective means of data augmentation. Zhu et al. [57] proposed training a generator on the server that can generate data conforming to the global data distribution. Subsequently, the trained generator is broadcasted to various edge devices for data augmentation, thereby enhancing performance. Unlike the former, Shin et al. [34] proposed a hybrid data augmentation technique called XorMixup. The core idea behind XorMixup is to gather encoded data samples from various edge devices. These data samples are decoded using only the data samples from each respective edge device. Due to non-IID data, it leads to bias in the global model and significant differences in accuracy among various edge devices. chiaro et al. [5] introduced the FL Enhance framework. Specifically, FL Enhance leverages cGANs locally trained at the server level to generate new data for data augmentation. In conclusion, data augmentation akin to data sharing, addresses the non-IID challenge in FL to notably enhance performance. Data augmentation, when uploading a small number of sample labels, introduces noise or employs encoding and decoding techniques to safeguard data privacy.

3.2   Process-Based

Generally, resolutions for the non-IID challenge, from an process perspective, primarily involve meta learning and multi-task learning.

Meta Learning. In a real-world scenario, due to the heterogeneity of data across various edge devices, the model aggregated by the central server during FL training may not perform as well as local individual models. Meta learning facilitates quick adaptation to unfamiliar situations,akin to tasks with a minimal quantity of additional data, making it a good complement to FL, making it a good complement to FL. Zhang et al. [49] proposed introducing meta learning into FL, thereby developing a federated meta learning model. The objective of this model is to train personalized models for each edge device, reducing the differences caused by non-IID data among the edge devices. Additionally, Yu et al. [47] introduced a method based on parallel integrated meta learning and FL algorithms, enhancing performance on non-IID data. See Fig. 4 training process involves both basic FL and meta FL. Firstly, some base models are trained in FL through cross validation. Subsequently, an ensemble model is trained using predictions from the validation set. At the same time, these models are trained in parallel and synchronized. Similarly, the suggestion presents a collaborative, group-oriented meta learning framework referred to as G-FML. This framework autonomously classifies edge devices into numerous groups, taking into account the similarity of their data distributions, and acquires user models through meta learning within each respective group [43]. Overall, the integration of meta learning with FL methods has significantly improved performance on non-IID data.You et al. [46] proposed addressing the non-IID data problem by adjusting gradient information, specifically by modifying past gradients of user models to optimize the server model. Additionally, in terms of privacy,

Zheng et al. [55] design a federated meta learning framework to address privacy concerns in both FL and meta learning scenarios.

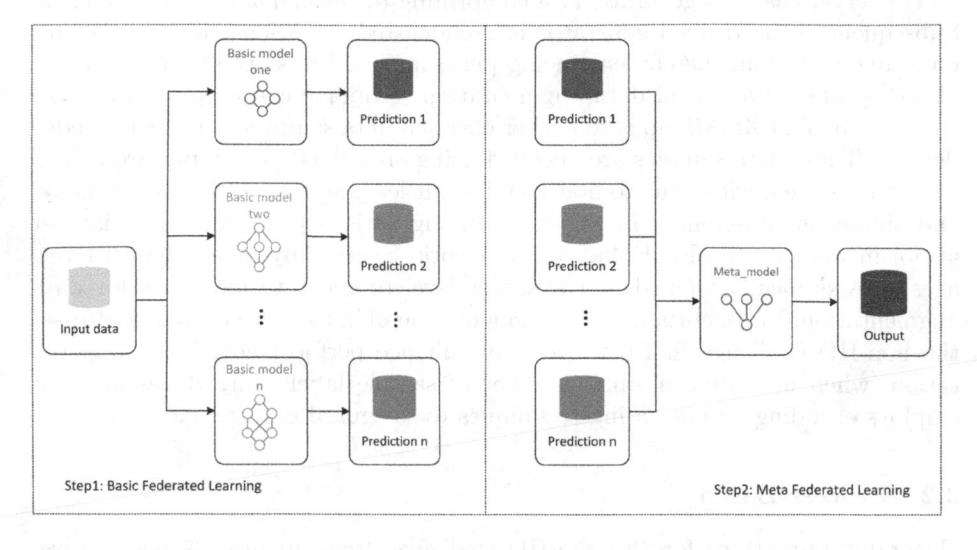

Fig. 4. The algorithmic structure combining meta learning and federated learning

Multi-task Learning. Addressing data privacy leakage has always been a challenging issue in multi-task learning. FL serves the purpose of privacy protection effectively, and the majority of FL objectives aim to train a global model without considering variations in tasks among different edge devices. Especially in heterogeneous data scenarios, the combination of FL with multi-task learning can effectively mitigate privacy concerns and lead to performance improvement. Mills et al. [29]presented was a multi-task FL approach that integrates non federated batch normalization layers into federated deep neural networks. MTFL allows users to train personalized models based on their own data, thereby achieving higher accuracy and faster convergence speed. Cheng et al. [4] proposed an adaptive algorithm that addresses the challenge of device heterogeneity. Dinh et al. [7] proposed a multi-task learning algorithm for general non-convex models is proposed, and it has been demonstrated to be significantly effective on non-IID data. Shu et al. [35] proposed a clustering FMTL to achieve multi-task teaching on non-IID data while improving communication efficiency and model accuracy. Zhang et al. [51]proposed an approach that call resilient FL. This approach leverages reinforcement learning in the context of FL, employing reinforcement learning for the weighted fusion of user models, rather than employing the traditional averaging fusion method. Overall, the objective of federated multi-task learning is to simultaneously learn different tasks from various devices and attempt to capture model relationships among them without privacy risks. Leveraging these model relationships allows each device's model to gain insights from other devices. Furthermore, the models learned for each device are always personalized.

4   Trends and Future Directions

Researchers have invested substantial efforts in addressing challenges related to non-IID data in FL. However, these challenges persist, including issues such as the degradation of model performance and the growing overhead between edge devices and servers. These problems remain a focal point for researchers in this field. In certain scenarios, privacy may be compromised to enhance FL model accuracy on non-IID data. Considering the array of challenges faced by FL, we provide an analysis of potential research directions that researchers may pursue in the future.

4.1   Privacy Protection

Privacy protection has always been the core principle of FL, allowing collaborative training without exposing the private data of local edge devices. However, in the context of addressing non-IID data, certain methods, such as Lin et al. [21] approach involving knowledge distillation and the exchange of model parameters among edge devices, may potentially lead to privacy breaches. Therefore, privacy protection remains a significant topic for future research in FL. The use of proxy datasets for knowledge distillation within FL has raised concerns about privacy exposure. Existing literature has proposed methods that mitigate non-IID data challenges while preserving privacy without the need for additional data. Differential Privacy [8] is primarily achieved through the introduction of random noise to ensure that the results of requests for publicly visible information do not leak individual privacy information. However, the introduction of random noise does impact the model's accuracy to some extent. In the context of non-IID data, striking a balance between privacy and model performance is poised to become one of the hot research directions in FL. Data sharing follows a similar rationale contributing a certain proportion of data can significantly enhance model performance. Defining quantitative standards for quantifying the extent of privacy disclosure and determining the maximum contribution of data are also prospective areas of focus in the future.

4.2   Communication Costs

Communication cost has consistently been a paramount consideration in FL research, as excessive communication costs lead to slow model training. Moreover, the level of communication cost serves as a crucial metric for assessing the efficacy of various methods in addressing non-IID data. The participating edge devices often possess limited computational power, making smaller models more accessible for a larger number of edge devices to engage in FL. Model pruning is a technique employed to reduce the complexity of neural network models by eliminating insignificantly weighted parameters to obtain lightweight models [13].

This approach effectively reduces communication costs without compromising model accuracy. Nevertheless, model pruning requires additional training data to be available on the server, which contradicts the principle of FL, where the central server does not possess data. Liu et al. [23] proposed an approach that combines FL with Single Learning split learning to concurrently address non-IID data in FL while enhancing communication efficiency. The future direction of research may revolve around how to employ various optimization algorithms for model pruning, how to perform model splitting effectively, and ultimately, how to enhance the performance of FL in the presence of non-IID data.

4.3 Personalized Federated Learning

Traditional FL involves collecting models from all edge devices to a central server, where model averaging and aggregation are performed. The aggregated model is then distributed back to the edge devices as the model for local training. However, in cases where the degree of non-IID data is substantial, the performance of the locally trained model on the received aggregated model may not be as effective as the model originally trained locally. Personalized FL places a greater emphasis on the historical knowledge of local models. In this context, Jin et al. [14] introduce self-knowledge distillation techniques, transferring knowledge from the previous round of locally trained models to the model received globally. This approach significantly addresses the challenges posed by non-IID data, preventing scenarios where the model diverges or fails to converge due to extreme data skewness. Presently, personalized FL research is gaining momentum, signifying a pivotal direction in resolving issues associated with non-IID data.

5 Summary

Federated Learning has gained widespread recognition and research interest due to its capability for distributed data processing, making it a potent framework for addressing collaborative training among edge devices in the real world. This paper introduces non-IID data and the challenges it presents to FL. Through a comprehensive analysis and synthesis of existing solutions, it becomes evident that FL continues to hold a promising future as research progresses. The paper also discusses the future directions of FL, aiming to provide valuable insights for scholars seeking to stay updated on the latest research advancements.

Acknowledgements. This work was supported by Youth Foundation Project of Hainan Natural Science Foundation(621QN211), National Natural Science Foundation of China (NSFC) (Grant No. 62162022, 62162024), the Major science and technology project of Hainan Province (Grant No. ZDKJ2020012), Hainan Provincial Natural Science Foundation of China (Grant No. 620MS021).

References

1. Abouelnaga, Y., Ali, O.S., Rady, H., Moustafa, M.: CIFAR-10: KNN-based ensemble of classifiers. In: 2016 International Conference on Computational Science and Computational Intelligence (CSCI), pp. 1192–1195. IEEE (2016)
2. Antunes, R.S., André da Costa, C., Küderle, A., Yari, I.A., Eskofier, B.: Federated learning for healthcare: systematic review and architecture proposal. ACM Trans. Intell. Syst. Technol. (TIST) **13**(4), 1–23 (2022)
3. Banabilah, S., Aloqaily, M., Alsayed, E., Malik, N., Jararweh, Y.: Federated learning review: fundamentals, enabling technologies, and future applications. Inf. Process. Manage. **59**(6), 103061 (2022)
4. Cheng, J., Luo, P., Xiong, N., Wu, J.: AAFL: asynchronous-adaptive federated learning in edge-based wireless communication systems for countering communicable infectious diseasess. IEEE J. Sel. Areas Commun. **40**(11), 3172–3190 (2022)
5. Chiaro, D., Prezioso, E., Ianni, M., Giampaolo, F.: FL-enhance: a federated learning framework for balancing Non-IID data with augmented and shared compressed samples. Inf. Fusion **98**, 101836 (2023)
6. Cohen, G., Afshar, S., Tapson, J., van Schaik, A.: EMNIST: anextension of MNIST to handwritten letters. arXiv preprint: arXiv:1702.05373 (2017)
7. Dinh, C.T., Vu, T.T., Tran, N.H., Dao, M.N., Zhang, H.: A new look and convergence rate of federated multitask learning with Laplacian regularization. IEEE Trans. Neural Netw. Learn. Syst. (2022)
8. Dwork, C.: Differential privacy. In: Bugliesi, M., Preneel, B., Sassone, V., Wegener, I. (eds.) Automata, Languages and Programming. Lecture Notes in Computer Science, vol. 4052, pp. 1–12. Springer, Berlin (2006). https://doi.org/10.1007/11787006_1
9. Hanzely, F., Richtárik, P.: Federated learning of a mixture of global and local models. arXiv preprint: arXiv:2002.05516 (2020)
10. Itahara, S., Nishio, T., Koda, Y., Morikura, M., Yamamoto, K.: Distillation-based semi-supervised federated learning for communication-efficient collaborative training with Non-IID private data. IEEE Trans. Mob. Comput. **22**(1), 191–205 (2021)
11. Jeong, E., Oh, S., Kim, H., Park, J., Bennis, M., Kim, S.L.: Communication-efficient on-device machine learning: federated distillation and augmentation under Non-IID private data. arXiv preprint: arXiv:1811.11479 (2018)
12. Jiang, J.C., Kantarci, B., Oktug, S., Soyata, T.: Federated learning in smart city sensing: challenges and opportunities. Sensors **20**(21), 6230 (2020)
13. Jiang, Y., et al.: Model pruning enables efficient federated learning on edge devices. IEEE Trans. Neural Netw. Learn. Syst. (2022)
14. Jin, H., et al.: Personalized edge intelligence via federated self-knowledge distillation. IEEE Trans. Parallel Distrib. Syst. **34**(2), 567–580 (2022)
15. Kairouz, P., et al.: Advances and open problems in federated learning. Found. Trends® Mach. Learn. **14**(1–2), 1–210 (2021)
16. Lakhan, A., et al.: Federated-learning based privacy preservation and fraud-enabled blockchain IoMT system for healthcare. IEEE J. Biomed. Health Inform. **27**(2), 664–672 (2022)
17. Li, L., Fan, Y., Tse, M., Lin, K.Y.: A review of applications in federated learning. Comput. Ind. Eng. **149**, 106854 (2020)
18. Li, Q., Diao, Y., Chen, Q., He, B.: Federated learning on Non-IID data silos: an experimental study. In: 2022 IEEE 38th International Conference on Data Engineering (ICDE), pp. 965–978. IEEE (2022)

19. Li, X., Huang, K., Yang, W., Wang, S., Zhang, Z.: On the convergence of FedAvg on Non-IID data. arXiv preprint: arXiv:1907.02189 (2019)

20. Li, X.C., Zhan, D.C., Shao, Y., Li, B., Song, S.: FedPHP: federated personalization with inherited private models. In: Oliver, N., Perez-Cruz, F., Kramer, S., Read, J., Lozano, J.A. (eds.) Machine Learning and Knowledge Discovery in Databases. Lecture Notes in Computer Science(), vol. 12975, pp. 587–602. Springer, Cham (2021). https://doi.org/10.1007/978-3-030-86486-6_36

21. Lin, T., Kong, L., Stich, S.U., Jaggi, M.: Ensemble distillation for robust model fusion in federated learning. In: Advances in Neural Information Processing Systems , vol. 33, pp. 2351–2363 (2020)

22. Liu, T., Ding, J., Wang, T., Pan, M., Chen, M.: Towards fast and accurate federated learning with Non-IID data for cloud-based IoT applications. J. Circuits, Syst. Comput. 31(13), 2250235 (2022)

23. Liu, X., Deng, Y., Mahmoodi, T.: Wireless distributed learning: a new hybrid split and federated learning approach. IEEE Trans. Wireless Commun. 22(4), 2650–2665 (2022)

24. Lo, S.K., Lu, Q., Wang, C., Paik, H.Y., Zhu, L.: A systematic literature review on federated machine learning: from a software engineering perspective. ACM Comput. Surv. (CSUR) 54(5), 1–39 (2021)

25. Ma, X., Zhu, J., Lin, Z., Chen, S., Qin, Y.: A state-of-the-art survey on solving Non-IID data in federated learning. Futur. Gener. Comput. Syst. 135, 244–258 (2022)

26. Mahini, H., Mousavi, H., Daneshtalab, M.: GTFLAT: game theory based add-on for empowering federated learning aggregation techniques. arXiv preprint: arXiv:2212.04103 (2022)

27. McMahan, B., Moore, E., Ramage, D., Hampson, S., y Arcas, B.A.: Communication-efficient learning of deep networks from decentralized data. In: Artificial Intelligence and Statistics, pp. 1273–1282. PMLR (2017)

28. McMahan, H.B., Moore, E., Ramage, D., y Arcas, B.A.: Federated learning of deep networks using model averaging, vol. 2, p. 2. arXiv preprint: arXiv:1602.05629 (2016)

29. Mills, J., Hu, J., Min, G.: Multi-task federated learning for personalised deep neural networks in edge computing. IEEE Trans. Parallel Distrib. Syst. 33(3), 630–641 (2021)

30. Qin, Z., Li, G.Y., Ye, H.: Federated learning and wireless communications. IEEE Wirel. Commun. 28(5), 134–140 (2021)

31. Sannara, E., Portet, F., Lalanda, P., German, V.: A federated learning aggregation algorithm for pervasive computing: evaluation and comparison. In: 2021 IEEE International Conference on Pervasive Computing and Communications (PerCom), pp. 1–10. IEEE (2021)

32. Sattler, F., Müller, K.R., Samek, W.: Clustered federated learning: model-agnostic distributed multitask optimization under privacy constraints. IEEE Trans. Neural Netw. Learn. Syst. 32(8), 3710–3722 (2020)

33. Sattler, F., Wiedemann, S., Müller, K.R., Samek, W.: Robust and communication-efficient federated learning from Non-IID data. IEEE Trans. Neural Netw. Learn. Syst. 31(9), 3400–3413 (2019)

34. Shin, M., Hwang, C., Kim, J., Park, J., Bennis, M., Kim, S.L.: XOR Mixup: privacy-preserving data augmentation for one-shot federated learning. arXiv preprint: arXiv:2006.05148 (2020)

35. Shu, J., et al.: Clustered federated multitask learning on Non-IID data with enhanced privacy. IEEE Internet Things J. 10(4), 3453–3467 (2022)

36. Song, J., Wang, W., Gadekallu, T.R., Cao, J., Liu, Y.: EPPDA: an efficient privacy-preserving data aggregation federated learning scheme. IEEE Trans. Netw. Sci. Eng. (2022)
37. Tanner, M.A., Wong, W.H.: The calculation of posterior distributions by data augmentation. J. Am. Stat. Assoc. **82**(398), 528–540 (1987)
38. Voigt, P., Von dem Bussche, A.: The EU General Data Protection Regulation (GDPR). A Practical Guide, vol. 10, no. 3152676, p. 10–5555, 1st Ed. Springer International Publishing, Cham (2017)
39. Wang, X., Garg, S., Lin, H., Kaddoum, G., Hu, J., Hossain, M.S.: A secure data aggregation strategy in edge computing and blockchain-empowered internet of things. IEEE Internet Things J. **9**(16), 14237–14246 (2020)
40. Wei, B., Li, J., Liu, Y., Wang, W.: Non-IID federated learning with sharper risk bound. IEEE Trans. Neural Netw. Learn. Syst. (2022)
41. Xu, J., Glicksberg, B.S., Su, C., Walker, P., Bian, J., Wang, F.: Federated learning for healthcare informatics. J. Healthc. Inf. Res. **5**, 1–19 (2021)
42. Xue, B., He, Y., Jing, F., Ren, Y., Jiao, L., Huang, Y.: Robot target recognition using deep federated learning. Int. J. Intell. Syst. **36**(12), 7754–7769 (2021)
43. Yang, L., Huang, J., Lin, W., Cao, J.: Personalized federated learning on non-IID data via group-based meta-learning. ACM Trans. Knowl. Discov. Data **17**(4), 1–20 (2023)
44. Yang, Z., Chen, M., Wong, K.K., Poor, H.V., Cui, S.: Federated learning for 6G: applications, challenges, and opportunities. Engineering **8**, 33–41 (2022)
45. Yin, X., Zhu, Y., Hu, J.: A comprehensive survey of privacy-preserving federated learning: a taxonomy, review, and future directions. ACM Comput. Surv. (CSUR) **54**(6), 1–36 (2021)
46. You, X., Liu, X., Jiang, N., Cai, J., Ying, Z.: Reschedule gradients: temporal Non-IID resilient federated learning. IEEE Internet Things J. **10**(1), 747–762 (2022)
47. Yu, H., Wu, C., Yu, H., Wei, X., Liu, S., Zhang, Y.: A federated learning algorithm using parallel-ensemble method on Non-IID datasets. Complex Intell. Syst., 1–13 (2023)
48. Yu, T., Bagdasaryan, E., Shmatikov, V.: Salvaging federated learning by local adaptation. arXiv preprint: arXiv:2002.04758 (2020)
49. Zhang, C., Yuan, X., Zhang, Q., Zhu, G., Cheng, L., Zhang, N.: Toward tailored models on private AIoT devices: federated direct neural architecture search. IEEE Internet Things J. **9**(18), 17309–17322 (2022)
50. Zhang, K., Song, X., Zhang, C., Yu, S.: Challenges and future directions of secure federated learning: a survey. Front. Comp. Sci. **16**, 1–8 (2022)
51. Zhang, W., et al.: R ^{2} Fed: resilient reinforcement federated learning for industrial applications. IEEE Trans. Ind. Inform. (2022)
52. Zhang, Z., Zhang, Y., Guo, D., Zhao, S., Zhu, X.: Communication-efficient federated continual learning for distributed learning system with Non-IID data. Sci. China Inf. Sci. **66**(2), 122102 (2023)
53. Zhao, L., Huang, J.: A distribution information sharing federated learning approach for medical image data. Complex Intell. Syst., 1–12 (2023)
54. Zhao, Y., Li, M., Lai, L., Suda, N., Civin, D., Chandra, V.: Federated learning with Non-IID data. arXiv preprint: arXiv:1806.00582 (2018)
55. Zheng, W., Yan, L., Gou, C., Wang, F.Y.: Federated meta-learning for fraudulent credit card detection. In: Proceedings of the Twenty-Ninth International Conference on International Joint Conferences on Artificial Intelligence, pp. 4654–4660 (2021)

56. Zhu, H., Xu, J., Liu, S., Jin, Y.: Federated learning on Non-IID data: a survey. Neurocomputing **465**, 371–390 (2021)

57. Zhu, Z., Hong, J., Zhou, J.: Data-free knowledge distillation for heterogeneous federated learning. In: International Conference on Machine Learning, pp. 12878–12889. PMLR (2021)

A Review of Relationship Extraction Based on Deep Learning

Guolong Liao[1,2], Xiangyan Tang[1,2(✉)], Tian Li[1], Li Zhong[1], and Pengfan Zeng[3]

[1] School of Computer Science and Technology, Hainan University, Haikou 570228, China
tangxy36@163.com
[2] Hainan Blockchain Technology Engineering Research Center, Haikou 570228, China
[3] Warrior Logistics, 450 S Denton Tap Rd, #2651, Coppell, TX 75019, USA

Abstract. Relation extraction is a key task in natural language processing. In recent years, deep learning techniques have been widely applied in relation extraction tasks. This paper systematically reviews relation extraction techniques based on deep learning, including the application of convolutional neural networks, recurrent neural networks and Transformer models. Firstly, it introduces the representative applications of these three models. Then, relation extraction methods based on the three deep learning models are comprehensively reviewed and compared. Finally, it discusses challenges faced by relation extraction tasks, including data sparsity, long-distance dependency, and outlooks new techniques like weakly supervised relation extraction. This paper provides a systematic overview of deep learning techniques for relation extraction, aiming to facilitate further research in this field.

Keywords: Relation Extraction · Deep Learning · CNNs · RNNs · Transformers

1 Introduction

Relation extraction is a major task in natural language processing (NLP), aiming to identify and classify predefined types of relations between entities from text. For example, in the sentence "Barack Obama was born in Illinois.", the goal of relation extraction is to identify the "place of birth" relation between the entities "Barack Obama" and "Illinois". Relation extraction is necessary for many NLP applications including information retrieval, question answering systems, and knowledge graph construction. The main approaches in this field can be categorized into kernel-based methods, distant supervision techniques, and deep learning based methods. However, these approaches face challenges such as data scarcity, ambiguity, and modeling of long-range relations. Therefore, future research directions include improving model robustness and generalization, as well as utilizing external knowledge and contextual information to enhance relation extraction performance [1].

H. Jin et al. (Eds.): IAIC 2023, CCIS 2058, pp. 73–84, 2024.
https://doi.org/10.1007/978-981-97-1277-9_6

2   Relation Extraction Methods Based on Deep Learning

In recent years, the application of deep learning in relation extraction has been extensively studied. Specifically, convolutional neural networks (CNNs), recurrent neural networks (RNNs), Transformer models and pre-trained models (such as BERT, GPT, etc.) have shown remarkable performance in relation extraction tasks.

Through a review of the literature, it is evident that the framework of entity relationship extraction methods based on deep learning can be roughly divided into five parts, as shown in Fig. 1.

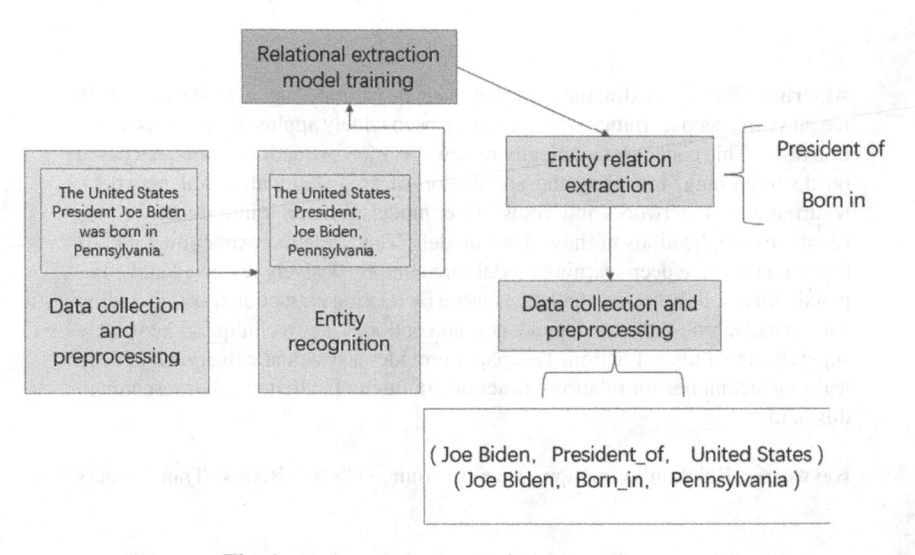

Fig. 1.   Entity relation extraction frame diagram.

The following section introduces entity relationship extraction methods based on CNNs, RNNs, Transformers and Pre-trained Models.

2.1   Relation Extraction Methods Based on CNNs

Convolutional neural networks (CNNs) have been widely applied in relation extraction tasks. CNNs can automatically learn semantic features from raw texts without complicated feature engineering.

The evolution of the famous CNN architecture for relationship extraction has gone through the following stages: sentence-based CNNs [2], piecewise CNNs [3, 4], multichannel CNNs [5], and hierarchical CNNs [6]. In the current state-of-the-art relation extraction models, advanced pre-trained CNN models like BERT have also been applied. Overall, CNNs have shown strong capabilities for relation extraction across various datasets and are commonly used as strong neural baselines.

2.2   Relation Extraction Methods Based on RNNs

Recurrent neural networks (RNNs) are a type of neural network that can process sequential data by recursively updating hidden states to model and encode sequences. They have played a significant role in the development of relation extraction techniques. Notably, RNNs have been utilized for sequence modeling and relation representation learning, with advancements in architectures like long short-term memory (LSTM) and gated recurrent units (GRU) addressing challenges such as vanishing and exploding gradients [7].

Sequence Modeling Based on RNNs. One of the early applications of RNNs in relation extraction was sequence modeling [8]. This approach involves extracting semantic features from sentences through sequential processing with RNNs. The steps include sequentially inputting words of a sentence into RNNs for recursive processing, utilizing the last hidden state to represent the semantic features of the entire sentence, and finally inputting these features into a classifier for relation classification.

Representation Learning is Based on the Relationship between Short Term Memory Network (LSTM) and Gated Cycle Unit (GRU). As the field progressed, RNNs, particularly architectures like LSTM and GRU, were leveraged for relation representation learning [9]. This approach aimed to capture long-distance dependencies by using LSTM or GRU networks. The procedure involves recursively processing the words in a sentence with LSTM or GRU, using the last hidden state to represent the semantic features of the entire sentence, and inputting these features into a classifier for relation classification. By exploiting the ability of LSTM and GRU to retain information about long-range context, this method effectively learns relation representations for sentences with complex linguistic structures and distant entities.

2.3   Relation Extraction Methods Based on Transformers and Pre-trained Language Models

Transformer is a neural network model based on self-attention mechanisms, suitable for processing sequential data. Self-attention mechanism, which is the core component of the Transformer model, establishes global dependencies in sequences, capturing semantic information from different positions. This can effectively extract sentence features, enhancing the performance of relation extraction tasks. Further performance improvement can be achieved by using a variant of self-attention called multi-head attention mechanism [10, 11]. This mechanism divides the input vectors into multiple subspaces and calculates the attention weights separately in each subspace.

Transformer Model: Enhancing Relation Extraction. The Transformer model, first proposed in the seminal paper "Attention Is All You Need" in 2017[51], marked a significant milestone in relation extraction. By employing self-attention mechanisms, Transformers can capture long-range dependencies and effectively model contextual information in sequences. This capability has greatly improved the performance of relation extraction tasks by enabling better understanding of semantic relationships between entities.

Pre-trained Models: BERT and GPT. In recent years, pre-trained models like BERT and GPT have revolutionized relation extraction tasks. BERT, introduced in 2018[10], is a transformer-based model pre-trained on a large corpus of text data. It captures contextual information from sentences and can be fine-tuned for various natural language processing tasks, including relation extraction. GPT, released in 2018 [52–55], is another influential pre-trained model that uses a transformer decoder architecture. It generates coherent and context-aware representations of text, which have proven to be valuable for relation extraction tasks.

In summary, both Transformer and pre-trained models have proven to be effective techniques for relation extraction. Their utilization of self-attention mechanisms, multi-head attention mechanisms, and fine-tuning strategies have significantly improved the performance of relation extraction tasks.

3 Analysis and Comparison of Relation Extraction Models Based on Deep Learning

3.1 Analysis and Comparison of Relation Extraction Models Based on CNNs

Since the introduction of convolutional neural networks, due to their sensitivity to local features, they have been widely applied in the field of relation extraction and have completely surpassed traditional methods in most metrics. Table 1 gives a comparison between traditional relation extraction methods and early CNNs-based methods, showing that CNNs-based methods can achieve higher scores [12].

Table 1. Comparison between traditional relation extraction methods and early CNNs-based methods.

Classifier	Example	F1
SVM	POS, stemming, syntactic patterns	60.1
SVM	word pair, words in between	72.5
RNN	POS, NER, WordNet	77.6
CNN + softmax	Words around word pair, WordNet	82.7

The table clearly demonstrates that early CNN-based models with end-to-end feature learning can surpass traditional models with manual feature engineering, leading to the prevalence of CNNs for relation extraction.

Currently, typical relation extraction models based on CNNs include: unidirectional CNN models, bidirectional CNN models, and multi-channel CNN models. The different CNN models have demonstrated the effectiveness of convolutional neural networks in learning semantic features for relation extraction tasks [13, 14]. However, CNNs are still weak in modeling long-distance dependencies and are prone to overfitting. The current research focuses are on how to improve the network architectures of CNN models, and how to combine them with attention mechanisms, graph neural networks and other models, to further improve the performance of relation extraction. Overall, relation extraction models based on CNNs, owing to their end-to-end feature learning capabilities, have become powerful models for relation extraction tasks. But CNN models still have room for improvement, and more research is needed on how to make them better at modeling text semantics, in order to play a bigger role in relation extraction.

3.2 Analysis and Comparison of Relation Extraction Models Based on RNNs

Although CNN-based models have achieved good results, CNNs rarely consider global features and temporal sequence information, especially the long-distance dependencies between entity pairs. On the other hand, RNNs can better address these issues and have achieved better results in experiments. Table 2 shows the comparison of F1 values of the relationship extraction model based on CNN and RNN in the dataset semeval2010-task8.

Table 2. The comparison of CNN-based and RNN-based methods.

Model	Encode	Dataset	F1
Nguyen et al. [15]	multi-sizes-filter-CNN	SemEval-2010 Task 8	82.8
Santos et al. [16]	fixed-size-filter-CNN	SemEval-2010 Task 8	84.1
Zhang et al. [17]	Bi-RNN	SemEval-2010 Task 8	82.5
Zhang et al. [18]	BLSTM	SemEval-2010 Task 8	84.3
Xu et al. [19]	Multi-Channel-RNN	SemEval-2010 Task 8	86.1

As the table shows, RNN-based models like BLSTM tend to achieve higher F1 scores than CNN models for relation extraction on this dataset.

Compared to CNN models, RNN models are better at learning global dependency features of sentences and can model sentence semantics more reasonably. However, RNN models have slower training processes and also suffer from vanishing/exploding gradient problems. Current experimental results of RNN models combined with attention mechanisms show that they can effectively improve relation extraction performance. Overall, both RNNs and CNNs provide powerful modeling capabilities for relation extraction and have their own advantages. Combining the characteristics of both to build models or ensemble their results can achieve better performance.

3.3 Analysis and Comparison of Relation Extraction Models Based on Transformers and Pre-Trained Models

With the development of pre-trained language models, especially the introduction of attention mechanisms like Transformer, the ability of neural network models to model long sequences has been greatly improved. This also provides new ideas for relation extraction tasks. At the same time, the complexity of current data sets is also increasing, and there are a lot of entity relationships overlapping, which also puts forward higher requirements for the current model framework.

The biggest advantage of Transformer-based relation extraction models is that they can sufficiently learn the contextual information of texts, especially the dependencies between entity pairs. The most representative of these is the Bert-based model. Table 3 below shows the F1 value comparison between them on dataset ACE05.

Table 3. The comparison of bert-based methods.

Model	Encode	ACE05 Ent, Rel, Rel +
TriMF [20]	BERT$_{base}$	87.6, 66.5, 62.8
PURE [21]	BERT$_{base}$	90.1, 67.7, 64.8
TableSeq [22]	ALBERT$_{XXLARGE}$	89.5, 67.6, 64.3
UniRE [23]	ALBERT$_{XXLARGE}$	90.2, ----, 66.0
PL-Marker [24]	ALBERT$_{XXLARGE}$	90.1, 73.0, 71.1

Ent refers to extracting entities correctly, Rel refers to extracting relationships correctly, and Rel + refers to extracting both entities and relationships correctly.

As shown in the table, PL-Marker achieves the highest Rel and Rel + F1 scores by incorporating prediction layer markers into ALBERT$_{XXLARGE}$ model.

Overall, the combination of Transformer architecture and large-scale pre-training has become a new direction for relation extraction research. Attention mechanisms help the model learn sentence-level semantics, while pre-training provides language prior knowledge. However, how to better tailor the model for the task and avoid overfitting still requires further research.

4   Challenges in Relation Extraction Based on Deep Learning

4.1   Data Sparsity Problem

Data sparsity refers to the fact that in relation extraction tasks, training data is often very limited and hardly covers all possible relations. This leads to insufficient model generalization capability in dealing with unseen relation types.

For example, in a commonly used benchmark dataset SemEval 2010 Task 8, there are only 10 relation types with about 10,000 training instances. The limited labeled data makes it difficult for models to generalize to other datasets and unseen relation types.

To alleviate the data sparsity problem, we suggest using data augmentation and distant supervision methods. Data augmentation methods include enhancing training data by using synonyms, antonyms, part-of-speech tags, etc. [25–33]. Distant supervision methods leverage relation information in knowledge bases to automatically annotate large-scale training data [34–37]. Although these two methods relieve the data sparsity problem to some extent, how to further solve the data sparsity problem remains an important future research direction in the relation extraction field.

4.2 Long Distance Dependence Problem

Long-distance dependency refers to relations that span across multiple sentences or passages in relation extraction tasks. This makes it difficult for models to capture semantic information in relations, thus hurting model performance.

For example, in document-level relation extraction, the two entities of a relation may appear in different paragraphs, requiring the model to infer the relation based on reasoning across long text distances.

Currently, there are two common solutions to address the long-distance dependency problem. The first is to introduce tree-structured modeling, which parses texts into syntactic trees for modeling [38, 39]. The second is to introduce attention mechanisms or iterative modeling, where attention mechanisms capture semantic information in relations by learning key information from texts [40, 41], and iterative modeling enhances model representation capability through multi-round learning [42]. It is worth noting that reference [42] converted the relation extraction task into a multi-round question answering problem based on the idea of multi-round iterations, and achieved state-of-the-art results on the ACE04, ACE05 and CoNLL04 datasets.

Inspired by the idea of incorporating tree-structured modeling, we propose a new approach of introducing graph-structured modeling, where graph models can parse texts into dependency graphs or co-occurrence graphs for modeling. Graph-structured modeling will clearly outperform tree-structured models, but at the same time increase model complexity. Therefore, how to effectively introduce graph-structured modeling for relation extraction remains an important future direction.

5 Trends in Relation Extraction Techniques Based on Deep Learning

5.1 Weakly-Supervised Relation Extraction Techniques

Relation extraction has long suffered from insufficient labeled data. Weakly-supervised relation extraction techniques use a small amount of labeled data to guide learning from large-scale unstructured text, thus receiving extensive attention. This section introduces weakly-supervised relation extraction methods based on deep learning.

To address the diversity of textual expressions, researchers have proposed CNN and RNN models with attention mechanisms, which focus on key words to improve the model's ability to identify relations. For example, the PCNN-ATT model achieved 80.1% F1 on the FewRel dataset [43]. Attention mechanisms allow models to pay attention to the most informative words and phrases when extracting relations, instead of treating all words equally. This improves performance in complex and diverse texts.

In addition to attention mechanisms, researchers have also explored using deep reinforcement learning for models to autonomously explore optimal strategies. The key idea is to train an agent that can learn relation extraction strategies through continuous trial-and-error interactions with the environment. Experiments show that the reinforcement learning framework significantly improves the performance of weakly-supervised relation extraction [44]. However, reinforcement learning suffers from low sample efficiency and training instability issues.

Moreover, the rapid development of knowledge graphs also provides new ideas for weakly-supervised relation extraction [45]. Using knowledge graph relations for distant supervision can automatically construct massive training data. However, this may also introduce noise which hurts model performance. Such methods have achieved state-of-the-art 82% F1 on the TACRED dataset.

Looking ahead, thoroughly exploring inherent textual correlations and achieving knowledge enhancement are beneficial directions for relation extraction. Improving evaluation metrics is also crucial. In summary, deep learning techniques have shown great potential in weakly-supervised relation extraction, but still face challenges in sample efficiency, noise control, etc. Continued research is needed to better leverage weakly labeled data.

5.2 Cross-Lingual Relation Extraction Techniques

With the explosive growth of multi-lingual textual data worldwide, cross-lingual relation extraction techniques have received extensive attention [46]. Their goal is to leverage models trained on high-resource languages to extract relations in low-resource languages. Main cross-lingual relation extraction methods based on deep learning are:

Multilingual word embeddings can represent vocabularies of different languages into the same semantic space, enabling effective cross-lingual transfer. However, constructing high-quality multilingual word embeddings is crucial for good cross-lingual transfer performance. Techniques like multi-task learning and corpus alignment can help build word embeddings that well capture cross-lingual semantics.

Adversarial learning frameworks have also been widely applied in cross-lingual relation extraction tasks. The key idea is to train feature extractors and relation classifiers in an adversarial way. By obtaining language-agnostic latent semantic feature representations through adversarial training, the performance of relation extraction in the target language can be significantly improved. However, adversarial training often suffers from instability issues and is difficult to converge.

Code-mixing techniques share bidirectional encoders across two languages to enable mutual learning of semantic information. Models like XLAN [47] allow semantic interaction between languages and outperform single-encoder structures in cross-lingual transfer. But balancing the contributions of different languages remains a challenge.

Currently, expanding relation extraction datasets for low-resource languages and designing transferable evaluation metrics are also key challenges. Techniques like crowd-sourcing and back-translation can help generate more labeled data. Composite metrics are needed to assess cross-lingual transfer performance.

In summary, cross-lingual relation extraction still faces challenges in building cross-lingual semantics, training stability, language balancing, etc. Continued research on deep learning techniques is important for its development.

5.3 Cross-Document Relation Extraction Techniques

Compared to single documents, cross-document relation extraction faces greater needs for semantic correlation and complex context modeling. In recent years, deep learning techniques have brought new opportunities for cross-document relation extraction tasks.

Methods based on hierarchical network structures significantly improve cross-document extraction performance by modeling semantic propagation across document groups, documents and sentences. This structure can reasonably represent contextual semantics at different granularities [48], balancing local and global correlations. However, designing an optimal hierarchical structure remains challenging.

Memory networks have also been extensively explored for cross-document relation extraction tasks. External memory can store document-level semantics [49], while internal memory focuses on current relations, gradually refining context representations over multiple reading rounds. The multi-hop architecture enhances the modeling capability. But multiple iterations of reading and writing also increase the training difficulty. Experiments have proven the effectiveness of memory network mechanisms in modeling complex cross-document relations.

In addition, by parsing texts into knowledge graphs, cross-document relations can be transformed into knowledge graph path reasoning problems. This method can unify document semantics, but path reasoning quality directly impacts extraction performance. Constructing accurate and complete knowledge graphs is non-trivial.

Currently, constructing high-quality cross-document relation datasets and designing composite metrics are key challenges. Lack of datasets hinders model development and evaluation. Overall, deep learning techniques have advanced the development of cross-document relation extraction [50], but continued research is still needed to improve modeling capability and training efficiency.

6 Conclusion

This paper comprehensively and systematically reviews relation extraction techniques based on deep learning. Our literature classification is based on a timeline, covering representative articles from 2015 to 2022. Currently, most relation extraction review articles classify based on relation extraction models, while we classify based on deep learning. Classifying based on deep learning can help us better understand the development process of relation extraction models from the bottom up, and better point out future directions. The main thread of this paper: This paper first introduces the definition and significance of relation extraction, as well as the technical system based on deep learning methods. It then discusses in detail relation extraction models based on convolutional neural networks, recurrent neural networks and Transformer, and compares the advantages and disadvantages of each model. Next, it analyzes the main challenges of deep learning in relation extraction, including data sparsity and long-range dependencies. Finally, it looks ahead at the development directions of relation extraction techniques such as weak supervised relation extraction.

In summary, deep learning provides a new perspective and models for the relation extraction task. But relation extraction still faces problems such as data quality and model optimization. Future research may focus on new technologies such as distant supervision, knowledge enhancement, and cross-lingual relation extraction, to further unleash the potential of deep learning in relation extraction. The summarization and outlook of this paper help better understand the modeling, application and development of deep learning in the field of relation extraction, and can also provide inspiration for subsequent research.

Acknowledgments. This work was supported by Youth Foundation Project of Hainan Natural Science Foundation (621QN211), National Natural Science Foundation of China (NSFC) (Grant No. 62162022, 62162024), the Major science and technology project of Hainan Province (Grant No. ZDKJ2020012), Hainan Provincial Natural Science Foundation of China (Grant No. 620MS021).

References

1. Zeng, W., Lin, Y., Liu, Z., et al.: Incorporating relation paths in neural relation extraction. arXiv preprint arXiv:1609.07479 (2016)
2. Liu, G., Fu, L., Yu, B., Cui, L.: Automatic recognition of parallel sentence based on sentences-interaction CNN and its application. In: 2022 7th International Conference on Computer and Communication Systems (ICCCS), pp. 245–250. IEEE, April 2022
3. Wen, H., Zhu, X., Zhang, L., Li, F.: A gated piecewise CNN with entity-aware enhancement for distantly supervised relation extraction. Inf. Process. Manage. **57**(6), 102373 (2020)
4. Li, Y., Ni, P., Li, G., Chang, V.: Effective piecewise CNN with attention mechanism for distant supervision on relation extraction task. In: 5th International Conference on Complexity, Future Information Systems and Risk, SciTePress. pp. 53–62, May 2020
5. Liu, Z., Huang, H., Lu, C., Lyu, S.: Multichannel CNN with attention for text classification. arXiv preprint arXiv:2006.16174 (2020)
6. Shimura, K., Li, J., Fukumoto, F.: HFT-CNN: learning hierarchical category structure for multi-label short text categorization. In: Proceedings of the 2018 Conference on Empirical Methods in Natural Language Processing, pp. 811–816 (2018)
7. Chung, J., Gulcehre, C., Cho, K., Bengio, Y.: Empirical evaluation of gated recurrent neural networks on sequence modeling. arXiv preprint arXiv:1412.3555 (2014)
8. Smirnova, E., Vasile, F.: Contextual sequence modeling for recommendation with recurrent neural networks. In: Proceedings of the 2nd Workshop on Deep Learning for Recommender Systems, pp. 2–9, August 2017
9. Zhang, Z., Cui, P., Zhu, W.: Deep learning on graphs: a survey. IEEE Trans. Knowl. Data Eng. **34**(1), 249–270 (2020)
10. Devlin, J., Chang, M.W., Lee, K., Toutanova, K.: Bert: Pre-training of deep bidirectional transformers for language understanding. arXiv preprint arXiv:1810.04805 (2018)
11. Kenton, J.D.M.W.C., Toutanova, L.K.: Bert: pre-training of deep bidirectional transformers for language understanding. In: Proceedings of naacL-HLT, vol. 1, p. 2, June 2019
12. Zeng, D., Liu, K., Lai, S., Zhou, G., Zhao, J.: Relation classification via convolutional deep neural network. In: Proceedings of COLING 2014, the 25th International Conference on Computational Linguistics: Technical Papers, pp. 2335–2344, August 2014
13. Basiri, M.E., Nemati, S., Abdar, M., Cambria, E., Acharya, U.R.: ABCDM: an attention-based bidirectional CNN-RNN deep model for sentiment analysis. Futur. Gener. Comput. Syst. **115**, 279–294 (2021)
14. Sun, K., Li, Y., Deng, D., Li, Y.: Multi-channel CNN based inner-attention for compound sentence relation classification. IEEE Access **7**, 141801–141809 (2019)
15. Nguyen, T.H., Grishman, R.: Relation extraction: perspective from convolutional neural networks. In: Proceedings of the 1st Workshop on Vector Space Modeling for Natural Language Processing, pp. 39–48, June 2015
16. Santos, C.N.D., Xiang, B., Zhou, B.: Classifying relations by ranking with convolutional neural networks. arXiv preprint arXiv:1504.06580 (2015)
17. Zhang, D., Wang, D.: Relation classification via recurrent neural network. arXiv preprint arXiv:1508.01006 (2015)

18. Zhang, S., Zheng, D., Hu, X., Yang, M.: Bidirectional long short-term memory networks for relation classification. In: Proceedings of the 29th Pacific Asia Conference on Language, Information and Computation, pp. 73–78, October 2015
19. Xu, Y., et al.: Improved relation classification by deep recurrent neural networks with data augmentation. arXiv preprint arXiv:1601.03651 (2016)
20. Shen, Y., Ma, X., Tang, Y., Lu, W.: A trigger-sense memory flow framework for joint entity and relation extraction. In: Proceedings of the Web Conference 2021, pp. 1704–1715, April 2021
21. Zhong, Z., Chen, D.: A frustratingly easy approach for entity and relation extraction. arXiv preprint arXiv:2010.12812 (2020)
22. Wang, J., Lu, W.: Two are better than one: Joint entity and relation extraction with table-sequence encoders. arXiv preprint arXiv:2010.03851 (2020)
23. Wang, Y., Sun, C., Wu, Y., Zhou, H., Li, L., Yan, J.: UniRE: a unified label space for entity relation extraction. arXiv preprint arXiv:2107.04292 (2021)
24. Ye, D., Lin, Y., Li, P., Sun, M.: Packed levitated marker for entity and relation extraction. arXiv preprint arXiv:2109.06067 (2021)
25. Shorten, C., Khoshgoftaar, T.M., Furht, B.: Text data augmentation for deep learning. J. big Data **8**, 1–34 (2021)
26. Bayer, M., Kaufhold, M.A., Reuter, C.: A survey on data augmentation for text classification. ACM Comput. Surv. **55**(7), 1–39 (2022)
27. Liu, P., Wang, X., Xiang, C., Meng, W.: A survey of text data augmentation. In: 2020 International Conference on Computer Communication and Network Security (CCNS), pp. 191–195. IEEE, August 2020
28. Wei, J., Zou, K.: EDA: easy data augmentation techniques for boosting performance on text classification tasks. arXiv preprint arXiv:1901.11196 (2019)
29. Karimi, A., Rossi, L., Prati, A.: AEDA: an easier data augmentation technique for text classification. arXiv preprint arXiv:2108.13230 (2021)
30. Rizos, G., Hemker, K., Schuller, B.: Augment to prevent: short-text data augmentation in deep learning for hate-speech classification. In: Proceedings of the 28th ACM International Conference on Information and Knowledge Management, pp. 991–1000, November 2019
31. Papanikolaou, Y., Pierleoni, A.: Dare: Data augmented relation extraction with gpt-2. arXiv preprint arXiv:2004.13845 (2020)
32. Wang, A., et al.: Entity relation extraction in the medical domain: based on data augmentation. Ann. Trans. Med. **10**(19) (2022)
33. Hu, X.: GDA: Generative Data Augmentation Techniques for Relation Extraction Tasks. arXiv preprint arXiv:2305.16663 (2023)
34. Su, P., Li, G., Wu, C., Vijay-Shanker, K.: Using distant supervision to augment manually annotated data for relation extraction. PLoS ONE **14**(7), e0216913 (2019)
35. Qin, P., Xu, W., Wang, W.Y.: Robust distant supervision relation extraction via deep reinforcement learning. arXiv preprint arXiv:1805.09927 (2018)
36. Zhou, Y., Pan, L., Bai, C., Luo, S., Wu, Z.: Self-selective attention using correlation between instances for distant supervision relation extraction. Neural Netw. **142**, 213–220 (2021)
37. Smirnova, A., Cudré-Mauroux, P.: Relation extraction using distant supervision: a survey. ACM Comput. Surv. (CSUR) **51**(5), 1–35 (2018)
38. Geng, Z., Chen, G., Han, Y., Lu, G., Li, F.: Semantic relation extraction using sequential and tree-structured LSTM with attention. Inf. Sci. **509**, 183–192 (2020)
39. Yang, T. et al.: Tree-capsule: tree-structured capsule network for improving relation extraction. In: Karlapalem, K., et al. Advances in Knowledge Discovery and Data Mining. PAKDD 2021. Lecture Notes in Computer Science(), vol 12714. Springer, Cham (2021). https://doi.org/10.1007/978-3-030-75768-7_26

40. Shen, Y., Huang, X.J.: Attention-based convolutional neural network for semantic relation extraction. In: Proceedings of COLING 2016, the 26th International Conference on Computational Linguistics: Technical Papers, pp. 2526–2536, December 2016
41. Yuan, Y., Zhou, X., Pan, S., Zhu, Q., Song, Z., Guo, L.: A relation-specific attention network for joint entity and relation extraction. In: International Joint Conference on Artificial Intelligence, January 2021
42. Li, X., et al.: Entity-relation extraction as multi-turn question answering. arXiv preprint arXiv: 1905.05529 (2019)
43. Liu, T., Wang, K., Chang, B., Sui, Z.: A soft-label method for noise-tolerant distantly supervised relation extraction. In: Proceedings of the 2017 Conference on Empirical Methods in Natural Language Processing, pp. 1790–1795, September 2017
44. Zhang, N., Deng, S., Sun, Z., Chen, X., Zhang, W., Chen, H.: Attention-based capsule networks with dynamic routing for relation extraction. arXiv preprint arXiv:1812.11321 (2018)
45. Alt, C., Gabryszak, A., Hennig, L.: TACRED revisited: A thorough evaluation of the TACRED relation extraction task. arXiv preprint arXiv:2004.14855 (2020)
46. Lample, G., Conneau, A.: Cross-lingual language model pretraining. arXiv preprint arXiv: 1901.07291 (2019)
47. Poeste, M., Müller, N., Arnaus Gil, L.: Code-mixing and language dominance: bilingual, trilingual and multilingual children compared. Int. J. Multiling. 16(4), 459–491 (2019)
48. Yao, Y.: CodRED: A cross-document relation extraction dataset for acquiring knowledge in the wild. In: Proceedings of the 2021 Conference on Empirical Methods in Natural Language Processing, pp. 4452–4472, November 2021
49. Wang, F., et al.: Entity-centered cross-document relation extraction. arXiv preprint arXiv: 2210.16541 (2022)
50. Wu, H., Chen, X., Hu, Z., Shi, J., Xu, S., Xu, B.: Local-to-global causal reasoning for cross-document relation extraction. IEEE/CAA J. Autom. Sinica 10(7), 1608–1621 (2023)
51. Vaswani, A., et al.: Attention is all you need. Advances in neural information processing systems, 30 (2022)
52. Radford, A., Narasimhan, K., Salimans, T., Sutskever, I.: Improving language understanding by generative pre-training (2018)
53. Radford, A., Wu, J., Child, R., Luan, D., Amodei, D., Sutskever, I.: Language models are unsupervised multitask learners. OpenAI blog 1(8), 9 (2019)
54. Brown, T., et al.: Language models are few-shot learners. Adv. Neural. Inf. Process. Syst. 33, 1877–1901 (2020)
55. Bubeck, S., et al.: Sparks of artificial general intelligence: early experiments with gpt-4. arXiv preprint arXiv:2303.12712 (2023)

BERT and LLM-Based Multivariate Hate Speech Detection on Twitter: Comparative Analysis and Superior Performance

Xiaohou Shi[1], Jiahao Liu[2(✉)], and Yaqi Song[1]

[1] China Telecom Corporation Limited Research Institute, Beijing, China
{shixh6,songyq11}@chinatelecom.cn
[2] Johns-Hopkins University, Baltimore, MD 21218, USA
jliu306@jh.com

Abstract. The detection of toxic and hate speech in online social media is becoming increasingly necessary due to its prevalence and the potentially harmful consequences it can cause. Previous research has demonstrated the vital role that machine learning and natural language processing models have in identifying inappropriate language. In this study, the aim is to assess the viability of BERT for accurately predicting multivariate classifications related to hate speech on Twitter. The analysis will be conducted using the Twitter hate speech dataset. BERT has demonstrated exceptional performance in numerous areas of NLP, making it a potentially superior alternative to traditional machine learning approaches. Experiments were performed on the same dataset using 1-layer BERT, 2-layers BERT, and logistic regression models for both training and prediction purposes. The results demonstrate that the 2-layer BERT produces an accuracy of 85%. Additionally, we incorporated transfer learning techniques by leveraging a Large Language model GPT-3 and data augmentation strategies to further enhance model performance. This experiment reached a higher accuracy of 88%. As this is a multivariate classification problem with an asymmetrical dataset, we anticipate BERT and GPT-3 will achieve greater accuracy for the binary classification problem of identifying hate speech. These findings enhance the comprehension of hate speech detection in online material and the implications of various modeling approaches.

Keywords: BERT · GPT-3 · Hate Speech Detection · Nature Language Process · Transfer learning · data augmentation · Large Language modeling · Fine-tuning · Social media

1 Introduction

In the last decade, the world has witnessed rapid growth in the number of users of social networks and online forums: about 1.91 billion people log onto Facebook daily [1], and around 6,000 tweets are tweeted on Twitter every second [2]. Users of these applications are composed of people from different cultures and educational backgrounds. These users may have contending viewpoints on many things, which is likely to lead to verbal

H. Jin et al. (Eds.): IAIC 2023, CCIS 2058, pp. 85–97, 2024.
https://doi.org/10.1007/978-981-97-1277-9_7

assaults using toxic language, drawing a series of problems. For example, hate speech can incite criminal activities or violent behaviors targeting others among young individuals. Furthermore, it inflicts harm on others, resulting in cyberbullying, which in turn can lead to depression and even suicide [3]. Since manually removing toxic language from the web can be very tedious, it is necessary to devise a model which detects such toxic language among social networks.

Social media companies have been consistently committed to eradicating hate speech and preventing its emergence as a disruptor of societal safety. Numerous scholars have dedicated their efforts to researching the automated detection of hate speech. Often, social media users do not explicitly employ hate-related terms for their attacks; instead, they resort to abbreviations or ordinary language, which further complicates the task of hate speech detection. To tackle this issue, we will use a free hate speech data set [4], in which toxic languages are divided into two categories: hate speech and offensive language. In Wikipedia, hate speech is defined as "any speech that attacks a person or group on the basis of attributes such as race, religion, ethnic origin, national origin, gender, disability" [5]. Offensive language is defined as text containing abusive slurs or derogatory terms. The input to our algorithm is posts in Twitter. We plan to use a neural network to classify them as hate speech, offensive language or neither.

2 Literature Review

2.1 Word2Vec

Word2Vec, an unsupervised learning algorithm, is integral in converting words into compact vectors within a continuous vector space [7–11]. It was introduced by introduced by Mikolov and colleagues in 2013. This inventive method will show semantic connections between words and improve the ability of processing languages. Word2Vec contains two main structures: Continuous Bag of Words (CBOW) and Skip-gram. Word2Vec has made significant contributions to improving hate speech prediction models in multiple significant ways [12]. With the improvement of semantic word meanings, Word2Vec makes the model more complex and recognizing subtle contents in hate speeches. Besides, machine learning models benefit from Word2Vec-generated vectors that include linguistic patterns and contextual information. The ability for vectors to distinguish hateful and non-hateful speech will be improved [13]. Word2Vec embeddings also demonstrate an impressive capacity to generalize meanings, with an invaluable resource for predicting hate speech in multiple languages.

2.2 Bert

Bidirectional Encoder Representations from Transformers (BERT) has significantly impacted natural language processing (NLP) since its introduction by Devlin et al. in 2018 [14]. BERT is a pre-trained model based on transformers, with rich contextual information in text. It is designed to comprehend the context of words in a sentence or document by considering the entire input sequence simultaneously. It uses a bidirectional approach, processing text in both left-to-right and right-to-left directions to capture dependencies and relationships between words efficiently. Its contextual awareness is especially useful in identifying hate speech that relies on subtle linguistic cues and context [15]. BERT's ability to show context and relationships between words enables it to identify hate speech. Besides, BERT models are pre-trained on extensive amounts of text data, which means that they are ideal for transfer learning. Refining these models for hate speech detection tasks enables them to utilize their extensive language knowledge, improving accuracy. With multilingual support: hate speech is present in several languages and dialects [16]. BERT's multilingual models can be customized to detect hate speech in different languages, improving the safety of online environments worldwide. Contextual Patterns means that hate speech may involve developing language patterns. BERT's flexibility in adapting to evolving linguistic trends allows it to identify emerging forms of hate speech [17].

2.3 Large Language Model

Over the past few years, advanced machine learning models, especially large language models like GPT-3, have emerged as leading solutions in the realm of natural language processing. OpenAI's GPT-3, a state-of-the-art model with 175 billion parameters, showcases capabilities beyond mere text generation, spanning tasks like translation, summarization, and even rudimentary reasoning [18]. Notably, the adaptability of such models has led researchers to explore their efficacy in hate speech detection. While traditional methods relied on explicit feature engineering and smaller datasets, the massive scale of models like GPT-3 allows them to discern intricate patterns within vast amounts of data, potentially offering improved accuracy in identifying and mitigating online hate speech. However, ensuring fairness and reducing biases in outputs remain key challenges.

3 Dataset and Features

Our available dataset is sourced from the Davidson-ICWSM dataset [5]. Davidson initially compiled a hate speech lexicon, which included words and phrases recognized by internet users as hate speech, and this lexicon was curated by Hatebase.org. They used the Twitter API to search for tweets containing the vocabulary and obtained a sample of tweets from 33,458 Twitter users. They extracted timelines for each user, resulting in a dataset of 85.4 million tweets. Subsequently, they randomly selected 25,000 tweets containing the vocabulary from this corpus and had them manually annotated by Crowd-Flower personnel [6]. These workers were tasked with categorizing each tweet into one of three categories: hate speech, offensive but not hate speech, and neither offensive nor hate speech. Davidson use CrowdFlower which drew a clear distinction between offensive language and hate speech, categorizing only those comments that escalated to racism, gender discrimination, and the like as hate speech.

In our data preprocessing process, we primarily focused on the tweet column, which contains the text of each tweet. This choice was made because our task's goal was to predict whether a tweet exhibited hateful or offensive tendencies through textual analysis. Other columns were not utilized because they demonstrated a high correlation with the labels, effectively functioning as a majority vote for classification. Regarding categorical labels, our datasets were imbalanced, with over 70% of the data labeled as offensive language. Handling missing data was not a concern for us as we did not encounter any missing data in our dataset. In terms of outliers, we did not address outliers explicitly. Since our input consisted of text data, detecting outliers in the traditional sense was challenging, and we assumed that there were no outliers in our dataset.

Given the imbalanced nature of our dataset, with over 70% of data labeled as offensive language, data augmentation becomes pivotal to enhance model robustness. For our textual data, we employed a technique called back-translation. This involves translating a tweet into another language (e.g., French or Spanish) using a machine translation tool, and then translating it back to English. The re-translated tweet often retains its original semantics but has slight lexical variations, effectively augmenting our dataset with "new" examples. Additionally, we considered synonym replacement, where random words in a tweet are replaced by their synonyms. This ensures model robustness against lexical diversity without altering the tweet's underlying sentiment [19]. Both these methods aim to increase the diversity of our training samples, especially in the minority class (hate speech), to alleviate the overfitting concerns that arise due to class imbalance.

4 Methods

4.1 Experiment Setup

In our experimental setup, two primary metrics were utilized to evaluate our methods: test accuracy and the confusion matrix. Test accuracy was a crucial evaluation metric due to the classification nature of our problem. It provided a clear indication of how effectively our models classified tweets into the categories of offensive language or hate speech. Additionally, the confusion matrix served as another valuable tool in our evaluation process. We utilized the methodology to detect pairs of classes that were prone to confusion. This provided important insights for comprehending the model's performance and locating potential areas of misclassification.

During the training of our models, diverse loss function options were investigated. Logistic regression employed logistic loss, and deep neural networks adopted cross-entropy loss. To address the issue of managing an imbalanced dataset, we conducted experiments with weighted loss schemes. These schemes assigned weights to each class that were inversely proportional to the class frequency. By doing so, the model was able to give greater significance to the minority class during training. This approach effectively helped to manage the effects of class imbalance.

Regarding our data splitting strategies, we utilized different ratios for our baseline and BERT models. For the initial model, an 80-20 split was executed, assigning 80% of the data for training and reserving the remaining 20% for testing. This division enabled us to efficiently train the initial model with the available dataset. On the other hand, the BERT model underwent an 80-10-10 split, with 80% of the data assigned for training, 10% for validation, and another 10% for testing purposes. This allocation guaranteed a separate validation set for tuning hyperparameters, a crucial aspect concerning BERT. Additionally, it upheld the continuity of data splitting through our baseline and BERT experiments.

4.2 Baseline

We selected logistic regression as our baseline model because of its simplicity, inter-pretability, and efficiency. Logistic regression's unambiguous probabilistic out-puts and feature contributions were ideal for establishing a reliable performance baseline. Additionally, it enabled us to devote more attention and resources to our more intricate BERT-based model, while still achieving dependable baseline out-comes.

During our analysis of the related literature, we observed the performances of different techniques applied to the same dataset or task. Specifically, S. S. Ali's research involving various statistical methods demonstrated logistic regression's exceptional accuracy of 0.904, affirming its potential as a baseline model.

Although logistic regression demonstrated impressive results, BERT was chosen as our primary method. BERT's outstanding performance in multiple natural language processing (NLP) tasks, showcased in various benchmarks, rendered it a compelling option for our project. Its capacity to apprehend contextual data and semantic subtleties in text coincided with our objective to enhance the detection of hate speech.

4.3 BERT Model

In our implementation, we utilized PyTorch-lightning for the BERT model, implement-ing backpropagation to update BERT's various parameters during training. For the base-line logistic regression model, we utilized Scikit-learn (Sklearn), which provides a sim-ple gradient descent-based training interface, streamlining our workflow. To enhance the performance of our models, we performed hyperparameter tuning for both BERT and logistic regression. We analyzed critical hyperparameters, such as training epochs, model architecture, dropout rate, and optimization approach.

It was observed that the model's performance presented sensitivity during the initial training stages, but became stable after approximately 15 epochs. Concerning BERT, we conducted experiments on the model's structure, explicitly varying the quantity of linear layers. Adding an additional linear layer from one layer to two layers produced a slight increase in accuracy in the test set. To prevent overfitting, we assessed various dropout rates, observing limited sensitivity to changes in this hyperparameter. Although selecting a suitable dropout rate could improve accuracy slightly, models with different dropout rates performed almost similarly given identical hyperparameters. Additionally, the optimization method we selected significantly impacted the model's ability to learn from the training data.

Overall, BERT proved to be more challenging to implement and train compared to logistic regression. This complexity primarily arose from BERT's extensive parameter count and the associated time and resource demands for training and hyperparameter tuning. However, its performance benefits, especially for NLP tasks like hate speech detection, justified the additional effort.

4.4 GPT-3 (Large Language Model)

In the experimental setup, we employed GPT-3 as the initial model, leveraging its extensive pre-trained language knowledge for the task of hate speech classification. The primary objective is to fine-tune GPT-3 to accurately identify hate speech while preserving its general language understanding capabilities.

The Fine-Tuning Strategy details include: 1. Freezing Lower Layers: To capitalize on GPT-3's vast pre-trained linguistic knowledge, we feezed the lower layers of the model, restricting fine-tuning to the upper layers, including the fully connected layers. This ensures that the model retains its generic language knowledge acquired during pre-training. 2. Lower Learning Rate: To prevent overfitting and maintain sensitivity to the pre-trained knowledge, we employed a lower learning rate during fine-tuning. This aids in ensuring that the model does not forget its previously learned information when adapting to the new hate speech classification task. 3. Exploring Different Fine-Tuning Strategies: We experimented with various fine-tuning strategies and learning rates (0.1, 0.5, 0.8) to discover the optimal performance configuration. This may involve exploring different top-level architectures, adjusting the depth of fine-tuning layers, and experimenting with different optimizers and hyperparameters.

By utilizing GPT-3 as the initial model and employing appropriate fine-tuning and data augmentation techniques, our aim is to construct an efficient model capable of robust hate speech detection. Through transfer learning, we seeked to harness the advantages of large-scale pre-trained language models to address this crucial issue effectively.

5 Results

The experiment results are presented in the Tables 1, 2 and 3. Table 1 shows the accuracy of our models. Tables 2 and 3 are the experiments of parameters, including optimization methods and dropout values.

Table 1. Results accuracy of Baseline, BERTS and GPT-3 on the different datasets.

Accuracy				
	Baseline	1-layer BERT	2-layers BERT	GPT-3
Original dataset	0.8063	0.8483	0.8571	0.8623
Weighted Loss dataset	0.698	0.8249	0.8429	/
Subset	0.6	0.7306	0.7583	/
Data-augmentation	0.8027	0.8546	0.8659	**0.8815**

We conducted three experiments to determine the impact of training on weighted, small and augmentation data sets. Our regression model indicates that it can attain 80% accuracy with the original data set, despite its imbalance. Due to overfitting on the original data set, the regression model was limited to predicting offensive language and not hate speech, as outlined in the mix-up matrix. Even when using a weighted dataset or

Table 2. Different Optimization Methods result accuracy of 2-layers BERT and GPT-3 on the augmentation dataset

	Optimization Methods		
	SGD	Adam	AdamW
2-layers BERT	0.7785	0.8571	0.8371
GPT-3	0.8127	**0.8815**	0.8622

Table 3. Different Dropout Values result accuracy of 2-layers BERT and GPT-3 on the augmentation dataset

	Dropout Values		
	0.1	0.5	0.8
2-layers BERT	0.8543	0.8644	0.8603
GPT-3	0.8704	**0.8815**	0.8791

subset, accuracy was not ideal and differed significantly from BERT, which is more stable in this regard. For the second experiment, we discovered that the accuracy of 2-layer BERT was superior to that of 1-layer BERT. To achieve this, we made adjustments to the 2-layer BERT by eliminating the dropout and linear layers in the second layer. These modifications were also applied to the unbalanced data set. Following the fine-tuning of the structure, the effects of 2-layer BERT were even more ideal. For transfer learning, GPT-3 excelled, especially on augmented data, showcasing impressive adaptability. Data augmentation significantly boosted model performance.

5.1 BERT Model

Our BERT-based model achieved an accuracy of 80% in comparison with related work in the field of hate speech detection, demonstrating a competitive performance, similar to other methods, including transformer-based approaches and various statistical methods. This comparison provides valuable context for the results, highlighting the effectiveness of deep learning methods for this task.

An important discovery we made was the rapid onset of overfitting during the BERT model training, as demonstrated by the divergence between the training and validation curves within a few epochs. The curves are shown in below Figs. 1 and 2. This indicated that the dataset size may not be adequate for effectively fine-tuning BWe noticed that both models were sensitive to specific hyperparameters like dropout values. Specifically, the model's performance was significantly affected by the number of training epochs, with stability achieved after about 15 epochs. The tendency are shown in the Figs. 3 and 4. The architectural choices, including the number of linear layers in the BERT model, had a minor effect on test set accuracy. The model exhibited limited sensitivity to variations in the dropout rate. Although an appropriate rate could increase accuracy slightly, its overall influence was minor. Additionally, selecting an appropriate optimizer significantly affected model performance, underscoring its importance.

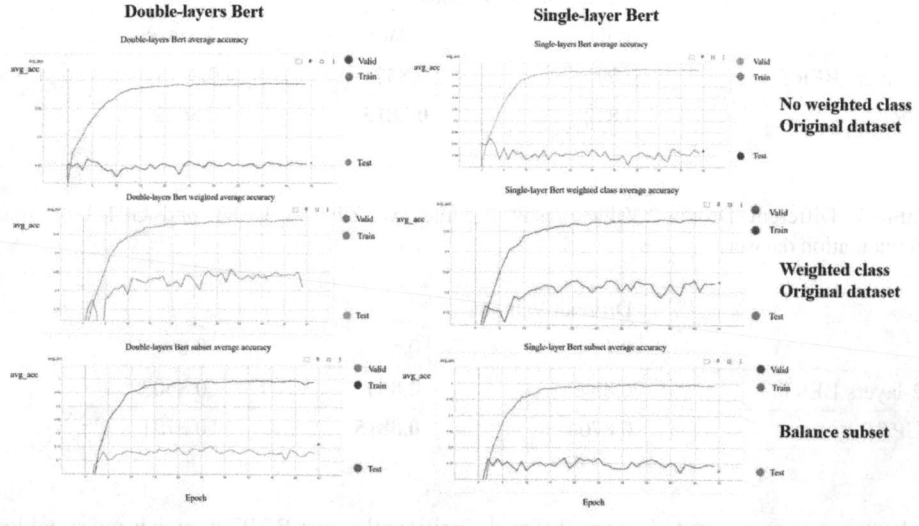

Fig. 1. Average accuracy growth curve of single-layer BERT and double-layers BERT on different datasets

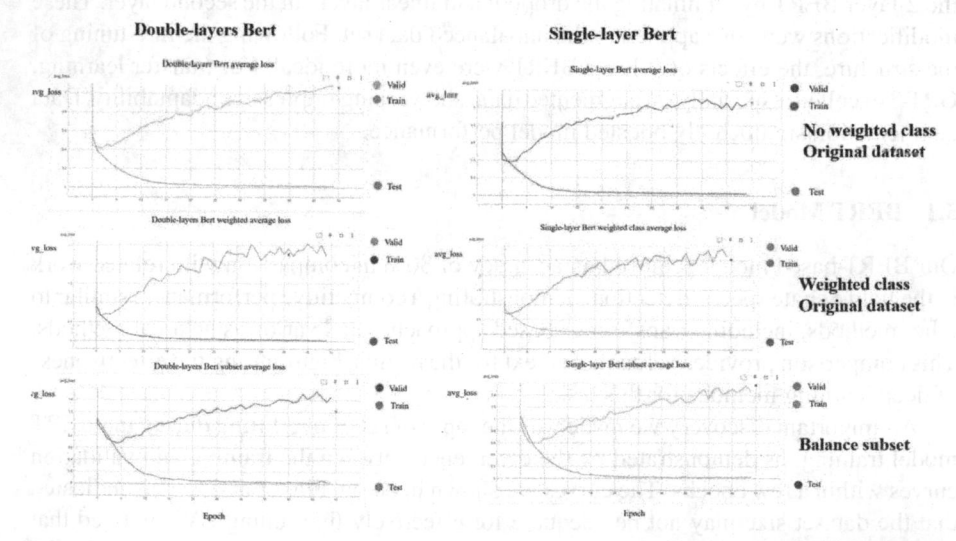

Fig. 2. Average loss growth curve of single-layer BERT and double-layers BERT on different datasets

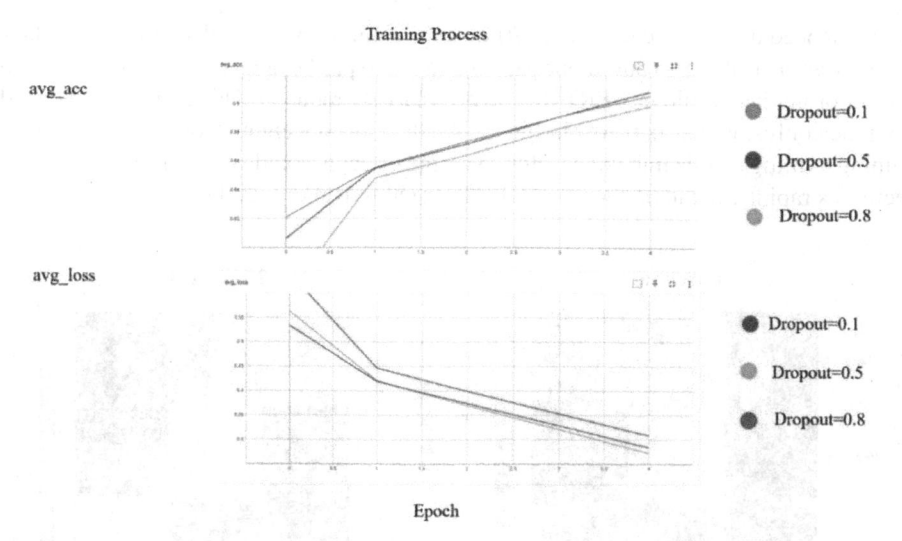

Fig. 3. Average accuracy and loss growth curve of double-layers BERT on different dropout values during training process

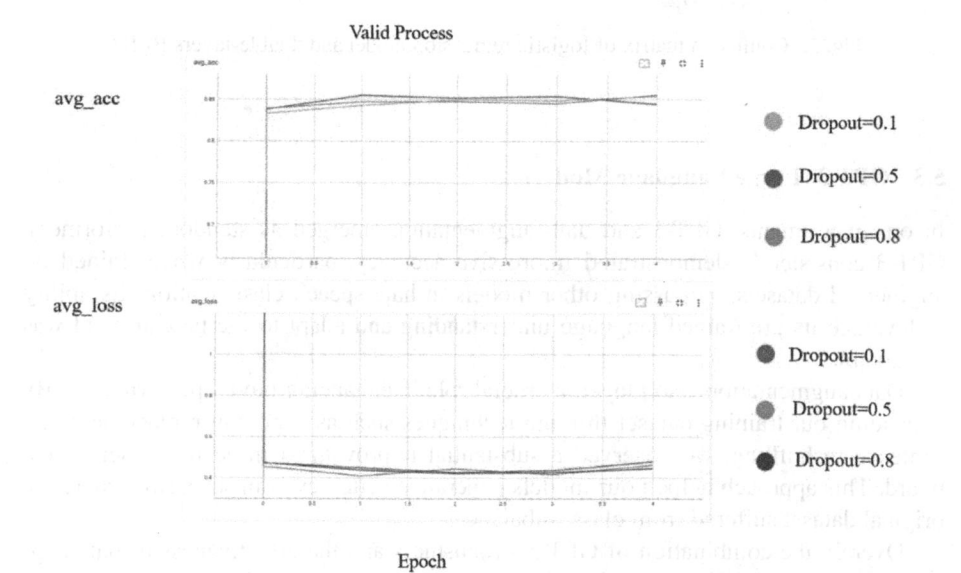

Fig. 4. Average accuracy and loss growth curve of double-layers BERT on different dropout values during valid process

5.2 Baseline

Our baseline logistic regression model demonstrated robustness and competitive performance, surpassing more intricate models such as Naive Bayes. Addressing the challenge of imbalanced data, where more than 70% of the samples were labeled as offensive language, was critical. This caused our baseline tends to predict a lot of offensive speech to hateful or neither, while in BERT this error is largely reduced. This could be observed from the confusion matrix below in Fig. 5. We developed weighted loss schemes during training to mitigate this imbalance. However, the effect is not ideal, and its accuracy rate decreases rapidly. It can be seen that BERT structure is more stable.

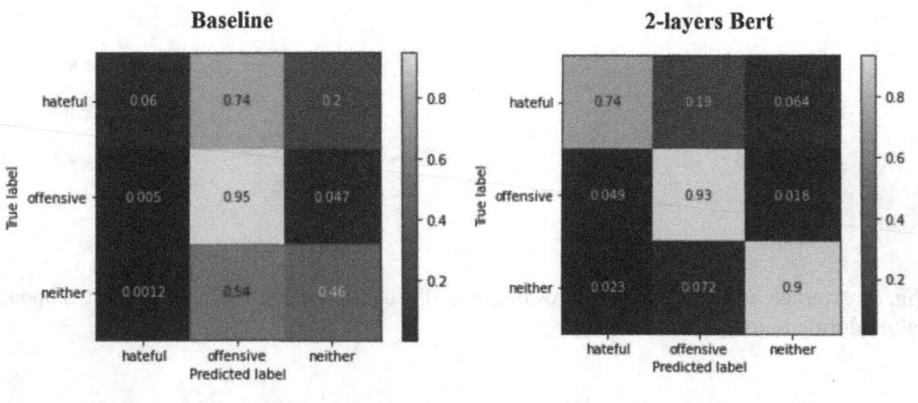

Fig. 5. Confusion matrix of logistic regression model and double-layers BERT

5.3 GPT-3 (Large Language Model)

In our experiments, GPT-3 and data augmentation emerged as standout performers. GPT-3 consistently demonstrated impressive accuracy, particularly when trained on augmented datasets, surpassing other models in hate speech classification. Its ability to leverage its pre-trained language understanding and adapt to the task at hand was remarkable.

Data augmentation also played a crucial role in enhancing model performance. By expanding our training dataset through techniques such as synonym replacement and sentence reshuffling, we observed a substantial improvement in accuracy across the board. This approach helped our models generalize better, even in scenarios where the original dataset suffered from class imbalance.

Overall, the combination of GPT-3's robustness and the effectiveness of data augmentation substantially boosted the accuracy of our hate speech classification models, underscoring the potential of advanced language models and data preprocessing techniques in addressing the challenges of hate speech detection.

5.4 Discussion

In our analysis of the results, we observed several key findings:

We conducted a comprehensive performance comparison between the baseline model (logistic regression) and two variations of the BERT classifier (1-layer and 2-layer).

The 2-layer BERT classifier outperformed the others, which was a surprising outcome. However, it was noteworthy that the 1-layer and 2-layer BERT models showed comparable performance, indicating that adding an extra layer did not significantly improve performance in this particular context.

We also investigated the management of class imbalance, analyzing various approaches, including situations where we did not address the imbalance at all. It was surprising to discover that the greatest accuracy was achieved without any specific imbalance handling. Nonetheless, we recognize that this outcome may have been impacted by the disproportionate character of our test sample. We underscored the significance of taking into account supplementary metrics to appraise model efficiency across diverse class distributions, even though the project's restrictions hindered our examination concerning this aspect.

Moreover, our scrutiny divulged that the optimization technique's selection had a weighty impact on model precision. The Adam optimizer surpassed alternatives such as SGD and AdamW, resulting in a noteworthy enhancement in performance. This inclination towards Adam coincides with trends in the industry, since it generally provides stability and efficiency for training deep neural networks.

6 Conclusion

In conclusion, our study demonstrates the effectiveness of BERT in hate speech detection through well-considered feature extraction and classification techniques. BERT leverages both domain agnosticism and domain-specific word embeddings by training on extensive data and fine-tuning on domain-specific datasets. The paper maintains a clear and comprehensible structure, elucidating technical abbreviations upon first use and adopting a formal, objective tone with precise vocabulary. Our experiments shed light on the impact of model structure, revealing that both 1-layer and 2-layer BERT models perform similarly. However, 2-layer BERT achieves remarkable accuracy when the second dropout and linear layers are removed, especially considering the dataset's inherent asymmetry.

We also delved into addressing class imbalance and noted that, surprisingly, the highest accuracy was achieved without any specific imbalance treatment. Nevertheless, we acknowledge the potential influence of the test set's composition on these findings, and additional metrics should be considered for varying class distributions.

Furthermore, our analysis underscores the significance of the optimization method choice. The Adam optimizer consistently outperforms other options, aligning with industry trends for stability and efficiency in deep neural network training.

Finally, our study extends its scope to consider the impact of data augmentation, transfer learning, and the role of GPT-3. These aspects collectively contribute to a more comprehensive understanding of hate speech detection in online content, emphasizing the diverse modeling approaches that enhance its accuracy and applicability.

References

1. Zephoria.com. https://zephoria.com/top-15-valuable-facebook-statistics/. Accessed 30 Oct 2022
2. Twitter usage statistics - internet live stats. https://www.internetlivestats.com/twitter-statistics/. Accessed 30 Oct 2022
3. Hinduja, S., Patchin, J.W.: Bullying, cyberbullying, and suicide. Arch. Suicide Res. 14(3), 206–221 (2010)
4. Hate speech and offensive language. https://data.world/thomasrdavidson/hate-speech-and-offensive-language. Accessed 30 Oct 2022
5. Hate speech – Wikipedia. https://en.wikipedia.org/wiki/Hate_speech. Accessed 30 Oct 2022
6. Davidson, T., Warmsley, D., Macy, M., Weber, I.: Automated hate speech detection and the problem of offensive language. In: Proceedings of the International AAAI Conference on Web and Social Media, vol. 11, no. 1, pp. 512–515 (2017)
7. Almeida, F., Xexéo, G.: Word embeddings: a survey. arXiv preprint arXiv:1901.09069 (2019)
8. Faris, H., Aljarah, I., Habib, M., Castillo, P.A.: Hate speech detection using word embedding and deep learning in the Arabic language context. In: ICPRAM, pp. 453–460, February 2020
9. Magu, R., Luo, J.: Determining code words in euphemistic hate speech using word embedding networks. In: Proceedings of the 2nd Workshop on Abusive Language Online (ALW2), pp. 93–100, October 2018
10. Saleh, H., Alhothali, A., Moria, K.: Detection of hate speech using BERT and hate speech word embedding with deep model. Appl. Artif. Intell. 37(1), 2166719 (2023)
11. Mikolov, T., Sutskever, I., Chen, K., Corrado, G.S., Dean, J.: Distributed representations of words and phrases and their compositionality. In: Advances in Neural Information Processing Systems, vol. 26 (2013)
12. Mikolov, T., Chen, K., Corrado, G., Dean, J.: Efficient estimation of word representations in vector space. arXiv preprint arXiv:1301.3781 (2013)
13. Ibrohim, M.O., Setiadi, M.A., Budi, I.: Identification of hate speech and abusive language on Indonesian Twitter using the Word2vec, part of speech and emoji features. In: Proceedings of the 1st International Conference on Advanced Information Science and System, pp. 1–5, November 2019
14. Devlin, J., Chang, M.W., Lee, K., Toutanova, K.: BERT: pre-training of deep bidirectional transformers for language understanding. arXiv preprint arXiv:1810.04805 (2018)
15. Shreyashree, S., Sunagar, P., Rajarajeswari, S., Kanavalli, A.: A literature review on bidirectional encoder representations from transformers. In: Inventive Computation and Information Technologies: Proceedings of ICICIT 2021, pp. 305–320 (2022)
16. Mozafari, M., Farahbakhsh, R., Crespi, N.: A BERT-based transfer learning approach for hate speech detection in online social media. In: Cherifi, H., Gaito, S., Mendes, J.F., Moro, E., Rocha, L.M. (eds.) COMPLEX NETWORKS 2019. SCI, vol. 881, pp. 928–940. Springer, Cham (2020). https://doi.org/10.1007/978-3-030-36687-2_77
17. Mozafari, M., Farahbakhsh, R., Crespi, N.: Hate speech detection and racial bias mitigation in social media based on BERT model. PLoS ONE 15(8), e0237861 (2020)
18. Gambäck, B., Sikdar, U.K.: Using convolutional neural networks to classify hate-speech. In: Proceedings of the First Workshop on Abusive Language Online, pp. 85–90 (2017)
19. Chiu, K.L., Collins, A., Alexander, R.: Detecting hate speech with gpt-3. arXiv preprint arXiv:2103.12407 (2021)
20. Rizos, G., Hemker, K., Schuller, B.: Augment to prevent: short-text data augmentation in deep learning for hate-speech classification. In: Proceedings of the 28th ACM International Conference on Information and Knowledge Management, pp. 991–1000, November 2019

21. Devlin, J., Chang, M.-W., Lee, K., Toutanova, K.: BERT: pre-training of deep bidirectional transformers for language understanding. arXiv preprint arXiv:1810.04805 (2018)
22. Brown, T., et al.: Language models are few-shot learners. Adv. Neural. Inf. Process. Syst. **33**, 1877–1901 (2020)
23. Lan, Z., Chen, M., Goodman, S., Gimpel, K., Sharma, P., Soricut, R.: Albert: a lite BERT for self-supervised learning of language representations. arXiv preprint arXiv:1909.11942 (2019)
24. Harris, C.R., et al.: Array programming with NumPy. Nature **585**(7825), 357–362 (2020). https://doi.org/10.1038/s41586-020-2649-2
25. Pedregosa, F., et al.: Scikit-learn: machine learning in Python. J. Mach. Learn. Res. **12**, 2825–2830 (2011)
26. T. pandas development team: pandas-dev/pandas: Pandas (2020). https://doi.org/10.5281/zenodo.3509134
27. Hunter, J.D.: Matplotlib: a 2D graphics environment. Comput. Sci. Eng. **9**(3), 90–95 (2007)
28. Paszke, A., et al.: Pytorch: an imperative style, high-performance deep learning library. In: Advances in Neural Information Processing Systems, vol. 32, pp. 8024–8035. Curran Associates, Inc. (2019). http://papers.neurips.cc/paper/9015-pytorch-an-imperative-style-high-performance-deep-learning-library.pdf
29. Wolf, T., et al.: Huggingface's transformers: state-of-the-art natural language processing. arXiv preprint arXiv:1910.03771 (2019)
30. Kingma, D.P., Ba, J.: Adam: a method for stochastic optimization. arXiv preprint arXiv:1412.6980 (2014)
31. Ma, E.: Nlp augmentation (2019). https://github.com/makcedward/nlpaug
32. Farooq, U., Rahim, M.S.M., Sabir, N., Hussain, A., Abid, A.: Advances in machine translation for sign language: approaches, limitations, and challenges. Neural Comput. Appl. **33**(21), 14357–14399 (2021)
33. Badjatiya, P., Gupta, S., Gupta, M., Varma, V.: Deep learning for hate speech detection in tweets. In: Proceedings of the 26th International Conference on World Wide Web companion, pp. 759–760, April 2017
34. Graves, A., Jaitly, N., Mohamed, A.R.: Hybrid speech recognition with deep bidirectional LSTM. In: 2013 IEEE Workshop on Automatic Speech Recognition and Understanding, pp. 273–278. IEEE, December 2013
35. Bisht, A., Singh, A., Bhadauria, H.S., Virmani, J., Kriti: Detection of hate speech and offensive language in twitter data using LSTM model. In: Recent Trends in Image and Signal Processing in Computer Vision, pp. 243–264 (2020)
36. Fazil, M., Khan, S., Albahlal, B.M., Alotaibi, R.M., Siddiqui, T., Shah, M.A.: Attentional multi-channel convolution with bidirectional LSTM cell toward hate speech prediction. IEEE Access **11**, 16801–16811 (2023)
37. Hakimov, S., Ewerth, R.: Combining textual features for the detection of hateful and offensive language. arXiv preprint arXiv:2112.04803 (2021)

Deep Learning for Protein-Protein Contact Prediction Using Evolutionary Scale Modeling (ESM) Feature

Lan Xu[✉]

School of Computer Science and Engineering, Nanjing University of Science and Technology,
200 Xiaolingwei Street, Nanjing, Jiangsu, China
121106010817@njust.edu.cn

Abstract. Protein-protein interactions (PPIs) are essential for various biological processes, and their binding sites provide important information for cell function and drug design. Traditional experimental methods for identifying these sites are expensive and time-consuming, prompting the emergence of computational forecasting tools. However, the performance of these tools tends to be limited due to single experimental training data and other limitations. We introduce a new hybrid deep neural network that fuses global sequence features with local influences, improving the traditional sequence sliding window strategy. Our method integrates a distance matrix between residue pairs to guide structural neighborhood screening, thereby increasing confidence in the results. In order to characterize the properties of proteins more comprehensively, we introduce protein language models into regular features, and design different attention mechanism models for feature learning in different dimensions. Experimental results show that our model performance has reached an advanced level and is ahead of other competing methods in several indicators.

Keywords: Protein-protein interactions (PPIs) · Hybrid deep neural network · Attention mechanism · Distance matrix · Structural neighborhood

1 Introduction

Protein-protein interactions (PPIs) are fundamental to the functioning of biological systems, playing a crucial role in essential processes such as cell signaling, metabolic pathways, and gene regulation [1]. Predictive studies of PPIs are of immense value in functional genomics, systems biology, and drug design, offering information into the complex networks within cells [2, 3]. While experimental methods have made significant advances in detecting PPIs, computational prediction methods remain vital due to their cost-effectiveness and time efficiency [4]. Proteins seldom work in isolation, they interact with other proteins to form complex networks that drive various biological functions [5]. Understanding the specific residues involved in protein-protein binding is key to unraveling cellular mechanisms, predicting disease-related proteins, and facilitating drug development. Over the past few decades, several biocomputational methods have emerged for predicting protein-protein binding residues. These methods can be broadly categorized into two types based on the type of features utilized: those relying on protein structure information and those leveraging protein sequence information. in this study, we aim to develop a novel computational approach for predicting PPI sites by integrating both protein sequence and structure information. By combining the strengths of these two types of features, we seek to enhance the accuracy and reliability of PPI site prediction. This research holds the potential to contribute valuable insights into cellular processes and promote the development of new therapeutics. Additionally, our proposed approach addresses the need for cost-effective and time-efficient methods in the field of protein-protein interaction prediction.

In the early stages, methods for predicting ligand-bound residues based on protein structure information dominate. This type of method can be divided into prediction method based on structure template matching, prediction method based on spatial geometry, and energy-based prediction method according to different calculation methods. Researchers in the field of bioinformatics generally believe that proteins with similar structures tend to have similar biological functions. This is also the source of ideas for prediction methods based on structure template matching, and excellent prediction methods based on structure template matching include 3DLigandSite [6], FINDSITE [7], and FunFOLD [8]. To identify ligand-binding residues in the protein to be tested, the prediction method based on structure template matching first uses protein structure alignment methods (MAMMOTH [9], TM-align [10], etc.) to evaluate the structural similarity between all labeled ligand binding site proteins and the protein to be tested. Then, based on the degree of structural similarity, the proteins at known ligand binding sites were sorted and screened, and several proteins were selected as templates, and the structural alignment information of these templates and the proteins to be tested was extracted. Finally, based on this alignment information, potential ligand binding residues in the protein to be tested are predicted according to specific ligand binding site identification rules.

The prediction method based on spatial geometry aims to identify the ligand binding region of the protein being tested by calculating specific geometric measures using its structural information. This method determines the binding site between the protein and the ligand. Studies on protein-ligand complexes in the Protein Data Bank (PDB) have revealed that small ligands often bind to concave regions on the protein surface, particularly the largest and deepest cavities. Energy-based prediction methods, on the other hand, focus on identifying empty regions on the protein surface that are favorable for ligand binding based on energy distribution. These methods typically involve designing a probe molecule and calculating the interaction energy between this probe and the surrounding protein atoms [11]. The goal is to locate ligand-binding residues in the protein being tested through this energy analysis.

However, the number of proteins with experimentally determined three-dimensional structures is much smaller compared to the number of proteins with known sequences. As a result, there is a significant portion of proteins where only sequence information is available, and their ligand binding sites cannot be predicted using structure-based methods. Thus, the prediction of protein-ligand binding sites directly from protein sequence information has gained considerable attention from researchers. In recent years, several prediction methods based on sequence information have emerged, including methods based on sequence template matching and machine learning algorithms utilizing sequence features.

Similar to the prediction method based on structure template matching described above, the prediction method based on sequence template matching mainly searches for one or more excellent sequence templates from the corresponding database based on the homology or similarity between protein sequences and sequences, and then predicts potential ligand binding sites in proteins based on the information of ligand binding sites corresponding to the searched sequence template. For example, a BLAST based prediction method is described in Literature [12], which uses the BLAST sequence alignment tool to select a protein with the highest sequence similarity and the lowest E-value from the protein data set of labeled binding site information as a template, and obtain the sequence alignment information between the protein template and the protein to be tested, and finally determine that all residues in the protein to be tested are binding sites on the alignment of the binding residues in the template protein.

The prediction method based on sequence template matching has a strong dependence on the homology of the sequence. When template proteins with high homology are not found, the confidence in the predicted ligand-bound residues is unreliable. In order to reduce the dependence of prediction methods on sequence homology and fully mine the hidden information in the data, relevant scholars try to construct a prediction model by introducing machine learning algorithms to identify protein-ligand binding residues. The main steps of the prediction method of machine learning based on sequence features are as follows: first, the effective information of the protein sequence (such as amino acid composition information, protein evolution information and secondary structure information based on sequence prediction) of the labeled ligand binding site is extracted, and then the feature vector of amino acid residues is constructed according to these information, the corresponding amino acid residue samples are generated, and then the training sample collection is constructed. Finally, an appropriate machine learning algorithm is used to train a predictive model on the training sample set, which is used to predict whether any amino acid residue of the protein to be tested is a ligand binding site.

Various machine learning techniques have found application in the field of protein-protein interaction prediction. Notable strategies include the use of random forests, as seen in IntPred [13], and the utilization of neural networks in SCRIBER [14] and ProNA2020 [15]. Recent advancements in the domain of bioinformatics have witnessed the incorporation of deep contextual learning. For instance, convolutional neural networks (CNNs) have been employed to investigate the contextual background of continuous residues [16]. DLPred [17] introduced a simplified long short-term memory (LSTM) network for multifaceted predictions, encompassing PPI sites and RSA. DeepPPISP [18] pioneered a CNN-based architecture, incorporating the entire protein chain to capture global information. MaSIF-site [19] leverages geometric deep learning techniques to extract surface fingerprints, relying solely on the structural aspects of proteins, which play a pivotal role in protein-protein interactions.

One notable advancement in this area is the Evolutionary Sequence Model (ESM) [20–22], a specialized language model tailored for biological sequences, particularly protein sequences. Trained on a vast corpus of protein sequence data, ESM leverages the Transformer architecture to capture complex biological features embedded within the sequences. Unlike general-purpose language models, ESM is optimized to understand the subtleties of protein structures and functions, making it highly effective for tasks such as structural prediction, functional annotation, and, importantly for this work, PPI prediction. Its ability to capture the intricate biological relationships within proteins allows for a more accurate and nuanced understanding of how proteins interact, providing a powerful tool for the computational exploration of biological systems.

In this research, we employed a deep learning network that incorporates semantic protein representations to provide predictions for protein-protein interaction sites, considering both sequence and structural aspects. Specifically, we approach the prediction of protein-protein interaction (PPI) sites as a binary classification task at the residue level. To achieve this, we combine evolutionary and structural data to create features for each residue node. This involves incorporating the semantic representation of ESM for proteins and calculating amino acid distances between pairs of residues to build distance matrices. Then, the multi-channel attention mechanism is used to realize the deep convolution framework to learn the advanced feature information of the labeled binding sites, and the sliding window strategy and distance matrix convolution are used to strengthen the influence of the context neighborhood and the structural neighborhood of the binding site sequence respectively. Through various evaluations, it is found that our hybrid neural network based on semantic representation is superior to other sequence- and structure-based methods, and can be derived to the rest of the relevant protein prediction problems.

2 Materials and Methods

The task of predicting binding sites for protein-protein interactions (PPI) is treated as a binary classification challenge, with our objective being the training of a deep learning convolutional neural network. This network learns from labeled residue binding sites to capture their distinctive features, predicts the probability of unlabeled nodes becoming binding sites, and divides nodes with a probability above the threshold into binding sites (positive), and nodes below the threshold into non-binding sites (negative).

2.1 Datasets

We utilized three commonly used public datasets: Data_186, Data_72 [23], and Data_164 [24], named based on the quantity of proteins. Data_186 is derived from known protein-protein complexes in the PDB, subjected to screening. Data_72 is based on the protein-protein docking benchmark set version 3.0 [25] and Data_164 is from newly annotated protein-protein complexes in the PDB. In this study, an amino acid is considered an interaction site if its absolute solvent accessibility is less than 1 $Å^2$ before and after protein binding; otherwise, it is a non-interaction site. To ensure a similar distribution of binding site proportions in the training and test sets, we merged these three datasets into an integrated dataset. Finally, redundant proteins with over 25% sequence similarity were removed using BLASTClust, resulting in 395 protein chains. In order to achieve cross-validation, prevent overfitting, and evaluate the model in time, we divide the training set, validation set and test set according to the ratio of roughly 8:1:1. Among them, the training set contains 316 protein chains, and the validation set and test set contain 40 and 39 protein chains, respectively. Statistics for these datasets can be found in Table 1, providing in-depth information.

Table 1. Statistical data of the training, validation, and test sets used in this study

Dataset	Interacting residues	Non-interacting residues	% Of interacting residues
Train	9923	53650	15.61
Valid	1328	6755	16.43
Test	1198	6656	15.25

2.2 Input Features

Feature extraction and selection is a key step in determining the effectiveness of convolutional models, and combined with previous excellent model examples, we decided to train our model using three sets of protein features: structural properties (DSSP and Distance_Map), evolutionary information (PSSM and HMM) and semantic expression (Esm_Representation). Considering that the interaction of local points cannot be ignored, we will also provide local features with a neighborhood length of 20 while entering global features.

2.2.1 Position-Specific Scoring Matrix and Hidden Markov Model

The Location-Specific Score Matrix (PSSM) has proven to be one of the widely adopted and highly effective protein sequence signatures, characterizing the frequency of twenty amino acids at each sequence site, demonstrating the evolutionary conservatism of the current site. The hidden Markov model can usually infer the implied state sequence based on the learned observation sequence, which can better express the protein sequence characteristics. Specifically, in this paper, we chose to use PSSM plus HMM to represent the evolutionary information of protein sequences. Among them, PSSM is generated by running (PSI-BLAST v2.10.1) to search for a sequence of queries against the UniRef90 database [26], which has three iterations and an E value of 0.001. HMM configuration files are generated by running HHblits v3.0.3 [27] to align the query sequence with the UniClust30 database [28] with the default parameters. In both PSSM and HMM, every amino acid is transformed into a 20-dimensional vector and then normalized to a fractional value between 0 and 1. This normalization is achieved using the formula below, with 'v' representing the initial feature value and 'Min' and 'Max' denoting the minimum and maximum values of this specific feature category as observed in the training set.

$$v_{norm} = \frac{v - Min}{Max - Min} \tag{1}$$

2.2.2 Secondary Structure and Accessible Surface Area

In order to get as close as possible to the three-dimensional structure of the protein to obtain a direct contact relationship from the perspective of data availability, we selected the secondary structure of the protein as the main structural characteristic, supplemented by solvent accessibility surface area and peptide backbone torsion angle (a measure of dihedral angle between adjacent residues). Specifically, it includes: 1) solitary thermal coding of the 9-dimensional representation of secondary structure characteristics, the first eight dimensions correspond to the eight distinct secondary structural states [3_{10}-helix (G), α-helix (H), π-helix (I), β-strand (E), β-bridge (B), β-turn (T), bend (S) and loop or irregular (L)] The last dimension represents the unknown; 2) The solvent accessibility surface area (ASA) is subsequently standardized into relative solvent accessibility (RSA) by scaling it with the maximum achievable ASA for the corresponding amino acid type; 3) The torsion angles PHI and PSI of the peptide backbone are converted to 4-dimensional eigenvectors using sine and cosine transforms. We have collectively named this 14-dimensional structural feature DSSP.

2.2.3 Evolutionary Scale Modeling Representation

Biological sequences, especially protein sequences, are characterized by patterns and changes that they accumulate over the course of evolution. To capture these complex relationships and patterns, a model that can handle long sequence dependencies is needed. The Transformer architecture has proven its effectiveness in natural language processing, so it is used for biological sequence data. ESM is based on the Transformer architecture, which consists of multiple self-attention layers that allow the model to establish relationships between different sequence locations. In this way, the model can capture complex interactions within proteins that are essential for their function and structure. And ESM was trained on a large amount of protein sequence data. In this way, it learns and understands the characteristics of evolutionary relationships, domains, and functional domains between protein families, which are essential for PPI prediction. The interaction between two proteins often depends on their structural and functional compatibility, and traditional PPI prediction methods often rely on hand-formulated features or simple sequence representations. In contrast, ESM provides a deep, rich representation of proteins capable of capturing more subtle information, which helps improve the accuracy of predictions. Combined with protein representation of ESM and other deep learning methods, end-to-end PPI prediction can be achieved without the need for manual feature formulation. Therefore, we use the semantic expression generated by Evolutionary Scale Modeling as a secondary feature to predict the contact between proteins, specifically, we extract the vector generated by the last hidden layer as the input feature, and because the dimension is too large, PCA is projected onto the principal component with a large feature value to achieve dimensionality reduction. Further, in order to obtain better results when processing local features, we will also stitch them into the above two types of features to achieve feature fusion.

2.2.4 The Distance_Map and Local Features

The affinity of a residue for a specific ligand is determined not just by its individual characteristics, but also by its surrounding environment. Consequently, several well-established models in the past have employed a sliding window approach to extract contextual information around the residue, enabling the capture of neighboring features at the current position. This approach is based on the assumption that adjacent residue sites in the sequence are structurally adjacent to each other, but this is not the case, and when we analyze the sequence information of the proteins with a measured 3D structure, we find that the structurally adjacent amino acids are not similar in the sequence, and some are even far apart. To this end, we propose an improved strategy, when looking for the neighborhood of the current site, we introduce distance_map matrix containing amino acid distance information, which is obtained in two steps: 1) according to the PDB file of the protein, we obtain the α carbon atomic coordinates of each amino acid residue, and then calculate the Euclidean distance between all residue pairs to form a distance matrix map; 2) To simplify this feature data, we convert values with distances less than 14 Å to 1 (except for those that are 0 in themselves), and convert values greater than or equal to 0. From this matrix, the sequence information of adjacent nodes of any node can be screened out, and then the collection of local features can be guided. The 2D distance matrix is a simplified representation of the tertiary structure of 3D proteins, and numerous investigations have demonstrated the effective reconstruction of protein tertiary structures through the analysis of 2D distance matrices [29, 30], the distance matrix encapsulates a wealth of structure-related information, with the ability to identify secondary structure (SS) elements through the examination of local contacts in contact diagrams. Moreover, the 2D distance matrix facilitates the detection of geometric features within protein structures, including domain boundaries and the relative positions between two SSs. This serves as a foundational underpinning for our proposed approach.

2.3 Architectural Design of the Network and Embedding of Features

In the hybrid neural network designed in this paper, we use two different network architectures to learn full-sequence features and neighborhood features respectively, providing diversified information at different scales and different levels, which can be optimized for their characteristics, so as to improve the ability of feature expression (Fig. 1).

The hybrid deep neural network architecture that we designed takes two types of inputs: structural neighborhoods and the entire protein sequence. For structural neighborhoods, we select 20 neighboring residue indices guided by the residue distance matrix. We then multiply their embedded features by certain weights (updated through backpropagation) and integrate them into the embedded features of the current residue position. Subsequently, these characteristics are fed into an LSTM + Attention network to derive a localized feature vector. When dealing with the entire protein sequence, we employ raw sequence feature embedding to merge three distinct feature categories. These amalgamated features are then channeled into an EPSA network to acquire a comprehensive feature vector for the global context. Finally, we combine these two feature vectors and input them into fully connected layers for prediction.

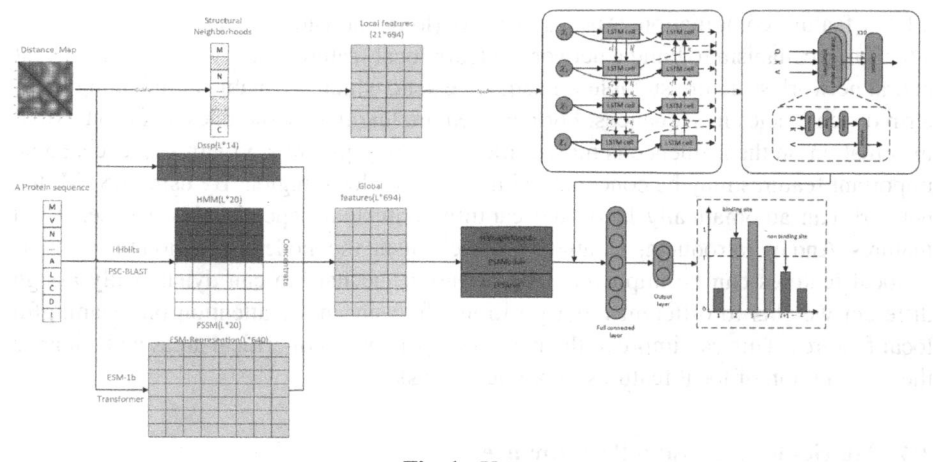

Fig. 1. Xyx

2.3.1 Global Feature Handling

In this part, our input is the global feature of a single protein chain with a characteristic dimension of L × 694D (L is the sequence length of the current protein chain, PSSM-20D, HMM-20D, DSSP-14D, Esm Representation-640D). A network model containing a highly efficient pyramid attention module (PSA [31]) was selected for training. PPI binding site prediction needs to consider the global characteristics of the entire protein sequence, because the interaction usually involves long-distance sequence dependencies, and the PSA module can interact with the sequence information at different locations through the self-attention mechanism to better capture the global context information of the protein sequence. And the module can learn the weights according to the similarity between different locations, weighted and combined more related features, enhance the importance of related features, and reduce the interference of unrelated features. Its implementation mainly goes through four steps: 1) the SPC module is used to segment the channels, and then multi-scale feature extraction (including multi-scale convolution and group convolution) is carried out for the spatial information on the feature map of each channel; 2) Use the SEWeight Module to extract the channel attention of different scale feature maps to obtain the channel attention vector at each different scale; 3) Use Softmax to re-calibrate the attention vector of multi-scale channels to obtain the attention weight after the interaction of the new multi-scale channel; 4) Multiply the recalibrated weights and corresponding feature maps by element, and output a feature map after attention weighting of multi-scale feature information. The multi-scale information representation ability of this feature map is richer, and the ability to characterize proteins is stronger.

2.3.2 Local Feature Handing

Under the guidance of the processed Distance_Map, we can obtain the adjacent site sequence number of any residue site on the protein chain, and considering the interaction between the residues, when extracting local features from the current site, we multiply the residue features in its neighborhood by a certain weight and also splice together to form a local feature combination. And use LSTM (long short-term memory) structure and attention mechanism to build a network to learn local features. LSTM [32] is a recurrent neural network structure suitable for sequence data, which is capable of modeling long-term dependencies in sequences. For local feature learning, sometimes local contextual information on the sequence is more critical, for example, in a protein sequence, some important features may be concentrated in a specific local region. By using LSTM, the network can automatically learn and capture contextual dependencies between local features. And by introducing an attention mechanism, the model's attention to key areas in local features can be improved. The attention mechanism can dynamically assign different weights to different input positions, focusing more attention on meaningful local features. This can improve the network's perception of critical areas and enhance the contribution of local features to prediction tasks.

2.4   Metrics for Assessing Performance

In the realm of PPI site prediction, we consider interaction sites as affirmative instances and non-interaction sites as adverse instances. In order to assess the efficacy of our model and other approaches for PPI site prediction, a set of five evaluation metrics were used [33]: accuracy (ACC), precision, recall, F_1 score, and the Matthews correlation coefficient (MCC).

$$Accuracy = \frac{TP+TN}{TP+TN+FP+FN} \tag{2}$$

$$Precision = \frac{TP}{TP+FP} \tag{3}$$

$$Recall = \frac{TP}{TP+FN} \tag{4}$$

$$F_1 = 2 \times \frac{Precision \times Recall}{Precision + Recall} \tag{5}$$

$$MCC = \frac{TP \times TN - FN \times FP}{\sqrt{(TP+FP) \times (TP+FN) \times (TN+FP) \times (TN+FN)}} \tag{6}$$

where TP is the number of deleterious mutations that are correctly predicted as positive; TN is the number of neutral mutations that are correctly predicted negative; FP is the number of neutral mutations that are incorrectly predicted as positive; FN is the number of deleterious mutations that are incorrectly predicted as negative, respectively.

2.5   Implementation Details

This experiment is based on the PyTorch framework, and in the training model we use the universal cross-entropy loss function, which is defined as follows:

$$Loss = -\frac{1}{n} \Sigma \left[y \log(y_{pred}) + (1-y) \log(1-y_{pred}) \right] \tag{7}$$

The optimizer also uses the common adaptive momentum (Adam), whose weight update expression is as follows:

$$\theta_t = \theta_{t-1} - \frac{\alpha}{\sqrt{\hat{v}_t} + \varepsilon} \hat{m}_t \tag{8}$$

The learning rate is 0.001 and the dropout rate is 0.3, the threshold is the value that makes the MCC metric maximum during training. When building a global feature training network, after many experiments, we constructed a structure containing four layers of PSA-Block. The first fully connected layer has an output of 1024 nodes, the second fully connected layer outputs 256 nodes, and finally outputs the probability value that the current node is a positive node. The whole process is to train the model on the training set, adjust the parameters of the model on the divided validation set, and adjust the model to the optimal state before overfitting. Finally, after sufficient training, it is evaluated on the test set.

3 Results and Discussions

3.1 Feature Ablation Experiments and Network Comparison

From the perspective of rationality of model construction strategy and feature selection, we will carry out some ablation experiments on the model, specifically, in order to explore the role of newly introduced feature Esm_Representation and the rationality of the Local feature fusion strategy, we will make four sets of controlled experiments: one is the result of running according to the above complete process, the second is to drop out Esm_Representation features, the third is to drop out local feature fusion, and the fourth is to replace the structural neighborhood under the guidance of the Distance_Map with a sequence context. And in order to highlight the correctness of our different neural network strategies for global and local feature training, we replace the network structure of global feature training with the same as local in the fifth set of experiments (Table 2).

Table 2. Four sets of experimental results

Group order	ACC	Precision	Recall	F_1	MCC
1	0.795	0.354	0.568	0.436	0.323
2	0.703	0.260	0.552	0.353	0.213
3	0.721	0.258	0.547	0.351	0.209
4	0.755	0.323	0.557	0.409	0.282
5	0.760	0.338	0.558	0.421	0.296

We mainly focus on ACC and MCC two indicators, the former represents the correct prediction rate of positive samples, the latter represents the overall prediction rate, from the experimental results it is not difficult to see that the ACC of control group one is significantly higher than that of control group two, three, slightly higher than control group four, which is in line with our original intention of experimental design, and the ACC of control group two is higher than that of control group three, the semantic expression characteristics of proteins account for a very heavy proportion of the prediction of positive samples, and it can be seen that the fusion of local neighborhood features can not be ignored. The results of control groups 1 and 4 tell us that neighborhood screening guided by distance matrix is more convincing than sequence context, but no matter what kind of local feature screening scheme has a positive effect on the effect of ACC. From the result analysis of MCC, the fusion of local features is the decisive factor, whether to consider the interaction of local residues will have more than 1/3 of the impact on the results, of course, semantic expression features also play an important role in it, and once again verify that the local influence considered from the structural perspective is more reasonable. In terms of network structure comparison, we can easily find that the first set of experimental results are relatively better than the fifth set of experimental results, which also means that EPSANet is higher than LSTM + Attention network in terms of learning ability of global sequence features.

3.2 Comparison with Competing Methods

In order to evaluate our methods more objectively and fairly, we selected six experimental methods for the same topic based on the same data set for comparison. Of these six methods, four are sequence-based prediction methods (ProNA2020, SCRIBER, DELPHI, and DeepPPISP) and two are structure-based prediction methods (SPPIDER and MaSIF-site). As shown in Table 3.

Table 3. Comparison of results

Method	ACC	Precision	Recall	F_1	MCC
ProNA2020	0.729	0.269	0.403	0.323	0.170
SCRIBER	0.672	0.258	0.572	0.356	0.209
DELPHI	0.701	0.262	0.553	0.356	0.211
DeepPPISP	0.675	0.254	0.546	0.347	0.175
SPRIDE	0.744	0.328	0.563	0.415	0.284
MaSIF-site	0.791	0.375	0.566	0.451	0.329
Our	0.795	0.354	0.568	0.436	0.323

Our model exhibits significant performance disparities compared to the four sequence-only prediction methods, which shows that the prediction accuracy of the semantic expression model of protein and the structural neighborhood screening scheme assisted by amino acid distance map is huge. Compared with structure-based prediction methods, our model is superior to SPRIDE and is on pair with MaSIF-site. The reason for this may be that the PPI site still mainly depends on the structural properties of the protein chain, and the basic feature combination selected in this model is basically based on sequence information, and more structural feature information should be incorporated to further improve performance.

4 Conclusion

In this study, we have made significant progress in predicting PPI residue junctions based on sequence information. We introduced new features and improved the strategy of neighborhood screening, which led to results that surpass existing model methods and even outperform some prediction methods based on structural information. However, while our model has achieved remarkable results, there is still room for further improvement.

Firstly, as the topic itself involves a binary classification problem with imbalanced data, we did not employ techniques like random undersampling to balance the model, which may have resulted in certain limitations in our model. Addressing this imbalance issue could potentially enhance the overall performance.

Secondly, when utilizing the distance matrix to identify structural neighborhoods, we applied a general preprocessing step (replacing non-zero values with 1). However, it is important to acknowledge that residues within the structural neighborhood may have varying degrees of influence on the current residues based on their respective distances. In future research, we can explore converting these distances into continuous values and weighting them differently when combining features.

Furthermore, we believe that the introduced protein semantic expression has broader applications beyond the scope of this study. It would be valuable to explore its potential in addressing various related topics and domains.

In conclusion, our research showcases promising advancements in PPI residue junction prediction using sequence information. By further addressing the data imbalance issue, refining the distance matrix preprocessing, and exploring new applications for protein semantic expression, we can continue to enhance the accuracy and overall capabilities of PPI prediction models. These improvements and future research directions will contribute to the broader understanding and exploration of protein interactions for various research applications.

References

1. Han, J.D., et al.: Evidence for dynamically organized modularity in the yeast protein-protein interaction network. Nature 430(6995), 88–93 (2004)
2. Li, M., Gao, H., Wang, J., Wu, F.X.: Control principles for complex biological networks. Brief Bioinform. 20(6), 2253–2266 (2019)
3. Wells, J.A., McClendon, C.L.: Reaching for high-hanging fruit in drug discovery at protein-protein interfaces. Nature 450(7172), 1001–1009 (2007)
4. Brettner, L.M., Masel, J.: Protein stickiness, rather than number of functional protein-protein interactions, predicts expression noise and plasticity in yeast. BMC Syst. Biol. 6, 128 (2012)
5. De Las Rivas, J., Fontanillo, C.: Protein–protein interaction networks: unraveling the wiring of molecular machines within the cell. Brief. Funct. Genom. 11(6), 489–496 (2012)
6. Wass, M.N., Kelley, L.A., Sternberg, M.J.: 3DLigandSite: predicting ligand-binding sites using similar structures. Nucl. Acids Res. 38(Web Server issue), W469–W473 (2010)
7. Brylinski, M., Skolnick, J.: A threading-based method (FINDSITE) for ligand-binding site prediction and functional annotation. Proc. Natl. Acad. Sci. USA 105(1), 129–34 (2008)

8. Roche, D.B., Tetchner, S.J., McGuffin, L.J.: FunFOLD: an improved automated method for the prediction of ligand binding residues using 3D models of proteins. BMC Bioinform. **12**, 160 (2011)
9. Ortiz, A.R., Strauss, C.E., Olmea, O.: MAMMOTH (matching molecular models obtained from theory): an automated method for model comparison. Protein Sci. **11**(11), 2606–2621 (2002)
10. Zhang, Y., Skolnick, J.: TM-align: a protein structure alignment algorithm based on the TM-score. Nucl. Acids Res. **33**(7), 2302–2309 (2005)
11. Xie, Z.R., Hwang, M.J.: Methods for predicting protein-ligand binding sites. Methods Mol. Biol. **1215**, 383–398 (2015)
12. Chen, K., Mizianty, M.J., Kurgan, L.: Prediction and analysis of nucleotide-binding residues using sequence and sequence-derived structural descriptors. Bioinformatics **28**(3), 331–341 (2012)
13. Northey, T.C., Baresic, A., Martin, A.C.R.: IntPred: a structure-based predictor of protein-protein interaction sites. Bioinformatics **34**(2), 223–229 (2018)
14. Zhang, J., Kurgan, L.: SCRIBER: accurate and partner type-specific prediction of protein-binding residues from proteins sequences. Bioinformatics **35**(14), i343-i353 (2019)
15. Qiu, J., et al.: ProNA2020 predicts protein-DNA, protein-RNA, and protein-protein binding proteins and residues from sequence. J. Mol. Biol. **432**(7), 2428–2443 (2020)
16. Xie, Z., Deng, X., Shu, K.: Prediction of protein-protein interaction sites using convolutional neural network and improved data sets. Int. J. Mol. Sci. **21**(2) (2020)
17. Buzhong Zhang, J.L., Quan, L., Chen, Y., Lü, Q.: Sequence-based prediction of protein-protein interaction sites by simplified long short-term memory network. Neurocomputing **357**, 86–100 (2019)
18. Zeng, M., Zhang, F., Wu, F.-X., Li, Y., Wang, J., Li, M.: Protein–protein interaction site prediction through combining local and global features with deep neural networks. Bioinformatics **36**(4), 1114–1120 (2019)
19. Gainza, P., et al.: Deciphering interaction fingerprints from protein molecular surfaces using geometric deep learning. Nat. Methods **17**(2), 184–192 (2020)
20. Rives, A., et al.: Biological structure and function emerge from scaling unsupervised learning to 250 million protein sequences. Proc. Natl. Acad. Sci. USA **118**(15) (2021)
21. Hong, Y., Lee, J., Ko, J.: A-Prot: protein structure modeling using MSA transformer. BMC Bioinform. **23**(1), 93 (2022)
22. Meier, J., Rao, R., Verkuil, R., Liu, J., Sercu, T., Rives, A.: Language models enable zero-shot prediction of the effects of mutations on protein function. bioRxiv (2021)
23. Murakami, Y., Mizuguchi, K.: Applying the Naive Bayes classifier with kernel density estimation to the prediction of protein-protein interaction sites. Bioinformatics **26**(15), 1841–1848 (2010)
24. Singh, G., Dhole, K., Pai, P.P., Mondal, S.: SPRINGS: prediction of protein-protein interaction sites using artificial neural networks. ResearchGate, April 2014
25. Howook Hwang, B.P., Mintseris, J., Janin, J., Weng, Z.: Protein–protein docking benchmark version 3.0, vol. 73, no. 3, pp. 705–709. WILEY Online Library (2008)
26. Suzek, B.E., Huang, H., McGarvey, P., Mazumder, R., Wu, C.H.: UniRef: comprehensive and non-redundant UniProt reference clusters. Bioinformatics **23**(10), 1282–1288 (2007)
27. Remmert, M., Biegert, A., Hauser, A., Soding, J.: HHblits: lightning-fast iterative protein sequence searching by HMM-HMM alignment. Nat. Methods **9**(2), 173–175 (2011)
28. Mirdita, M., von den Driesch, L., Galiez, C., Martin, M.J., Soding, J., Steinegger, M.: Uniclust databases of clustered and deeply annotated protein sequences and alignments. Nucl. Acids Res. **45**(D1), D170–D176 (2017)

29. Vassura, M., Margara, L., Di Lena, P., Medri, F., Fariselli, P., Casadio, R.: Reconstruction of 3D structures from protein contact maps. IEEE/ACM Trans. Comput. Biol. Bioinform. **5**(3), 357–367 (2008)
30. Yang, J., Jin, Q.Y., Zhang, B., Shen, H.B.: R2C: improving ab initio residue contact map prediction using dynamic fusion strategy and Gaussian noise filter. Bioinformatics **32**(16), 2435–2443 (2016)
31. Hu Zhang, K.Z., Lu, J., Zou, Y., Meng, D.: EPSANet: an efficient pyramid squeeze attention block on convolutional neural network. arXiv:2105.14447 [cs.CV] (2021)
32. Bommidi, B.S., Kosana, V., Teeparthi, K., Madasthu, S.: Hybrid attention-based temporal convolutional bidirectional LSTM approach for wind speed interval prediction. Environ. Sci. Pollut. Res. Int. **30**(14), 40018–40030 (2023)
33. Yuan, Q., Chen, J., Zhao, H., Zhou, Y., Yang, Y.: Structure-aware protein-protein interaction site prediction using deep graph convolutional network. Bioinformatics **38**(1), 125–132 (2021)

FedTag: Towards Automated Attack Investigation Using Federated Learning

Mu Chen[1,2,3](✉), Zaojian Dai[2,3], Yong Li[2,3], and Ziang Lu[2,3]

[1] Department of Informatics and Communication Engineering, Xiamen University, Xiamen 361005, China
[2] State Grid Smart Grid Research Institute Co., Ltd., Nanjing 210003, China
[3] State Grid Laboratory of Power Cyber-Security Protection and Monitoring Technology, Nanjing 210003, China
luziang@geiri.sgcc.com.cn

Abstract. With the popularity of microservice systems in the industry, the security problems faced by microservice systems have received more and more attention, and the security of their systems is usually reflected in traces, logs, and some monitoring information, which record inter-service interactions and intra-service behaviors, respectively. Existing attack detection methods require a lot of manual intervention to label the data and have a large overhead in constructing the graph. In this paper, we propose a new federated learning-based attack investigation system, FedTag, which consists of three components: data preprocessing, model training, and attack investigation. The data preprocessing package fuses three kinds of log data from different sources, and the model training phase trains the BERT model in the form of joint learning to generate a pre-trained model, which is then fine-tuned to obtain a downstream classifier. In the attack investigation phase, FedTag first flags and embeds detected anomalous log queries then extracts events related to the attack investigation point and labels them as malicious events. Eventually, the attack story can be reconstructed by generating a causal graph or reporting all suspicious events to achieve traceability of attack links. Final experimental results show that FedTag displays TPR and FPR metrics comparable to AirTag when compared to the centralized model. Specifically, for S1–S6 single-host data, federated learning shows less than 1% drop and S2 achieves 100% TNR, outperforming AirTag for M1–M6 multi-host data, federated learning shows only about 1% drop. However, compared to centralized anomaly detection, FedTag not only detects log anomalies but also protects the privacy of each client's log data.

Keywords: Microservices · attack detection · data fusion · federated learning

1 Introduction

Cybersecurity is one of the more serious problems in the world today. Hackers launch frequent attacks on industrial systems, and with the popularity of

H. Jin et al. (Eds.): IAIC 2023, CCIS 2058, pp. 112–126, 2024.
https://doi.org/10.1007/978-981-97-1277-9_9

microservice systems in the industry, the security issues faced by microservice systems are receiving more and more attention [1]. Microservice systems running dozens to thousands of services on different machines and operating in highly uncertain and dynamic environments face a wide variety of security threats. OWASP (open web application security project) suggests threat scenarios that microservices may face such as theft of private data, unauthorized access, and injection attacks [13]. Therefore, engineers overseeing these industrial systems need newer and more accurate methods to detect and trace cyber attacks injected into their microservices.

In microservice systems, the interactions between individual services are monitored by using distributed tracing, and in this way, attacks are identified. Distributed trace plays a vital role in analyzing and monitoring the execution and there are several open-source frameworks for tracing the sequence of service instance calls under microservice architecture such as Jaeger, skywalking, etc. The generated trace describes the execution process (i.e., invocation chain) of a service instance's request [2]. Also, log messages are widely used by developers to record the behavior of each service for debugging errors and root cause analysis. Metrics represent multi-dimensional metrics data provided by the application code as it runs, common runtime dimensional metrics are CPU utilization, service response time, etc., which has a long history in IT monitoring and is widely used by engineers, along with logs and trace are used together with logs and traces to detect if a system is not performing as expected.

Trace is a view of the entire chain of requests in a distributed system, span represents the view within different services in the entire chain, and span combined is the view of the entire trace. In the whole request invocation chain, the request will always carry the traceID to the downstream service, each service will also generate its internal spanID for generating its internal invocation view, and together with the traceID passed to the downstream service. traceID in the whole request invocation chain remains unchanged, so the log can be queried by traceID for all logs recorded by the system during the entire request. An industrial distributed tracing system can link log messages of the same trace by injecting traceID and spanID into the log messages generated by different service instances [3].

With the wide application of deep learning in various domains, existing research has also applied deep learning in system attack detection. The aim is to build an end-to-end attack story from off-the-shelf audit logs through a framework. The initial traceability graph construction method is backtracker [10], which is the classic host-side traceability graph construction method, based on the system logs to define process dependencies on the host side and dependencies between processes and files, etc., respectively, thus constructing the graph, but the method mainly mines the causal dependencies of system behaviors from the system level and does not take into account the application layer semantics, based on which, we propose an end-to-end traceability framework OmegaLog, Omegalog [4] merges application event logs with system logs to generate a universal origin graph (UPG). However, all of these previous traceability methods are performed on a single host, while a complete attack chain usually spans multiple hosts, so some more recent work [17] has constructed attack paths spanning multiple hosts as causal graphs and applied different graph optimization techniques, but even so, these graphs are still very large and difficult to interpret in practice. ATLAS [8] is currently the most advanced deep learning-based attack research framework, which constructs causal relationships for given logs and attack events and iterates over the nodes in the graph using the given attack events to categorize them as malicious or benign. However, the method still has some limitations, such as it still requires a lot of human intervention to label the data, and there is a large overhead when constructing the graph. There exists another method, AirTag [9], which uses log messages in the initial analysis and converts them to causal graphs only when needed, avoiding manual labeling.

This paper presents FedTag, a new federated learning-based attack investigation system that consists of three components: data preprocessing, model training, and attack investigation. The data preprocessing package fuses three kinds of log data from different sources, such as Firefox, DNS, and different security events. This is followed by model training, where FedTag labels the log files on each client separately, trains the BERT model in the form of federated learning to generate a pre-trained model, and then fine-tunes it to obtain downstream classifiers. Finally, in the attack investigation phase, FedTag first labels and embeds detected anomalous log queries, and then extracts events related to the attack investigation point and labels them as malicious. FedTag can reconstruct the attack story by generating a causal graph or reporting all suspicious events to achieve traceability of the attack links.

The data preprocessing package fuses three types of log data from different sources, and the model training phase trains the BERT model in the form of federated learning to generate a pre-trained model [5], which is then fine-tuned to obtain downstream classifiers. In the attack investigation phase, FedTag first flags and embeds detected anomalous log queries then extracts events related to the attack investigation point and labels them as malicious events. Eventually, the attack story can be reconstructed by generating a causal graph or reporting all suspicious events to achieve traceability of attack links.

Final experimental results show that FedTag displays TPR and FPR metrics comparable to AirTag when compared to the centralized model. Specifically, for S1–S6 single-host data, federated learning shows less than a 1% drop and S2 achieves 100% TNR, outperforming AirTag for M1–M6 multi-host data, federated learning shows only about a 1% drop. However, compared with centralized anomaly detection, FedTag not only detects log anomalies but also protects the privacy of each client's log data.

In summary, this paper makes the following contributions:

- Fusing three kinds of log data from different sources, such as Firefox, DNS, and different security events.
- Training BERT models in the form of joint learning to generate pre-trained models

- The method not only detects attacks but also enables traceability of attack links.
- FedTag not only detects log anomalies but also protects the privacy of each client's log data.

2 Background

2.1 Trace and Log

In large-scale distributed systems, microservice systems make extensive use of distributed tracing to analyze and monitor user requests for instantiated services [3]. A user's request to an instantiated service instance triggers a series of service calls, which are represented by traces, each of which is called a span. In link tracing, a trace is the complete path of a request or an operation from start to finish. It covers the invocation relationships and performance information of all relevant components in a distributed system. Specifically, Trace contains a series of spans, each representing an instance of an invocation or operation. A span records information such as the start time, end time, elapsed time, operation type, etc. of the instance. By combining multiple spans, a complete Trace can be composed, describing the flow of requests through the distributed system. As the Fig. 1 represents an example of a user request service, a trace consists of a set of tree-structured spans, each corresponding to a service call. The trace contains five spans, all of which have a parent span except spanA, which starts the synchronization calls corresponding to spanB and spanC in turn.

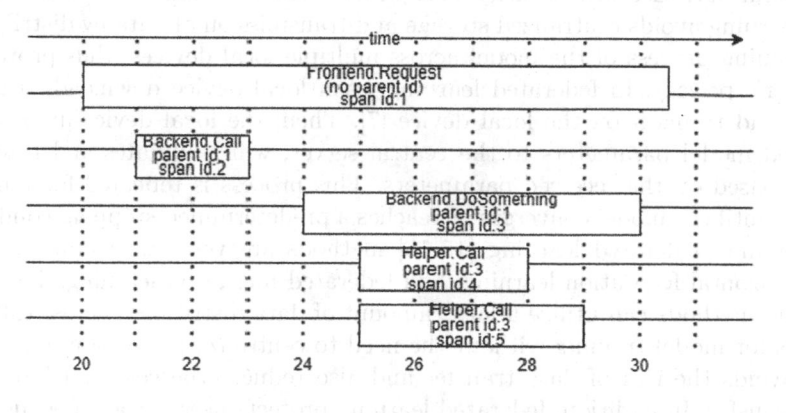

Fig. 1. Trace and log.

2.2 Metrics

Metrics represent the multi-dimensional metrics data provided by the application code running, each metric data usually consists of a time series and a set of tagged key-value pairs of structured data, common running dimensions metrics are: service response time, HTTP requests, CPU utilization, memory usage, disk read/write size, JVM memory usage, etc. [11]; metrics are used to measure the trend of performance, consumption, efficiency, and many other software attributes over time, efficiency, and many other software attributes over time. They allow engineers to monitor the evolution of a set of measurements (e.g., CPU or memory usage, request duration, latency, etc.) through alerts and dashboards. In its most basic form, a metrics data point is made up of three components: a metric name, a timestamp of the time when the data point was collected, and a measurement value represented by a number.

Prometheus is an open-source monitoring and alerting system based on a temporal database that is well-suited for monitoring Kubernetes clusters [6]. The core of the monitoring system is the Metric metrics. The basic principle of Prometheus is to periodically capture the state of the monitored component via HTTP protocol, any component can be accessed as long as it provides the corresponding HTTP interface for monitoring. No SDK or other integration process is required. This is ideal for virtualized environment monitoring systems such as VM, Docker, Kubernetes, etc.

2.3   Federal Learning

In the context of everyone's attention to data rights and data security, one of the major challenges facing AI is the problem of data silos, and the emergence of federated learning as a new machine learning paradigm can achieve the purpose of solving the data silos as well as the privacy protection problem without exposing the data of all parties. In traditional machine learning, the data usually needs to be centralized on a central server for training, which may involve user privacy issues and may also face security risks in data transmission. In contrast, federated learning avoids centralized storage and transmission of data by distributing the training process of the model across multiple local devices, thus protecting the user's privacy. In federated learning, each local device downloads a global model and trains it on the local device [7]. Then, the local device uploads the updated model parameters to the central server, which updates and fuses the model based on the received parameters. This process is repeated for multiple rounds until the model converges or reaches a predetermined stopping condition.

Common federated learning [15,16] methods are vertical federation learning, horizontal federation learning, and federated migration learning. The three learning methods can utilize a large amount of data distributed across different devices for model training without the need to centralize the data in one place. This avoids the risk of data transfer and also reduces the cost and latency of data transfer. In addition, federated learning protects user privacy because user data does not leave the local device (Fig. 2).

3   Methodology

3.1   Overview

As shown in Fig. 3, in order to achieve traceability of attack links, we propose a new federated learning based attack investigation system FedTag, which consists

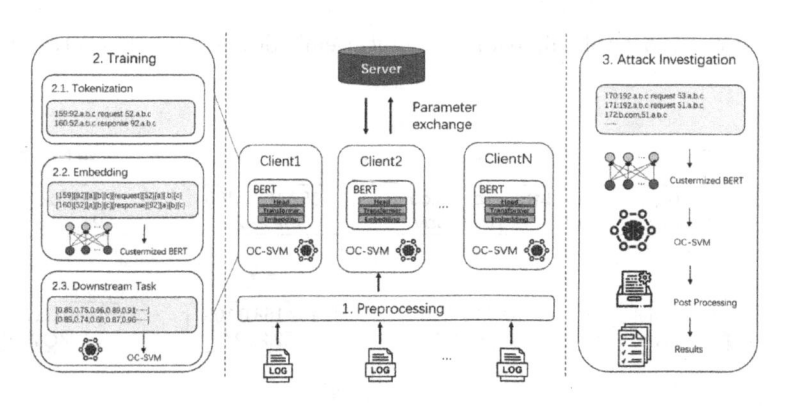

Fig. 2. Overview of FedTagX

of three components: data preprocessing, model training, and attack investigation. Firstly, data preprocessing involves sorting log entries and merging log data from different sources, such as FireFox, DNS, and different security events. Each client preprocesses their private non homologous log data separately. The second step is the model training process. FedTag tokenizes log files on each client separately, trains the BERT model in the form of federated learning to generate pre trained models, and then fine-tuning to obtain downstream classifiers. We have designed a new marker that utilizes the semantic syntax of specific domains in different source log files. Meanwhile, due to the presence of a large amount of benign data, we use one class support vector machine (OC-SVM) as our downstream classifier to overcome the problem of data imbalance. We use existing frequency based heuristic methods to filter out false positives generated by using single class support vector machines and improve accuracy. Finally, during the attack investigation phase, FedTag first marks and embeds the detected abnormal log queries, and then extracts events related to the attack investigation point and marks them as malicious events. FedTag can reconstruct attack stories by generating causal diagrams or reporting all suspicious events, achieving traceability of attack links.

3.2   Data Preprocessing

Preprocessing involves merging log data from different sources to maintain the correlation of the log data. We merge logs from different sources based on their timestamps and relationships. Referring to the preprocessing methods of ATLAS and AIRTAG, clustering and merging logs of the same behavior from DNS, firefox, and syslog, respectively, to facilitate the capture of causal relationships between the same behavior.

3.3   Model Training

We train a model through federated learning to capture the causal relationships between log events and achieve the restoration of attack chains. Our training includes three steps: tokenization, semantic embedding, and downstream task training.

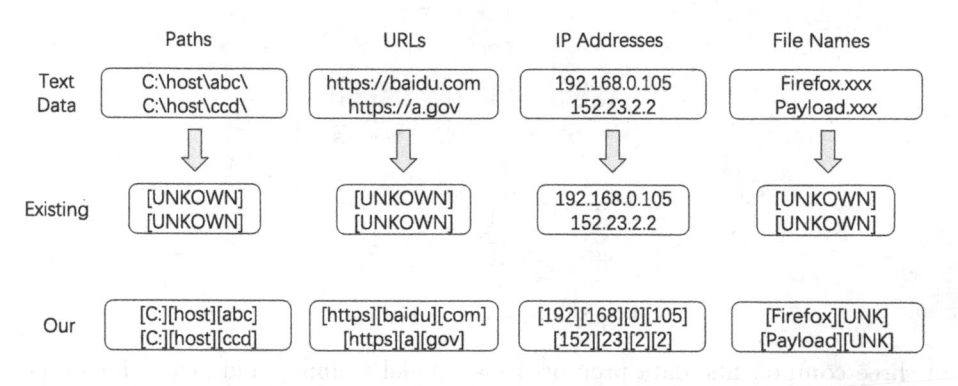

Fig. 3. Comparsion of Existing Tokenizer and Our Solution.

Tokenization. Tokenization is the process of breaking down text data into smaller units known as tokens. For instance, in the case of the advanced natural language processing model BERT, tokenization plays a pivotal role in its success. BERT's tokenization involves splitting input text into discrete tokens, which can encompass words, subwords, or even individual characters. Notably, BERT introduces the [CLS] and [SEP] tokens to demarcate sentence boundaries at the beginning of the input text. It employs a vocabulary of approximately 30,000 tokens, including various specialized tags like ## ed, #ly, and ##ing, which convey tense, part-of-speech information, and more. BERT relies on the Word-Piece Tokenizer to further break down text into subword units. For example, a word like "unhappiness" might be segmented into ["un", "## happiness"]. This approach facilitates the handling of unknown words and complex word forms within the vocabulary. Complex words are represented by their stems along with one or more specific markers. As an example, the word "embedding" corresponds to four markers: [em], [#bed], [#ding], and [#s]. Additionally, there's a special marker, [UNKNOWN], to signify any unknown words.

We have designed a new tokenizer that takes into account the specific semantic context of log events. Our tokenizer splits text using predefined characters (such as "." and "//"). Initially, we scan each field's value in every log entry, for instance, searching for URLs, and then split them accordingly. Therefore, compared to traditional tokenizers that tokenize all text in one go, our approach is more granular. When log entries contain multiple fields, we perform multiple splits. It's worth noting that we can still reuse existing tokenization tools for most words in the logs. All domain-specific tokenization tools are designed to break down the names of recorded entities, such as paths, URLs, IP addresses, and temporary file names, into smaller units. We use predefined separators within these names to achieve word-level splitting and then apply tokenization to each word. Figure 3 illustrates some examples. For instance, a path like C://host//abc

is split into tokens [C:], [host], and [abc] by recognizing special slash symbols, implying a directory structure. Importantly, for words like "host," "abc," and "ccd," we directly employ the BERT tokenizer. URLs can be seen as combinations of network protocols, domain names, ports, and paths. For temporary files like Firefox.xxx, FedTag also tokenizes them into [Firefox] and [xxx], making it easier to associate files with their corresponding processes. To handle various log formats, we leverage the LogStash [12] log parsing framework.

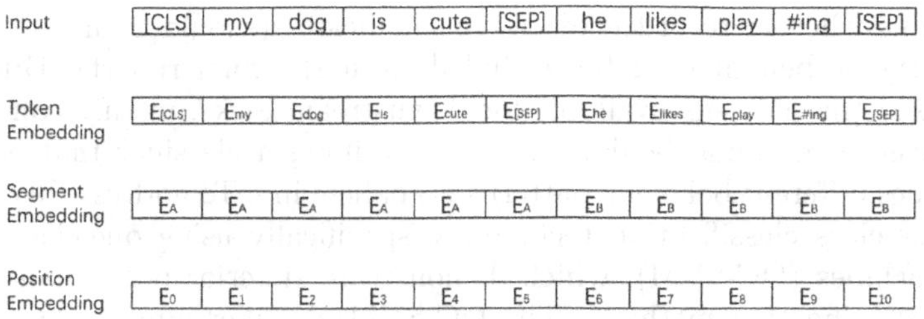

Input	[CLS]	my	dog	is	cute	[SEP]	he	likes	play	#ing	[SEP]
Token Embedding	$E_{[CLS]}$	E_{my}	E_{dog}	E_{is}	E_{cute}	$E_{[SEP]}$	E_{he}	E_{likes}	E_{play}	$E_{\#ing}$	$E_{[SEP]}$
Segment Embedding	E_A	E_A	E_A	E_A	E_A	E_A	E_B	E_B	E_B	E_B	E_B
Position Embedding	E_0	E_1	E_2	E_3	E_4	E_5	E_6	E_7	E_8	E_9	E_{10}

Fig. 4. Embeddings in BERT.

Embedding. After tokenizing logs from various sources, we employ federated learning to train embeddings. Specifically, we leverage the Bidirectional Transformer-based BERT model for natural language processing tasks. To tailor it for our specific context, we use BERT as the initialization model and retrain it using our unlabeled log data.

In FedTag, we closely adhere to the design principles of the BERT model, which comprises three crucial embedding layers. As depicted in Fig. 4, the ultimate input embedding combines token embeddings, segment embeddings, and positional embeddings. In BERT, each sentence is regarded as a segment, implying that tokens within the same sentence share identical segment embeddings, also referred to as sentence embeddings. However, our application context differs from traditional sentences, as there is no inherent notion of sentences. Instead, we treat each event as a unit, where all tokens belonging to the same event share the same segment embeddings. Similar to BERT, we also employ Masked Language Modeling (MLM) and Next Sentence Prediction (NSP, referred to as Next Segment/Event Prediction in FedTag) to train our model.

MLM is founded on the idea of replacing tokens or segments in logs with a special placeholder token [MASK], and during training, we utilize contextual information, such as neighboring tokens and segments, to predict the specific value of [MASK]. For example, we mask the request token 192.a.b.c and attempt to predict the exact content of [MASK] using 192.a.b.c [MASK] 152.a.b.c. On the other hand, NSP focuses on predicting whether the second sentence in a given sentence pair is a subsequent sentence to the first. BERT takes pairs of sentences, such as A: 192.a.b.c requests 152.a.b.c and B: 152.a.b.c responds 192.a.b.c, as input and predicts True, indicating that sentence B follows sentence A.

Downstream Task Training. After model training, the model has acquired the ability to analyze causal relationships among different source log markers, making it applicable to various downstream tasks. Typically, we incorporate pre-trained models (such as linear classifiers) into the model and employ a small amount of labeled data for supervised learning during training. For instance, BERT can be utilized to learn positive-negative relationships in sentiment analysis and create a model through classifier fine-tuning to identify emotional vocabulary in text. By fine-tuning the model on task-specific datasets, better performance can be achieved in downstream tasks. In attack investigations, we need to determine whether recorded events are malicious (should be included in the final report) or benign (can be excluded from the final report). Due to the imbalance in the datasets available for downstream tasks, mainly consisting of or exclusively containing benign patterns, training a classifier that effectively distinguishes different behavior patterns is challenging. To address this issue, we employ one-class classification techniques, specifically using one-class Support Vector Machines (OC-SVM), which demonstrate superior performance among one-class classification methods. The OC-SVM classifier aims to learn benign patterns and classify samples that significantly differ from benign patterns as suspicious. As illustrated in Fig. 5, OC-SVM learns benign behavior from unlabeled training logs (i.e., benign training data) and strives to find a decision boundary suitable for the training data. During anomaly detection, OC-SVM classifies input data that falls outside the decision boundary as malicious inputs (as indicated by the red dots). Besides training on benign data, OC-SVM can also be trained on datasets that are not entirely clean, as long as the majority of the data is benign. Since OC-SVM learns the patterns of the majority of activities, it naturally filters out the very few malicious activities once the model converges.

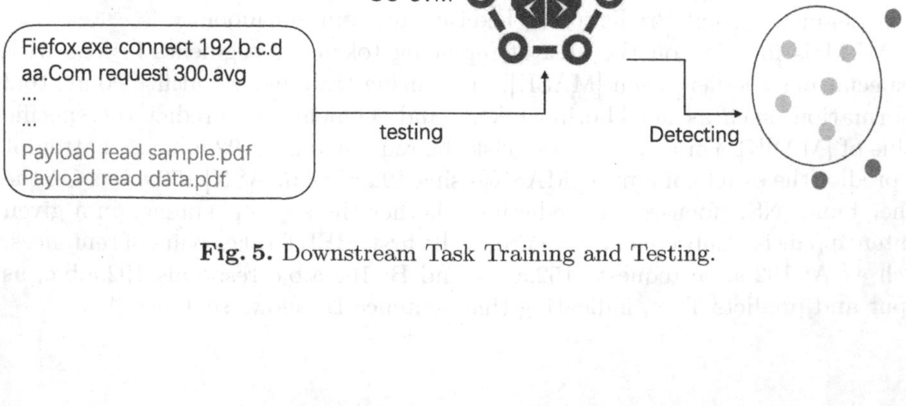

Fig. 5. Downstream Task Training and Testing.

3.4   Attack Investigation

The aim of the attack investigation is to meticulously reconstruct the entire narrative of the attack, leveraging the information contained in the provided audit logs. FedTag employs pre-trained models and OC-SVM to identify and precisely locate all potential suspicious events, thereby enabling the comprehensive reconstruction of the entire attack storyline. We employ heuristic techniques based on frequency-based analysis to effectively sift through and eliminate false positives that may arise from the one-class support vector machine. Following this, FedTag proceeds to create a causal graph encapsulating the reported suspicious events within concise, small-scale reports. This innovative approach significantly reduces computational overhead compared to traditional graph construction methods that rely on all available logs. Figure 6 provides a visual representation of FedTag's approach to constructing this causal graph based on the ATLAS methodology. More specifically, FedTag begins by generating an initial attack graph for log entries classified as malicious by OC-SVM. Given that FedTag only reports single events rooted in log text, it's possible for the aforementioned graph construction process to yield isolated graph components, as depicted in Fig. 6-A. To address this issue, FedTag actively associates these disconnected graph components, ultimately providing a seamless and comprehensive causal graph. Within each individual graph, FedTag extends its coverage by identifying nodes adjacent to other graphs within the original log text events, as indicated by the yellow nodes in Fig. 6-B. When two components share a common node (represented by a red node), FedTag consolidates these disjoint graph components by merging the necessary nodes and their corresponding edges, as illustrated in Fig. 6-C. FedTag iterates through this process until all isolated graph components are seamlessly connected. Notably, during this iterative process, FedTag's primary focus remains on identifying the first matching event between any two disjointed graph components. This approach streamlines the construction process, requiring the creation of the initial graph only once and connecting each pair of disjointed components just once, eliminating the need for redundant graph reconstruction.

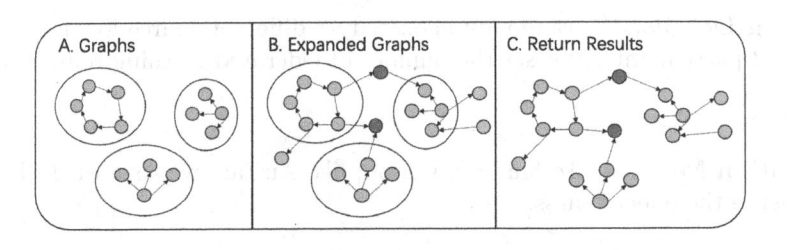

Fig. 6. The process of graph reconstruction. (Color figure online)

4   Experiments

4.1   Experimental Settings

Datasets. We utilized the public datasets provided by ATLAS (S1 to S4 and M1 to M6), where attacks S-1 to S-4 were executed on a single host, and M-1 to M-6 were conducted across multiple hosts. For each multi-host attack, simulations were performed on two hosts, with the second host serving as the target for lateral movement. Following the completion of these attacks, ATLAS collected audit logs within a 24-h window. Details of the datasets, including dataset names, attack types, attack targets, the number of known attack indicators (such as attacker IPs), original log sizes, the ratio of nodes involved in benign and malicious activities, and the ratio of malicious activities to all activities in each column, are presented in Table 1.

Table 1. Overview of the datasets

Dataset	Attack Type	Target	#Clue	Size (MB)	%Overlap	%Malicious
S1	Web compromise	BS	3	382	78.68%	6.46%
S2	Malvertising	BS	3	1015	84.00%	4.30%
S3	Spam campaign	MS	6	522	76.96%	12.12%
S4	Pony campaign	MS	5	449	84.65%	16.17%
M1	Web compromise	BS	6	711,102	83.8%	4.06%
M2	Phishing	BS	5	671,112	85.33%	13.68%
M3	Malvertising	BS	4	336,138	86.12%	11.60%
M4	Monero miner	Win	6	533,91	83.69%	3.80%
M5	Pony campaign	MS	5	726,113	87.36%	5.33%
M6	Spam campaign	MS	6	551,142	79.13%	4.18%

Training Details. We randomly allocated all different-source log data, with a total of 3 participants, and set the number of federated learning rounds to N = 20.

Evaluation Metrics. To fair comparison, like AirTag, we also use TPR, FPR to measure the effectiveness.

4.2   Results

Overall Results. The overall results of the model comparison on the same dataset are presented in Fig. 7, with model results from AirTag [9] cited for comparison. From Fig. 6, it can be observed that, compared to the centralized

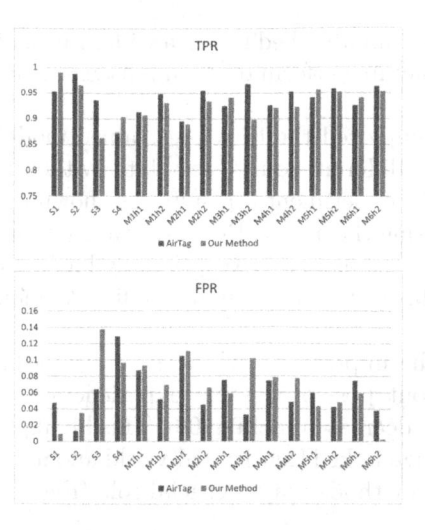

Fig. 7. Overall Results

model, FedTag shows TPR and FPR metrics that are on par with AirTag. Specifically, for the S1–S6 single-host data, federated learning exhibits a decrease of less than 1%, with S2 achieving a TNR of 100%, surpassing AirTag. For the M1–M6 multi-host data, federated learning shows only around a 1% decrease. However, in contrast to centralized anomaly detection, FedTag not only detects log anomalies but also preserves the privacy of each client's log data, an achievement that centralized learning cannot attain.

Table 2. Result of Ablation Study

Dataset	Methods	TNR	TPR	FPR	FNR
S1	w/o post	0.98	0.98	0.02	0.02
	w/o pretraining	0.45	0.92	0.08	0.55
	ours	0.98	0.99	0.01	0.01
S2	w/o post	1	0.94	0.06	0
	w/o pretraining	0.03	0.99	0.01	0.97
	ours	1	0.97	0.03	0
S3	w/o post	0.77	0.86	0.14	0.23
	w/o pretraining	0.46	0.9	0.1	0.54
	ours	0.77	0.95	0.05	0.23
S4	w/o post	0.99	0.17	0.83	0.01
	w/o pretraining	0.58	0.88	0.12	0.42
	ours	0.99	0.9	0.1	0.01

Ablation Study. We evaluated FedTag in a wider range of scenarios, measuring factors that could impact its performance, and discussing the potential sensitivity of FedTag.

First, we directly employed the Bert pre-trained model provided on the official website to embed different-source log data without involving the logs in Bert's pre-training. The experimental results are shown in Table 2. Compared to our FedTag method, using the official Bert pre-trained model directly can achieve decent TPR; however, it leads to an exceptionally high false positive rate, reaching as high as 0.97. This would result in a significant amount of manual work in real-world scenarios.

Then, we tested the importance of the post-processing stage. As shown in Table 2, FedTag without post-processing experiences a noticeable decrease in TPR, with the highest decrease reaching 0.6. This highlights the essential role of the post-processing stage in FedTag. In real-world anomaly detection, frequency-based post-processing methods play a crucial role (Fig. 8).

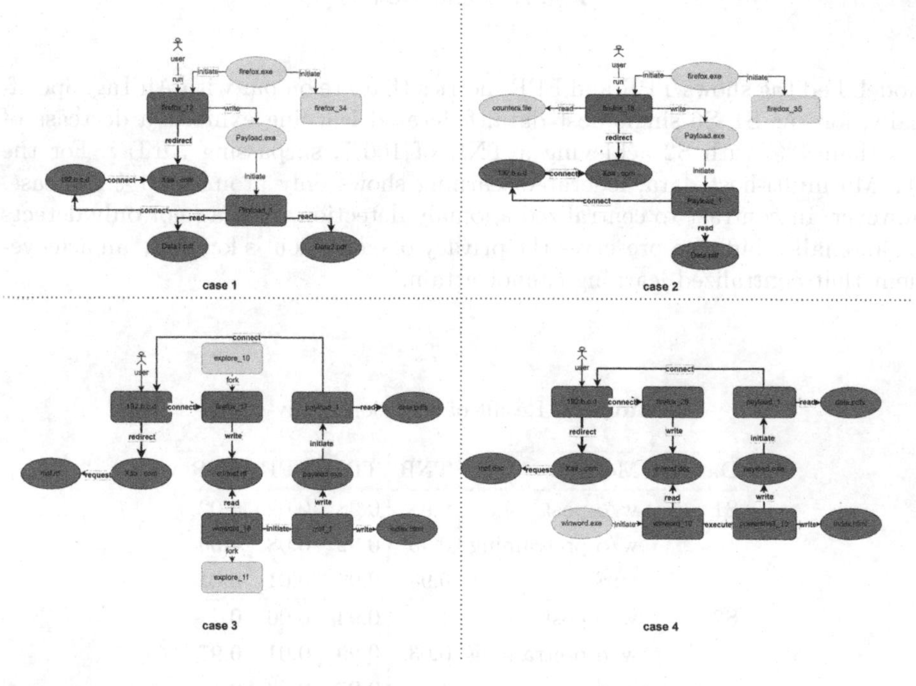

Fig. 8. Investigation Cases by FedTag. (Color figure online)

4.3   Case

In this section, we present and discuss the results of FedTag's construction of causal graphs from four attack examples. In the attack case, we present Fig. 7, the red color indicates the attack chain, and the added edges indicate that they were added by connecting disconnected components during the reconstruction of the graph. During the construction of the causal graph, the original attack subgraph was too large, so we manually simplified it during the graphing, showing only the key attack steps and omitting many other false positives.

Case 1: This attack exploits CVE-2015-5122. The vulnerability installs allow remote attackers to execute arbitrary code or cause a denial of service (memory corruption) via a designed link. During the attack, the user runs a process and clicks on a link that is redirected to a malicious website resolved by 192.b.c.d. The attacker can execute arbitrary code or cause a denial of service (memory corruption). The attacker exploited CVE-2015-5122 through this link, corrupted the browser plugin, and wrote a payload program payload.exe on the victim's computer. The process Payload1 derived from the execution of this program scanned the victim's system for files, established a connection to the attacker, and uploaded all pdf files.

Case 2: This attack exploits the CVE-2015-3105 vulnerability, which is similar to the vulnerability in case 1 and allows an attacker to execute arbitrary code or cause a denial of service (memory corruption) via an unspecified vector.

Case 3: This attack exploits the CVE-2017-11882 vulnerability. The vulnerability allows an attacker to pass through objects in memory that cannot be handled correctly. In this attack, a user opens a malicious email that contains a link to a malicious website that is resolved by 192.b.c.d. The user opens a malicious email that contains a link to a malicious website. The user requested and downloaded a malicious file, msf.rtf, which was read with Winword_18. The msf.rtf then builds msf_1, which writes a payload program, payload.exe, and replaces the benign website page in the victim's host with a malicious index.html. The attacker executes payload.exe, initializes payload_1, scans the pdf file, and receives the pdf file.

Case 4: This attack exploits the CVE-2017-0199 vulnerability. When a user opens a document that contains the exploit code for this vulnerability, the malicious code downloads and executes a Visual Basic script that contains PowerShell commands.

5 Conclusion

In this paper, we present FedTag, a new federated learning-based attack investigation system. It fuses three kinds of log data from different sources, trains a BERT model in the form of federated learning to generate a pre-trained model, and then performs fine-tuning to obtain downstream classifiers. The detected anomalous log queries are labeled and embedded during attack identification, and then the events related to the attack investigation points are extracted and labeled as malicious events. Eventually, the attack story can be reconstructed by generating a causal graph or reporting all suspicious events to achieve traceability of attack links. The final experimental results show that FedTag displays TPR and FPR metrics on par with AirTag. However, FedTag not only detects log anomalies but also protects the privacy of each client's log data.

Acknowledgements. Please place your acknowledgments at the end of the paper, preceded by an unnumbered run-in heading (i.e. 3rd-level heading).

References

1. Zhou, X., Peng, X., Xie, T., et al.: Fault analysis and debugging of microservice systems: industrial survey, benchmark system, and empirical study. IEEE Trans. Softw. Eng. **47**(2), 243–260 (2018)
2. Zhang, C., Peng, X., Sha, C., et al.: DeepTraLog: trace-log combined microservice anomaly detection through graph-based deep learning. In: Proceedings of the 44th International Conference on Software Engineering, pp. 623–634 (2022)
3. Peng, X., Zhang, C., Zhao, Z., et al.: Trace analysis based microservice architecture measurement. In: Proceedings of the 30th ACM Joint European Software Engineering Conference and Symposium on the Foundations of Software Engineering, pp. 1589–1599 (2022)
4. Hassan, W.U., Noureddine, M.A., Datta, P., et al.: OmegaLog: high-fidelity attack investigation via transparent multi-layer log analysis. In: Network and Distributed System Security Symposium (2020)
5. Hassan, W.U., Bates, A., Marino, D.: Tactical provenance analysis for endpoint detection and response systems. In: 2020 IEEE Symposium on Security and Privacy (SP), pp. 1172–1189. IEEE (2020)
6. Hossain, M.N., Sheikhi, S., Sekar, R.: Combating dependence explosion in forensic analysis using alternative tag propagation semantics. In: 2020 IEEE Symposium on Security and Privacy (SP), pp. 1139–1155. IEEE (2020)
7. Milajerdi, S.M., Gjomemo, R., Eshete, B., et al.: HOLMES: real-time APT detection through correlation of suspicious information flows. In: 2019 IEEE Symposium on Security and Privacy (SP), pp. 1137–1152. IEEE (2019)
8. Alsaheel, A., Nan, Y., Ma, S., et al.: ATLAS: a sequence-based learning approach for attack investigation. In: 30th USENIX Security Symposium (USENIX Security 21), pp. 3005–3022 (2021)
9. Ding, H., Zhai, J., Nan, Y., et al.: AIRTAG: towards automated attack investigation by unsupervised learning with log texts. In: 32nd USENIX Security Symposium (USENIX Security 23), pp. 373–390 (2023)
10. King, S.T., Chen, P.M.: Backtracking intrusions. In: Proceedings of the Nineteenth ACM Symposium on Operating Systems Principles, pp. 223–236 (2003)
11. Grandini, M., Bagli, E., Visani, G.: Metrics for multi-class classification: an overview. arXiv preprint arXiv:2008.05756 (2020)
12. Paganini, P.: Phishing campaigns target us government agencies exploiting hacking team flaw CVE-20155119 (2015). https://securityaffairs.co/wordpress/38707/cyber-crime/phishing-cve-2015-5119.html. Accessed 06 Jun 2020
13. Logstash (2022). https://www.elastic.co/cn/logstash
14. Jiang, G., Mohandas, R., Leathery, J., Berry, A., Galang, L.: CVE-20170199: In the Wild Attacks Leveraging HTA Handler (2017). https://www.fireeye.com/blog/threat-research/2017/04/cve-2017-0199-hta-handler.html. Accessed 06 Jun 2020
15. Li, P., et al.: Multi-key privacy-preserving deep learning in cloud computing. Futur. Gener. Comput. Syst. **74**, 76–85 (2017)
16. Yang, Q., Liu, Y., Chen, T., Tong, Y.: Federated machine learning: concept and applications. ACM Trans. Intell. Syst. Technol. (TIST) **10**(2), 1–19 (2019)
17. Ul Hassan, W., Bates, A., Marino, D.: Tactical provenance analysis for endpoint detection and response systems. In: Proceedings of the IEEE Symposium on Security and Privacy (2020)

Deep Learning Based SQL Injection Attack Detection

Zhang Pan[1], Qianli Huang[2], Ziqing Tian[2], Ying Liu[1], Jiapeng Lou[2], Yongguang Gong[2], and Zhiqiang Wang[2,3(✉)]

[1] The Air Traffic Control Bureau of Civil Aviation Administration of China, Beijing 100022, China
[2] Beijing Electronic Science and Technology Institute, Beijing 100070, China
wangzq@besti.edu.cn
[3] State Information Center, Beijing 100045, China

Abstract. The malicious behavior of SQL injection attack is one of the major threats to database-containing web applications, which has been widely concerned by Internet users and developers. SQL injection attack is an attacker implanting malicious SQL statements into legitimate user inputs, which leaks or tampers with the information of the database system, thus leading to serious security problems. To solve the above security problems, this paper carries out research on SQL injection attack detection technology based on deep learning, and designs and implements an attack detection tool, which provides an efficient means of security protection and helps to improve the security of database management systems. In this paper, URL text is transformed into computer-recognizable vectors by decoding the URL samples, feature vectorization and other operations. Based on this, the vectors are trained using four neural networks, MLP, LSTM, CNN and TextCNN. The experimental results show that the models using BertMLP and BertCNNs algorithms achieve 98.92% and 99.38% accuracy on the test set, respectively. The analysis of the experimental results shows that the accuracy rate obtained by training the classifiers using deep learning is 5% higher than that obtained using traditional machine learning, where the model using the BertMLP algorithm has fewer false positives and the model using the BertCNN algorithm has a higher overall performance.

Keywords: SQL injection attack detection · Deep learning · Neural network

1 Introduction

1.1 Research Background and Significance

With the increasing popularity of Internet applications and the generalization of data collection and storage, database security is a growing concern. According to the report of Imperva, a data security company, the number of Web security vulnerabilities is

---

Z. Pan and Q. Huang—Contributed equally to this work and should be considered co-first authors.

H. Jin et al. (Eds.): IAIC 2023, CCIS 2058, pp. 127–141, 2024.
https://doi.org/10.1007/978-981-97-1277-9_10

increasing year by year, and there is at least 1 vulnerability in almost every Web application, and SQL injection attack is one of the most common and harmful attacks [1]. Attackers can construct malicious SQL statements to trick programs to perform illegal operations, including stealing sensitive data and modifying database contents. Therefore, SQL injection attacks have become a problem that cannot be ignored in the field of Internet security. SQL injection attacks are harmful in the following ways [2]: data leakage, data tampering, attacking other hosts, and destabilizing websites.

With the continuous development of deep learning technology breakthroughs, the types of data that can be processed are increasingly rich, and the computer arithmetic power is increasingly enhanced, the application of deep learning technology to solve the problem of detecting SQL injection attacks has become a widely publicized idea. Deep learning can automatically detect and classify SQL injection attacks by learning and modeling a large amount of data, thus improving the accuracy, real-time and effectiveness of network security protection. Therefore, in the context of the gradual maturation of deep learning technology, it is of great theoretical and practical application value to carry out the design and realization of SQL injection attack detection tools based on deep learning [3].

Traditional SQL injection attack detection techniques usually rely on specific rules, which need to be constantly updated and modified, and it is easy for attackers to circumvent the detection of these rules by some simple means. In contrast, deep learning-based SQL injection attack detection techniques have the following advantages [4]: more adaptive, after the introduction of deep learning technology into SQL injection attack detection technology, it can automatically learn the characteristics of SQL injection attacks by learning and modeling the data, without the need for explicit rules, and can be more adaptive, with better generalization ability; more accurate, deep learning technology introduced into the SQL injection attack detection technology can fully explore the underlying data features, and optimize the model to improve detection accuracy; more robust, traditional SQL injection attack detection techniques can only detect known attacks and cannot cope with unknown attacks. Deep learning-based SQL injection attack detection techniques can be learned and modeled to have better robustness against unknown attacks.; and reduced maintenance costs, for traditional SQL injection attack detection techniques, it is necessary to constantly update the rules, and the maintenance cost is high. The SQL injection attack detection technology based on deep learning technology does not require human intervention, only need to continuously accumulate data, and then only need to carry out model training. In conclusion, SQL injection attack detection technology based on deep learning has many advantages, which can effectively improve the ability of network security protection and is an important research direction in the field of network security in the future.

1.2   Current Status of the Research

At present, many scholars at home and abroad use machine learning and deep learning methods to detect SQL injection attacks. Next, we briefly introduce a few latest researches in recent years.

Zhang [4] introduced a new approach to detect the vulnerability of SQL injection in PHP code. Various machine learning algorithms have been trained and evaluated, such as

Random Forest, Logistic Regression, SVM, Multilayer Perceptron (MLP), Long Short Term Memory (LSTM) and Convolutional Neural Networks (CNN). Zhang found that CNN offered the best accuracy of 95.4%.

Sivasangari et al. [9] detected SQL injection attacks with AdaBoost. In this research, the data are transformed into stumps, and the stumps are classified as either the weaker ones with a lower weight than the output, or the stronger ones that provide the largest proportion of the total output. Experiments show that this algorithm can detect injection attacks with high accuracy and efficiency.

Daset al. [10] proposed a method to classify a dynamic SQL query into an attack or an ordinary one, based on a Web profile prepared in the training stage. Simple Bayes, SVM and analytic tree are adopted to classify the data. The total detection rate of the two sets of data sets is 91% and 90%.

Deep learning has attracted a lot of attention in recent years due to the fact that it doesn't require the researchers to observe and deal with the data and perform complex features. Moreover, it has been widely applied in SQL injection testing.

Tanget et al. [11] proposed a method to detect SQL injection attacks. Firstly, they analyze a lot of SQL injected data to extract the related features, then train various neural network models, such as MLP, LSTM and so on. Experiments indicate that MLP is superior to LSTM in detection.

Gandhi et al. [12] proposed a hybrid CNN-BiLSTM approach to detect SQL injection attacks. They compare the different kinds of machine learning algorithms for detecting SQL injection attacks in detail. The CNN-BiLSTM method provides approximately 98% precision in comparison with other methods.

Zhang et al. proposed a method to detect SQL injection in the context of network traffic. In this approach, only the target features needed for the training of the model are selected with Deep Belief Network (DBN) models. They also tested the performance of LSTM, CNN, and MLP. According to experiment results, the precision of DBN reached 96%.

Overall, the current machine learning-based methods require high a priori knowledge from the researcher, and the manually defined features can cover a limited number of SQL injection scenarios. Therefore, on the one hand, the detection effect depends heavily on the accuracy of the researcher's predefinition of the features. On the other hand, it cannot learn well from the constantly evolving new injection attack features. Nevertheless, the continuous maturity and frequent application of deep learning technology in the field of NLP provide a new research direction for SQL injection detection. Therefore, in this paper, we will do an in-depth research on the SQL injection detection method based on deep learning. And the construction of its input data, i.e., the way of sample vectorization is also discussed and studied together.

1.3 Research Content and Main Work

This paper mainly researches the SQL injection attack detection technology based on deep learning and completes the design and realization of the detection tool. According to the actual application scenarios and requirements, MLP, CNN, TextCNN, LSTM and Bert are utilized to build a deep network to train the classifier, and the deployment is completed. At the same time, we used and evaluated the tool in the real environment and verified the effectiveness of the tool in detecting attacks in real scenarios and investigated the impact of different algorithmic models on the detection effect of the tool. The research in this paper focuses on URL analysis to detect whether it contains the risk of SQL injection attacks, and this research combines practical problems with state-of-the-art technology to make a more meaningful exploration.

1.4 Thesis Organization

This paper is divided into a total of four chapters, which mainly study the SQL injection detection technology based on deep learning, exploring three major parts of the technology, namely, data preprocessing, deep learning algorithm model design and tool system design, and experimental validation. The main contents of each chapter are as follows:

The Sect. 1 is an introduction, which describes the research background and research significance of this paper, introduces the main work and innovations of this paper, and then describes the thesis organization.

The Sect. 2 is the design and implementation, which designs and implements a specific detection tool based on deep learning SQL detection technology and provides a comprehensive introduction to the implementation details of the four specific modules contained therein.

Section 3 is the experimental environment and evaluation of experimental results, which compares the performance effects of different vectorization methods, different deep learning algorithms, and deep learning and machine learning on the same dataset according to the evaluation criteria adopted, and explains the experimental results and makes suggestions to the user users on the use of the tool.

Section 4 is the summary and outlook, which summarizes the main work and main results of this paper.

2   Deep Learning Based SQL Injection Detection Tool Design

In this chapter, a detection tool will be built based on the technical foundation to facilitate users to make judgments according to their needs.

2.1   Detection Tool Architecture

The design of the detection tool is divided into four modules, namely, data preprocessing module, sample vectorization module, deep learning model module, and tool design module, and the framework diagram is shown Fig. 1.

The data preprocessing module decodes and reduces the samples of the dataset, tags them and separates them into words; the sample vectorization module takes the processed text data as input and outputs feature vectors with semantic features; the deep learning model module trains the feature vectors to train classifiers that can differentiate whether or not the URLs contain SQL injection attacks; and finally, it deploys classifiers on the back-end and designs the front-end interface to compose a complete detection tool.

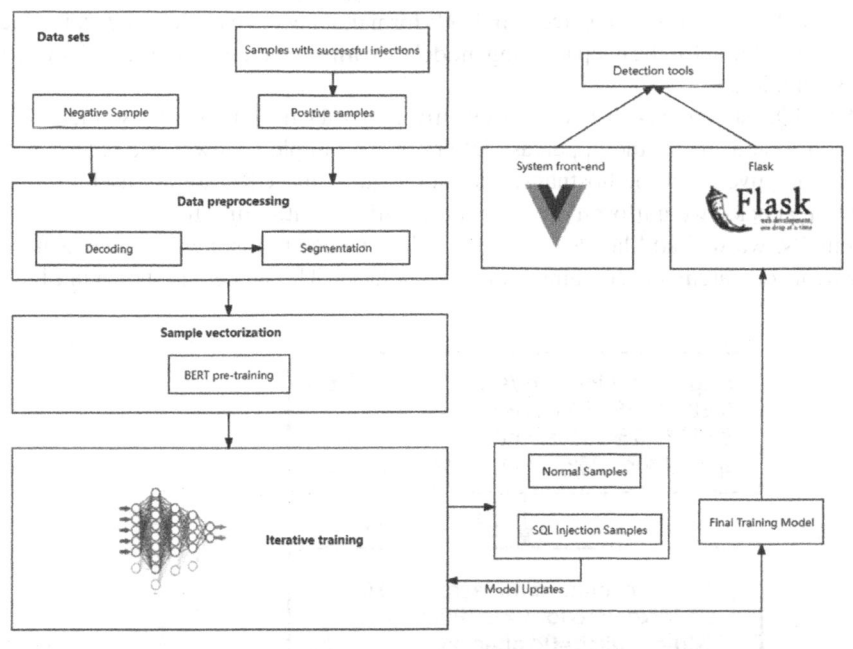

Fig. 1. Schematic diagram of the design structure of the deep learning-based SQL injection detection tool

2.2 Data Preprocessing

Data preprocessing is crucial to the training results of deep learning. It can be said with certainty that the quality of data preprocessing and data quality directly determines the highest level of detection results of the trained classifiers, and every deep learning algorithm model optimization is just a continuous attempt to reach this highest level. Therefore, in this subsection, the preprocessing steps of SQL injected sample data will be elaborated in detail.

In this paper, the work of the data preprocessing module includes decoding and reducing the encoded data, labeling, and word splitting operations.

2.2.1 Data Decoding and Reduction and Tagging

URL encoding refers to escaping non-ASCII characters or special characters in a URL to ensure its transmission and parsing correctness on the Internet. Since the Internet uses ASCII encoding as the basic format for transmitting characters, if there are special characters to be transmitted, they need to be converted into the corresponding ASCII code by adding the % symbol prefix at the end of the URL and then connecting to the ASCII code value (e.g., %20 for space). Common characters that need to be translated include spaces, metacharacters (& +, #, etc.), internationalized domain names, and so on.

Since the dataset we collected is in URL format, it needs to be decoded and reduced to normal text first for the deep learning model to learn valid features of the SQL injection attack samples.

For SQL statements, it is case-insensitive, so after decoding, we use the lower () function to convert all the uppercase letters in the sample to lowercase letters. At the same time, given that the hostname and pathname in the URL are not useful for SQL injection attacks, we uniformly replace these with the "http://u" string.

Finally, we will add labels to each sample sentence to distinguish them as normal statements or statements containing injection attacks. The process is shown in Fig. 2.

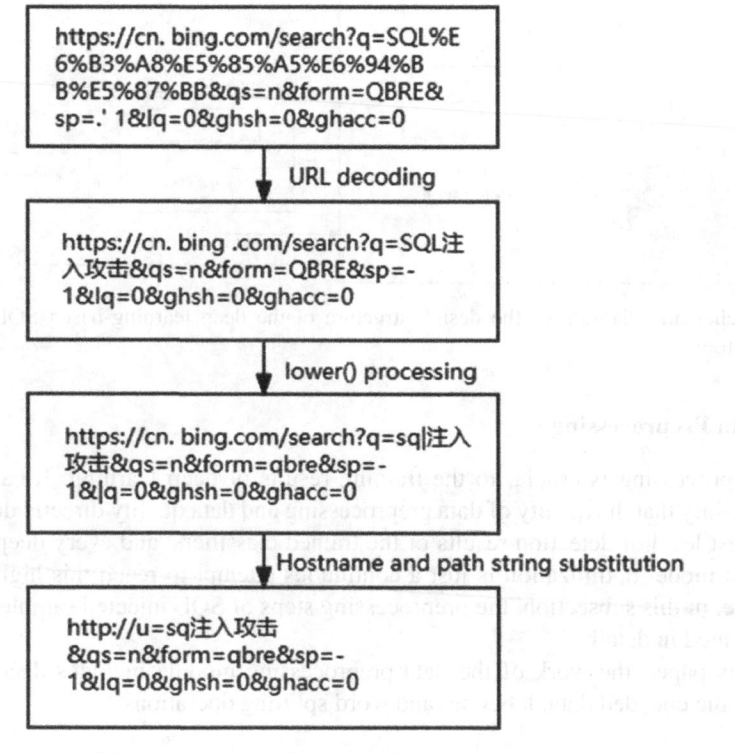

Fig. 2. Data Decoding Reduction and Tagging Codes

2.2.2 Segmentation

Since we subsequently chose the Bert algorithm for text vectorization, this paper uses the BertTokenizer tool to split the text of the decoded URLs into words. BertTokenizer uses the WordPiece model to split the input text into words, splitting each URL into several parts and labeling them with special symbols to achieve better linguistic processing effect. The example of word splitting used in this paper is shown in Fig. 3.

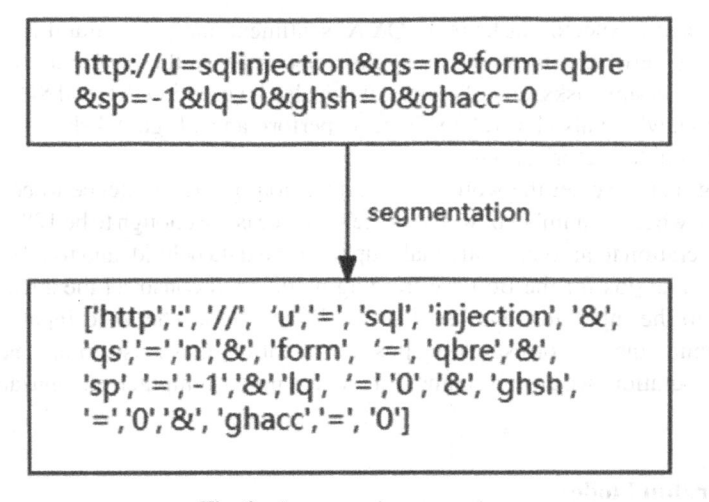

Fig. 3. Segmentation example

2.3 Sample Vectorization

The approach to sample vectorization taken in this paper is more specific, using the Bert algorithm for the process. The Bert-base-uncased model is a natural language processing model based on the Transformer Encoder architecture, released by Google Inc. in 2018, which consists of a lowercase (uncased) version of the base model for English subject words, which has 12 layers, 768 hidden units, and 110M parameters. Bert-base-uncased was developed by using the Bert-base-uncased model on a large corpus and

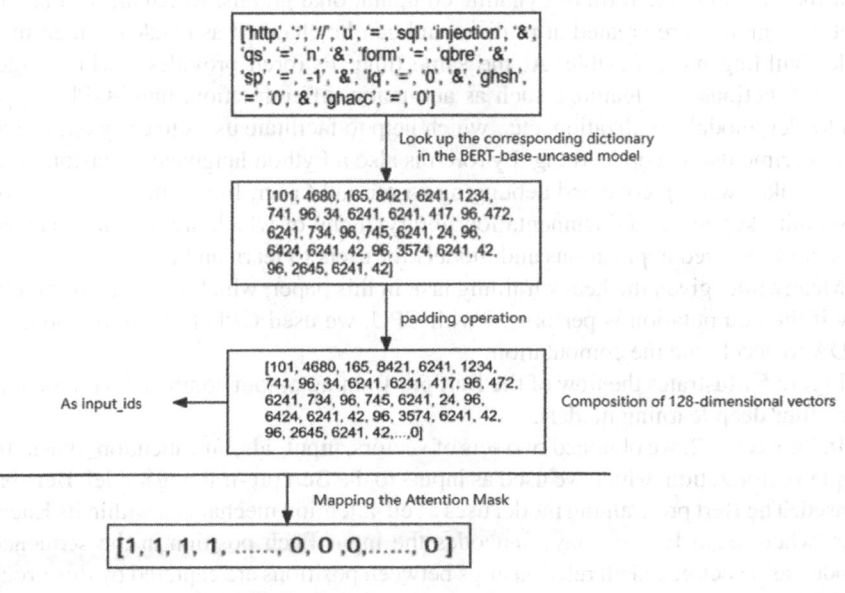

Fig. 4. Sample vectorization process

then fine-tuned to specific tasks (e.g., Q&A, sentiment analysis, natural language reasoning, etc.), Bert-base-uncased can significantly improve the performance of natural language processing tasks, and of course, it can also be used in URL and SQL statement vectorization, which also has relatively good performance. Figure 4 shows the detailed process of sample vectorization.

Among them, we set the word vector dimension of each sentence to be 128 at the longest, and when the number of words in the sentence is not enough to be 128, we take the Padding operation to automatically make up 0 for the data behind, and use the generated vectors as input_ids for the Bert pre-training model, and construct the attention_mask according to the input_ids, and use 1 to mark the vector The valid input data in the Padding complement 0 corresponding position attention_mask also complement 0, after the above operation, we convert a line of text data into a computer-recognizable feature vector.

2.4 Algorithm Mode

This part is the most important and core module of the whole detection system. Here we only explain the engineering part of the algorithm implementation.

Currently, there are a variety of deep learning frameworks in the industry, including TensorFlow, PyTorch, Keras, Caffe, MXNet, and so on. Each framework has its advantages and applicable scenarios, for example, TensorFlow has been widely used in machine learning and artificial intelligence projects in various fields, Keras is easy to get started and compatible with TensorFlow and PyTorch, and PyTorch is characterized by ease of use, dynamic computational graphs, and good flexibility.

Or the engineering practices in this paper, we have chosen the PyTorch framework. PyTorch is an open-source deep learning framework developed and maintained by Facebook Inc. It adopts the form of dynamic computational graphs, which means that computational graphs are created at runtime and can be changed as needed, which makes model building more flexible. At the same time, PyTorch provides a wide range of built-in functions and features, such as automatic differentiation, multi-GPU support, data loader, model serialization, etc., which help to facilitate users to carry out research and experiments in deep learning. PyTorch is also a Python language extension library, which makes writing code and debugging easier and faster. In addition, it has a strong community support and documentation library, many of which are available as source codes for advanced applications and models for users to learn and use.

Meanwhile, given the heavy training task in this paper, which would be particularly slow if the computation is performed with CPU, we used GPU to train the model and CUDA to accelerate the computation.

Figure 5 illustrates the flow of the Bert pre-training output combined with the inputs of the four deep learning models.

In Subsect. 2.2, we obtained two sets of vectors, input_ids, and attention_mask, from sample vectorization, which we used as inputs to the Bert pre-training model, Bert-base-uncased. The Bert pre-training model uses a self-attention mechanism within its Encoder layer, where each Encoder layer encodes the input Each position in the sequence is encoded as a vector, and all relationships between positions are captured by this process. In addition to the input layer, this pre-trained model contains 12 Encoder layers, and we

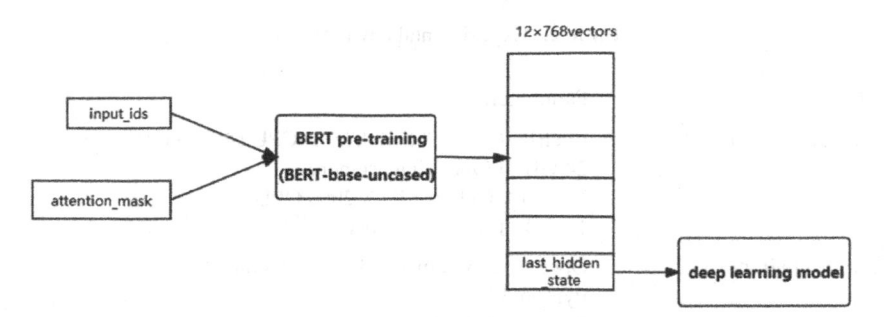

Fig. 5. Bert pre-training output combined with deep learning model flow

select the last layer vector last_hidden_state as the final textual representation to be fed into the subsequent deep learning model to construct the classifier.

3 Simulation Experiment and Evaluation

3.1 Dataset

The dataset used in this paper is partly from the open-source Web attack detection project dataset [5] and partly from the HttpParams Dataset [6] dataset, which contains different kinds of attacks, and we use the SQL injection attack with the "sqli" identifier. The combined dataset is shown in Table 1.

Table 1. Sample profile of the dataset

Dataset	Normal request data	SQL attack	Total
HttpParams Dataset	8547	5574	14121
Other Github projects	29953	33913	63866
(grand) total	38500	39487	77987

3.2 Experimental Environment

To verify the effectiveness of the SQL injection attack detection scheme and tools proposed in this paper, a high-performance experimental environment is built on a personal PC. The specific experimental environment information is shown in Table 2.

Table 2. Experimental environment

Platform	Parameters
Hardware environment	Intel(R) Core (TM) i7-8750H CPU @ 2.20 GHz CPU 24 GB of Operating Memory NVIDIA GeForce RTX 2060 GPUs 12 GB of Video Memory
Software environment	Operating system Windows 10 64 bit Python 3.7 Pycharm 2020.3 HBuider X 3.3.5 Pytorch 1.11 CUDA component Sklearn Transformers Matplotlib Flask 2.3.2

3.3 Evaluation Indicators

There is a need to use reasonable evaluation metrics to evaluate and compare the effectiveness of classifier models trained by neural networks. Common evaluation metrics for deep learning are Accuracy, Precision, Recall, F1-score, ROC curve and AUC value, etc. [7]. In this paper, the first four evaluation metrics are used, and their calculation formulas and descriptions are given next.

1. Accuracy: Accuracy is the most used evaluation index in classification models, indicating the ratio of the number of samples correctly predicted by the model to the total number of samples. The formula is:

$$Acc = (1 - \frac{err\_sum}{sum}) \times 100\% \tag{1}$$

   err_sum denotes the number of samples that were misclassified, and sum denotes the number of all samples. There are some shortcomings in the accuracy metric: on the one hand, the accuracy may be distorted for datasets with unbalanced categories; on the other hand, the accuracy cannot reflect the misclassification of the classifier. So, we need other metrics for comprehensive evaluation.

2. Precision. Precision indicates the proportion of samples with positive predictions that are actually true positive examples. The formula is:

$$Pre = \frac{TP}{TP + FP} \times 100\% \tag{2}$$

   TP denotes the number of true samples among the samples that were predicted as positive examples, and FP denotes the number of false samples that were incorrectly predicted as positive examples.

3. Recall. Recall indicates the proportion of the truly positive samples that are correctly predicted as positive by the model. The formula is:

$$Rec = \frac{TP}{TP + FN} \times 100\% \tag{3}$$

 TP denotes the true number of samples that were predicted to be positive examples, and *FN* denotes the number of samples that were actually positive examples but were predicted to be negative examples.

4. F1-score: The F1-score is the reconciled average of Precision and Recall, combining the performance of both. The calculation formula is:

$$F1 = 2 \times \frac{Pre \times Rec}{Pre + Rec} \tag{4}$$

 The F1-score is used to synthesize Precision and Recall are two criteria that contain contradictory meanings.

3.4 Analysis of Experimental Results

3.4.1 Impact of Different Sample Vectorization Methods

Word2Vec is an algorithm developed by Google for converting text into vector. The algorithm maps words to vectors in a high-dimensional space such that words with similar meanings are close together in this space. Word2Vec mainly consists of Skip-gram, a continuous skip-gram model, and CBOW, a continuous bag-of-words model. Skip-gram model predicts the target word based on the target word, and the network

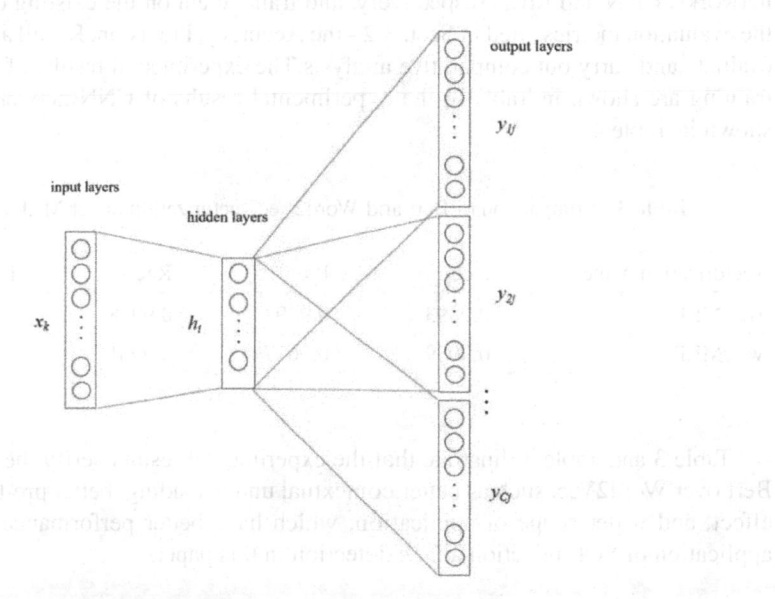

Fig. 6. Skip-gram network structure

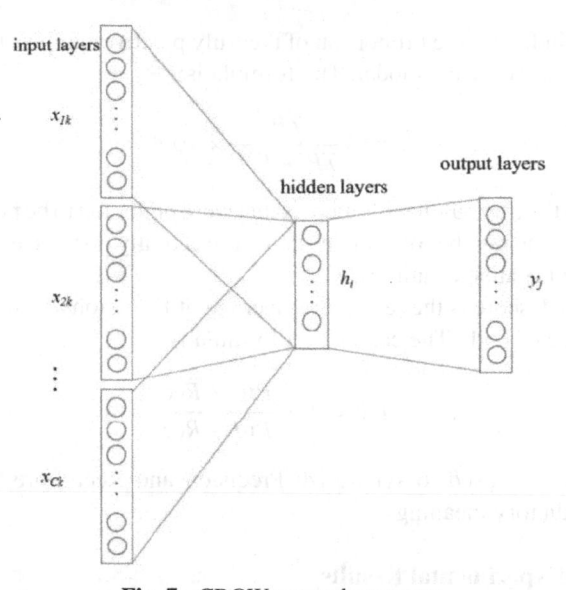

Fig. 7. CBOW network structure

structure is shown in Fig. 6. CBOW model predicts the target word based on the context word, and the network structure is shown in Fig. 7.

In order to compare the ability of the two algorithms, Bert model and Word2Vec model, to capture the semantic features of the samples and the effect of vectorization, this subsection uses these two vectorization methods in combination with two neural networks, CNN and MLP, respectively, and trains them on the existing datasets, using the evaluation metrics cited in Sect. 3.2 - the Accuracy, Precision, Recall and F1-score to evaluate and carry out comparative analysis. The experimental results of MLP network training are shown in Table 3; the experimental results of CNN network training are shown in Table 4.

Table 3. Comparison of Bert and Word2Vec vectorization under MLP network

vectorization name	Acc	Pre	Rec	F1-score
BertMLP	0.9893	0.9892	0.9938	0.9893
w2vMLP	0.9089	0.9677	0.9961	0.9669

Table 3 and Table 4 illustrate that the experimental results verify the advantages of Bert over Word2Vec, such as better contextual understanding, better pre-training model effect, and wider scope of application, which have better performance results in the application of SQL injection attack detection in this paper.

Table 4. Comparison of Bert and Word2Vec vectorization under CNN network

vectorization name	Acc	Pre	Rec	F1-score
BertMLP	0.9956	0.9998	0.9912	0.9955
w2vMLP	0.8956	0.9967	0.9465	0.9565

3.4.2 Comparison of the Effectiveness of Different Deep Learning Models

In order to compare the training effectiveness of MLP, CNN, TextCNN, and LSTM on sample vectors, this subsection utilizes the PyTorch framework to construct the network and train it, and also uses the evaluation metrics cited in Sect. 3.2 - Accuracy, Precision, Recall Recall and F1-score for validation and comparative analysis. The experimental results on the validation set are shown in Table 5, and the experimental results on the Test set are shown in Table 6.

Table 5. Comparison of different neural networks on the validation set

Algorithm name	Acc	Pre	Rec	F1-score
BertMLP	0.9881	0.9686	0.9906	0.9981
BertCNN	0.9956	0.9998	0.9912	0.9955
BertTextCNN	0.8613	0.7837	0.9982	0.8613
BertLSTM	0.8663	0.7902	0.9974	0.8663

Table 6. Comparison of different neural networks on test set

Algorithm name	Acc	Pre	Rec	F1-score
BertMLP	0.9892	0.9893	0.9938	0.9893
BertCNN	0.9938	0.9939	0.9875	0.9939
BertTextCNN	0.8331	0.7496	0.9990	0.7496
BertLSTM	0.8376	0.7583	0.9965	0.7583

In view of the experimental results, both LSTM and TextCNN are eliminated due to their poor performance, and MLP and CNN can be selected according to the level of concern about false alarms as well as missed alarms; if fewer false alarms are required, MLP is selected; for the rest of the cases, CNN with higher comprehensive performance can be selected.

3.4.3 Comparison with Common Machine Learning Methods

In reference [8], SVM, Ensemble Boosted Trees, and Linear Discriminant are the three machine learning algorithms that have been used to do SQL injection detection, and with reference to the feature extraction approach in reference, the test is conducted in the existing dataset and compared by Accuracy as an evaluation criterion. The results are obtained as shown in Table 7.

Table 7. Comparison with classical machine learning algorithms

	MLP	CNN	SVM	EBT	LD
Acc	0.9981	0.9956	0.9381	0.9370	0.9358

All three machine learning algorithms end up with accuracy rates between 93% and 94%, which is also a superior performance, but about 5% less accurate than the MLP or CNN models. So, the upper limit of accuracy that can be achieved by using a proper deep learning algorithm is higher than using machine learning when it comes to detecting whether a URL contains an SQL injection attack or not.

4 Conclusions

In this paper, we design a SQL injection detection model based on deep learning, utilize the automatic feature extraction and high extensibility of deep learning to overcome the dilemma, and design a detection tool based on the detection model to achieve effective detection of SQL injection attacks. The main results of the work are shown as follows:First, in this paper, text vectorization is performed on preprocessed data using Bert method to extract semantic features, and four deep learning models, MLP, CNN, LSTM, and TextCNN, are used to train classifiers.Secondly, on the basis of detection techniques, this article designs a SQL injection detection tool, and finally gives the tool results to show.Third, the article conducts comparative experiments and analyzes the experimental results, concluding that the Bert method is better than the Word2Vec method, the deep learning model is better than the traditional machine learning model, and puts forward pertinent suggestions for the advantages and disadvantages of the two types of deep learning models, CNN and MLP, and how to choose them in practical applications.

Conflict of Interest. The authors declare that they have no competing financial interests exist.

Funding. This paper's work financially was supported by the Fundamental Research Funds for the Central Universities (3282023052, 3282023013), China Postdoctoral Science Foundation funded project (2019M650606).

References

1. Liu, H.: Security threats and countermeasures in computer network databases. Comput. Program. Skills Maintenance **12**, 154–156 (2020)
2. Tajpour, A., Shooshtari, J.J.Z.: Evaluation of SQL injection detection and prevention techniques. In: Second International Conference on Computational Intelligence, Communication Systems and Networks, CICSyN 2010, Liverpool, UK, 28–30 July 2010. IEEE (2010)
3. Alghawazi, M., Alghazzawi, D., Alarifi, S.: Detection of SQL injection attack using machine learning techniques: a systematic literature review. Mach. Learn. Data Anal. Cyber Secur. **2**, 764–777 (2022)
4. Luo, C.: A research on SQL injection and XSS attack detection based on deep learning. Chinese Academy of Engineering Physics (CAEP) (2020)
5. Skykami. [EB/OL]. [10 May 2023]. https://github.com/skykami/detect_traffic_sqlias_CNN
6. Morzeux. HttpParams Dataset. [EB/OL]. [10 May 2023]. https://github.com/Morzeux/HttpParamsDataset
7. Hasan, M., Balbahaith, Z., Tarique, M.: Detection of SQL injection attacks: a machine learning approach. In: Proceedings of the 2019 International Conference on Electrical and Computing Technologies and Applications (ICECTA). IEEE (2020)
8. Bu, S., Ge, X.: Analysis and research on principles and detection techniques of common high-risk web vulnerabilities. Electron. Softw. Eng. (2), 243–244.0 (2021)
9. Sivasangari, A., Jyotsna, J., Pravalika, K.: SQL injection attack detection using machine learning algorithm. In: 2021 5th International Conference on Trends in Electronics and Informatics (ICOEI), pp. 1166–1169. IEEE Press (2021)
10. Das, D., Sharma, U., Bhattacharyya, D.K.: Defeating SQL injection attack in authentication security. An experimental study. Int. J. Inf. Secur. 1–22 (2019)
11. Tang, P., Qiu, W., Huang, Z., et al.: Detection of SQL injection based on artificial neural network. Knowl.-Based Syst. (190) (2020)
12. Gandhi, N.: A CNN-BiLSTM based approach for detection of SQL injection attacks. In: Proceedings of the 2021 International Conference on Computational Intelligence and Knowledge Economy (ICCIKE), Dubai, United Arab Emirates, pp. 378–383. IEEE Xplore (2021)

RSCC: Robust Semi-supervised Learning with Contrastive Learning and Augmentation Consistency Regularization

Xinran Jing and Yongli Wang(✉)

Nanjing University of Science and Technology, Nanjing 210000, China
{jingxr,yongliwang}@njust.edu.cn

Abstract. Semi-supervised learning (SSL) can effectively take advantage of unlabeled data. Aiming at the poor performance of existing SSL methods in the case of only a very small number of labels and the problem of pseudo-label confirmation bias, we propose a novel SSL method, RSCC, which combines the powerful representation learning capability of contrast learning with augmentation consistency regularization methods and introduces a symmetric cross-entropy learning to mitigate the impact of noisy pseudo-labels on model performance. RSCC consists of two key steps. We first perform self-supervised pre-training on unlabeled data using contrast learning to extract meaningful representations from the data, and then perform tuning training based on SSL methods of augmentation consistency regularization and symmetric cross-entropy learning. We conduct rich experiments, which show that RSCC achieves state-of-the-art accuracy on multiple datasets, such as CIFAR-10 and CIFAR-100, especially when labeled data is extremely scarce. This underscores its cutting-edge and effective performance.

Keywords: Semi-supervised learning · Contrastive Learning · Consistency Regularization

1 Introduction

Recently, numerous data-driven models have demonstrated remarkable performance. However, in reality, it is extremely hard to get big amounts of labeled data, especially in highly specialized domains where it is even more expensive to do so. As a result, the autonomous and efficient learning from unlabeled or sparsely labeled data has become a focal point of research. In response, semi-supervised learning (SSL) has emerged as a potent solution to address this issue, offering promising avenues for tackling the problem.

SSL plays a pivotal role in the field of deep learning, effectively harnessing limited labeled data to enhance model performance. By incorporating a substantial amount of unlabeled data, SSL enables models to acquire more robust feature representations, consequently improving their performance and generalization capabilities [1]. This makes the model more resistant to interference and able to cope with noise and changes in the data. Furthermore, SSL offers an effective way to address issues of imbalanced data label distributions, as unlabeled data provides supplementary support. Nevertheless, as a general trend, the fewer labeled instances available, the less effective SSL becomes. Consequently, the challenge of achieving effective SSL with an extremely limited number of labeled instances has emerged as a significant research topic. Self-supervised learning allows for the direct utilization of abundant unlabeled data in training neural network models, automatically extracting data features [2]. Once the model completes its self-training, the learned feature representations can be fine-tuned or employed in transfer learning for various machine learning tasks [3]. These feature representations provide a new ray of hope for addressing the problem of limited labeled data [4]. If SSL is regarded as a downstream task of self-supervised learning, it becomes possible to fully harness the high-quality data representations extracted by self-supervised learning models and further fine-tune the model for specific tasks. Theoretically, this approach can yield more desirable results with extremely few labeled instances [5]. This method combines the strengths of both self-supervised learning and SSL, offering a promising avenue to address the challenges posed by data limitations and label scarcity.

In SSL, pseudo-labeling is one of the most mainstream and commonly used methods. However, Low-quality pseudo-labels not only fail to enhance model performance but can also lead to error accumulation, thereby undermining learning performance [6]. Existing methods often employ confidence thresholds to filter pseudo-labels [7]. However, setting confidence thresholds too high can result in the selection of fewer pseudo-labels, reducing the model's learning efficiency and potentially causing overfitting or underfitting. Conversely, setting confidence thresholds too low can lead to the selection of numerous low-quality pseudo-labels, impairing model performance. Balancing the quality and quantity of pseudo-labels is a critical issue worthy of in-depth investigation.

To address the above issues, we present a novel SSL method called RSCC. RSCC combines the robust representation learning capabilities of contrastive learning with data augmentation consistency regularization and introduces symmetric cross-entropy learning to alleviate the impact of noisy pseudo-labels. RSCC primarily consists of two key steps. First, we employ contrastive learning for self-supervised pretraining, extracting meaningful representations from the data. Subsequently, we fine-tune the model using a combination of augmentation consistency regularization and symmetric cross-entropy in the SSL phase. Specifically, inspired by MoCo V2 [8], we first construct a comparative learning network with momentum encoder for pre-training. Subsequently, two distinct data augmentations are applied to data samples. Pseudo-labels are generated from the predictions of weakly augmented samples, and losses are enforced on strongly augmented sample predictions, thereby facilitating consistency regularization learning. Additionally, we introduce the symmetric cross-entropy learning method, denoted as SL, as an enhancement to the conventional cross-entropy loss. This is done to mitigate the adverse effects resulting from confirmation bias in pseudo-labels. On standard datasets such as CIFAR-10 and CIFAR-100, RSCC achieves state-of-the-art performance, especially when there are only very few labels available.

2 Related Work

2.1 Self-supervised Data Representation Learning

Self-supervised learning constitutes a crucial branch of deep learning, predicated on the fundamental concept of learning from unlabeled data. Notably, contrastive learning stands out as a prominent representative. In recent years, the field of contrastive learning has witnessed the emergence of numerous outstanding methods, including BYOL [9], SimCLR [10], SwAV [11], MoCo [12], SimSiam [13], and others. These contrastive learning methods have not only achieved remarkable results in the domain of self-supervised learning but have also had a positive impact on SSL. Their high-quality data representations and feature learning capabilities provide robust support for SSL.

Contrastive learning methods can often be conceptualized as a dictionary lookup problem. In theory, it is essential to ensure that this dictionary is sufficiently large to store rich visual representation information. Only when the dictionary is large enough can the inherent features of matching images be more easily recognized. In traditional end-to-end methods, the size of the dictionary is typically aligned with the batch size. However, due to limitations in computational resources, the choice of batch size cannot be excessively high, leading to decreased learning efficiency and convergence challenges for the model. The introduction of a Memory bank has addressed this issue by storing represen-tations of all samples before model training. However, after updating the parameters of the encoder in a training batch, only the sample representations used as "keys" in the current batch are updated. Consequently, the consistency of the Memory bank is relatively poor.

He et al. proposed the MoCo method, effectively balancing the size of the dictio-nary and its consistency [12]. The key innovation of MoCo lies in the introduction of a momentum encoder to encode representations in the dictionary. This momentum encoder maintains a moving average of the model parameters, enabling it to create more con-sistent and robust data representations. By comparing the representations generated by the current encoder with those generated by the momentum encoder, MoCo effectively learns the similarities and differences between images. This process allows the model to capture complex patterns and features in the data, thereby generating high-quality learning representations. In MoCo v2 [8], the authors further improved the model's per-formance by incorporating projection heads and employing stronger image augmentation methods.

2.2 SSL

SSL achieves impressive results in the field of image classification [14–17]. SSL commonly employs pseudo-labeling or consistency regularization methods. Regarding pseudo-labeling, Lee et al. selected the class with the highest predicted probability as pseudo-labels [18], but this approach can lead to overconfident pseudo-labels. Arazo et al. suggested using network predictions to generate soft pseudo-labels, a simple and effective approach that mitigates the impact of overconfident pseudo-labels [19]. Shi et al. approached the problem from a new perspective by introducing confidence on unlabeled image samples to estimate unreliable labels [20], and this method has also demonstrated strong competitiveness.

Simultaneously, research on consistency regularization methods is flourishing. Laine et al. introduced the concept of temporal ensembling, which maintains an exponential moving average of label predictions for each training sample and penalizes inconsistent predictions [21]. However, Tarvainen et al. found that this approach can be quite cumbersome on large scale datasets and proposed the Mean Teacher model to address this issue [22]. Subsequently, Ke et al. pointed out that Mean Teacher restricts the learning of the student model and proposed the Dual Student model, utilizing a stable constraint for knowledge exchange and furthering consistency learning [23]. In addition, other types of consistency learning methods have also made significant advancements. Miyato et al. introduced a method based on virtual adversarial loss to assess the local smoothness of prediction distributions for unlabeled data [24]. Xie et al. explored how to employ advanced data augmentation for consistency regularization learning, bringing significant improvements to numerous visual tasks [25].

Moreover, many advanced algorithm models combine both methods, such as Fix-Match [7], UPS [26], Dash [27], FlexMatch [28], ConMatch [29], and others. Additionally, there are some graph-based SSL methods that make full use of the topological structure and similarity information in the data, aiding the model in better capturing the data's distribution and relationships during the learning process [30, 31]. However, it's worth noting that these methods may involve complex procedures.

2.3 The Issue of Pseudo-label Confirmation Bias

As an integral part of SSL, pseudo-labeling faces several challenges, and one of them is the issue of confirmation bias. Therefore, researchers have been actively working on developing new methods to enhance the quality of pseudo-labels. One of the simplest, most effective, and commonly used approaches is based on model predictions. This method typically involves training an initial model on the limited labeled data and then predicting a class distribution for each unlabeled sample, assigning it a pseudo-label. However, this method is susceptible to the uncertainty in model predictions. As a result, researchers have proposed various techniques to filter and correct pseudo-labels, including threshold filtering [7], ensemble methods [21, 32, 33], and model uncertainty estimation [22, 25]. Threshold filtering methods aim to filter out predictions with high uncertainty. However, setting the confidence threshold too high results in fewer pseudo-labels, affecting model learning efficiency. Conversely, setting the threshold too low introduces noisy pseudo-labels, degrading model performance. Striking a perfect balance between the quality and quantity of pseudo-labels is not straightforward. Taking the representative FixMatch model, which employs threshold filtering, as an example, Li et al. identified asymmetric noise in their pseudo-labels due to batch structure reconstruction, leading to confirmation bias [6]. Ensemble methods enhance the quality of pseudo-labels by combining predictions from multiple models. However, they often come with high computational costs, especially when dealing with large-scale datasets. Additionally, ensemble methods typically require careful tuning of multiple hyperparameters, adding an extra workload. Using ensemble methods on large datasets may demand substantial computational resources and time, necessitating a trade-off between benefits and costs. Incorporating model uncertainty estimation in pseudo-label can improve the stability of pseudo-labels. However, model uncertainty estimation is not always perfectly accurate, and inaccurate estimates can lead to low-quality pseudo-label generation. Furthermore, employing model uncertainty estimation requires more complex model architectures and may necessitate more labeled data and smaller batch sizes to maintain model robustness.

Currently, various pseudo-label learning methods exhibit a degree of confirmation bias, particularly in the early stages of model training. Deep neural networks tend to fit noise, resulting in lower loss values. Consequently, the confirmation bias in pseudo-labels gradually accumulates over multiple training batches. Methods to improve the quality of pseudo-labels have become increasingly complex. This paper approaches the problem from a different perspective, focusing on robust SSL with noisy pseudo-labels, which involves tolerating a certain level of noisy pseudo-labels to enhance model generalization. Wang et al. proposed the Symmetric Cross-Entropy Learning (SL) method [34]. This approach symmetrically enhances cross-entropy (CE) using the noise-robust Reverse Cross-Entropy (RCE) to address this issue. In this paper, SL is introduced into SSL to mitigate the impact of pseudo-label confirmation bias.

3   Method

The structure of RSCC is illustrated in Fig. 1. RSCC primarily consists of two crucial steps: (1) Self-Supervised Pretraining Based on Contrastive Learning. In this step, the model first applies various image augmentations to the unlabeled data to create positive and negative sample pairs. Subsequently, a contrastive loss is employed to adjust the similarity between positive and negative sample pairs. The objective of this step is to extract meaningful data representations. (2) SSL Fine-Tuning with Augmentation Consistency Regularization and Symmetric Cross-Entropy Learning. In this step, RSCC combines augmentation consistency regularization and symmetric cross-entropy learning methods. The goal of this step is fine-tuning based on specific tasks.

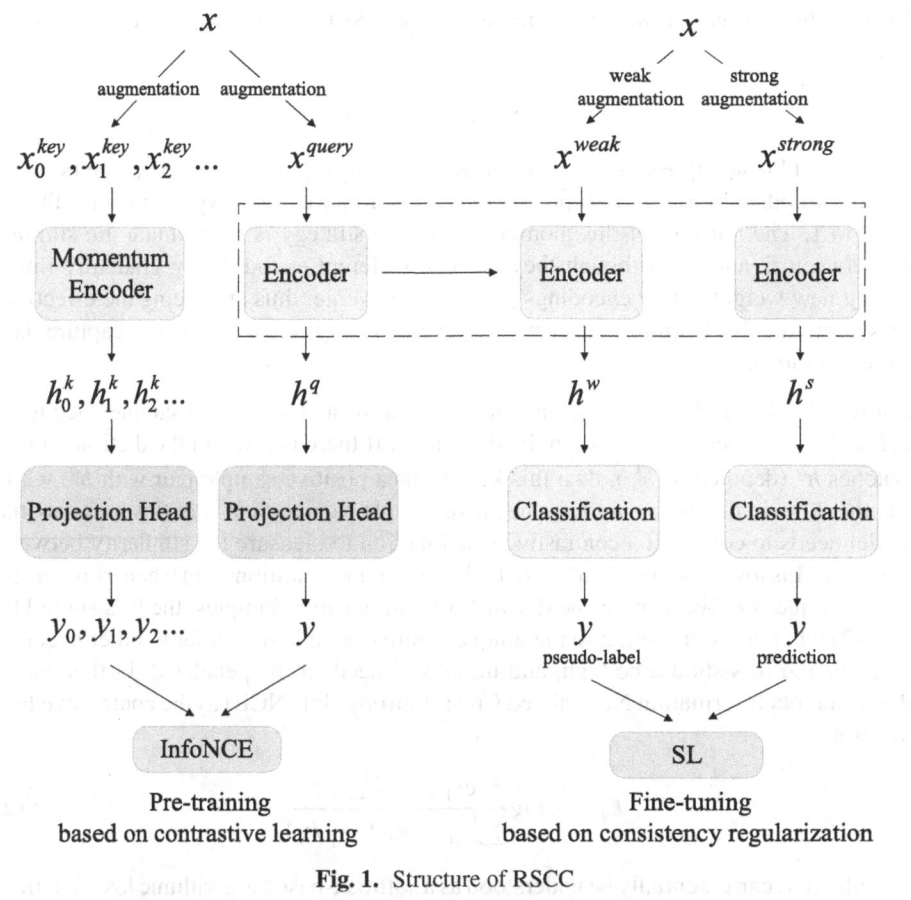

Fig. 1. Structure of RSCC

3.1 Self-supervised Pretraining Based on Contrastive Learning

To reduce the dependency on a large number of labeled data, improve model learning efficiency and performance, RSCC initiates self-supervised learning on unlabeled data. In this section, taking inspiration from MoCo V2 [8], RSCC constructs a network for self-supervised pretraining based on contrastive learning, composed of four components: data augmentation, encoder, projection head, and contrastive loss. Regarding the storage of data representations, RSCC employs a queue-based approach. Each time a new image is processed, its representation is added to the queue, while the oldest element in the queue is removed, ensuring that the queue's size remains constant. This operation helps mitigate the impact of older encoders on key consistency loss, thus improving the effectiveness of contrastive learning.

Encoder and Update Strategy. To enhance the consistency of data representations, RSCC utilizes two encoders: f_q and f_k. f_q is responsible for encoding input images into feature representations, which will be used for subsequent contrastive learning tasks. f_k is used to encode the representations in the queue, facilitating contrastive learning.

To smoothly update the weights of the encoders, RSCC employs a momentum update strategy as:

$$\varphi_k \leftarrow n\varphi_k + (1 - n)\varphi_q \tag{1}$$

In Eq. (1), φ_k represents the parameters of f_k, φ_q represents the parameters of f_q, and φ_q is updated through backpropagation. n is a momentum hyperparameter that is close to 1. The purpose of this momentum update strategy is to maintain the similarity between f_k and f_q. Although they are two different encoders, by gradually introducing new weights, their encodings gradually converge, thus enhancing the effectiveness of contrastive learning. This smoothing update helps the model better capture data representations.

Contrastive Loss. We define an encoded query as h^q and a series of sample encodings as $h_0^k, h_1^k, h_2^k \ldots$ serving as keys in the dictionary. If there is a key in the dictionary that matches h^q (denoted as h_+^k), then this key forms a positive sample pair with h^q, while all other keys in the dictionary form negative sample pairs with h^q. . In this scenario, the model needs to construct a contrastive loss function to measure the similarity between samples. This loss function should satisfy the following conditions: (1) when h^q is similar to the unique positive sample and dissimilar to all negative samples, the loss should be low; (2) when h^q is dissimilar to the unique positive sample or similar to other negative samples, the loss should be high, and the model needs to be penalized. In this paper, RSCC adopts Information Normalized Cross-Entropy (InfoNCE) as the contrastive loss function:

$$\ell_q = -log \frac{\exp(h^q \cdot h_+^k / \gamma)}{\sum_{c=0}^{C} \exp(h^q \cdot h_i^k / \gamma)} \tag{2}$$

InfoNCE can essentially be understood as a softmax-based logarithmic loss function, where the classifier's objective is to classify h^q as h_+^k. In Eq. (2), γ is a temperature hyperparameter primarily used to adjust the model's discrimination between sample pairs.

3.2 Robust SSL Fine-Tuning Based on Augmentation Consistency

The SSL step of RSCC inherits the pre-trained encoder from contrastive learning, then combines it with the augmentation consistency regularization and symmetric cross-entropy learning methods to improve its performance. For a given dataset, we provide the following definitions: $X = \{(x_c, p_c) : c \in (1, \ldots, C)\}$ represents a set of C labeled samples, where x_c denotes training samples and p_c represents their respective one-hot labels; $U = \{u_c : c \in (1, \ldots, \mu C)\}$ is a collection of μC unlabeled samples, where μ is a hyperparameter determining the relationship between labeled and unlabeled samples; $p(y|x)$ denotes the class probability distribution obtained by inputting x into the model; $H(p, q)$ represents the cross-entropy between two probability distributions p and q; $M(\cdot)$ and $m(\cdot)$ represent strong augmentation and weak augmentation operations.

Consistency Regularization Loss. RSCC utilizes the model's output on weakly augmented images to obtain pseudo-labels, which are then applied to strongly augmented images of corresponding unlabeled samples, followed by enforcing a loss. This embodies the concept of consistency regularization. The model's original loss function consists of two standard cross-entropy loss terms: ℓ_l and ℓ_u. ℓ_l represents the supervised loss function:

$$\ell_l = \frac{1}{C}\sum_{c=1}^{C} H(p_c, p(y|m(x_c))) \tag{3}$$

For unlabeled samples, we first predict the class probability distribution for their weakly augmented images, denoted as q_m. Then, we choose the class with the highest probability as the pseudo-label, denoted as $\hat{q}_m$. Subsequently, we construct a cross-entropy loss using the pseudo-label $\hat{q}_m$ and the class probability distribution of the strongly augmented version of u_c. ℓ_u is as follows:

$$\ell_u = \frac{1}{\mu C}\sum_{c=1}^{\mu C} \text{if}(\max(q_m) \geq \tau)\text{H}(\hat{q}_m, p(y|\text{M}(u_c))) \tag{4}$$

In Eq. (4), τ is a hyperparameter, which represents the confidence threshold. Specifically, the model retains the pseudo-label only when the maximum class probability corresponding to a sample exceeds τ. Finally, the complete loss function is given by:

$$\updownarrow = \updownarrow_l + \lambda \ell_u \tag{5}$$

In Eq. (5), λ is a hyperparameter that represents the weight of the unsupervised loss.

Robust Loss with Noisy Pseudo-labels. In this work, we recognize that there is a certain bias in the confirmation of pseudo-labels, especially in the early stages of model training, and these biases accumulate over multiple training batches. Therefore, we introduce Symmetric Cross-Entropy learning (SL) in the process of consistency regularization to address overfitting on "easy" classes and underfitting on "hard" classes. Specifically, SL introduces the concept of Symmetric Cross-Entropy (SCE):

$$SCE = CE + RCE \tag{6}$$

In Eq. (6), the reverse cross-entropy (RCE) is derived from CE in reverse, and the formula is given by:

$$\ell_{rce} = -\sum_{c=1}^{C} q \log p \tag{7}$$

To avoid computational issues due to p possibly being 0 in Eq. (7), we define $\log 0 = G$, where G is a constant less than zero. This setting is similar to the pruning technique. The formula for symmetric cross-entropy (SCE) is given as follows:

$$\ell_{sce} = \ell_{ce} + \ell_{rce} \tag{8}$$

To maintain the good convergence of CE and the robustness of RCE to noise, SL introduces two hyperparameters, α and β, as weights for ℓ_{ce} and ℓ_{rce}, so the loss of SL is given as follows:

$$\ell_{sl} = \alpha \ell_{ce} + \beta \updownarrow_{rce} \tag{9}$$

Therefore, the unsupervised loss function during the SSL fine-tuning process in RSCC is defined as:

$$\ell_u^* = \frac{1}{\mu C} \sum_{c=1}^{\mu C} \text{if}(\max(q_m) \geq \tau)(\alpha \text{H}(\hat{q}_m, p(y|\text{M}(u_c))) + \beta \text{H}(p(y|\text{M}(u_c)), \hat{q}_m))$$

(10)

Hence, the loss function for RSCC SSL is given as:

$$\ell_{RSCC} = \mathbb{I}_l + \lambda \ell_u^*$$

(11)

4 Experiment

4.1 Datasets

We conducted a comprehensive evaluation of RSCC's performance, primarily focusing on two popular standard datasets: CIFAR-10 and CIFAR-100. We considered different scenarios with varying quantities of labeled data. We paid particular attention to RSCC's performance with very limited labeled data. To achieve this, we randomly selected 1, 2, 4, 8 and 25 labeled data from each class in CIFAR-10 for our experiments. In CIFAR-100, we also randomly selected 4, 25 and 100 labeled data from each class to evaluate the performance of RSCC.

4.2 Implementation Details

To ensure a fair comparison, we used the same backbone network structure in our experiments. Specifically, we employed WRN28-2 and WRN28-8 as base networks for CIFAR-10 and CIFAR-100, respectively. We used standard stochastic gradient descent (SGD) with a momentum of 0.9 as the optimizer. The cosine learning rate decay schedule was defined as $\eta = \eta_0 \cos(\frac{7\pi w}{W})$, where η_0 was the initial learning rate (set to 0.03), W (set to 2^{20})was the total training steps, and w was the current training step. For RSCC's hyperparameters, we set $n = 0.9999$, $\gamma = 0.2$, $\tau = 0.94$, $\lambda = 1$, $\mu = 7$ and $G = -4$. For CIFAR-10, we used $\alpha = 0.1$ and $\beta = 1.0$, while for CIFAR-100, $\alpha = 6.0$ and $\beta = 0.1$. Regarding data augmentation, during the contrastive learning pretraining phase, RSCC employed random cropping, random color distortion, random horizontal flipping, random grayscale transformation, and blur augmentation. During the robust SSL fine-tuning phase, for weak augmentation, RSCC used random cropping, translation, and random horizontal flipping. For strong augmentation, RSCC utilized RandAugment.

4.3 Main Results

We used a range of state-of-the-art and classic SSL methods for comparison. These algorithms include MixMatch [32], FixMatch [7], CoMatch [5], SimMatch [35], FreeMatch [36], and FullMatch [37]. These algorithms represent the cutting-edge research in the field, and this comparison helps us evaluate the performance of the RSCC under various conditions more comprehensively. It allows us to get a deep understanding of the advantages and limitations of RSCC and provides a strong foundation for future research.

Table 1. Comparison of Method Accuracy (%) on CIFAR-10

Dataset	CIFAR-10				
#Label	10	20	40	80	250
MixMatch	34.24 ± 7.06	46.84 ± 10.63	59.90 ± 11.76	80.79 ± 1.28	88.97 ± 0.85
FixMatch	75.21 ± 7.65	82.32 ± 9.77	86.12 ± 3.53	92.06 ± 0.88	94.90 ± 0.67
CoMatch	76.33 ± 6.45	87.67 ± 8.47	93.09 ± 1.39	93.97 ± 0.62	95.69 ± 0.33
SimMatch	77.36 ± 7.93	87.39 ± 6.29	94.40 ± 1.37	94.67 ± 0.68	95.16 ± 0.39
FreeMatch	91.93 ± 4.24	93.32 ± 2.52	**95.10 ± 0.44**	95.63 ± 0.03	**95.90 ± 0.02**
FullMatch	85.83 ± 6.25	91.73 ± 4.92	94.11 ± 1.01	94.85 ± 0.46	95.36 ± 0.12
RSCC (Ours)	**92.16 ± 4.83**	**93.84 ± 2.69**	95.03 ± 0.67	**95.68 ± 0.59**	95.81 ± 0.24

Table 2. Comparison of Method Accuracy (%) on CIFAR-100

Dataset	CIFAR-100		
#Label	400	2500	10000
MixMatch	32.41 ± 0.66	60.24 ± 0.48	72.22 ± 0.29
FixMatch	57.45 ± 1.76	71.97 ± 0.16	77.80 ± 0.12
CoMatch	59.98 ± 1.11	72.99 ± 0.21	78.17 ± 0.23
SimMatch	62.19 ± 2.21	**74.93 ± 0.32**	**79.42 ± 0.21**
FreeMatch	62.02 ± 0.42	73.53 ± 0.20	78.32 ± 0.03
FullMatch	59.42 ± 1.40	73.06 ± 0.40	78.56 ± 0.10
RSCC (Ours)	**62.86 ± 1.03**	74.82 ± 0.33	79.11 ± 0.38

Table 1 and Table 2 displays the accuracy comparison between RSCC and other baseline methods on CIFAR-10 and CIFAR-100. It is evident that RSCC consistently achieves top-tier performance across various levels of labeled data. Remarkably, RSCC excels in scenarios with extremely limited labels, achieving an accuracy of 91.98% on CIFAR-10 with only 10 labels, 93.84% with 20 labels, and 62.86% on CIFAR-100 with 400 labels, surpassing prior methods. RSCC also maintains accuracy very close to the best-performing methods on CIFAR-10 and CIFAR-100 across other label quantities. These results underscore the effectiveness and state-of-the-art performance of RSCC.

4.4 Ablation Study

Module Effectiveness Analysis. RSCC consists of three parts: a pre-training method based on contrastive learning, a SSL method based on consistency regularization, and SL. The SSL based on consistency regularization serves as the foundation. To assess the impact of the contrastive learning pre-training and SL components on the enhancement of the consistency regularization-based SSL method in RSCC, extensive ablation experiments were conducted on CIFAR-10 with different label quantities (10, 20, 40, 80, 250). The experimental results are shown in Table 3. As indicated in Table 3, representation learning through pre-training based on contrastive learning significantly improved the accuracy of the consistency regularization-based SSL method. Moreover, the enhancement became more noticeable with fewer labels (accuracy increased by 6.66% with 40 labels, 9.82% with 20 labels, and 15.52% with 10 labels). This demonstrates that pre-training based on contrastive learning is crucial for RSCC to achieve high accuracy with extremely limited labels. The introduction of SL alleviated the impact of pseudo-label confirmation bias in SSL, leading to a noticeable improvement in the accuracy of the consistency regularization-based SSL method (7.94%, 4.11%, 3.59%, 1.05%, and 0.35% improvements in the cases of 10, 20, 40, 80, and 250 labels, respectively). This experimental result confirms the effectiveness of the contrastive learning pre-training method and SL in RSCC, leading to outstanding performance on CIFAR-10.

Table 3. Impact of Different Modules on the Accuracy of RSCC in CIFAR-10 with Varying Numbers of Labels

Label Amount	10	20	40	80	250
Consistency Regularization	75.21 ± 7.65	82.32 ± 9.77	86.12 ± 3.53	92.06 ± 0.88	94.90 ± 0.67
Consistency Regularization and Contrastive Learning	90.73 ± 6.36	92.14 ± 8.92	92.78 ± 5.94	93.84 ± 1.03	95.41 ± 0.91
Consistency Regularization and SL	83.15 ± 7.28	86.43 ± 8.52	89.71 ± 4.19	93.11 ± 0.95	95.25 ± 0.83
RSCC	92.16 ± 4.83	93.84 ± 2.69	95.03 ± 0.67	95.68 ± 0.59	95.81 ± 0.24

Analysis of the Relationship Between SL and Confidence Thresholds. We studied robust SSL with noisy pseudo-labels, specifically by introducing the SL to tolerate some level of noisy pseudo-labels for better model generalization. Therefore, we proposed a hypothesis: after introducing SL into the model, can we relax the threshold filtering conditions, i.e., decrease the value of the confidence threshold? (The original FixMatch paper's experimental results suggest that, for 250 labeled data in CIFAR-10, the model has the lowest error rate with a confidence threshold of 0.95). To test this hypothesis, we evaluated RSCC's accuracy under different confidence threshold values, still using 250 labeled data on CIFAR-10. The results, as shown in Fig. 2, reveal that the RSCC without SL achieved the highest accuracy with a confidence threshold of 0.94, while the RSCC with SL achieved the highest accuracy with a confidence threshold of 0.93.

This indicates that the introduction of SL allows for a slight reduction in the confidence threshold. However, this effect is not particularly significant, and relatively higher confidence thresholds still play a important role. The introduction of SL can improve the performance of the model by tolerating fewer noisy pseudo-labels based on high confidence thresholds.

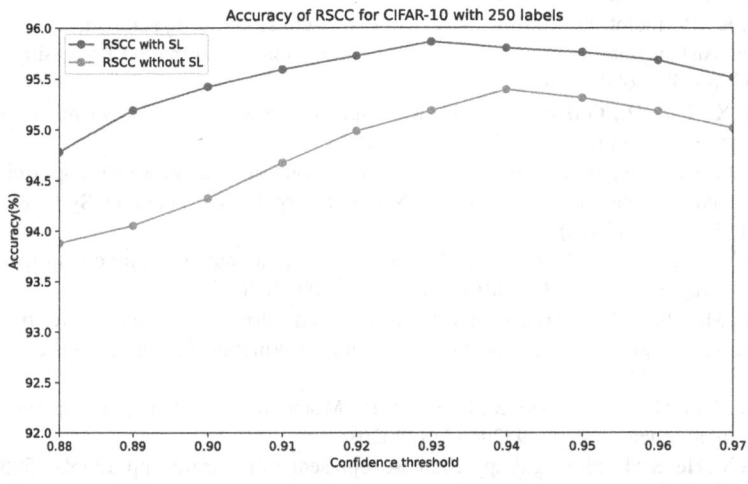

Fig. 2. Results of RSCC accuracy on CIFAR-10 with 250 labeled data as the confidence threshold varies.

5 Conclusion

We have introduced a novel SSL method, RSCC, it combines the powerful representation learning capability of contrastive learning with data augmentation-based consistency regularization and introduces symmetric cross-entropy learning to mitigate the negative impact of low-quality pseudo-labels on model performance. Extensive experiments on CIFAR-10 and CIFAR-100 with varying numbers of labels demonstrate the outstanding performance of RSCC, particularly in scenarios with few labeled samples. Thus, RSCC provides an effective solution for SSL tasks, and we hope the "unsupervised pre-training plus task-specific SSL fine-tuning" learning paradigm proposed in this paper become a trend in future deep learning development.

References

1. Grandvalet, Y., Bengio, Y.: Semi-supervised learning by entropy minimization. In: Advances in Neural Information Processing Systems, vol. 17 (2004)
2. Jing, L., Tian, Y.: Self-supervised visual feature learning with deep neural networks: a survey. IEEE Trans. Pattern Anal. **43**, 4037–4058 (2020)
3. Zoph, B., Ghiasi, G., Lin, T., et al.: Rethinking pre-training and self-training. In: Advances in Neural Information Processing Systems, vol. 33, pp. 3833–3845 (2020)

4. Bachman, P., Hjelm, R.D., Buchwalter, W.: Learning representations by maximizing mutual information across views. In: Advances in Neural Information Processing Systems, vol. 32 (2019)
5. Li, J., Xiong, C., Hoi, S.C.: Comatch: semi-supervised learning with contrastive graph regularization, 9475–9484 (2021)
6. Li, C., Tu, R., Zhang, H.: Reimplementation of FixMatch and investigation on noisy (pseudo) labels and confirmation errors of FixMatch (2020)
7. Sohn, K., Berthelot, D., Carlini, N., et al.: Fixmatch: simplifying semi-supervised learning with consistency and confidence. In: Advances in Neural Information Processing Systems, vol. 33, pp. 596–608 (2020)
8. Chen, X., Fan, H., Girshick, R., He, K.: Improved baselines with momentum contrastive learning. arXiv preprint arXiv:2003.04297 (2020)
9. Grill, J., Strub, F., Altché, F., et al.: Bootstrap your own latent-a new approach to self-supervised learning. In: Advances in Neural Information Processing Systems, vol. 33, pp. 21271–21284 (2020)
10. Chen, T., Kornblith, S., Norouzi, M., Hinton, G.: A simple framework for contrastive learning of visual representations. In: PMLR, pp. 1597–1607 (2020)
11. Caron, M., Misra, I., Mairal, J., et al.: Unsupervised learning of visual features by contrasting cluster assignments. In: Advances in Neural Information Processing Systems, vol. 33, pp. 9912–9924 (2020)
12. He, K., Fan, H., Wu, Y., Xie, S., Girshick, R.: Momentum contrast for unsupervised visual representation learning, pp. 9729–9738 (2020)
13. Chen, X., He, K.: Exploring simple Siamese representation learning, pp. 15750–15758 (2021)
14. Pham, H., Dai, Z., Xie, Q., Le, Q.V.: Meta pseudo labels, pp. 11557–11568 (2021)
15. Xie, Q., Luong, M., Hovy, E., Le, Q.V.: Self-training with noisy student improves imagenet classification, pp. 10687–10698 (2020)
16. Yalniz, I.Z., Jégou, H., Chen, K., Paluri, M., Mahajan, D.: Billion-scale semi-supervised learning for image classification. arXiv preprint arXiv:1905.00546 (2019)
17. French, G., Mackiewicz, M., Fisher, M.: Self-ensembling for visual domain adaptation. arXiv preprint arXiv:1706.05208 (2017)
18. Pseudo-Label DL. The simple and efficient semi-supervised learning method for deep neural networks, pp. 1–6 (2013)
19. Arazo, E., Ortego, D., Albert, P., O'Connor, N.E., McGuinness, K.: Pseudo-labeling and confirmation bias in deep semi-supervised learning. In: IEEE, pp. 1–8 (2020)
20. Shi, W., Gong, Y., Ding, C., Tao, Z.M., Zheng, N.: Transductive semi-supervised deep learning using min-max features, pp. 299–315 (2018)
21. Laine, S., Aila, T.: Temporal ensembling for semi-supervised learning. arXiv preprint arXiv: 1610.02242 (2016)
22. Tarvainen, A., Valpola, H.: Mean teachers are better role models: weight-averaged consistency targets improve semi-supervised deep learning results. In: Advances in Neural Information Processing Systems, vol. 30 (2017)
23. Ke, Z., Wang, D, Yan, Q., Ren, J., Lau, R.W.: Dual student: breaking the limits of the teacher in semi-supervised learning, pp. 6728–6736 (2019)
24. Miyato, T., Maeda, S., Koyama, M., Ishii, S.: Virtual adversarial training: a regularization method for supervised and semi-supervised learning. IEEE Trans. Pattern Anal. **41**, 1979–1993 (2018)
25. Xie, Q., Dai, Z., Hovy, E., Luong, T., Le, Q.: Unsupervised data augmentation for consistency training. In: Advances in Neural Information Processing Systems, vol. 33, pp. 6256–6268 (2020)

26. Rizve, M.N., Duarte, K., Rawat, Y.S., Shah, M.: In defense of pseudo-labeling: an uncertainty-aware pseudo-label selection framework for semi-supervised learning. arXiv preprint arXiv: 2101.06329 (2021)
27. Xu, Y., Shang, L., Ye, J., et al.: Dash: semi-supervised learning with dynamic thresholding. In: PMLR, pp. 11525–11536 (2021)
28. Zhang, B., Wang, Y., Hou, W., et al.: Flexmatch: boosting semi-supervised learning with curriculum pseudo labeling. In: Advances in Neural Information Processing Systems, vol. 34, pp. 18408–18419 (2021)
29. Kim, J., Min, Y., Kim, D., et al.: Conmatch: semi-supervised learning with confidence-guided consistency regularization. In: Avidan, S., Brostow, G., Cissé, M., Farinella, G.M., Hassner, T. (eds). ECCV 2022. LNCS, vol. 13690, pp. 674–690. Springer, Cham (2022). https://doi. org/10.1007/978-3-031-20056-4_39
30. Ju, W., Luo, X., Qu, M., et al.: TGNN: a joint semi-supervised framework for graph-level classification. arXiv preprint arXiv:2304.11688 (2023)
31. Jiang, B., Chen, S., Wang, B., Luo, B.: MGLNN: semi-supervised learning via multiple graph cooperative learning neural networks. Neural Netw. **153**, 204–214 (2022)
32. Berthelot, D., Carlini, N., Goodfellow, I., et al.: Mixmatch: a holistic approach to semi-supervised learning. In: Advances in Neural Information Processing Systems, vol. 32 (2019)
33. Li, J., Socher, R., Hoi, S.C.: Dividemix: learning with noisy labels as semi-supervised learning. arXiv preprint arXiv:2002.07394 (2020)
34. Wang, Y., Ma, X., Chen, Z., et al.: Symmetric cross entropy for robust learning with noisy labels, pp. 322–330 (2019)
35. Zheng, M., You, S., Huang, L., et al.: SimMatchV2: semi-supervised learning with graph consistency, pp. 16432–16442 (2023)
36. Wang, Y., Chen, H., Heng, Q., et al.: Freematch: self-adaptive thresholding for semi-supervised learning. arXiv preprint arXiv:2205.07246 (2022)
37. Chen, Y., Tan, X., Zhao, B., et al.: Boosting semi-supervised learning by exploiting all unlabeled data, pp. 7548–7557 (2023)

Enhanced Prototypical Network for Few-Shot Named Entity Recognition

Tianwen Huang[1] , Mingming Zhang[2], Kai Liu[2], Xianhui Li[3,4],
and Yongli Wang[1(✉)]

[1] Nanjing University of Science and Technology, Nanjing 210000, China
yongliwang@njust.edu.cn
[2] State Grid Jiangsu Electric Power Co., Ltd., Information and Communication Branch,
Nanjing 210024, China
[3] Jiangsu Ruizhong Data Co., Ltd., Nanjing 210012, China
[4] State Grid Electric Power Research Institute Co., Ltd., Nanjing 211106, China

Abstract. Few-shot Named Entity Recognition (NER) aims to perform NER tasks with limited data. Prototypical networks have already demonstrated excellent performance in addressing issues in few-shot settings. Traditional prototypical networks face issues such as inaccurate representation of O-class entities and poor distribution of entity prototypes. This paper presents an Enhanced Prototypical Network (EPN) for few-shot Named Entity Recognition, mainly including O-class subclass clustering module and feature consistency evaluation module. EPN divides O-class entities into multiple subclasses, each with its own prototype, to represent O-class entities more accurately. It also optimizes the distribution of entities in the feature space for better entity classification. Experimental results show that the model proposed in this paper improves performance on mainstream datasets.

Keywords: Few-shot NER · Enhanced Prototypical Network · Subclass Clustering · Feature Consistency

1 Introduction

Named Entity Recognition (NER) plays a pivotal role in Natural Language Processing (NLP), aiming to identify and classify named entities within text. Previous research has predominantly employed deep neural network methods [1–4], which have exhibited remarkable performance in NER tasks. However, these methods heavily rely on extensive annotated corpora. In certain scenarios, acquiring a substantial amount of annotated data can be quite challenging. Even with access to large-scale annotated corpora, the issue of rare entities is inevitable. Therefore, addressing NER in the context of few-shot learning has become a focal point of current research. The few-shot learning task is delineated as follows (Fig. 1).

For few-shot Named Entity Recognition, an effective approach is metric learning [5]. Metric learning aims to learn similarity or distance metrics between data points to better

H. Jin et al. (Eds.): IAIC 2023, CCIS 2058, pp. 156–170, 2024.
https://doi.org/10.1007/978-981-97-1277-9_12

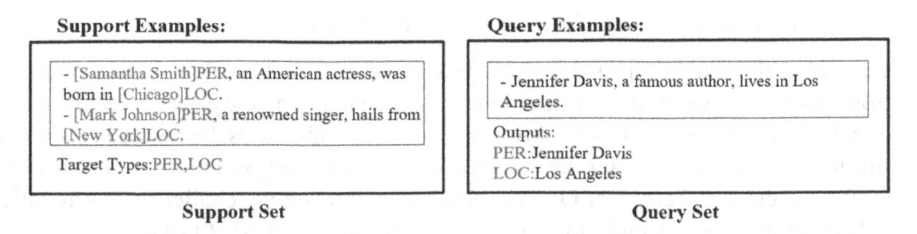

Fig. 1. A 2-way 2-shot NER task. Identify with fewer samples

handle tasks such as classification, clustering, dimensionality reduction, and information retrieval. Prototypical networks are a form of metric learning, and their primary concept involves learning a prototype for each class. This prototype represents the average embedding for each entity category, and classification is performed by calculating the distance or similarity between the sample and the prototype. When prototypical networks are applied to few-shot recognition, they can effectively utilize structural information of common entities to label uncommon entities.

Currently, there are some issues with few-shot Named Entity Recognition based on prototypical networks. Traditional prototypical networks typically calculate class prototypes by averaging the embeddings of entities in the support set for each class. However, Yoon and Hou et al. [6] demonstrated that this prototype representation leads to tightly clustered entities in the embedding space, resulting in classification errors. Additionally, there may be cases where entities of the same class have distant distances in the feature space, which can also introduce errors. For O-class, the entities of O-class are semantically noisy. Creating a single O-class prototype to represent all O-class entities is considered inappropriate. Unlike entity classes, O-class entities lack uniformity in semantics. Entities labeled as O in a dataset may correspond to different semantic spaces. This is because different categories from various datasets may be labeled as O-class in the current dataset. Entities from these different semantic spaces should not be uniformly represented by a single prototype.

To address the aforementioned issues, this paper proposes an Enhanced-Prototype-Network (EPN) approach for few-shot Named Entity Recognition. The main idea of this method is as follows:

First, this paper introduces O-class subclass clustering module, which represents O-class prototypes through the learning of multiple subclasses. Drawing inspiration from unsupervised clustering, O-class entities are clustered based on their latent semantics, dividing them into multiple subclasses, each of which can be represented by a prototype. Subclass prototypes are then jointly trained with target prototypes. Furthermore, this paper introduces feature consistency evaluation module to optimize the distribution of entities in the feature space. Through backpropagation of the loss function, network parameters are updated, making homogeneous entities with similar features tend to cluster together, while entities with weaker feature relationships disperse. This way, the model presented in this paper can comprehensively explore the similarity and dissimilarity features among entities.

The primary contributions of this paper are as follows:

1. The paper introduces an O-class subclass clustering module. Drawing inspiration from unsupervised clustering principles, by leveraging semantic information from the target classes, it clusters O-class entities into multiple subclasses. Each subclass shares similar semantic features and can be represented by a prototype. This innovation helps to more accurately represent O-class entities, addressing the challenge of semantic polysemy, and consequently improving the performance of NER tasks.
2. The paper introduces a feature consistency evaluation module. By introducing a new learning task, the model is trained to optimize the distribution of entities in the feature space. This enables the model to identify shared features among entities of the same class as well as distinguish the differing features among entities of different classes, thereby enhancing its feature extraction capability.
3. Performance improvements were achieved across multiple datasets. Through multiple tests on various datasets, the method proposed in this paper consistently demonstrated significant improvements over mainstream methods according to standard evaluation metrics. This indicates the robustness and effectiveness of the method across different domains and task settings.

2 Related Work

2.1 Few-Shot Learning

Few-shot learning is a crucial subfield in machine learning and deep learning, focusing on how to construct and train models when training data is extremely limited. These models need to learn from a very small number of training examples to be able to perform accurate classification or inference on unseen data [8]. In few-shot learning, some typical algorithms include: Optimization-based methods [9]. These methods typically attempt to adapt to small sample data by optimizing a loss function to achieve better generalization performance; Metric learning-based methods [10, 11]: These methods concentrate on learning similarity or distance metrics between data points to better handle small-sample classification problems; Reinforcement learning-based methods [12]: These methods use reinforcement learning techniques to train models to perform well in small-sample scenarios. Few-shot learning initially found wide applications in the field of image classification [13] and has gradually gained prominence in natural language processing. For example, it has been used in tasks such as relation extraction, text classification [14, 15], question-answering systems [16], providing effective approaches for handling few-shot problems in these tasks.

2.2 Prototypical Network

The prototypical network [10] studied in this paper belongs to the category of metric learning-based methods and is widely applicable to tasks such as image classification, text classification [15], and named entity recognition. The training approach of the prototypical network involves sampling a series of tasks from the source domain to simulate tasks in the target domain. Each task consists of a support set and a query set, and the model is trained on each task. The prototypical network makes predictions based on the similarity between the support set and the query set. The core idea of a prototypical network is to achieve classification tasks by learning prototypes for categories, where it is expected that prototypes for entities of the same class are clustered together in the semantic space. However, in named entity recognition, there is a special class of entities known as O-class entities, which themselves encompass multiple semantics. Therefore, they should not be clustered together in the semantic space, contradicting the principles of a prototypical network. Some research attempts to address this issue. Fritzler et al. [17] introduced a hyperparameter b_0 as a parameter for distance similarity and optimized it during training, effectively treating O-class as a single entity. However, this approach treats O-class as a whole, ignoring the potential presence of multiple semantics within O-class entities. On the other hand, Yang et al. [18] argued that O-class entities do not have a unified semantics. They adopted a nearest-neighbor approach, using the entity closest in distance to the query set as the representation of the query sample. This method, however, overlooks potential similar semantic information within O-class entities. These studies suggest that when dealing with O-class entities, careful consideration of how to handle their diversity and semantic richness is necessary to fully leverage this information to improve model performance.

2.3 Few-Shot NER

For the task of few-shot named entity recognition, there has been substantial prior work. Fritzler et al. [17, 19] applied the prototypical network ProtoBERT to the problem of few-shot named entity recognition and conducted an evaluation of various network architectures. Their findings demonstrate that the prototypical network performs exceptionally well in the context of few-shot scenarios. Yang et al. [18] introduced the NNShot model, which utilizes a pre-trained named entity recognition model from the source domain as a feature extractor to generate context-relevant representations for each word. During testing, it assigns labels to test words by calculating their similarity. They also proposed StructShot, which includes structured decoding steps, using an abstract transition matrix to capture dependencies between labels. Finn et al. [9] introduced the model-agnostic meta-learning (MAML) approach, which is a meta-learning algorithm. Its core concept involves training the initial parameters of a model to enable it to rapidly adapt to new tasks with minimal data and gradient updates. Katiyar et al. [20] proposed the ContaiNER model, which adopts a contrastive learning approach. It utilizes Gaussian embeddings and a loss based on KL divergence to enhance the model's ability to distinguish between different entities, aiming to maximize the distance between entities in the embedding space. Ding et al. [21] introduced the Big Prototypes model, which utilizes tensor fields to model prototypes from a geometric perspective. It abstracts category information using hyperspheres in the feature space and involves two sets of learnable parameters. Wang et al. [22] proposed a distillation method for few-shot named entity recognition tasks, utilizing large-scale unlabeled datasets. Cui et al. [23] introduced an approach that effectively leverages the BART [24] model for guidance. Hou et al. [25] extended Tap-Net [6] and proposed a Conditional Random Field (CRF) network suitable for few-shot scenarios, presenting the L-TapNet+CDT model. Ma et al. [26] and Athiwaratkun et al.

[27] formulated the few-shot named entity recognition task as a machine comprehension problem and a generation problem, respectively.

3 Model

The proposed Enhanced Prototype Network in this paper consists of two key modules: O-class subclass clustering module and feature consistency evaluation module. The O-class subclass clustering module aims to achieve a finer-grained division of O-class entities, categorizing them into multiple subclasses, each with similar semantic features and represented by a single prototype. The feature consistency evaluation module primarily adjusts the distribution of embedded features through heterogeneous and homogeneous evaluation. Heterogeneous evaluation helps increase the dissimilarity between different prototypes, making them more dispersed in the feature space. Homogeneous evaluation aids in bringing the embeddings of the same-class entities closer to each other, making them more compact in the feature space. The model architecture diagram is shown in Fig. 2.

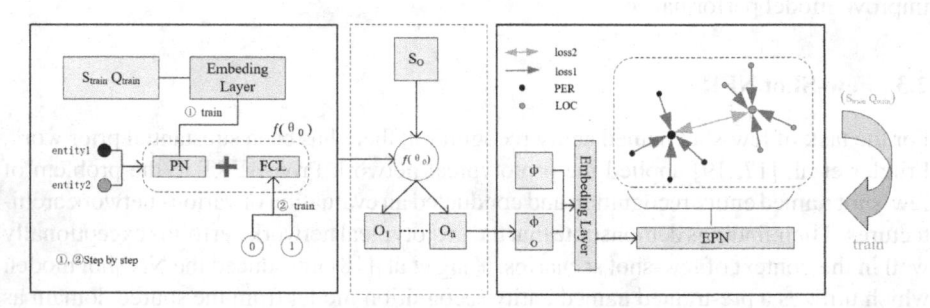

Fig. 2. The architecture diagram of our model EPN.

3.1 Prototypical Network

For a task $T = \{S, Q\}$, where S represents the support set and Q represents the query set. Define the set of target classes as C. For any class $c_k \in C$, all entities of class c_k are extracted from the support set S, resulting in $S_k = \{x_1^k, x_2^k, x_3^k, \ldots, x_n^k\}$. The prototype of c_k is represented as the mean of the embedding vectors of all entities x_i^k in c_k.

$$\phi_k = \frac{1}{|S_k|} \sum_{x_i^k \in S_k} f_\theta\left(x_i^k\right) \tag{1}$$

where f_θ represents the embedding layer function, and in this paper, BERT [28] model is used as the embedding layer. $|S_k|$ denotes the number of entities in class c_k. Then the collection of prototypes for all entity classes is represented as:

$$\phi = \{\phi_1, \phi_2, \ldots, \phi_n\} \tag{2}$$

For the non-entity class O, we define $S_o = \{x_1^o, x_2^o, x_3^o \ldots, x_n^o\}$. Following the approach of Yang et al. [18], in this paper, each O-class entity's contextual embedding is individually represented as a prototype:

$$\phi_0 = \{\phi_{o_1}, \phi_{o_2}, \ldots, \phi_{o_n}\} = \{f_\theta(x_1^o), f_\theta(x_2^o), \ldots, f_\theta(x_n^o)\} \tag{3}$$

ϕ_0 represents the collection of all O-class prototypes.

The collection of all prototypes is represented as:

$$\Phi = \phi \cup \phi_0 = \{\phi_0, \phi_1, \phi_2, \ldots, \phi_n\} \tag{4}$$

For any sample q_i in the query set Q, first normalize the vector representation using $L2$ normalization, and then compute its Euclidean distance $d(q_i, \phi_j)$ to all prototypes in Φ as follows:

$$d(q_i, \phi_j) = \||f_\theta(q_i) - \phi_j\|_2^2 \tag{5}$$

For O-class entities, we select the minimum distance from all O-class prototypes. In other words, for each O-class entity, calculate the distance to all O-class prototypes and choose the prototype with the shortest distance as the representative prototype for that entity:

$$d(q_i, \phi_0) = \operatorname*{argmin}_{\phi_j \in \phi_0} d(q_i, \phi_j) \tag{6}$$

Finally, we use the softmax function to convert the calculated similarities into a probability distribution. The softmax function will assign probabilities to each prototype based on the similarities, and these probabilities will represent the likelihood of a given entity belonging to a particular class or category. This way, we can obtain a probability distribution over the prototypes for each query sample:

$$\hat{y}_{i,j} = \frac{\exp^{-d(q_i, \phi_j)}}{\sum_{k=0}^{n} \exp^{-d(q_i, \phi_k)}} \tag{7}$$

Finally, the loss function is defined as follows:

$$l(\theta_1) = -\sum_{k=1}^{|Q|} \sum_{j=0}^{n} y_{i,j}^k \log \hat{y}_{ij}^k \tag{8}$$

where, y_i is the one-hot vector of the true label for q_i.

3.2 O-Class Subclass Clustering Module

This paper draws inspiration from unsupervised clustering concepts. In the context of the same task, there will be similarity in feature space and interaction between categories. In natural language processing tasks, textual data is often represented as vectors in the feature space. These vectors capture the semantic and syntactic features of the text. Different entity categories may share similar features or have mutual influences in the feature space. In the same task, if categories A and B exhibit similarity or interaction in the feature space, then when supervised training is conducted on category A, it adjusts the model's feature representation to better capture the features of category A. This adjustment not only affects category A but also impacts category B in the feature space. This is because the model attempts to find a feature representation in the feature space that suits category A. This naturally leads to the clustering of task- or semantically-related category B in the feature space, facilitating the better capture of its relevant features. In essence, this phenomenon is driven by the similarity in the feature space and the semantic associations between categories. This similarity and semantic correlation result in the clustering of unlabeled categories in the feature space, as described in the literature [7], which provides strong support for this phenomenon.

Hence, assuming the mapping function defined in the process in Sect. 3.1 is denoted as $\tilde{f}_\theta$, while clustering the target classes, entities within class O will also cluster based on the task or semantic information. Therefore, we design a fully connected neural network layer to determine whether two entities cluster. Since class O and the target class coexist in the same feature space, we train the neural network using the target class and then apply it to class O. The fully connected network is defined as follows:

$$P_{i,j} = W\left(\tilde{f}_\theta(x_i), \tilde{f}_\theta(x_j)\right) + b \tag{9}$$

where W represents the weight matrix, b is the bias matrix, and $P_{i,j}$ represents the judgment result. 1 indicates belonging to the same class, while 0 indicates not belonging to the same class. Training the network on the target classes enables it to possess the ability to judge whether two prototypes belong to the same class.

Based on the value of $y_{i,j}$, the loss is defined as follows:

$$l = \begin{cases} -\log(P_{i,j}), y_{i,j} = 1 \\ -\log(1 - P_{i,j}), y_{i,j} = 0 \end{cases} \tag{10}$$

Finally, the loss function of the model is defined as:

$$l(\theta) = \frac{1}{N^2} \sum_{i=1}^{N} \sum_{j=1}^{N} (-(y_{i,j} * \log(P_{i,j}) + (1 - y_{i,j}) * \log(1 - P_{i,j}))) \tag{11}$$

N represents the total number of entities in the target class, and $y_{i,j}$ indicates whether x_i and x_j belong to the same prototype, where 1 represents they are the same, and 0 represents they are not.

Through the aforementioned training process, the model obtains p_θ. Using p_θ, all samples in S_o are clustered, resulting in multiple subclasses, following the logic below:

Algorithm	Subclass Clustering Algorithm
Input	Collection of O-class entity samples: $S = \{x_1, \ldots, x_T\}$
	Network p_θ, Number of samples: T
Steps	*Begin*
1	Initialize the prototype list **prototypes** with a length of 0.
2	*for* $i \leftarrow \{1, \ldots, T\}$ *do*
	// Calculate the number of prototypes.
3	$len \leftarrow LEN(prototypes)$
4	$flag \leftarrow$ *False* // Flag variable.
5	*for* $j \leftarrow \{1, \ldots, len\}$
	// AVG represents taking the average of all entities
6	$\overline{prototypes_j} \leftarrow AVG(prototypes_j)$
7	*if* $p_\theta(x_i, \overline{prototypes_j})$
8	$prototypes_j \leftarrow prototypes_j \cup x_i$
9	$flag \leftarrow$ *True*
10	*break*
11	*end if*
12	*end for*
13	*if not* $flag$
14	Create a new prototype $\widehat{prototype}$ containing the sample x_i.
15	$prototypes \leftarrow prototypes \cup \widehat{prototype}$ //Merge it into the prototype collection.
17	*end if*
18	*end for*
Output	The results of the subclass Clustering: *prototypes*

Finally, we obtain $n\prime = len(prototypes)$ subclass, and each subclass is represented by the average embedding of all entities within it. These represent the prototypes of the O-class subclasses. We use the new prototypes $\phi_0\prime$ to replace the original ϕ_0.

$$\phi_0' = \{\phi_{o1}', \phi_{o2}', \phi_{o3}', \ldots, \phi_{o_{n'}}'\} \tag{12}$$

The final definition for the new prototype set is given by:

$$\Phi = \phi_0' \cup \phi == \{\phi_0', \phi_1, \phi_2, \ldots, \phi_n\} \tag{13}$$

Subsequently, the training of the prototypical network is performed using the new prototype set.

3.3 Feature Consistency Evaluation Module

Based on the prototypical network in the first section and the new prototype set obtained in the second section, two new models were designed to ensure the reasonable distribution of entities in the embedded space.

Heterogeneity Evaluation Module. The core idea behind the heterogeneity evaluation module is to maximize the distance between different prototypes, ensuring that the prototypical network effectively captures the differing features among prototypes. This is achieved by calculating the distances between different prototypes and subsequently computing the average of these distances. The model's objective is to maximize this average distance, defined as the loss function $l(\theta_2)$, to reasonably spread out the distances between prototypes of different classes. The calculation process is as follows:

$$\overline{d}_\phi = \frac{\sum_{i=1}^{N}\sum_{j=1}^{n}(\phi_i - \phi_j)^2}{n^2} \tag{14}$$

$$l(\theta_2) = -d_\phi \tag{15}$$

Here, N represents the total number of prototypes. The goal of $l(\theta_2)$ is to maximize the distance between different prototypes.

Homogeneity Evaluation Module. The core idea of the homogeneity assessment module is to encourage the distribution of entities from the same class to be more compact in the feature space. This objective is achieved by defining a loss function, denoted as $l(\theta_3)$, that optimizes the distance between entities within the same prototype, making them cluster more effectively.

The way to achieve this goal is by first calculating the average distance between entities. Then, the training objective of the model is to minimize this average distance, ensuring that entities of the same class aggregate more closely around the prototypes. This process helps improve the model's ability to recognize shared features among entities of the same class. The calculation formula is as follows:

$$\overline{d} = \frac{\sum_{i=1}^{N}\sum_{j=0}^{K}\sum_{m=0}^{n}\left(f'_\theta\left(x_j^i\right) - f'_\theta(x_m^i)\right)^2}{(N*K)^2} \tag{16}$$

$$l(\theta_3) = \overline{d} \tag{17}$$

K represents the number of entities owned by a prototype, N represents the total number of all prototypes, and x_j^i represents the jth entity of the ith prototype.

Finally, based on the training objectives of the prototypical network and the two new models proposed in this paper, the loss function of the EPN is defined as follows:

$$l(\theta) = l(\theta_1) + l(\theta_2) + l(\theta_3) \tag{18}$$

4 Experiments

4.1 Dataset

This paper evaluates the performance of the EPN (Enhanced Prototypical Network) model on two datasets: the FEW-NERD dataset [19] and the Cross-Dataset [25] dataset (Table 1).

The FEW-NERD dataset is constructed from Wikipedia and includes approximately 18,000 sentences, 66 fine-grained entity types, and 8 coarse-grained entity types. These

Table 1. Dataset Introduction

Dataset	Domain	Sentences	Classes
FEW-NERD	Wikipedia	188.2k	66
CoNLL03	News	20.7k	4
GUM	Wiki	3.5k	11
WNUT	Social	5.6k	6
OntoNotes	Mixed	159.6k	18

coarse-grained entity types encompass various fine-grained entity types. The FEW-NERD dataset uses two segmentation criteria: INTRA and INTER. The INTRA criterion divides the data based on coarse-grained entity types to ensure that entities of different coarse-grained entity types are present in the training, validation, and test sets. The INTER criterion, on the other hand, segments the data based on fine-grained entity types to ensure that fine-grained entity types do not overlap in different tasks but share the same coarse-grained entity type.

In this study, we adopted the N-way K-shot task setting, which is a common approach in few-shot learning. For each task in the training set, we selected N target classes, and for each target class, we sampled K samples to create a support set for the task $S_{train} = \{x_i, y_i\}_{i=1}^{N*K}$. Additionally, we sampled $K\prime$ examples to form a query set for the task $Q_{train} = \{x_j, y_j\}_{j=1}^{N*K\prime}$. Since contextual information is crucial for sequence labeling tasks, sentence-level sampling was necessary. However, ensuring a strict K-shot setup during the sampling process was challenging because a single sentence might contain multiple entities. Therefore, we adopted the N-way K–2K shot setting as suggested by Ding et al. [19].

The Cross-Dataset comprises datasets from four different domains, including CoNLL03 (news), GUM (Wikipedia), WNUT (social), and OntoNotes (mixed).

4.2 Evaluation Metrics

In accordance with the standards defined in relevant literature, Named Entity Recognition (NER) performance is generally evaluated using three metrics: Precision (P), Recall (R), and F1-score (F1). Their calculation formulas are as follows:

$$P = \frac{correct}{correct + incorrect} \tag{19}$$

$$R = \frac{correct}{correct + missing} \tag{20}$$

$$F = \frac{2PR}{P + R} \tag{21}$$

In the formulas, *correct* represents the number of entities recognized and correctly classified by the model. *incorrect* refers to the number of entities recognized by the model but classified incorrectly. *missing* Indicates the number of entities that the model should have recognized but failed to do so, meaning it represents entities that were missed or not recognized by the model.

Compared to precision (P) and recall (R), the F1 score is a more comprehensive metric for assessing the model's performance. Therefore, in this study, the F1 score is adopted as the primary evaluation metric.

4.3 Baselines

The paper compares the proposed model with several mainstream few-shot named entity recognition methods. For the FEW-NERD dataset, the model is compared to BERT, Big Prototypes, NNShot, StructShot, and CONTAINER. For the Cross-Dataset dataset, the model is compared to SimBERT [25], TransferBERT [25], Matching Network (MN) [31], ProBERT [17], and L-TapNet+CDT [25].

4.4 Parameter Settings

The paper follows the method proposed by Hou et al. [25] and uses BERT-base as the model's embedding layer with a dimension of 768. The maximum sequence length is set to 64. The optimization is performed using the Adam optimizer [30]. A dropout rate of 0.1 is applied, and the learning rate is set to 1e-5. The training batch size is set to 4.

In the data set configuration, following Ding et al.'s [25] publicly available data, the sampling for the FEW-NERD dataset is as follows: It contains a total of 20,000 training tasks, 1,000 validation tasks, and 5,000 testing tasks. Each task follows the N-way K–2K shot setting. For the CrossDataset experiments, two of the datasets are randomly selected for training data, while the remaining two are used with one as the validation set and the other as the test set. In the construction of tasks, different strategies are applied for 5-shot and 1-shot. In the 5-shot setting, it includes 200 training tasks, 100 validation tasks, and 100 testing tasks. In the 1-shot setting, the OntoNotes dataset contains 400 training tasks, 200 validation tasks, and 100 testing tasks. The others all consist of 200 training tasks, 100 validation tasks, and 200 testing tasks.

4.5 Results and Analysis

Based on the experimental results in Table 2, it was observed that ProtoBERT performs poorly under the 1–2 shot benchmark but shows better performance under the 5–10 shot benchmark. This is mainly because in cases with relatively fewer samples, ProtoBERT cannot accurately represent semantic features through averaging embeddings. Big Prototypes are able to learn the general distribution patterns of different entity types, thus significantly improving performance even in cases where very little information is shared between the training and testing sets. Under the INTRA benchmark, in the 1–2 shot setting, the ContainNER model outperforms this study's model. This may be due to the fact that ContainNER, through contrastive learning and large-scale training in the source domain, has acquired substantial domain knowledge, which guides the identification of domain-specific data. In addition, under the INTER benchmark, StructShot performs best in the 1–2 shot setting, thanks to its nearest neighbor classifier, which better explores the connections between similar entities.

In summary, the model studied in this paper performs exceptionally well in most aspects, particularly on benchmarks with larger sample sizes. Our designed O-class subclass clustering module, along with feature consistency evaluation module, makes more comprehensive use of semantic information within O-class entities, enhancing the model's ability to recognize entity similarity and dissimilarity, leading to significant improvements. However, it's worth noting that the performance improvement of this study's model is relatively smaller in the 1–2 shot setting. This might be attributed to two main factors. First, this paper defines more training tasks requiring more sample support. In situations with relatively few samples, noise interference might become more pronounced, making it difficult for the tasks to be adequately learned, thus resulting in average performance. Second, even though prototype enhancement is employed, this study uses an averaging embedding approach to construct prototypes. This means that, even with prototype enhancement, when the sample size is too small, the constructed prototypes might have issues with semantic accuracy, making it challenging for enhanced prototypes to demonstrate their advantages. This is also a limitation of prototypical networks in the few-shot NER (Table 3).

Table 2. F1 scores on Few-NERD for both inter and intra settings. The best results are in bold.

Models	INTRA				INTER			
	1–2 shot		5–10 shot		1–2 shot		5–10 shot	
	5-way	10-way	5-way	10-way	5-way	10-way	5-way	10-way
ProtoBERT	23.45	19.76	41.93	34.61	44.44	39.09	58.80	53.97
Big Prototypes	32.26	24.02	50.88	42.46	52.09	44.26	65.59	60.73
NNShot	31.01	21.88	35.74	27.67	54.29	46.98	50.56	50.00
StructShot	35.92	25.38	38.83	26.39	**57.33**	**49.46**	57.16	49.39
ContainNER	**40.34**	**33.84**	53.70	47.49	55.95	48.35	61.83	57.12
EPN	37.23	24.71	**58.77**	**48.21**	56.21	46.99	**69.12**	**62.03**

In summary, the model EPN presented in this paper performs admirably in most cases, especially on benchmarks with a larger number of samples, demonstrating significant performance improvements. This indicates the effectiveness of the model's design and approach for the task of few-shot named entity recognition.

Table 3. F1 scores on CrossDataset. The best results are in bold.

Models	1-shot				5-shot			
	CoNLL03	GUM	WNUT	OntoNotes	CoNLL03	GUM	WNUT	OntoNotes
SimBERT	19.22	6.91	5.18	13.99	32.01	10.63	8.20	21.14
TransferBERT	4.75	0.57	2.71	3.46	15.36	3.62	11.08	**35.49**
MN	19.50	4.73	17.23	15.06	19.58	5.58	6.61	8.08
ProtoBERT	32.49	3.89	10.68	6.67	50.06	9.54	17.26	13.59
L-TapNet+CDT	**44.30**	**12.04**	**20.80**	**15.17**	45.34	11.65	23.30	20.95
EPN	22.87	9.23	12.24	13.25	**54.82**	**20.12**	**25.76**	26.54

5 Conclusion

This paper introduces an Enhanced Prototypical Network, which includes O-class sub-class clustering module and feature consistency evaluation module. This approach helps address the polysemy of O-class entities and avoids the issues associated with using a unified prototype to represent O-class entities. Additionally, the new model improves the classification performance of the prototypical network and enhances the model's feature extraction capabilities. The testing on the FEW-NERD and Cross-Dataset datasets demonstrates that the model proposed in this paper achieves satisfactory results.

While this study has made significant progress in most cases, there is room for improvement, particularly in the 1–2 shot scenarios where performance gains are relatively small. Future research could explore more sophisticated data augmentation techniques and more powerful model designs to enhance the model's performance on this challenging benchmark. Additionally, we encourage the application of the prototypical network innovations from this study to other natural language processing domains, such as relation extraction and text classification, to validate their generality and practicality. This will contribute to further advancing research and applications in the fields of few-shot learning and NLP.

Acknowledgements. This article has been awarded by the State Grid Corporation Technology Guide Project (5700-202218185A-1-1-ZN).

References

1. Huang, Z., Xu, W., Yu, K.: Bidirectional LSTM-CRF models for sequence tagging. arXiv preprint arXiv:1508.01991 (2015)
2. Ma, X., Hovy, E.: End-to-end sequence labeling via bi-directional LSTM-CNNS-CRF. arXiv preprint arXiv:1603.01354 (2016)
3. Lample, G., Ballesteros, M., Subramanian, S., et al.: Neural architectures for named entity recognition. arXiv preprint arXiv:1603.01360 (2016)
4. Peters, M.E., Ammar, W., Bhagavatula, C., et al.: Semi-supervised sequence tagging with bidirectional language models. arXiv preprint arXiv:1705.00108 (2017)
5. Bellet, A., Habrard, A., Sebban, M.: A survey on metric learning for feature vectors and structured data. arXiv preprint arXiv:1306.6709 (2013)

6. Yoon, S.W., Seo, J., Moon, J.: Tapnet: neural network augmented with task-adaptive projection for few-shot learning. In: International Conference on Machine Learning, pp. 7115–7123. PMLR (2019)
7. Koch, G., Zemel, R., Salakhutdinov, R.: Siamese neural networks for one-shot image recognition. In: ICML Deep Learning Workshop, vol. 2, no. 1 (2015)
8. Huisman, M., Van Rijn, J.N., Plaat, A.: A survey of deep meta-learning. Artif. Intell. Rev. **54**(6), 4483–4541 (2021)
9. Finn, C., Abbeel, P., Levine, S.: Model-agnostic meta-learning for fast adaptation of deep networks. In: International Conference on Machine Learning, pp. 1126–1135. PMLR (2017)
10. Snell, J., Swersky, K., Zemel, R.: Prototypical networks for few-shot learning. In: Advances in Neural Information Processing Systems, vol. 30 (2017)
11. Liu, K., Liu, W., Ma, H., et al.: Generalized zero-shot learning for action recognition with web-scale video data. World Wide Web **22**(2), 807–824 (2019)
12. Wei, J., Zou, K.: EDA: easy data augmentation techniques for boosting performance on text classification tasks. arXiv preprint arXiv:1901.11196 (2019)
13. Miller, E.G., Matsakis, N.E., Viola, P.A.: Learning from one example through shared densities on transforms. In: Proceedings IEEE Conference on Computer Vision and Pattern Recognition. CVPR 2000 (Cat. No. PR00662), vol. 1, pp. 464–471. IEEE (2000)
14. Geng, R., Li, B., Li, Y., et al.: Dynamic memory induction networks for few-shot text classification. arXiv preprint arXiv:2005.05727 (2020)
15. Sun, S., Sun, Q., Zhou, K., et al.: Hierarchical attention prototypical networks for few-shot text classification. In: Proceedings of the 2019 Conference on Empirical Methods in Natural Language Processing and the 9th International Joint Conference on Natural Language Processing (EMNLP-IJCNLP), pp. 476–485 (2019)
16. Wang, J., Wang, C., Qiu, M., et al.: KECP: knowledge enhanced contrastive prompting for few-shot extractive question answering. arXiv preprint arXiv:2205.03071 (2022)
17. Fritzler, A., Logacheva, V., Kretov, M.: Few-shot classification in named entity recognition task. In: Proceedings of the 34th ACM/SIGAPP Symposium on Applied Computing, pp. 993–1000 (2019)
18. Yang, Y., Katiyar, A.: Simple and effective few-shot named entity recognition with structured nearest neighbor learning. arXiv preprint arXiv:2010.02405 (2020)
19. Ding, N., Xu, G., Chen, Y., et al.: Few-NERD: a few-shot named entity recognition dataset. arXiv preprint arXiv:2105.07464 (2021)
20. Das, S.S.S., Katiyar, A., Passonneau, R.J., et al.: CONTaiNER: few-shot named entity recognition via contrastive learning. arXiv preprint arXiv:2109.07589 (2021)
21. Ding, N., Chen, Y., Cui, G., et al.: Few-shot classification with hypersphere modeling of prototypes. arXiv preprint arXiv:2211.05319 (2022)
22. Wang, Y., Mukherjee, S., Chu, H., et al.: Meta self-training for few-shot neural sequence labelling. In: Proceedings of the 27th ACM SIGKDD Conference on Knowledge Discovery and Data Mining, pp. 1737–1747 (2021)
23. Cui, L., Wu, Y., Liu, J., et al.: Template-based named entity recognition using BART. arXiv preprintarXiv:2106.01760 (2021)
24. Lewis, M., Liu, Y., Goyal, N., et al.: BART: denoising sequence-to-sequence pre-training for natural language generation, translation, and comprehension. arXiv preprint arXiv:1910.13461 (2019)
25. Hou, Y., Che, W., Lai, Y., et al.: Few-shot slot tagging with collapsed dependency transfer and label-enhanced task-adaptive projection network. arXiv preprint arXiv:2006.05702 (2020)
26. Ma, J., Yan, Z., Li, C., et al.: Frustratingly simple few-shot slot tagging. In: Findings of the Association for Computational Linguistics: ACL-IJCNLP 2021, pp. 1028–1033 (2021)
27. Athiwaratkun, B., Santos, C.N., Krone, J., et al.: Augmented natural language for generative sequence labeling. arXiv preprint arXiv:2009.13272 (2020)

28. Devlin, J., Chang, M.W., Lee, K., et al.: BERT: pre-training of deep bidirectional transformers for language understanding. arXiv preprint arXiv:1810.04805 (2018)
29. Ming, H., Yang, J., Jiang, L., et al.: Few-shot nested named entity recognition. arXiv preprint arXiv:2212.00953 (2022)
30. Kingma, D.P., Ba, J.: Adam: a method for stochastic optimization. arXiv preprint arXiv:1412.6980 (2014)
31. Vinyals, O., Blundell, C., Lillicrap, T., et al.: Matching networks for one shot learning. In: Advances in Neural Information Processing Systems, vol. 29 (2016)

Regularized DNN Based Adaptive Compensation Algorithm for Gateway Power Meter in Ultra-High Voltage Substations

Yonggui Wang, Xiao Feng[✉], Wenjing Li, and Tengfei Dong

State Grid Information Telecommunication Group Co., LTD., Beijing 100000, China
19801116781@163.com

Abstract. The development of the power Internet of Things (IoT) has made the data of ultra-high voltage substation metering critical information sources for supporting power energy scheduling and market transactions. However, as key data acquisition devices, the operation performance of ultra-high voltage substation energy meters exhibits instability under different environmental conditions, and even small metering errors can lead to significant discrepancies in settlement. This issue directly impacts the overall safety and stability of the power system. Therefore, this paper considers the influence of environmental factors such as temperature, humidity, and air quality on substation energy metering and proposes an adaptive compensation algorithm for ultra-high voltage substation energy meter based on regularized deep neural networks (DNN) to enhance metering accuracy. Finally, the effectiveness of this algorithm is validated through simulation experiments.

Keywords: Ultra-high Voltage · Artificial Intelligence · DNN · Power IoT

1 Introduction

With the continuous progress and development of society, the scale of the electrical power system has been expanding. The data from ultra-high voltage substation energy meters contain valuable information and have become a crucial basis for supporting power scheduling, electricity market transactions, and economic and technical assessments in the power industry. Furthermore, through in-depth analysis of the metering data at ultra-high voltage substations, we can gain insights into customers' electricity consumption patterns, thus better understanding their behavior. The metering accuracy of ultra-high voltage substation energy meters is easily affected by environmental factors, and even small metering errors can lead to significant discrepancies in settlement. This not only poses challenges to the safety and stability of energy meters but also has far-reaching implications for the fairness and equity of metering transaction settlements. Therefore, achieving high-precision metering with ultra-high voltage substation energy meters has become an indispensable cornerstone for the stable operation of the power system.

However, various factors in the usage environment can directly affect the normal operation of gateway energy meters. For example, environmental temperature changes

H. Jin et al. (Eds.): IAIC 2023, CCIS 2058, pp. 171–181, 2024.
https://doi.org/10.1007/978-981-97-1277-9_13

can cause changes in the performance parameters of the internal components of the gate meter, such as thermoelectric potential, resistance, capacitance, inductance, etc., thereby causing changes in the frequency, amplitude, phase, and other characteristics of the circuit operation. In high humidity environments, the characteristics of the insulation layer material of the equipment will decrease, and the inductance and resistivity of the circuit will also undergo different changes. In high-altitude areas, the air pressure is relatively low and the air density is relatively thin, which may affect the electrical and mechanical properties of circuit components. If the air quality is poor, the dust in the environment will gradually accumulate on various precision components of the gateway meter over time. When the amount of dust particles attached to the components reaches a certain level, the operating efficiency of the meter will be affected, and even there will be poor heat dissipation or flashover. All of the above factors can cause significant measurement errors and even permanent damage to the gateway meter, bringing economic and safety risks to society.

Therefore, high-precision measurement of gateway meters has become an important link in power grid management. This paper proposes an adaptive compensation algorithm for gateway energy meters under the influence of multi-dimensional environmental factors based on regularized deep neural network (DNN). The algorithm takes into account a wide range of environmental factors, including temperature, humidity, and air quality, and performs adaptive compensation for the measurement of gateway energy meters. The objective is to rectify measurement errors attributed to environmental factors and enhance the overall safety and stability of the power system.

2 Related Works

Research has shown that for every 10 degrees Celsius increase in temperature, the failure rate of electronic devices will double, while the reliability of systems composed of electronic components will correspondingly decrease by 50% [1]. Especially with the rapid development of modern computer and communication technology, gateway energy meters have gradually developed into a new generation of energy meters with intelligent chips such as central processing unit (CPU) as the core, and the impact of environmental factors is also increasing [2].

Therefore, many scholars have conducted in-depth research on the impact of temperature and other factors on various aspects of smart energy meters. Journal articles [3] proposed a method for measuring errors in electric energy meters due to changes in environmental temperature. One is to consider temperature and phase changes, and then use compensation devices for error correction. The other is to use the meter itself to adjust the device and consider temperature coefficient to eliminate errors. Journal articles [4] conducted experiments on the accuracy of single-phase smart energy meters based on the working principle of the meters. Journal articles [5] proposed two temperature compensation methods to improve the measurement accuracy of three-phase energy meters effectively, in response to the problem of accumulated system errors caused by temperature drift of various components due to temperature differences when operating in different temperature ranges. Journal articles [6] proposed a consistency analysis and optimization scheme for the measurement accuracy of intelligent energy meters, and took single-phase intelligent energy meters as the research object, considering temperature, tolerance design, temperature compensation, and other aspects. Using Matlab simulink software for simulation modeling, various uncertain factors were considered, thus comprehensively improving the measurement accuracy of intelligent energy meters. At the same time, Yuan Ruiming and Ye Xuerong also used simulation software for modeling and established temperature mapping relationships between the measurement chip and key components in the measurement circuit that affect the measurement results based on the obtained temperature data [7, 8].

However, most of the existing work only considers the impact of temperature on electricity meters, without comprehensively considering the impact of various environmental factors, which has certain limitations in practical applications. Therefore, this article comprehensively considers various environmental factors and proposes a new compensation algorithm based on regularized DNN to further reduce the measurement error of gateway energy meters.

3   System Model

3.1   Working Principle of Gateway Power Meter

The basic working principle of a standard electric energy meter is shown in Fig. 1. Firstly, the load current and load voltage are sampled in real-time, and then the converted current and voltage are sent to the digital processing system (DSP) through their respective analog-to-digital converters. The DSP digital processing system can perform real-time U and I calculations to calculate the actual parameters. Then, the control frequency generator generates high electrical energy pulses, which is obtained by a frequency divider for use in calibrating standard meters. The high and low frequency electrical energy pulses are output to the interface, and the microcontroller and DSP communicate in real-time. The power consumption is displayed through the liquid crystal display (LCD), thereby displaying the output [9].

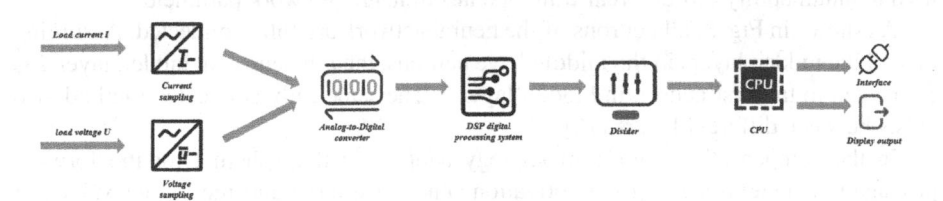

Fig. 1. Working principle of gateway power meter.

3.2   Temperature Compensation Principle

The schematic diagram of adaptive compensation for the gateway electricity meter using a neural network is shown in Fig. 2. The model comprises two components: the electricity meter model and the neural network compensation model.

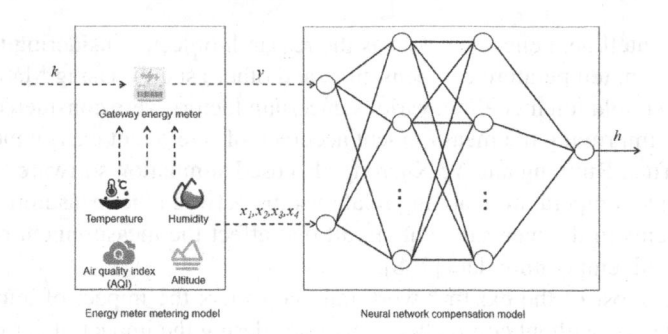

Fig. 2. Schematic diagram of adaptive compensation.

where h represents the output energy value compensated by the neural network, and the mathematical model of the gateway energy meter is

$$y = f(k, x_1, x_2, x_3, x_4) \tag{1}$$

where k is the target electrical energy parameter to be measured, x_1 is the temperature parameter of the environmental influencing factor, x_2 is the humidity parameter, x_3 is the altitude parameter, x_4 is the environmental quality parameter, and y is the energy meter output.

4   Regularized DNN Based Adaptive Compensation Algorithm

DNN is a multi-layer perceptron architecture that is essentially the same as Multi Layer Perception (MLP). When classifying from different depths of a deep neural network (DNN), the entire network structure can be categorized into three layers: the input layer, hidden layer, and output layer. Utilizing neural networks for regression analysis enables the attainment of high accuracy in intricate nonlinear mappings and ensures high maintainability through real-time updates of neural network parameters.

As shown in Fig. 3, all neurons of the neural network are fully connected. Assuming that all the hidden layers in the middle have neurons, the i th neuron in hidden layer 2 is connected to the first neuron in hidden layer 3. The same fully connected method also exists between different layers [10].

In the compensation algorithm strategy adopted in this scheme, for the forward propagation stage, assuming the activation function is $\tau(k)$ and the output values of each layer are χ_n^i, the input parameters and environmental parameters of the electricity meter that need to be compensated are all variables of the input layer. Using the same approach as the perceptron, the output of the first layer is calculated and gradually transmitted to the output layer, which is the proposed forward propagation algorithm for compensation based on DNN.

In the forward propagation stage, we read electrical energy value K and vectorize it before it is transmitted to the network. The parameter feature extraction is performed in the hidden layer to obtain the corresponding matrix W and bias vector b. Assuming the existence of n identical neurons in the hidden layer, and matrix passed in from input layer

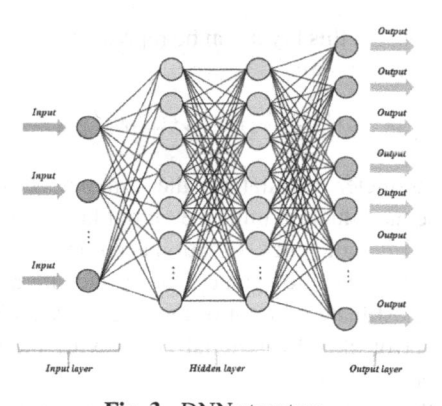

Fig. 3. DNN structure.

will be passed to each neuron. After completing the calculation of the corresponding matrix A and bias vector b, it will be output to the next layer, where x_1 respresent temperature, x_2 respresent humidity, x_3 respresent altitude, x_4 respresent environmental quality.

Starting from the input layer, the output of the hidden layers $\chi_1^2, \chi_2^2, ..., \chi_{n-1}^2, \chi_n^2$ can be expressed as:

$$
\begin{cases}
\chi_1^2 = \tau(k_1^2) = \tau(\omega_{11}^2 x_1 + \omega_{12}^2 x_2 + \omega_{13}^2 x_3 + \omega_{14}^2 x_4 + \omega_{15}^2 K + b_1^2) \\
\chi_2^2 = \tau(k_2^2) = \tau(\omega_{21}^2 x_1 + \omega_{22}^2 x_2 + \omega_{23}^2 x_3 + \omega_{24}^2 x_4 + \omega_{25}^2 K + b_2^2) \\
... \\
\chi_{n-1}^2 = \tau(k_{n-1}^2) = \tau(\omega_{(n-1)1}^2 x_1 + \omega_{(n-1)2}^2 x_2 + \omega_{(n-1)3}^2 x_3 + \omega_{(n-1)4}^2 x_4 + \omega_{(n-1)5}^2 K + b_{(n-1)}^2) \\
\chi_n^2 = \tau(k_n^2) = \tau(\omega_{n1}^2 x_1 + \omega_{n2}^2 x_2 + \omega_{n3}^2 x_3 + \omega_{n4}^2 x_4 + \omega_{n5}^2 K + b_n^2)
\end{cases}
\tag{2}
$$

where $\tau(k)$ is the activation function, and ω_{mn}^i is the weight between the n th neuron in the $i-1$ th layer and the m th neuron in the i th layer, b_n^i is the bias corresponding to the n th neuron in the i th layer.

Similarly, the output of the second hidden layer χ_1^3 can be expressed as:

$$
\chi_1^3 = \tau(k_1^3) = \tau(\omega_{11}^3 \chi_1^2 + \omega_{12}^3 \chi_2^2 + ... + \omega_{1(n-1)}^3 \chi_{(n-1)}^2 + \omega_{1n}^3 \chi_n^2 + b_1^3)
\tag{3}
$$

where the algebraic representation $\omega_{11}^3 \chi_1^2 + \omega_{12}^3 \chi_2^2 + ... + \omega_{1(n-1)}^3 \chi_{(n-1)}^2 + \omega_{1n}^3 \chi_n^2$ within the activation function can be written as $\sum_{p=1}^{n} \omega_{1p}^3 \chi_p^2$.

Therefore, for the j th neuron in the i th layer, the output is:

$$
\chi_j^i = \tau(k_j^i) = \tau(\sum_{p=1}^{n} \omega_{jp}^i \chi_p^{i-1} + b_j^i)
\tag{4}
$$

In the process of forward propagation, the previous algebraic output methods can be induced and represented through matrix form. If the i th and $i+1$ th layer networks have m and n nodes respectively, the weight ω of the $i+1$ th layer network can be represented by matrix $E_{n \times m}^{i+1}$, and the bias of this layer can be represented by b^{i+1}. The relevant node output of the $i + 1$ layer network are

$$\chi^{i+1} = \tau(E^{i+1}\chi^i + b^{i+1}) \tag{5}$$

After completing the parameter calculation, the parameters can be passed from the current layer to the subsequent hidden until the output layer is reached.

To address the issue of insufficient estimation accuracy, it is necessary to introduce a loss function for the backpropagation stage. During each iteration of training, the loss function can compare the estimated output of the original sample with the real dataset, and minimize expected error, thereby gradually achieving the adjustment of network parameters for each layer.

The introduction of the loss function requires problem constraints to achieve the goal of optimizing prediction. The commonly used solution is the Gradient Descent method [11, 12], which solves in the direction of the fastest gradient descent to obtain a local minimum or global minimum. We select the Cross entropy function to derive the loss, with minimizing the loss function as our goal.

$$L(E, b, x, y) = -\frac{1}{n} \sum_{\chi^{i-1}} [yLn\chi^i + (1-y)Ln(1-\chi^i)] \tag{6}$$

Once the loss function is established, the parameters E, b for each layer can be iteratively optimized using the gradient descent method.

For the output layer, the loss function is:

$$L(E, b, x, y) = -\frac{1}{n} \sum_{j=1}^{n} [yLn(\tau(E^i\chi^{i-1} + b^i)) + (1-y)Ln(1 - \tau(E^i\chi^{i-1} + b^i))] \tag{7}$$

where y represents expected output, and χ^i is actual output of the i th layer neuron. For derivation of (7), we can obtain the Eq. (8):

$$\frac{\partial L(E, b, x, y)}{\partial \chi^i} = -\frac{1}{n} \sum_{\chi^{i-1}} [\frac{1}{\tau(k^{i-1})(1 - \tau(k^{i-1}))} y - \frac{1}{1 - \tau(k^{i-1})}] \tag{8}$$

In addition, using θ^i to represent the gradient of layer i, i can be obtained by taking the derivative of the linear relationship $k = \sum \omega_i x_i + b$ through the cross entropy loss function:

$$\theta^i = -\frac{1}{n} \sum_{\chi^{i-1}} [y - \chi^i] = \chi^i - y \tag{9}$$

The gradient of E^i and b^i can be obtained:

$$\theta_E^i = \frac{\partial L(E, b, x, y)}{\partial A^i} = \frac{\partial L(E, b, x, y)}{\partial k^i} \frac{\partial k^i}{\partial A^i} = \theta^i \frac{\partial k^i}{\partial A^i} = (\chi^i - y)\chi^{i-1}$$

$$\theta_b^i = \frac{\partial L(E, b, x, y)}{\partial b^i} = \frac{\partial L(E, b, x, y)}{\partial k^i} \frac{\partial k^i}{\partial b^i} = \theta^i \frac{\partial k^i}{\partial b^i} = \chi^i - y \tag{10}$$

In the equation, the gradients of E^i and b^i are represented by θ_E^i and θ_b^i, respectively, and the update of parameters can be expressed as:

$$E^i = E^i - \alpha \frac{\partial L(E, b, x, y)}{\partial A^i} = E^i - \alpha(\chi^i - y)\chi^{i-1}$$
$$b^i = b^i - \alpha \frac{\partial L(E, b, x, y)}{\partial b^i} = b^i - \alpha(\chi^i - y)$$

(11)

where α is the preset learning rate, gradually diminishes as the increase of training iterations. During update process of the above parameters, they are affected by error $\chi^i - y$, so the correlation weight and bias update speed show a positive correlation with the size of the error.

After solving the gradient of the output layer, it is necessary to gradually solve the gradients of each layer in the previous hidden layer, in order to achieve reverse parameter updates for the entire network. If the inactive output of layer s is k^s, then its gradient is expressed as θ^s:

$$\theta^s = \frac{\partial L(E, b, x, y)}{\partial k^s} = (\frac{\partial k^i}{\partial k^{i-1}} \frac{\partial k^{i-1}}{\partial k^{i-2}} \cdots \frac{\partial k^{s+1}}{\partial k^s})^T \frac{\partial L(E, b, x, y)}{\partial k^i}$$

(12)

According to the above derivation, the linear representation of layer $s + 1$ is:

$$k^{s+1} = E^{s+1}\chi^s + b^{s+1} = E^{s+1}\tau(k^s) + b^{s+1}$$

(13)

Therefore, for the derivation of (3), it can be obtained that:

$$\frac{\partial k^{s+1}}{\partial k^s} = E^{s+1} diag(\tau(k^s))$$

(14)

Since the backpropagation process updates layer by layer from the output layer forward, when solving the gradient of layer s, the gradient θ^{s+1} of layer $s+1$ has been obtained, and the gradient θ^s of layer s shown in Eqs. (3–14) can be obtained:

$$\theta^s = \frac{\partial L(E, b, x, y)}{\partial k^s} = (\frac{\partial k^{s+1}}{\partial k^s})^T \frac{\partial L(E, b, x, y)}{\partial k^{s+1}} = (\frac{\partial k^{s+1}}{\partial k^s})^T \theta^{s+1}$$

(15)

By introducing (13) into (14) and combining (10) and (13), the E^s and b^s gradients of layer s can be obtained:

$$E_E^s = \frac{\partial L(E, b, x, y)}{\partial k^s} \frac{\partial k^s}{\partial E^s} = (\chi^{s-1}E^{s+1} diag(\tau(k^s)))^T \theta^{s+1}$$
$$E_b^s = \frac{\partial L(E, b, x, y)}{\partial k^s} \frac{\partial k^s}{\partial b^s} = (E^{s+1} diag(\tau(k^s)))^T \theta^{s+1}$$

(16)

After deduction, the algorithm recurses based on gradients from the output layer, and can update the linear relationships and biases of each layer layer layer by layer. After parameter updates, more effective prediction accuracy can be achieved.

Furthermore, to solve the common overfitting question in neural network training, we introduce the L2 regularization scheme, which adjusts the weight parameters without affecting the update of bias b. According to the loss function mentioned earlier, there is an Eq. (17)

$$L(E, b) = -\frac{1}{n}[yLny^p + (1 - y)Ln(1 - y^p)] \tag{17}$$

After L2 regularization, the loss function is updated to:

$$L(E, b) = -\frac{1}{n}[yLny^p + (1 - y)Ln(1 - y^p)] - \frac{\lambda}{n}\sum_{i=2}^{l}\omega_i \tag{18}$$

where λ is a hyperparameter that needs to be adjusted according to the actual situation when updating the weight parameters and bias, while ω_i is the vector of the weights of each layer.

Combined with Eq. (11), E^i is updated to:

$$E^i = E^i - \alpha(\chi^i - y)\chi^{i-1} - \alpha\lambda E^i \tag{19}$$

At this point, we have achieved adaptive compensation based on regularized DNN under the influence of multi-dimensional environmental variables.

5   Simulation and Analysis

We have developed a software validation platform to assess the algorithm's efficacy. Select 5000 test data under different environmental factors, with 80% being the training dataset and 20% being the validation dataset. Each data contains five elements x_1, x_2, x_3, x_4, k, including temperature variation range of -45 °C - + 65 °C, humidity variation range of 20–90% RH, altitude variation range of 50–2000 m, and air quality index (AQI) variation range of 0–350. In this article, we set the incentive function $\tau(x) = sigmoid(x)$ with a learning rate of 0.001.

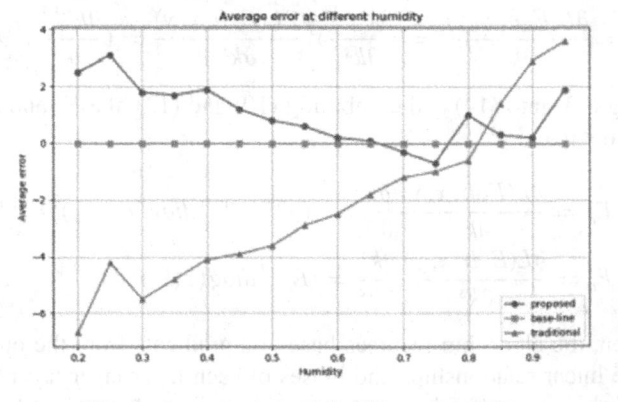

Fig. 4. Average error with different humidity.

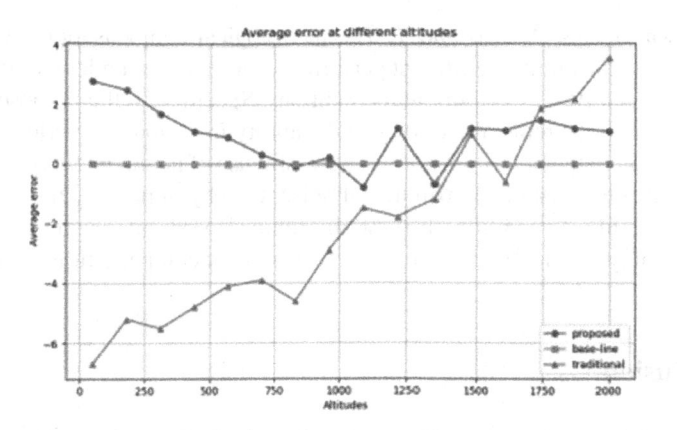

Fig. 5. Average error with different altitudes.

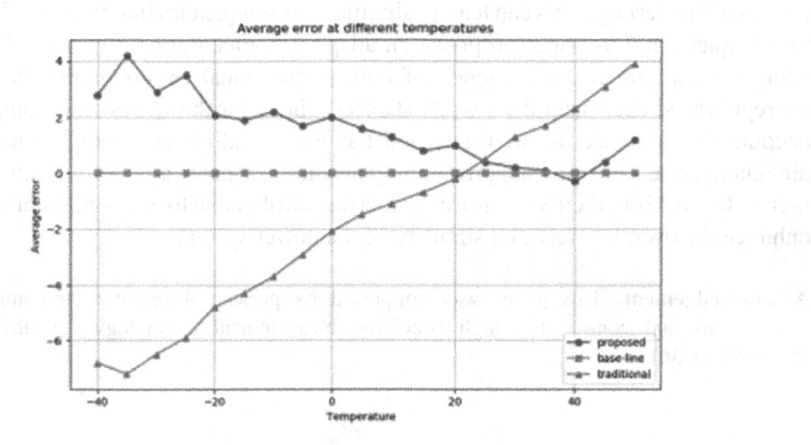

Fig. 6. Average error with different temperatures.

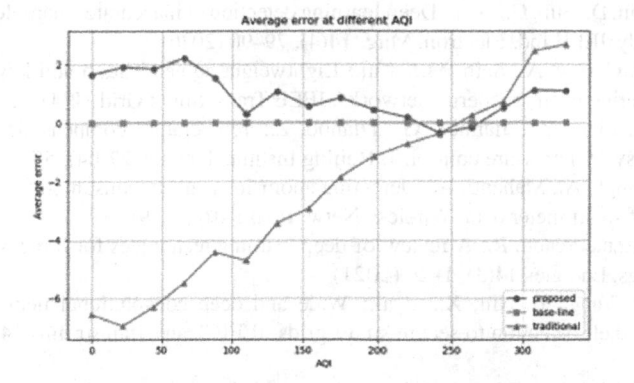

Fig. 7. Average error with different environmental AQI.

AS shown in Figs. 4, 5, 6 and 7, The metrological compensation algorithm proposed in this paper exhibits significant performance advantages under humidity, temperature, altitude, and AQI environmental conditions. Specifically, the proposed algorithm achieves higher measurement accuracy and stability in various scenarios. This advantage is partly due to the new algorithm's successful use of neural networks to reduce the impact of sensor errors and environmental noise, thereby improving the consistency of measurement results. In addition, the new algorithm also exhibits excellent reproducibility and can stably provide high-precision measurement compensation results under all conditions.

6   Conclusion

As key data acquisition devices, the operation performance of ultra-high voltage substation energy meters exhibits instability under different environmental conditions, and even small metering errors can lead to significant discrepancies in settlement. In response to this question, this paper proposes an adaptive compensation algorithm for gateway energy meters under the influence of multi-dimensional environmental factors based on regularized deep neural network (DNN). The algorithm takes into comprehensive account the influence of environmental factors, including temperature, humidity, and air quality, and performs adaptive compensation for gateway energy meter measurements. Its goal is to rectify measurement errors attributable to environmental factors and enhance the overall safety and stability of the power system.

Acknowledgement. This paper was supported by project of research and application of autonomous and controllable high precision measurement technology for uhv substation (546809220036).

References

1. Liu, M., Liu, D., Sun, G., et al.: Deep learning detection of inaccurate smart electricity meters: a case study. IEEE Ind. Electron. Mag. **14**(4), 79–90 (2020)
2. Kumar, P., Gurtov, A., Sain, M., et al.: Lightweight authentication and key agreement for smart metering in smart energy networks. IEEE Trans. Smart Grid **10**(4), 4349–4359 (2019)
3. Huien, G., Liang, C., Jianhua, G., Dianbo, Z.: Temperature compensation algorithm for hydraulic system pressure control. J. Beijing Institut. Technol. **27**(04), 556–563 (2018)
4. Sial, A., Singh, A., Mahanti, A.: Detecting anomalous energy consumption using contextual analysis of smart meter data. Wireless Netw. **1**(5), 1–16 (2019)
5. Runge, J., Zmeureanu, R.: A review of deep learning techniques for forecasting energy use in buildings. Energies **14**(3), 1–26 (2021)
6. Zheng, Z., Yang, Y., Niu, X., et al.: Wide and deep convolutional neural networks for electricity-theft detection to secure smart grids. IEEE Trans. Industr. Inf. **14**(4), 1606–1615 (2018)
7. Aslam, Z.: A combined deep learning and ensemble learning methodology to avoid electricity theft in smart grids. Energies **13**(21), 5599 (2020)
8. Zhang, J., Wang, Z., Meng, J., et al.: Boosting positive and unlabeled learning for anomaly detection with multi-features. IEEE Trans. Multimedia **21**(5), 1332–1344 (2019)

9. Xie, L., Yiqun, Z., Xuejun, Y.: Design and application of standard energy meters based on class 0.02 installation. Electr. Technol. Softw. Eng. **1**(02), 240 (2016)
10. Xia, L., Li, Z.: A new method of abnormal behavior detection using LSTM network with temporal attention mechanism. J. Supercomput. **77**(4), 3223–3241 (2021)
11. Himeur, Y., Alsalemi, A., Bensaali, F., et al.: A novel approach for detecting anomalous energy consumption based on micro-moments and deep neural networks. Cogn. Comput. **12**(6), 1381–1401 (2020)
12. Punmiya, R., Choe, S.: Energy theft detection using gradient boosting theft detector with feature engineering-based pre-processing. IEEE Trans. Smart Grid **10**(2), 2326–2329 (2019)

Dynamic Occlusion Expression Recognition Based on Improved GAN

Minchao Liang[1] , Mingming Zhang[2], Kai Liu[2], Xianhui Li[3,4], and Yongli Wang[1(✉)]

[1] Nanjing University of Science and Technology, 210000 NanJing, China
liangminchao1998@163.com, yongliwang@njust.edu.cn
[2] Information Communication Branc,State Grid Jiangsu Electric Power Co., Ltd., 210024 Nanjing, China
[3] Jiangsu Ruizhong Data Co., Ltd., 210012 Nanjing, China
[4] State Grid Electric Power Research Institute Co., Ltd., 211106 Nanjing, China

Abstract. In order to address the issue of local occlusion in practical dynamic expression recognition, this paper first introduces a facial restoration network that combines Vision Transformer (ViT) and GAN. This network can accurately identify missing facial features and perform detailed and efficient restoration. Secondly, for the task of expression recognition, a more robust dynamic expression recognition network is trained by cascading ViT with a Two-Stream CNN, effectively leveraging ViT's feature extraction capability and the Two-Stream CNN's ability to acquire spatio-temporal features. Finally, by combining these two networks, we can efficiently recognize dynamically occluded expressions. A multitude of experiments demonstrate that the facial image restoration network trained on the CelebA and VGG Face2 datasets outperforms other networks in handling small and medium occlusions. Expression recognition experiments on AFEW and MMI datasets show that this paper's expression recognition network achieves an accuracy of 54.95% and 81.2%, respectively, for dynamic expression recognition, surpassing mainstream networks. Moreover, the restoration network outperforms mainstream networks in addressing occlusions and provides an average accuracy improvement of 5.34% in occluded expression recognition.

Keywords: dynamic facial expression recognition · Two-Stream CNN · convolutional neural network · generative adversarial networks · cascaded network

1 Introduction

Facial expression recognition has gained substantial momentum as a prominent research area within computer vision. Its significance extends across various domains and applications. The ability to recognize facial expressions empowers computer systems to comprehend human emotions and emotional states, thereby enhancing human-computer interactions. This advancement holds

H. Jin et al. (Eds.): IAIC 2023, CCIS 2058, pp. 182–197, 2024.
https://doi.org/10.1007/978-981-97-1277-9_14

immense potential in applications such as virtual assistants, emotion-aware technologies, and the enhancement of user experiences in safe-driving scenarios. Additionally, it finds utility in automating pain assessment and other medical applications.

Partial feature loss in partially occluded images reduces the accuracy and effectiveness of expression recognition. An effective way to address the occlusion problem is to complement the lost features as comprehensively as possible. Traditional face completion methods include full variational methods [1] and block matching methods [2], but these methods heavily rely on the unobscured parts. In contrast, Zhang et al. [3] proposed a cascading approach that utilizes multiple deep regression networks and denoising autoencoders to obtain a robust deep model for handling partial occlusions, automatically restoring the true appearance of the occluded parts.

In addition to encoder-based feature completion, generative adversarial networks (GANs) have also proven effective in image restoration. GANs operate on the principle of a 'two-player game' between generative and discriminative models, and they have seen broad applications across various fields in recent years [4]. For example, in 2016, Pathak et al. developed the Context-Encode GAN [5], where the generative model's context encoder is essentially an AlexNet [6]. The GAN network leverages the features learned by the encoder and compares them with the original features through the mutual reinforcement of the generative and discriminative models, resulting in more realistic complemented images. Dynamic expression recognition has to ensure real-time recognition while considering the temporal relationship between neighboring frames. Various interfering factors, such as illumination changes, pose variations, and random occlusions, pose challenges for dynamic expression recognition, making it more prone to occlusion issues compared to static expression recognition. Among these issues, random occlusion is a primary concern in dynamic expression recognition [7].

In this paper, for the above mentioned problem of dynamic expression recognition with occlusion in natural environment, we address the problem that traditional generative adversarial networks are unable to maintain the consistency and naturalness of global features of an image when generating localized images. For the face recovery generation task, we design an improved GAN network model. In the network, the ViT module with attention mechanism is used to detect the missing features of the face image and verify the feature generation, and in the generative network, the ability to perceive the local and global features of the image is improved. Secondly, a cascade neural network is constructed for dynamic expression recognition, which first extracts the spatial and temporal flow features of facial expressions by ViT, and then inputs these features into Two-Stream CNN to combine the two features for facial expression recognition.

In conclusion, our main contributions in this paper are summarized as follows:

- We have investigated and proposed an improved GAN model for face image recovery generation task, which addresses the inability of most GAN-based image generation networks to achieve uniformity and natural fidelity of the generated image in image localized feature generation.

- Within the network architecture, a ViT module is incorporated to verify the absence of facial features before proceeding with the generation of local facial attributes and feature prediction. In the generative network, the ViT is integrated to jointly detect both local and global features within the generated image.
- By cascading ViT with Two-Stream CNN, a powerful dynamic expression recognition network is trained by fully utilizing the feature extraction capability of ViT and the spatio-temporal information acquisition capability of Two-Stream CNN, and combined with the occlusion restoration network can efficiently perform dynamic occlusion expression recognition.
- The network proposed in this paper is experimented using several open-source datasets (e.g. MMI, CelebA and VGG Face2), and its efficiency is verified under a variety of facial occlusion levels.

2   Related Work

Early research on facial expression recognition focused on recognizing expressions from static images. These methods effectively extracted spatial information but could not capture information from movements. Recently, several studies have attempted to capture dynamic changes in facial physical structures from continuous frames based on handcrafted features or deep learning methods (e.g., LBP-TOP, HOG 3D, STM-ExpLet, and DTAGN).

Nowadays, expression recognition goes into deep learning stage [8], AlexNet [6], GooleNet [9], ResNet [10] have achieved good results in expression recognition, for the expression occlusion problem in recognition, Zhang et al. [11]based on symmetric transformation for eye occlusion, which is better for recognizing partially fixed occlusion, and Zhang et al. converted Gabor facial templates into template-matching distance features, and the generated feature vectors are robust to minor occlusions like eyes and mouth corners. However, all these methods only recognize expressions on static images, but none of them pay attention to the randomness of the occluded parts, so it is a worthwhile research direction to complement the randomly occluded expressions and then recognize the dynamically occluded expressions.

Generative Adversarial Networks (GAN) stand as a focal point of current research in the domain of image generation. GANs are renowned for their capability to produce high-quality samples using an innovative zero-sum game and adversarial training paradigm. In the realms of feature learning and feature representation, GANs outperform conventional machine learning algorithms, marking a significant advancement. Their substantial success in the domain of sample generation attests to their prowess [12]. There are two parts in GAN, Generator and Discriminator. The Generator is mainly used to learn the real image distribution to make the image generated by itself more realistic, in order to fool the discriminator [13]. The discriminator, on the other hand, needs to discriminate the received images as true or false. The structure of the classical GAN model is shown in Fig. 1.

The GAN model is composed of two parts: the first part is the generator G, which receives a random noise Z and generates a picture from this random noise; the second part is the discriminator D, which identifies the authenticity of the image, and determines whether a picture is "real" or not. Throughout the process, the generator tries to produce more realistic images as much as possible, while the discriminator tries to recognize the authenticity of the image. Over time, the generator and the discriminator play an adversarial game until the two networks reach a final dynamic equilibrium state, where the discriminator cannot recognize whether the image generated by the generator is real or not.

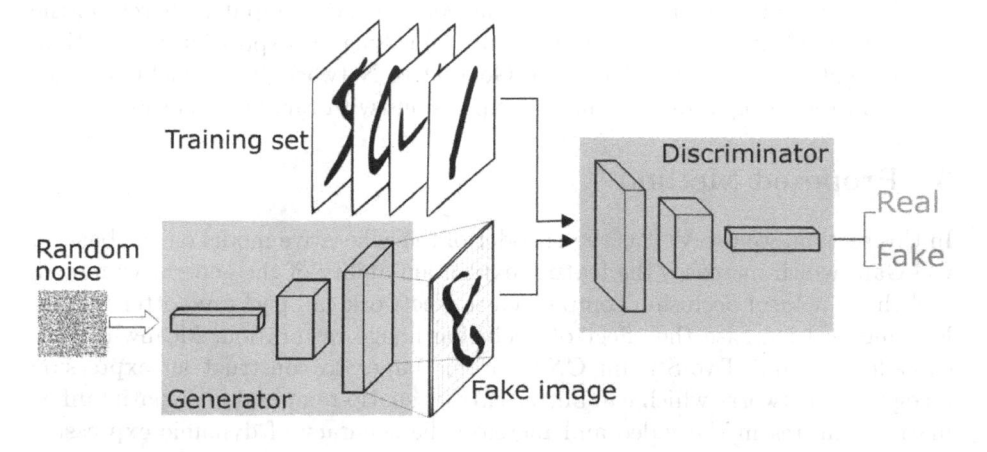

Fig. 1. Classical GAN architecture

Zhang and colleagues introduced a multi-face prior search network (MFP-SNet) designed to extract information from various face priors to tackle the challenge of blind face recovery (BFR) [14]. Ge and collaborators presented the Identity Diversity GAN (ID-GAN) [15]. This innovative model combines a CNN face recognizer with a GAN, enabling the generation of authentic and identity-preserving images. It achieves this by utilizing CNN for feature reconstruction and GAN for visual reconstruction. Xu et al. [16]proposed a GAN-based feed-forward network designed to generate high-quality face images. This network seamlessly integrates facial attributes like identity and expression by utilizing a 3D a priori and extracting distinct facial attribute information. Initially, convolutional networks were effective in processing static images, but with the increasing GPU processing power and the development of more sophisticated neural network architectures, recurrent neural networks (RNNs) have emerged as a more suitable choice for processing dynamic sequence data of arbitrary length. In the 2018 ACII conference, Sun et al.'s recurrent neural network model achieved a recognition rate 23.81% points higher than the MEC2017 baseline [17]. Traditional CNN and RNN approaches have overlooked the fact that dynamic expressions contain two essential types of information: spatial location information and temporal motion information. Typically, emotion recognition in expression recognition relies on a single spatial video frame. However, the formation of an expression is an ongoing process, and a single video frame can only represent the object's state at that particular moment, unable to effectively capture its motion information. Consequently, valuable information is missing, leading to relatively low emotion recognition rates. To more comprehensively exploit all the available information in the video and utilize the previously neglected temporal motion information, a pivotal network structure was introduced by Simonyan and Zisserman in 2014 [18]: the Two-Stream Convolutional Neural Network (CNN). This network leverages spatial information in the video as input for one network and temporal motion information as input for another network. The two networks are then integrated to create a unified network with dual inputs. Therefore, the Two-Stream CNN is particularly well-suited for dynamic expression recognition. When combined with an Adversarial Generative Network, it can yield superior results in the recognition of dynamic expressions with random occlusions.

3   Proposed Method

In this section, we use ViT as the encoder of the generative model on the basis of CC-Gan, which increases the feature extraction ability of the generative model, and the new local occlusion complementary network can perform better feature learning and increase the effect of occlusion image restoration. Meanwhile, we cascade ViT and Two-Stream CNN in this paper to construct an expression recognition network, which can better extract spatio-temporal and spatial information features in the video and increase the accuracy of dynamic expression recognition.

3.1   Facial Restoration Generative Network

In the realm of image generation, generating facial features presents a more intricate challenge compared to generating global features entirely from scratch. This is because facial feature generation necessitates not only creating the missing components of the image but also preserving the overall coherence and natural appearance of the facial features. This is particularly crucial since subtle changes in facial expressions hold significant value in expression recognition. It is widely acknowledged that convolutional computation may not excel in capturing comprehensive global feature information. Therefore, relying solely on a generative network model based on GANs does not yield the desired outcomes. The use of Vision Transformer (ViT) for facial image feature extraction has some features and advantages: ViT is able to capture the global information of the whole image rather than just local features, ViT utilizes the self-attention mechanism, which enables the model to adaptively focus on features in different regions of the image and after ViT is trained on large-scale datasets, ViT can provide information about the features in different regions of the image due to the self-attention mechanism [19]. Properties, ViT can provide interpretable information about the relationship between different regions in the image, which helps to understand the decision-making process of the model. The structure of ViT is shown in Fig. 3. In this paper network, the global features of the face are extracted using the ViT network and the missing parts of the features are detected under the multilayer perceptron and the multicast self-attention computation, which are used to repairing and generating the missing parts of the face. The module consists of alternating multi-head self-attention layers and MLP. Layernorm is applied before each block and residual join is performed after each block (Fig. 2).

The MLP consists of two fully connected layers sandwiched by a GELU activation function:

$$\mathbf{z}_0\& = \left[\mathbf{x}_{\text{class}}\,; \mathbf{x}_p^1 \mathbf{E}; \mathbf{x}_p^2 \mathbf{E}; \cdots ; \mathbf{x}_p^N \mathbf{E}\right] + \mathbf{E}_{\text{pos}}\, , \& \mathbf{E} \in \mathbb{R}^{\left(P^2 \cdot C\right) \times D}, \mathbf{E}_{\text{pos}} \in \mathbb{R}^{(N+1) \times D} \tag{1}$$

$$\mathbf{z}_\ell'\& = \text{MSA}\left(\text{LN}\left(\mathbf{z}_{\ell-1}\right)\right) + \mathbf{z}_{\ell-1}, \&\&\ell = 1 \ldots L \tag{2}$$

$$\mathbf{z}_\ell\& = \text{MLP}\left(\text{LN}\left(\mathbf{z}_\ell'\right)\right) + \mathbf{z}_\ell', \&\&\ell = 1 \ldots L \tag{3}$$

$$\mathbf{y}\& = \text{LN}\left(\mathbf{z}_L^0\right)\&\& \tag{4}$$

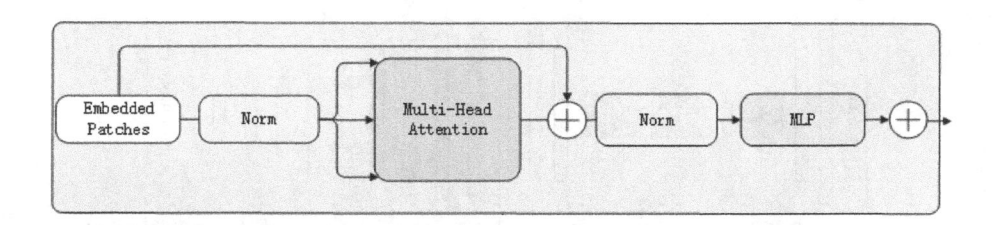

Fig. 2. Transformer module architecture

In this paper, we use the CC-GAN framework and perform a reconstruction of both the generator and discriminator components. The network structure is illustrated in Fig. 3. Within the generator section, we primarily utilize ResNet as the core network architecture and opt for dilation convolution as a replacement for traditional upsampling and downsampling operations. The size of the perceptual field of view significantly impacts the generation of image texture features during the image generation process. To address this, we integrate dilation convolution to reduce redundant convolution operations while concurrently expanding the perceptual field. Resnet avoids image generation overfitting expressions by using multiple residual blocks, which generates more detailed image textures.

The complementary and real images of the generated model are used simultaneously as inputs to the discriminative model, and the adversarial network is generated based on the condition, which treats the real image as the condition, and the higher judgment value is obtained only when the complementary image matches the real image more closely. The two types of images of size (128, 128, 1) are passed through the VGGNet in CC-Gan to output a one-dimensional result, the discriminative model consists of eight convolutional layers with convolutional kernel size of 3×3 and one fully connected layer. The final image generation network is shown in Figure 4.

3.2   Cascade Expression Recognition Network

After completing the local occlusion complementation, we also need to carry out expression recognition to verify the effectiveness of occlusion complementation. Dynamic expressions generally contain two aspects of information, one is spatial location information, and the other is temporal motion information. Generally, expression recognition is only a single spatial video frame for emotion recognition, but the formation of expression is ultimately a process, a single video frame can only represent the state information of the object at this time, and its motion information we can not be effectively utilized, the two parts of the useful information but a part of it is missing, which leads to a low rate of recognition of the emotion, and the Two-Stream CNN utilizes the motion information to the dynamics, it utilizes the spatial deep convolutional network to classify still images, and then utilizes the temporal deep convolutional network to train and

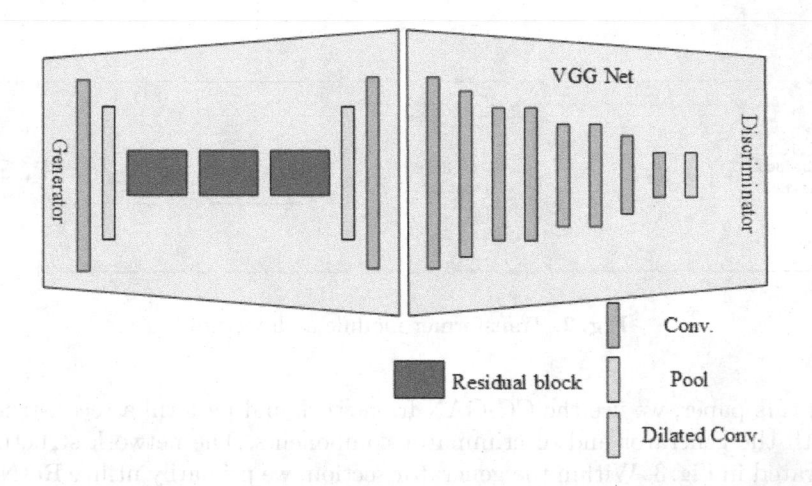

Fig. 3. Facial restoration generative network architecture

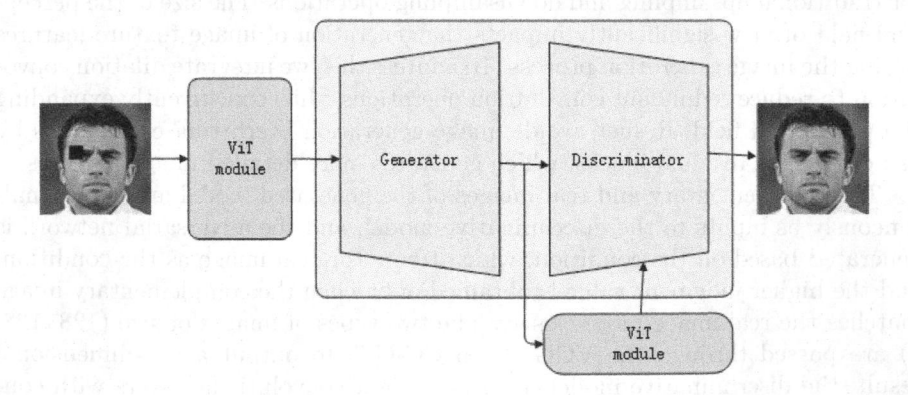

Fig. 4. The improved GAN framework

classify the optical flow information of the action video, and finally fuses the two branches softmax scores, and it shows that for the temporal flow network, the effect of using the optical flow information as an input is much higher than the recognition rate of using a series of frames as an input. In this paper, we have cascaded the ViT and the Two-Stream convolutional neural network in a cascade, and the constructed cascade dynamic expression recognition network is better at spatio-temporal information processing, which can optimize the expression recognition results.

Before the emergence of Two-Stream Convolutional Neural Networks (CNNs), there were also many works that tried to apply Convolutional Neural Networks (CNNs) to the task of video motion recognition, but the results were not as good as those previous methods that hand-designed the shallow features. CNNs are good at learning localized features, and it is difficult to learn the temporal-based movement patterns. Two-Stream CNNs first extract the dynamic information well in advance, that is, to get the optical flow information, and then let the CNNs learn the mapping relationship between optical flow and labels, which is equivalent to not letting the CNN learn the motion features directly, but extracting the motion features, that is, the optical flow information, and then letting the CNN learn the extracted motion features. The dual-stream neural network structure is mainly divided into two parts: the spatial stream network and the temporal stream network.

Time flow networks use the optical flow map between sequences of frames as input information, when the object undergoes motion, it will leave a series of image information on the retina, like light flowing through, hence the name. Optical flow passes through the pixels in an image sequence in the time domain and the correlation between neighboring frames to find out the correspondence that exists between the previous frame and the current frame, and then calculates the object between neighboring frames. The motion information of the object between neighboring frames is then calculated. In this paper, we adopt the $TV - L^1$ method to extract the optical flow, and input several consecutive frames of the corresponding videos to get the optical flow maps between the neighboring frames of these videos.

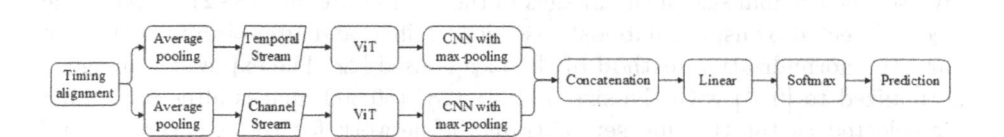

Fig. 5. Cascade FER Network architecture

The spatial network uses a single frame from each moving video as input information. Because it takes a process to produce expressions of joy, anger, sadness and hatred, in principle, every frame in the video can be used as an input to the spatial network. However, in order to better express people's emotional information, it is necessary to try to select the frame in the video with clear face images and the largest change in expression as the input to the network. Traditional spatial networks use CNN as the convolutional layer, although CNN is easy to train, but lacks the global interaction between pixels, so it is necessary to use the transformer to obtain the global information. ViT can get the global information, but it requires a large amount of computation, so it is very important to design a lightweight ViT model. Existing lightweight networks are generally based on CNN, but CNN networks have two drawbacks, one is that it only has local sensory field so it is difficult to model the global information, and the second is that the weights are invariant in inference and difficult to apply flexibly. Therefore, in this paper, ViT is introduced to mitigate these two drawbacks, and a hybrid ViT and CNN network architecture is used for spatial feature extraction, which is shown in Fig. 5.

4   Experiments and Results

The experiments were performed on a 64-bit Microsoft Windows 11 system with an Intel ® Core ™ i9-12900HX CPU (a) 2.30 GHz. The GPU was an NVIDIA GEFORCE GTX4060 with 6 GB of video memory. Using a Pytorch-based deep learning platform.

4.1   Facial Restoration Generative Network

The improved GAN constructed in this paper is compared with CC-Gan for experiments to explore the effect of the constructed local occlusion complementary network on the performance of expression repair, and the complementary network is verified for expression repair by comparing the Generative loss.

Datasets. In this paper, three datasets, CelebA [20], VGG Face2 [21], are selected for evaluation and experimentation.

CelebA dataset consists of 202599 face images, the number of samples is large and 10177 individuals with different emotional expressions are sufficiently diverse, the original size of the images in the dataset are all 178×218, firstly, the face is detected by using Adaboost cascade classifier, and after obtaining the face part, the normalization method of the face is used [22] The expression image is normalized to [0, 1] with the size of 128×128. 150,000 images after processing are selected as the training set to train the network for the experiment, and 52,599 images for the test set. Iterative training on CelebA's dataset 200 cycles, 2523 iterations per cycle, each batch size is set to 8, the initial local binary mask Mask is set to 48, and the spatial dimensions of the hidden variable z is 100. Gradient decay in training is optimized for loss using Adam's algorithm, and the initial learning rate l_r is 1×10^{-4}, where the parameter $\beta1$ is set to 0.5, $\beta2$ to 0.999, and ε to 1×10^{-8}.

The VGG Face2 dataset comprises around 3.31 million images categorized into 9131 distinct groups, with each category representing a different individual's identity. The dataset features images with an average resolution of 137×180 pixels. These images were collected from Google Image Search and exhibit substantial diversity in terms of pose, age, lighting conditions, ethnicity, and occupation. To prepare the data for analysis, we apply the same preprocessing methods as described earlier.

Results. In order to verify the robustness of the local occlusion complementary network in facial restoration in this paper, the size of the occlusion region is set to 1/2, 1/4, 1/8 for the experiments on CelebA, VGG Face2, respectively.Figure 6 shows the experimental comparison results of the facial complementary network using the parallel structure and the restoration map without the parallel structure, from left to right, respectively. Comparison results of experiments on face complementation networks with and without the parallel structure, from left to right are the occlusion image, the restored image without the parallel structure, the restored image with the parallel structure, and the original image, respectively. The experimental results show that the restoration effect is more realistic for small to medium degree of occlusion like 1/8 to 1/4, and the overall loss is stable at 0.07–0.1 on CelebA dataset, and 0.07–0.09 on VGG Face2 dataset.

However, 1/2 occlusion produces partial distortions in the generated image and discontinuities in the boundary between the repaired part and the unoccluded part,Figure 7 showing that the combined loss of 1/2 occlusion compared

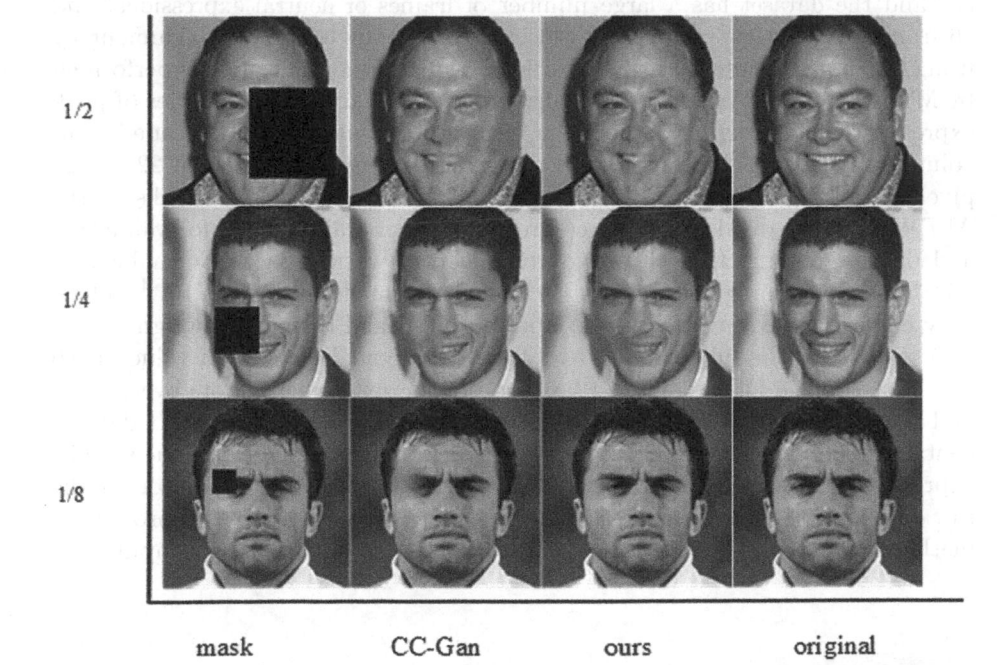

Fig. 6. Different occlusion repair process

to other degrees of occlusion on the CelebA and VGG Face2 datasets has a more than 0.2 difference, the reason for the excessive loss is that a large portion of the facial features are occluded, and the discriminative model improperly trades off the large number of features extracted by the encoder, which makes the overall loss rate for facial expression recovery too large.

4.2 Cascade Expression Recognition Network

In order to verify the effect of the occlusion-complementary network in this paper on the final expression recognition effect, the performance effects of the cascade expression recognition network constructed in this paper on unoccluded expression recognition and occluded expression recognition are explored separately. Firstly, we do experiments on AFEW and MMI [23] with unobscured expressions to analyze the performance of the cascade network on different datasets. Then experiments with occlusion are performed on the MMI dataset to analyze the contribution of occlusion complementation to dynamic expression recognition.

Datasets. In this setction, AFEW and MMI are used as the datasets, AFEW is selected from movie and TV series clips, and the dataset has 7 emotional states: anger, disgust, fear, happiness, neutrality, sadness, and surprise. 1426 segments of the AFEW dataset are collected, and the dataset is 720×576 pixels, and the dataset has a large number of frames of neutral expressions, and 16 frames of peak expressions are selected for face detection and alignment by using the dynamic time warping algorithm, and their normalization is performed by MTCNN. Time Warping (DW) algorithm is used to select 16 frames of peak expression for face detection and alignment using Multi-masked Cascaded Convolutional Network (MTCNN), and at the same time normalized to 224×224 pixel size for input. In order to make the experimental results more realistic, the AFEW data are divided into three groups: 773 for training, 373 for validation and 653 for testing. In each convolutional layer, we use Batch Normalization (BN) with a learning rate of 1×10^{-4} and weight decay of 5×10^{-4}, add a fully connected layer to the fully connected layer, and then add a weight decay of 5×10^{-4} to the fully connected layer. A Dropout layer with a value of 0.6 was added to the fully connected layer to prevent experiments from overfitting, and $L2$ regularization was performed in the SoftMax layer. The MMI dataset contains 2900 videos from 32 subjects, of which there are 205 frontal views. The expression sequences in MMI peak near the middle, and the resolution of the raw data is 720×576 pixels in size. In this paper, we use the same preprocessing method as Sect. 4.1 to normalize the data to 224×224 pixel size for input.

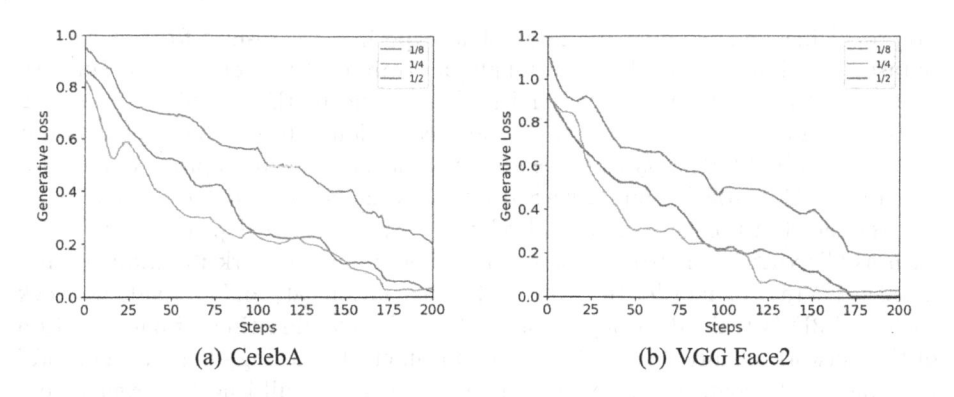

(a) CelebA   (b) VGG Face2

Fig. 7. Different occlusion of loss change

Results of Unmasked Dynamic Expression Recognition. In this experiment, 10-fold cross-validation is used, and the AFEW and MMI data sets are divided into 10 groups, 9 of which are used as the training set, and the other 1 group is used as the test set, and the average score of the 10 experimental results is taken at last. This ensures that the experimental results are real and effective, and not disturbed by inter-sample errors. Figure 8 is the recognition rate of the cascade expression recognition network constructed in this paper in the AFEW and MMI datasets randomly selected for one training, after 100000 iterations of training, the final result of the AFEW gets 52.12% recognition rate, and the MMI gets 80.31% recognition rate. In the case of GPU accelerated computing AFEW iteration 100 000 times spent a total of 297 min, MMI spent 256 min, after 100 000 iterations of the loss function are lower than 0.01 and the change has been stabilized, the recognition results are also sufficient to converge.

In order to compare the performance of the cascade network proposed in this paper, comparison experiments of training time and recognition rate are conducted with other networks. The comparison experiments all use the same

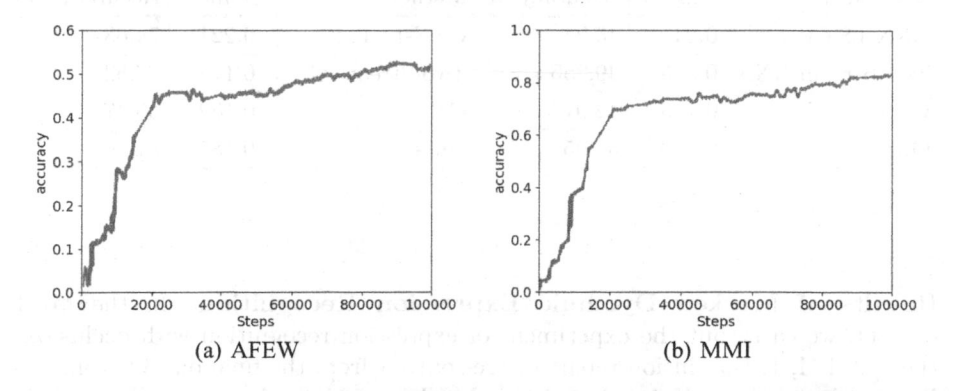

(a) AFEW   (b) MMI

Fig. 8. Change of accuracy

preprocessing and training methods. The experimental results for the AFEW dataset are shown in Table 1. Although the training time for a single network such as C3D is only 0.185 s for one iteration, due to the insufficient learning of the spatio-temporal information, the recognition rate on the same amount of data is only 42.91%, as the cascade network increases in depth, the recognition rate of CNN-LSTM and Two-Stream CNN increases to 45.24% and 42,91%, respectively, but the time spent is also longer, especially the depth of the network such as CNN-LSTM is too large, and the time of one network iteration reaches 0.243 s. Therefore, in order to improve the recognition rate, only a single network such as C3D is trained in depth for 0.185 s. The training time of one iteration of the network is only 0.185 s due to the insufficient learning of spatio-temporal information. Recognition rate only the depth increase will lead to slower recognition speed and increase the computational cost of the computer. In this paper, the ViT module deepens the depth of the network, which is better for feature learning, but the recognition time increases, and the experimental results show that the cascade of ViT and Two-Stream CNN improves the recognition rate to 54.95% on the basis of ensuring the stability of the training time.

Similarly, the experiments on the MMI dataset iterated once on a single network C3D in 0.152 s, with a recognition rate of only 69.47%, CNN-LSTM and Two-Stream CNN recognition rate increased by 5.16 and 8.05% points, but the time increased by 0.069 s and 0.026 s. The cascade network in this paper has the highest recognition rate of 81.20%, but the time is close to that of Two-Stream CNN, which is only 0.187 s. The method to improve the recognition rate is generally to increase the depth of the network, but the cost of network operation will also increase, and the structural topology of this paper not only improves the ability of feature extraction, but also ensures that the recognition speed will not surge.

Table 1. Accuracy and time comparison

(a) AFEW			(b) MMI		
Method	Time/s	Accuracy/%	Method	Time/s	Accuracy/%
CNN-LSTM	0.243	45.24	CNN-LSTM	0.221	74.63
Two-Stream CNN	0.205	49.256	Two-Stream CNN	0.178	77.52
C3D	0.185	42.91	C3D	0.152	69.47
Ours	0.213	54.95	Ours	0.187	81.20

Results of Masked Dynamic Expression Recognition. In the MMI dataset, we carry out the experiment of expression recognition with occlusion, the 1/2, 1/4, 1/8 occlusion repair images output from the final output from the local occlusion network int Sect. 4.1are batch input into the cascade expression recognition network in Sect. 4.2, and finally we get the comparison of the experimental results such as the Table 2, as the occlusion area decreases the recognition rate is gradually improving, the optimization rate also shows an upward trend, and the increasing trend of the optimization rate is gradually smooth. The recognition rate of the repaired images under the three occlusion conditions reaches 63.12%, 68.39%, 80.25%, with an average recognition rate of 70.59%, which is 3.03, 6.05, 6.93% points higher than that of the unrepaired broken images, and the average recognition rate is 5.34% points higher, so it is necessary to repair the broken images and then carry out expression recognition. Therefore, it is necessary to repair the damaged images and then recognize the expressions.

Table 2. Different occlusion accuracy comparison

Mask	Repair accuracy/%	No rpair accuracy/%	Upgrade/%
1/2	63.12	60.09	3.03
1/4	68.39	62.34	6.05
1/8	80.25	73.32	6.93
Average	70.59	65.25	5.34

From Table 2, we can see that the 1/8 occlusion repair recognition rate is relatively high, above 80%, and the recognition optimization rate is above 6%, and the 1/4 occlusion repair recognition rate is lower than the 1/8 occlusion repair recognition rate, but the optimization rate is considerably higher than that of the 1/2 occlusion, by The optimization rate is much higher than that of 1/2 masking, which is 3.02% points higher. Although the recognition rate of 1/2 occlusion is improved, the optimization rate is only 3.03% points. It can be seen that the loss rate of large-area occlusion restoration is too high, so the restored image loses too many features related to expression, and the contribution rate of its expression recognition is not high. Comprehensive experimental results, this paper's method for small and medium degree of occlusion, through the complementation of the expression recognition effectiveness significantly improved.

5   Conclusions

In this paper, we propose a generative adversarial framework using ViT in a complementary network, which reduces the effect of occlusion on expressions to some extent, and the network is faster and more stable. In this paper, the content loss of local occlusion complementary network is lower than CC-Gan, in which small and medium degree of random occlusion such as 1/8 to 1/4 is repaired better than 1/2 large occlusion, for large occlusion, the generative model and discriminative model need to be balanced in the allocation of network layers.

The constructed cascade expression recognition network performs expression recognition with different occlusion complements on the MMI dataset, and finds that occlusion repair contributes an average of 5.34% points to expression recognition, of which the optimization rates of 1/8 and 1/4 occlusion repair are 6.93 and 6.05% points, respectively, and there is also an optimization rate of 3.03% points for 1/2 occlusion, which is not optimally repaired.

The cascaded expression recognition network constructed in this paper also has high recognition performance for no occlusion, and the recognition rate is not only higher than that of a single network structure such as C3D, but also better than that of a cascaded network such as CNN-LSTM, and the highest recognition rate reaches 54.95 and 81.20% points on the AFEW and MMI datasets, respectively.

In future work, since the current network adds depth and does not improve the temporal part significantly we will try other DNN structures (3DCNN, LRCN, etc.) to improve the temporal part. We will also do more tests on other facial expression databases, especially in the field environment.

Acknowledgements. This article has been supported by State Grid Corporation Technology Guide Project(5700-202218185A-1-1-ZN).

References

1. Chan, T.F., Shen, J.: Mathematical models for local nontexture inpaintings. SIAM J. Appl. Math. **62**(3), 1019–1043 (2002)
2. Barnes, C., Shechtman, E., Finkelstein, A., Goldman, D.B.: Patchmatch: a randomized correspondence algorithm for structural image editing. ACM Trans. Graph. **28**(3), 24 (2009)
3. Zhang, J., Kan, M., Shan, S., Chen, X.: Occlusion-free face alignment: deep regression networks coupled with de-corrupt autoencoders. In: 2016 IEEE Conference on Computer Vision and Pattern Recognition (CVPR) (2016)
4. Lee, J., Kim, S., Kim, S., Park, J., Sohn, K.: Context-aware emotion recognition networks. In: 2019 IEEE/CVF International Conference on Computer Vision (ICCV) (2020)
5. Pathak, D., Krahenbuhl, P., Donahue, J., Darrell, T., Efros, A.A.: Context encoders: feature learning by inpainting. In: 2016 IEEE Conference on Computer Vision and Pattern Recognition (CVPR) (2016)
6. Egils, A., Tomasz, S., Maie, B., Dorota, K.: Audiovisual emotion recognition in wild. Mach. Vis. Appl. **30**, 975–985 (2018)
7. Wang, K., Peng, X., Yang, J., Meng, D., Qiao, Y.: Region attention networks for pose and occlusion robust facial expression recognition. IEEE Trans. Image Process. **29**, 4057–4069 (2020)
8. Jain, D.K., Zhang, Z., Huang, K.: Multi angle optimal pattern-based deep learning for automatic facial expression recognition. Pattern Recogn. Lett. **139**, 157–165 (2020)
9. Mollahosseini, A., Chan, D., Mahoor, M.H.: Going deeper in facial expression recognition using deep neural networks. IEEE (2016)
10. He, K., Zhang, X., Ren, S., Sun, J.: Deep residual learning for image recognition. IEEE (2016)
11. Jianming, Z., Xiaocui, Z.: Processing method of facial expression images under partial occlusion. Comput. Eng. Appl. **47**(3), 170–173 (2011)

12. Li, Y., Liu, S., Yang, J., Yang, M.H.: Generative face completion. In: 2017 IEEE Conference on Computer Vision and Pattern Recognition (CVPR) (2017)
13. Goodfellow, I.J., et al.: Generative adversarial nets (2014)
14. Zhang, P., Zhang, K., Luo, W., Li, C., Wang, G.: Blind face restoration: benchmark datasets and a baseline model (2022)
15. Ge, S., Li, C., Zhao, S., Zeng, D.: Occluded face recognition in the wild by identity-diversity inpainting. IEEE Trans. Circ. Syst. Video Technol. **30**(10), 3387–3397 (2020)
16. Xu, Z., et al.: Facecontroller: controllable attribute editing for face in the wild. In: Proceedings of the AAAI Conference on Artificial Intelligence, pp. 3083–3091 (2022)
17. Sun, M.C., Hsu, S.H., Yang, M.C., Chien, J.H.: Context-aware cascade attention-based rnn for video emotion recognition. In: 2018 First Asian Conference on Affective Computing and Intelligent Interaction (ACII Asia) (2018)
18. Simonyan, K., Zisserman, A.: Two-stream convolutional networks for action recognition in videos. In: Ghahramani, Z., Welling, M., Cortes, C., Lawrence, N., Weinberger, K.Q. (eds.) Advances in Neural Information Processing Systems, vol. 27. Curran Associates Inc. (2014)
19. Dosovitskiy, A., et al.: An image is worth 16×16 words: transformers for image recognition at scale. In: International Conference on Learning Representations (2021)
20. Liu, Z., Luo, P., Wang, X., Tang, X.: Deep learning face attributes in the wild. In: Proceedings of International Conference on Computer Vision (ICCV) (2015)
21. Parkhi, O.M., Vedaldi, A., Zisserman, A.: Deep face recognition. In: Proceedings of the British Machine Vision Conference 2015 (2015)
22. Zhu, X., Lei, Z., Yan, J., Yi, D., Li, S.Z.: High-fidelity pose and expression normalization for face recognition in the wild. In: 2015 IEEE Conference on Computer Vision and Pattern Recognition (CVPR) (2015)
23. Pantic, M., Valstar, M., Rademaker, R., Maat, L.: Web-based database for facial expression analysis. In: 2005 IEEE International Conference on Multimedia and Expo, p. 5 (2005)

Contrastive Learning Based on AMR Graph for Logic Reasoning

Yan He[✉] [iD] and Yongli Wang [iD]

Nanjing University of Science and Technology, NanJing 210000, China
heyan@njust.edu.cn

Abstract. In the field of Machine Reading Comprehension (MRC), existing models have already surpassed human average performance in many tasks such as SQuAD. In recent years, more challenging MRC datasets have been introduced, such as ReClor and LogiQA datasets. These datasets place a greater emphasis on evaluating the logical reasoning abilities of models. To enhance the model's logical reasoning capabilities, we propose the AMR-CL method, a contrastive learning pretraining approach based on AMR (Abstract Meaning Representation) logical graphs. We employ an AMR parser to construct AMR logical graphs that represent the semantic information implied in the text. Then, we enhance the logical relationships in the AMR graph based on logical predicates and perform logical expansion using the principle of logical equivalence. We create logically consistent positive examples and logically inconsistent negative examples using logical equivalences for data augmentation. Contrastive learning is applied to help models better understand logical information within the text. We conducted experiments on two logical reasoning datasets, ReClor and LogiQA, and the results confirm the effectiveness of our method.

Keywords: Machine reading comprehension · Logic reasoning · Contrastive learning

1 Introduction

Machine Reading and Comprehension (MRC) is one of the core tasks in the field of Natural Language Processing (NLP) and has found extensive applications in information retrieval, recommendation systems, intelligent question-answering, and other task scenarios. In the field of Natural Language Processing, Machine Reading Comprehension is defined as follows: given one or more documents, a question, and multiple answer choices. The task is to use machines to answer questions based on the provided context, testing the model's ability to understand natural language.

In recent years, with the vigorous development of pretraining language models, many of these models have shown exceptional performance on popular reading comprehension datasets such as SQuAD [1] and RACE [2]. Examples of these models include BERT [3], RoBERTa [4], XLNet [5], GPT-3 [6], and more. However, early machine reading comprehension datasets were relatively straightforward, often requiring answers to be

© The Author(s), under exclusive license to Springer Nature Singapore Pte Ltd. 2024
H. Jin et al. (Eds.): IAIC 2023, CCIS 2058, pp. 198–211, 2024.
https://doi.org/10.1007/978-981-97-1277-9_15

located in the original text through fuzzy string matching. They lacked in-depth assess-ments of a model's comprehension abilities. The current research trend in machine reading comprehension is to decompose comprehension abilities into a series of skills, such as numerical reasoning [7], common-sense reasoning [8], and logical reasoning [9, 10]. Logical reasoning is highly challenging as it involves making inferences based on the semantic relationships between events or facts in a text. To advance this field, more challenging datasets have been introduced, such as the ReClor [9] dataset and the LogiQA [10] dataset. Figure 1 provides an example from the ReClor dataset, which is a typical logical reasoning problem comprising a context, a question, and four answer options, with only one correct answer. In such tasks, the model needs to extract the logical relationships between statements within the given context, combine this with the question, and also extract the logical structure of each answer option to demon-strate its validity, ultimately arriving at the correct answer. In this example, based on the inferred reasoning process $\neg s5 \rightarrow s1 \rightarrow s2 \rightarrow s3$ obtained from the context, option B is determined to be the correct answer.

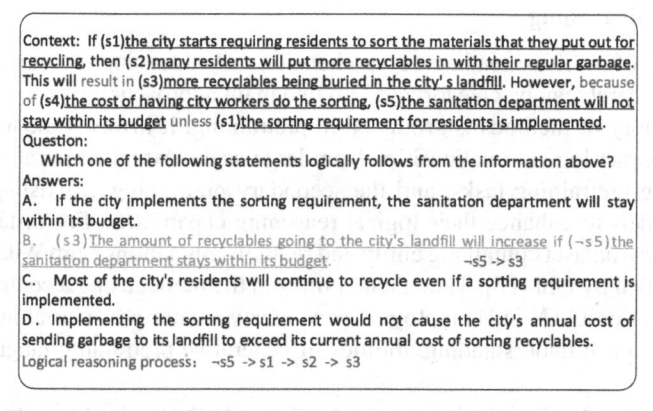

Fig. 1. An example of logical reasoning from the ReClor dataset.

Recent research has shown some progress on benchmark datasets. DAGN [15] uses Elementary Discourse Units (EDUs) as the fundamental reasoning unit and leverages graph networks to learn discourse features within a passage. However, the event graphs constructed by this approach lack the necessary logical structures for complex logical reasoning and may have lower graph construction accuracy. LReasoner [16] employs simple logical rules for logical enhancement. While these methods can effectively utilize logical rules, they may not enable the model to learn intricate logical structures. MERIt [12] introduced a counterfactual data augmentation method, where entity positions in context and options are swapped to construct negative samples for contrastive learning pre-training. However, sentences constructed in this manner may not necessarily exhibit logical relationships.

In order to leverage the rich semantic and logical structures present in both the context and options, we adopted the use of Abstract Meaning Representation (AMR) [11]. AMR is an advanced semantic abstraction that represents the semantic structure of a sentence as a single-rooted directed acyclic graph, allowing different sentences with similar semantics to share the same AMR graph. We applied logical equivalence laws to construct logically consistent positive instances and logically inconsistent negative instances for data augmentation. This approach enables the model to learn the logical structure of the text through contrastive learning. The main contributions of this paper are summarized as follows:

1. We adopted the method of constructing AMR logic graphs, using logical identifiers and logical equivalence laws to augment and extend the logical relationships, so that the logic graphs contain the semantic information and logical structure of the text.
2. We constructed logically consistent positive samples and logically inconsistent negative samples using logical equivalence laws, and used contrast learning to learn logical differences.
3. We demonstrate the effectiveness of our method by significant improvements on two datasets, Reclor and LogiQA.

2 Related Work

2.1 Logical Reasoning

In the context of machine reading comprehension tasks focusing on logical reasoning, the current research can be categorized into two main approaches:

One category of methods is grounded in pretraining. It involves the use of heuristic rules to extract logical relationships from large-scale textual corpora, the design of corresponding pretraining tasks, and the secondary pretraining of existing pretrained language models to enhance their logical reasoning capabilities. For instance, MERIt [12] generated a dataset containing entity and relation annotations from Wikipedia based on data provided by Qin [13]. They then optimized the model using a contrastive learning approach. LogiGAN [14] employs a generator-validator adversarial mechanism to enhance the logical understanding abilities of generative pretraining language models such as T5.

Another category of methods focuses on optimizing the model during the fine-tuning stage. These methods use various techniques to divide text into different granular logical units and introduce symbolic inference rules to define logical relationships between these units. Currently, most proposed methods aim to enhance the model's logical reasoning capabilities from this perspective. DAGN [15] utilizes Elementary Discourse Units (EDUs) as fundamental reasoning units and employs graph networks to learn discourse features within a passage. LReasoner [16] introduces a context expansion framework that incorporates logical equivalence laws, including the law of contrapositive and the law of transitivity. Logiformer [17] introduces a dual-stream architecture that includes grammar and logic branches to address the challenge of modeling long-distance dependencies between logical units. AdaLoGN [18] adaptively infers logical relationships to expand event graphs, enabling mutual reinforcement and iterative enhancement between neural and symbolic reasoning.

2.2 Abstract Meaning Representation

Abstract Meaning Representation (AMR) [11] is a semantic representation in natural language processing. It abstractly represents the semantic structure of a sentence as a rooted directed acyclic graph, using a graph structure to extract the sentence's semantic structure information. Sentences with the same fundamental meaning are assigned the same AMR graph. AMR graphs utilize concept nodes to represent entities, time, attributes, states, and other information. The edges between nodes represent semantic relationships between concept nodes. AMR graphs emphasize the intrinsic meaning of a sentence, focusing on semantically relevant sentence components while ignoring aspects such as voice and tense. Figure 2 provides an example of an AMR graph.

Sentence: I saw Joes's dog,which was running in the garden.

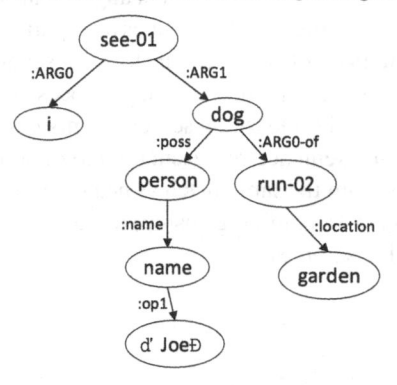

Fig.2. Example of an AMR graph.

In Fig. 2, entity concepts can be "dog", "person", event predicates "see-01", and "run-02". In AMR, entities, time, attributes, and states are introduced as variables. When an entity plays multiple roles within a sentence, the corresponding variable can be reused in the graph symbols. Relationships in the AMR graph begin with ":" and follow Proposition Bank annotation conventions such as ":ARG0", ":ARG1", and ":ARG2". According to [19], the AMR graph contains approximately 100 types of relationships, including general semantic relationships (":age", ":location"), quantity relationships (":quant"), date entity relationships (":month", ":season"), and more.

In recent years, significant progress has been made in the field of semantic parsing using AMR. [21] employs an attention-based Long Short-Term Memory network (Att-BiLSTM) to encode the input source text sequence. Simultaneously, they utilize a Graph Recursive Network (GRN) AMR encoder to encode the corresponding AMR graph. This approach supplements the source text with semantic structural information from the AMR graph and is applied to tasks such as translation and summarization. [22] integrates AMR graphs with external knowledge sources, utilizing AMR's entity nodes to achieve entity linking and graph fusion in common-sense reasoning and question-answering tasks. [23] combines small AMR graphs from multiple short sentences into a larger graph using co-reference and co-occurrence techniques. This method is employed in multi-hop reasoning question-answering tasks.

2.3 Contrastive Learning

Contrastive Learning (CL) is a commonly used method for constructing self-supervised tasks. The training objective in contrastive learning aims to minimize the distance between positive sample pairs while maximizing the distance between positive and negative sample pairs. Contrastive learning methods were initially applied in the field of computer vision and have achieved success with models like Moco and SimCLR [24]. The core of contrastive learning lies in how positive and negative samples are constructed to learn to distinguish their representations. For instance, SimCLR generates positive sample pairs by applying different data augmentations (e.g. cropping, flipping) to the same image, while treating all other images as negative samples. The loss function encourages the representations of positive pairs to be closer and those of negative pairs to be farther apart. In the context of natural language processing, SimCSE [25] introduced a contrastive learning method for learning sentence embeddings. It does so by using two dropout layers on the same sentence representation to obtain a positive pair, while considering other sentences from the same batch as negative samples. This approach learns sentence representations by maximizing cosine similarity between positive samples and minimizing it for negative samples.

3 Method

3.1 Logical Graph Construction

AMR Graph Generation. An AMR graph can be denoted as G = (V,E), where V is the set of vertices, and E is the set of edges, including core semantic roles (ARG0, ARG1, etc.), non-core semantic roles (age, location, etc.), logic relations (condition, polarity, etc.), modifiers (mod), and more. During the construction of the AMR graph, an AMR parser is used to parse the text into an AMR graph. This way, the AMR graph captures essential semantic information from the text sentences, helping the model better utilize the semantic and logical structures in the text. To enhance the interaction between the context, question, and options when constructing the AMR graph, a global root node is added to the unified AMR graph. Additionally, neuralcoref is employed to parse co-occurrence or co-reference to integrate AMR sentence subgraphs.

Logical Relation Identification. To better recognize logical relationships in text sentences and subsequently generate AMR graphs that include these relationships, we conducted an analysis of text passages in logical reasoning machine reading comprehension datasets. From this analysis, we found that relationships between logical units (such as entities or events) can be broadly categorized into five types: premise, consequence, negation, contrast, and coordination. These relationships are typically expressed in text through indicator words. Hence, taking reference from [26], we categorized and summarized common logical relationship indicator words. For specific details, please refer to Table 1.

As seen in Table 1, we categorize logical relationships into five types: premise, consequence, negation, contrast, and coordination. In these categories:

Table 1. Summary of logical relationship identifiers

Type	Identifiers
Premise	according to, as a result of, as indicated by, as long as, based on that, because, because of, by virtue of, considering, considering that, depend on, due to, for the reason that, given that, if and only if, in that, in view of, inasmuch as, may be inferred from, now that, on account of, on the grounds that, only if, or the sake of, owing to, rely on, seeing that, thanks to
Consequence	accordingly, as a conclusion, as a consequence, as a result, as it turns out, because of this, by such means, conclude that, conclusively, consequently, entail that, eventually, for this reason, hence, imply that, in conclusion, in order that, in this manner, in this way, infer that, it follows that, on that account, prove that, result in, show that, so that, suggest that, that being so, that is why, thereby, therefore, thus, to that end, whence, wherefore
Negation	aren't, barely, cannot, can't, couldn't, didn't, doesn't, don't, few, hardly, hasn't, haven't, isn't, little, merely, neither, never, no longer, nobody, none of, not, nothing, nowhere, rarely, scarcely, seldom, unable, wasn't, weren't, without, won't, wouldn't
Contrast	although, but, conversely, despite, even, even if, even though, however, in contrast, in spite of, instead of, nevertheless, nevertheless, nonetheless, on the contrary, otherwise, rather, reckles, regardless of, though, unless, whereas, yet
Coordination	additionally, afterward, also, and, as well, besides, both, either, further, furthermore, in addition, like-wise, mean-while, meantime, moreover, next, nor, on the other hand, or, similarly, simultaneously

- Premise refers to the "cause" in a cause-and-effect relationship. Common indicator words include "because of", where the statement following the indicator word serves as the reason or explanation for the preceding statement.
- Consequence indicates the "effect" in a cause-and-effect relationship. Common indicator words include "in order that", where the statement following the indicator word represents the result of the preceding statement.
- Negation alters the semantics of a statement. Common indicator words include "hardly", where adding this word to a sentence significantly changes the meaning.
- Contrast signifies a shift in the statement between sentences. Common indicator words include "however", where the meaning of the two sentences before and after the indicator word changes.
- Coordination is the most common type of relationship between statements, where two statements are presented in parallel or progression. Common indicator words include "in addition", where the sentences connected by the indicator word express coordinated or parallel facts.

Logical Expansion. In addition to the explicit logical relationship indicator words extracted from the context and options, there are implicit logical relationships that are implied and need to be inferred and expanded through logical reasoning. We perform logical expansion on the generated AMR graphs based on the principles of logical equivalence (Table 2).

Table 2. Law of logical equivalence

Name	Formulas
Commutative Law	$(A \wedge B) \iff (B \wedge A)$
Implication Law	$(A \to B) \iff (\neg A \vee B)$
Contrapositive Law	$(A \to B) \iff (\neg B \to \neg A)$
Double Negation Law	$\neg\neg A \iff A$
Category Inference Law	$(\forall A, f(A)) \iff (\neg \exists A, \neg f(A))$

Among these, the application of logical laws involves several principles:

- Commutative Law: Applied in AMR graphs with coordination relationships, where the left and right branches of subgraphs are exchanged.
- Implication Law: Used in two scenarios. Firstly, in AMR graphs with premise or consequence relationships, these can be replaced with negation relationships alongside coordination relationships. Secondly, in AMR graphs with coordination relationships, premise relationships can be used as replacements, while negation relationships in the premise are removed.
- Contrapositive Law: Utilized in conditional reasoning to infer the sufficiency and necessity of conditions. It involves swapping the order of the premises or conclusions in AMR graphs and adding or removing negation relationships at both ends.
- Double Negation Law: Applied to sentences with negation indicator words.
- Category Inference Law: Used for reasoning if a specific entity concept belongs to a particular category, such as "dog $\subseteq$ animal". This type of inference is achieved by using quantifiers like "all", "whole", "both", and so on.

3.2 Data Augmentation

Our training samples are often derived from real text data, representing objective facts from the real world. However, when constructing incorrect options, it is usually done by directly replacing entities. As a result, the content described in these incorrect options often conflicts with the objective facts of the real world. When pretraining language models using a substantial amount of unlabeled text data, the models learn and memorize common knowledge from the real world mentioned in the training data. As a result, during answer prediction, the model can rely on the direct judgment of whether the options align with common knowledge from the real world to arrive at the correct answer, rather than explicitly comparing the logical relationships between the context and options.

To encourage the model to pay more attention to the consistency of logical relationships between the context and options, we employ a data augmentation method that constructs positive instances consistent with the logic of the context and negative instances that are inconsistent. This augmentation enhances the model's logical reasoning capability. As outlined in Sect. 3.1, logical equivalence principles can be used to generate graphs that are both logically consistent and inconsistent with the original AMR graph.

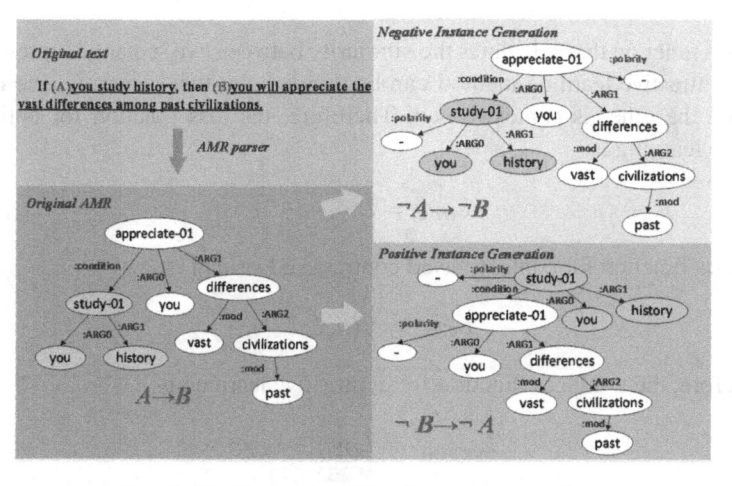

Fig. 3. Positive and negative instance generation

Figure 3 illustrates an example using the sentence "If you study history, then you will appreciate the vast differences among past civilizations". In this example, an AMR parser generates the original AMR graph, followed by the construction of positive and negative instances using logical equivalence principles for data augmentation. In particular, positive instances are constructed using the contrapositive law, while negative instances are created using the "then" relation. The AMR graph is then encoded using a Graph Convolutional Neural Network (R-GCN).

3.3 Contrastive Learning Pretraining

Contrastive Learning (CL) helps the model differentiate between logical differences in sentences by mapping two similar representations closer while expanding the mapping distance between two different representations. Each sentence is trained with both a positive sample (logically equivalent) and a negative sample (logically non-equivalent). For an instance x, a positive sample x + and a negative sample x- can be represented as follows:

$$distance\big(f(x), f(x^{+})\big) \ll distance\big(f(x), f(x^{-})\big) \tag{1}$$

Here, x is an instance sentence, x + is a positive sample where x and x + are logically equivalent, and x- is a negative sample where x and x- are not logically equivalent. The function f is an encoder for learning semantic representations, and the distance function represents the mapping distance between two semantic representations. The expected semantic representation distance between x and x + is expected to be closer than the distance between x and x-. Therefore, the formula for contrastive learning can be written as:

$$L = L(x, x^+, x^-) = -\sum \log\left(\frac{e^{s(x,x^+)}}{e^{s(x,x^-)}}\right)$$

(2)

where s is a function that calculates the similarity between two semantic representations.

Our contrastive learning method can be divided into two scenarios: one is option-based, and the other is context-based. Therefore, the loss function for option-based contrastive learning is:

$$L_{OBC} = L(c, a, A^-)$$

(3)

The loss function for context-based contrastive learning is:

$$L_{CBC} = L(a, c, C^-)$$

(4)

Therefore, the total loss function for contrastive learning is:

$$L_{contrast} = L_{OBC} + L_{CBC}$$

(5)

3.4 Answer Prediction

The purpose of the answer prediction module is to select the correct answer among the four options. For the AMR logical graphs corresponding to the four options, joint updates are performed using a Graph Convolutional Neural Network (R-GCN) and text encoding (Etext). All graph nodes are added to the alignment positions of the original text encoding vectors. Finally, a linear layer is used for the ultimate prediction.

$$\hat{y} = ReLU(linear(E_{AMR} + E_{text}))$$

(6)

$$L_{answer} = CrossEntropy(\hat{y}, y)$$

(7)

Therefore, during the training process, the total loss is given by:

$$L = \alpha L_{answer} + (1 - \alpha)L_{contrast}$$

(8)

In this equation, α represents a hyperparameter.

4 Experiments

4.1 Datasets

In this paper, experiments and analysis were conducted on two datasets, both of which are multiple-choice reading comprehension datasets that require logical reasoning abilities.

Reclor is a single-choice reading comprehension dataset from standardized tests such as the Graduate Management Admission Test (GMAT) and the Law School Admission Test (LSAT) in the United States. After filtering and selection, it comprises 6,138 data instances that test logical reasoning abilities. These data are randomly divided into 4,638 for training, 500 for development, and 1,000 for testing. The dataset contains 6,138 questions. The test set is categorized into simple and challenging test sets based on 24 logical reasoning types.

LogiQA is a dataset from the Chinese National Civil Service Examination, designed to assess the critical thinking and logical reasoning abilities of civil service candidates. After filtering and translation by experts into English, the dataset contains 8,678 questions. It is randomly split into training, development, and testing sets with an 8:1:1 ratio. To obtain the correct answers in these two datasets, logical reasoning and analysis based on the passage, question, and options are necessary.

4.2 Experimental Details

For our model, we employ state-of-the-art AMR parsers [27] and aligners [28] to obtain AMR graphs. Subsequently, we use Roberta-large [4], ALBERT-xxlarge [29], and DeBERTa-v2-xxlarge [30] as baselines. We set the learning rate to $\{1e-5, 2e-5, 3e-5\}$, with a warm-up rate of 0.1. By setting gradient accumulation, we use batch sizes of $\{16, 24, 32\}$, and the maximum number of epochs is set to 10. The hyperparameter α is chosen from $\{0.3, 0.5, 0.7\}$, and we select Adamw [31] as the optimizer with a weight decay rate of 0.01.

4.3 Result Comparison

The experimental evaluation metrics on the ReClar dataset and LogiQA dataset are based on accuracy, calculated as:

$$Accuracy = \frac{Number\ of\ questions\ answered\ correctly}{Total\ number\ of\ questions} \# \tag{9}$$

We compared our model with several baseline models and human performance. The baseline models for comparison include RoBERTa [4], ALBERT [29], and DeBERTa [30]. Additionally, we compared the experimental results with state-of-the-art methods such as DAGN [15], LReasoner [16], Logiformer [17], MERIt [12], IDOL [26], among others. Both datasets have average scores from graduate students on randomly selected test samples, which can be used as human performance benchmarks to compare with the model's performance.

Table 3 presents the experimental results on the ReClor dataset and the LogiQA dataset. From the results, we can analyze the following points:

1. The model performs well on RoBERTa, ALBERT, and DeBERTa.

2. The model shows significant improvement in performance on the ReClor difficult set, indicating a substantial enhancement in the model's logical reasoning ability.

Certainly, we also conducted experiments targeting different inference types within the ReClor dataset, as shown in Fig. 4. It can be observed that our model has shown

Table 3. The overall results on ReClor and LogiQA.

Model/Dataset	ReClor				LogiQA	
	Dev	Test	Test-E	Test-H	Dev	Test
RoBERTa	62.6	55.6	75.5	40.0	35.0	35.3
DAGN	65.2	58.2	76.1	44.1	35.5	38.7
LReasoner	66.2	62.4	**81.4**	47.5	38.1	40.6
AdaLoGN	65.2	60.2	79.3	45.1	39.9	40.7
MERIt	67.8	60.7	79.6	45.9	42.4	41.5
Logiformer	68.4	63.5	79.1	**51.3**	42.2	42.6
IDOL	**70.2**	**63.9**	-	-	42.5	41.8
AMR-CL	69.5	63.7	79.2	50.8	**42.6**	**42.7**
ALBERT	69.1	66.5	76.7	58.4	38.9	37.6
LReasoner	73.2	70.7	81.1	62.5	41.6	41.2
MERIt	73.2	·71.1	**83.6**	61.5	43.9	**45.3**
IDOL	**74.6**	70.9	-	-	44.7	43.8
AMR-CL	73.6	**71.2**	83.4	**62.6**	**44.9**	43.5
DeBERTa	74.4	68.9	83.4	57.5	44.4	41.5
LReasoner	74.6	71.8	83.4	62.7	**45.8**	43.3
MERIt	78.0	73.1	**86.2**	**64.4**	40.0	38.9
AMR-CL	**78.4**	**73.2**	86.0	64.2	44.8	**44.5**
Human	-	63.0	57.1	67.2	-	86.0

improvements across various inference types, especially in the "Implication", "Conclusion" and "Most Strongly Supported" categories. These types of questions often entail the prompt: "Of the following statements, which one is most strongly supported by the information above?" and place a greater emphasis on the model's comprehension of contextual semantic structures.

4.4 Ablation Experiments

To evaluate the effectiveness of the two main enhancement techniques in the model, namely, AMR graph construction and data augmentation based on logical equivalence, we conducted ablation experiments. The results of the ablation experiments are presented in Table 4.

The data in the table clearly indicates that both the AMR graph construction and the logic equivalence-based data enhancement modules significantly improved the model's performance. Additionally, it is evident that among the five logical equivalence laws, the laws of implication, contrapositive, and double negation make significant contributions to enhancing the model's logical reasoning ability.

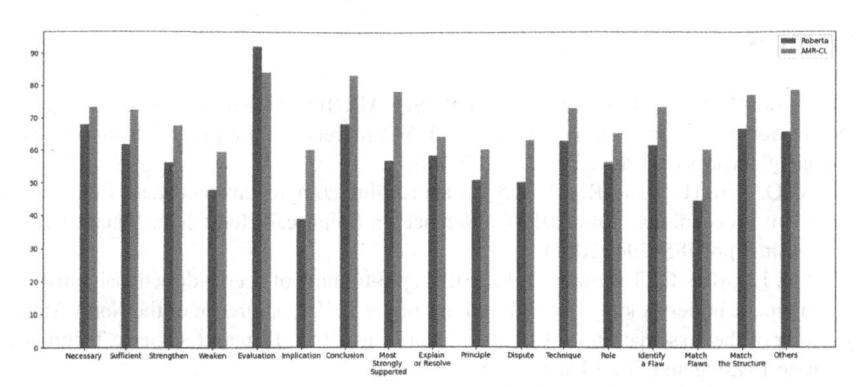

Fig. 4. Results of experiments with different types of reasoning

Table 4. Results of ablation experiments

Model/Dataset	ReClor			
	Dev	Test	Test-E	Test-H
AMR-CL	78.4	73.2	86.0	64.2
- AMR logical graph	71.2	68.4	84.2	62.8
- logical indicator	75.5	70.5	85.2	63.1
Logical equivalence				
- Commutative Law	78.3	73.1	86.0	64.1
- Implication Law	77.4	72.1	85.9	63.6
- Contrapositive Law	77.2	71.7	85.8	63.5
- Double Negation Law	76.9	71.4	85.6	63.5
- Category Inference Law	78.2	73.0	86.0	64.0

5 Conclusion

This paper presents a contrastive learning pretraining method based on AMR logical graphs for the task of logical reasoning in machine reading comprehension. We construct AMR logical graphs to represent the semantic and logical structure in the text, enhance logical relations based on logical trigger words, and employ logical equivalence principles for logical expansion. Subsequently, we use logical equivalence principles to create logically consistent positive samples and logically inconsistent negative samples for data augmentation, thereby assisting the model in learning logical differences through contrastive learning. We conducted experiments on two logical reasoning datasets, ReClor and LogiQA, and the results confirm the effectiveness of this approach.

References

1. Rajpurkar, P., Zhang, J., Lopyrev, K., et al.: SQuAD: 100,000+ questions for machine comprehension of text. In: Proceedings of the 2016 Conference on Empirical Methods in Natural Language Processing, pp. 2383–2392 (2016)
2. Lai, G.Q., Liu, H., et al.: RACE: large-scale reading comprehension dataset from examinations. In: Proceedings of the 2017 Conference on Empirical Methods in Natural Language Processing, pp. 785–794 (2017)
3. Kenton, J.D.M.W.C., Toutanova, L.K.: BERT: pre-training of deep bidirectional transformers for language understanding. In: Proceedings of the 2019 Conference of the North American Chapter of the Association for Computational Linguistics: Human Language Technologies, Volume 1 (Long and Short Papers), pp. 4171–4186 (2019)
4. Liu, Y., Ott, M., Goyal, N., et al.: Roberta: a robustly optimized bert pretraining approach. arXiv preprint arXiv:1907.11692 (2019)
5. Yang, Z., Dai, Z., Yang, Y., et al.: XLNet: generalized autoregressive pretraining for language understanding. In: Advances in Neural Information Processing Systems, vol. 32 (2019)
6. Brown, T., Mann, B., Ryder, N., et al.: Language models are few-shot learners. Adv. Neural. Inf. Process. Syst. **33**, 1877–1901 (2020)
7. Dua, D., Wang, Y., Dasigi, P., et al.: DROP: a reading comprehension benchmark requiring discrete reasoning over paragraphs. In: Proceedings of the 2019 Conference of the North American Chapter of the Association for Computational Linguistics: Human Language Technologies, Volume 1 (Long Papers), 2368–2378 (2019)
8. Huang, L., Le Bras, R., Bhagavatula, C., et al.: Cosmos QA: machine reading comprehension with contextual commonsense reasoning. In: Proceedings of the 2019 Conference on Empirical Methods in Natural Language Processing and the 9th International Joint Conference on Natural Language Processing (EMNLP-IJCNLP), 2391–2401 (2019)
9. Yu, W., Jiang, Z., Dong, Y., et al.: ReClor: a reading comprehension dataset requiring logical reasoning. Int. Conf. Learn. Representations (2019)
10. Liu, J., Cui, L., Liu, H., et al.: LogiQA: a challenge dataset for machine reading comprehension with logical reasoning. In: Proceedings of the Twenty-Ninth International Conference on International Joint Conferences on Artificial Intelligence, pp. 3622–3628 (2021)
11. Banarescu, L., Bonial, C., Cai, S., et al.: Abstract meaning representation for sembanking. In: Proceedings of the 7th Linguistic Annotation Workshop and Interoperability with Discourse (2013)
12. Jiao, F., Guo, Y., Song, X., et al.: MERIt: meta-path guided contrastive learning for logical reasoning. Find. Assoc. Comput. Linguist. ACL 2022, 3496–3509 (2022)
13. Qin, Y., Lin, Y., Takanobu, R., et al.: ERICA: Improving entity and relation understanding for pre-trained language models via contrastive learning. In: Joint Conference of the 59th Annual Meeting of the Association for Computational Linguistics and the 11th International Joint Conference on Natural Language Processing, ACL-IJCNLP 2021. Association for Computational Linguistics (ACL), pp. 3350–3363 (2021)
14. Pi, X., Zhong, W., Gao, Y., et al.: LogiGAN: learning logical reasoning via adversarial pretraining. In: Advances in Neural Information Processing Systems (2022)
15. Huang, Y., Fang, M., Cao, Y., et al.: DAGN: discourse-aware graph network for logical reasoning. In: Proceedings of the 2021 Conference of the North American Chapter of the Association for Computational Linguistics: Human Language Technologies, pp. 5848–5855 (2021)
16. Wang, S., Zhong, W., Tang, D., et al.: Logic-driven context extension and data augmentation for logical reasoning of text. Find. Assoc. Comput. Linguist. ACL 2022. 1619–1629 (2022)

17. Xu, F., Liu, J., Lin, Q., et al.: Logiformer: a two-branch graph transformer network for inter-pretable logical reasoning. In: Proceedings of the 45th International ACM SIGIR Conference on Research and Development in Information Retrieval, pp. 1055–1065 (2022)
18. Li, X., Cheng, G., Chen, Z., et al.: AdaLoGN: adaptive logic graph network for reasoning-based machine reading comprehension. In: Proceedings of the 60th Annual Meeting of the Association for Computational Linguistics (Volume 1: Long Papers), pp. 7147–7161 (2022)
19. Kingsbury, P., Palmer, M.: From TreeBank to PropBank. In: Proceedings of the Third International Conference on Language Resources and Evaluation (LREC'02) (2002)
20. Zhou, J., Naseem, T., Astudillo, R.F., et al.: AMR parsing with action-pointer transformer. In: Proceedings of the 2021 Conference of the North American Chapter of the Association for Computational Linguistics: Human Language Technologies, pp. 5585–5598 (2021)
21. Song, L., Gildea, D., Zhang, Y., et al.: Semantic neural machine translation using AMR. Trans. Assoc. Comput. Linguist. 7, 19–31 (2019)
22. Kapanipathi, P., Abdelaziz, I., Ravishankar, S., et al.: Leveraging abstract meaning representation for knowledge base question answering. In: Findings of the Association for Computational Linguistics: ACL-IJCNLP 2021, pp. 3884–3894 (2021)
23. Bai, X., Chen, Y., Song, L., et al.: Semantic representation for dialogue modeling. In: Proceedings of the 59th Annual Meeting of the Association for Computational Linguistics and the 11th International Joint Conference on Natural Language Processing (Volume 1: Long Papers), pp. 4430–4445 (2021)
24. Chen, T., Kornblith, S., Norouzi, M., et al. A simple framework for contrastive learning of visual representations. In: International Conference on Machine Learning, pp. 1597–1607. PMLR (2020)
25. Gao, T., Yao, X., Chen, D.: SimCSE: simple contrastive learning of sentence embeddings. In: Proceedings of the 2021 Conference on Empirical Methods in Natural Language Processing, pp. 6894–6910 (2021)
26. Xu, Z., Yang, Z., Cui, Y., et al.: IDOL: indicator-oriented logic pre-training for logical reasoning. In: Findings of the Association for Computational Linguistics: ACL 2023, pp. 8099–8111 (2023)
27. Bevilacqua, M., Blloshmi, R., Navigli, R.: One SPRING to rule them both: symmetric AMR semantic parsing and generation without a complex pipeline. In: Proceedings of the AAAI Conference on Artificial Intelligence, vol. 35, issue 14, pp. 12564–12573 (2021)
28. Blodgett, A., Schneider, N.: Probabilistic, structure-aware algorithms for improved variety, accuracy, and coverage of AMR alignments. In: Proceedings of the 59th Annual Meeting of the Association for Computational Linguistics and the 11th International Joint Conference on Natural Language Processing (Volume 1: Long Papers), 3310–3321 (2021)
29. Lan, Z., Chen, M., Goodman, S., et al.: ALBERT: a Lite BERT for Self-supervised learning of language representations. In: International Conference on Learning Representations (2019)
30. He, P., Liu, X., Gao, J., et al.: DeBERTa: decoding-enhanced BERT with disentangled attention. In: International Conference on Learning Representations (2020)
31. Loshchilov, I., Hutter, F.: Decoupled weight decay regularization. In: International Conference on Learning Representations (2018)

Automated Detection and Recognition of Wild Dolphin Behaviors Using Deep Learning

Jiahua Lin[1], Duan Gui[3], Quan Xie[3], Xundao Zhou[4,5(✉)],
and Yunxiao Shan[1,2(✉)]

[1] The School of Artificial Intelligence, Sun Yat-sen University, Zhuhai, Guangdong,
People's Republic of China
shanyx@mail.sysu.edu.cn
[2] Shenzhen Institute, Sun Yat-sen University, Shenzhen, Guangdong,
People's Republic of China
[3] The Southern Marine Science and Engineering Guangdong Laboratory, Zhuhai,
Guangdong, People's Republic of China
[4] The Fifth Research Institute of Ministry of Industry and
Information Technology (MIIT), Zhuhai, People's Republic of China
zhouxundao@ceprei.biz
[5] Key Laboratory of MIIT for Intelligent Products Testing and Reliability, Zhuhai,
People's Republic of China

Abstract. We presented a deep convolutional neural network approach that was able to automatically detect and recognize the behavior of wild dolphins. This study used a deep learning approach to automatically identify and analyze three visu-ally distinct behaviors (exiting behavior, wandering behavior, and entering be-havior) in dolphins. Using data directly from video recordings by both drones and handheld cameras, an action recognition model trained on this data achieved high levels of accuracy (0.99 for exiting, 0.92 for wandering, and 0.88 for entering). The method is the first to achieve visual action recognition of dol-phins behavior, opening new possibilities for using large datasets in dolphins research.

Keywords: Deep Learning · Convolutional Neural Networks · Dolphin Behavior Recognition

1 Introduction

Advancements in data collection technology, including drones, cameras, and hydrophones, now allow for the capture of dolphin behavior with unprecedented detail. This enables the measurement of changes at both individual and population levels, as well as alterations in dolphin behavior that may span extended time periods and large spatial scales. However, the training and human labor required to process vast quantities of video data still limit the scale and depth of behavioral analyses. The application of AI-based behavior recognition can expedite research on cetacean animal behavior, offering a potent tool for detailed investigations into dolphin behavior and life habits.

The prevailing research on dolphin detection predominantly centers on the classification of dolphin species and dolphin detection. In the field of acoustics, Luo et al. [7]. introduced the K-nearest neighbor method to distinguish between different dolphin species by identifying their unique acoustic click signals. In the field of image recognition, Mo et al. [10] utilize the YOLOv4 algorithm to construct a real-time detection model for the dorsal fin of the Chinese white dolphin. However, there remains an absence of studies addressing long-term dolphin behavioral patterns. The goal of this paper is to enable automated detection and recognition of behaviors in wild dolphin videos. Behavioral recognition based on deep learning has so far mostly been studied in the laboratory under constraints or using still images only, and has not yet been shown to work well for unconstrained video footage from the wild. Detecting behavior in wild dolphins in videos recorded from the sea surface is a major challenge-behaviors are often hard to detect because they are obscured by motion blur, occlusions, sea surface reflections, low resolution, or lighting conditions. If successful, these tools will enable us to explore a large number of questions in the field of dolphin behavioral ecology and dolphin conservation research. Data were obtained through long-duration automated recognition of behaviors in cross-comparisons of groups of dolphins in various contexts. These detailed behavioral data are also an important component of conservation research: They allow us to study how human activities may disrupt dolphin behaviors (i.e. migratory patterns, foraging, reproduction) and provide concrete data for assessing threats and levels of endangerment. This study used deep learning methods to automatically recognize and analyze three visually distinct behaviors (exiting the water, wandering, and entering the water) of dolphins. Chinese white-finned dolphins

For the first time, a method has been developed to recognize visual actions in dolphin behavior, paving the way for new opportunities in zoological research and conservation studies that utilize large data sets. The Chinese white dolphin, which prefers nearshore distribution, primarily resides within a 15 km radius from the coast. This species is an ideal test subject for recognizing dolphin behavior due to their presence in regions such as China, Guangdong, Guangxi, Hong Kong, Macao, Fujian, Taiwan, Hainan, and Zhejiang along the coast [3]. The population in the Pearl River Estuary may be the largest white dolphin population identified worldwide, and of course the largest in China, with an estimated number of about 2000 [4]. With complex sociality and behavioral flexibility, more abundant behaviors can be observed and recorded. Our goal is to identify entering, wandering and exiting behaviors. We randomly sampled several instances of dolphins and visualized them in Fig. 1.The framework for automatically identifying wild Chinese white dolphins' behavior consists of two stages: dolphin identification and key point detection and visual behavior recognition. First, through SAM on the video to do the segmentation of individual dolphins on the image to determine the location of the dolphin in space. Then through long-term changes in key points of a dolphin's posture to further determine the current behavior of the dolphin. LSTM [11] models are used respectively at this stage.

SAM can also be used to provide a preview mechanism to filter out the sequences of dolphin behaviors that need to be labeled by human annota-

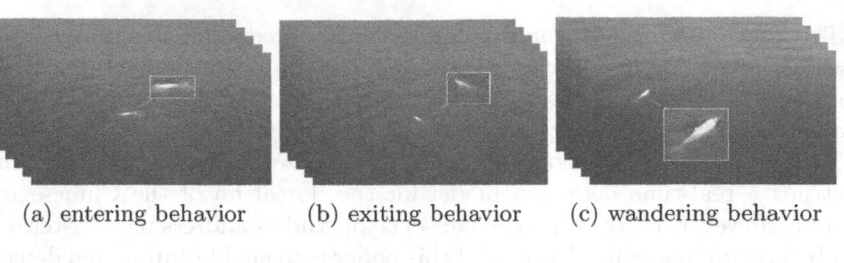

(a) entering behavior   (b) exiting behavior   (c) wandering behavior

Fig. 1. Dolphin behavior

tors, thereby greatly reducing the time required to collect annotations. This is achieved through a semantic segmentation model using grounded-SAM that can identify short video sequences where dolphins are likely to occur and segment out the images of the dolphins in Fig. 2. Human annotators then only need to verify if the sequence contains dolphin behavior. Key points are then directly output via an improved deeplabcut for the LSTM network to perform dolphin behavior recognition. This method is able to automatically recognize when and where fine movements (such as emergence, cruising, and entry behavior) occur. The behavioral prediction component consists of a deep LSTM model that uses both vision and key point predictions for behavior. Data was collected by the authors and their team on trips out at sea. Given the richness of behavioral data, automated parsing can facilitate a more nuanced understanding of specific dolphin behaviors. This approach represents the inaugural instance of automatic identification of wild dolphin behavior through key point and visual analysis.

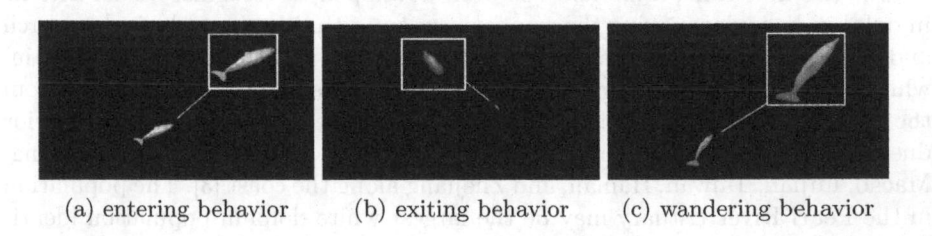

(a) entering behavior   (b) exiting behavior   (c) wandering behavior

Fig. 2. After grounded-SAM dolphin behavior

2 Dolphin Dataset

The Dolphin Behavioural Data Set is a comprehensive collection of information on the behavior of dolphins. This data set provides an in-depth analysis of their movements, social interactions, and other key behaviors. It serves as a valuable resource for researchers studying marine biology, psychology, and ethology. The detailed observations and measurements captured within this dataset enable scientists to gain a deeper understanding of these intelligent creatures and their natural habitats.

2.1   Collecting and Annotating

The majority of Chinese white dolphins are typically found in the Pearl River Estuary [9]. We conduct 20 boat-line surveys in the Jiangmen and Zhuhai Port of Zhuhai-Macao Bridge area to take images and videos of wild dolphin samples. The videos were taken under different weather conditions, with a total duration of 31.5 h. We cut the collected dolphin videos into image frames and extract segmented images of each frame of dolphins by semantic segmentation using the SAM-grounded model. To train the dolphin pose estimation model, we obtain key point predictions of dolphins and manually correct incorrect key points or add missing key points. Dolphin Key Point Annotation. In the first stage, we manually browse through the video and select 4192 frames containing dolphins. We use deeplabcut to annotate the key points of each frame in Fig. 3. Then, we split 4000 frames into 3192 frames and 1000 frames for model training/validation. Finally, we train a dolphin pose estimation model on this initial subset for the second annotation stage.

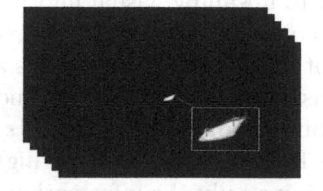

Fig. 3. Key points of dolphin pose estimation

Dolphin behavior annotation. We define three behaviors of dolphins based on the content of the preliminary dolphin ethology and key points of dolphins. Specifically, we first obtain the dolphin key point data generated by the dolphin pose estimation model. Note that these dolphin key point data can be both true positive (e.g., dolphin) and interfering objects (e.g., waves, sunlight, ship). In addition, there is also a problem of missing true positives (dolphin). To solve this problem, we manually check the prediction results, filter false-positive labels through the characteristics of continuity of dolphin key points in time series, and consider missing true positives. This strategy enables us to more accurately realize the behavior prediction of dolphins. For long time series, only the image frames with head key points or with head key points and blowhole key points are annotated as emergence behavior.

2.2   Dataset Statistics

The final accumulation of only 4002 images contains dolphins (one image can contain one or more dolphins), and the number of dolphins accounts for only

16% of the total frames, which indicates that wild dolphins are rare. Because the body size of wild dolphins is small, they will be disturbed by objects with similar sizes.

3   Detection Algorithms and Systems

We adopt the off-the-shelf SAM-grounded [5] semantic segmentation large model algorithm for dolphin detection. Specifically, we fine-tune a Grounded-Segment-Anything (GSA) model on the dolphin dataset, where GSA models are a combination of several deep learning models, including EfficientNet [6], DINO [12] and SAM. Using EfficientNet as the image encoder, DINO is used for feature extraction, and SAM is used for segmentation1. The GSA model segments the input image into dolphin objects, then each object is divided into different numerical blocks to save the generated dolphin mask blocks to generate txt files. The deeplabcut [8] model was then used to extract feature frames of different poses of the dolphins from the video, followed by manual annotation of body parts such as head, blowhole, dorsal fin, tail. A deep residual neural network (ResNet [2]) was used as a feature extractor which is pre-trained on ImageNet dataset [1] and has strong object recognition capabilities. A deconvolution layer is added after the output layer of ResNet to upsample visual information and generate spatial probability density maps. For each body part, its probability density map represents the "confidence" with which the part appears at a certain pixel location. The network was trained using annotated data to fine-tune it so that it assigns high probabilities to locations of annotated body parts and low probabilities elsewhere. Subsequently, data from body parts with high and low probabilities is filtered and simplified, retaining only the information of each body part present in the current frame. Utilizing semantic key point data to enhance the training of the behavior recognition model. The long-term memory characteristics of the LSTM network enable recognition of dolphin behavior within the current 10 frames through LSTM network training. The method flow structure is shown in Fig. 4. The average accuracy has achieved a high level (exiting behavior: 0.99, wandering behavior: 0.92, entering behavior: 0.88), marking the first instance of visual motion recognition of dolphin behavior.

4   Experiments

4.1   Training and Inference

Since these detectors have different training/inference strategies, we illustrate their settings separately below. The initial section focuses on the training and inference of the Deeplabcut model, which employs ResNet-50 as its backbone network. During the training phase, the model undergoes 10,000 epochs. In terms of pose estimation during the inference stage, data is filtered and simplified based on high and low probability body part positions. Body part position information

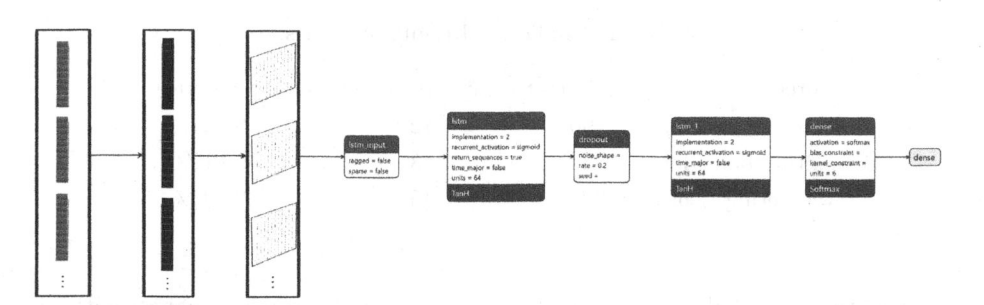

Fig. 4. The method flow structure

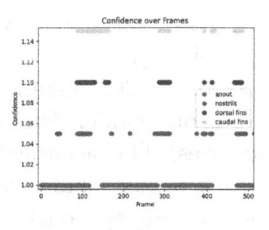

Fig. 5. The distribution of keys per frame

is retained if it has a confidence score exceeding 0.8, while any data with a confidence score below 0.8 is excluded (Fig. 5).

In the second part of the study, behavior recognition was employed. The training and inference processes utilized the LSTM model to match each frame's current behavioral big category with the dolphin mask data on a one-to-one basis. The model underwent training for a total of 1000 epochs.

4.2   Result

For the entering behavior recognition, we analyzed 40.2 h of video, which contains 102 action sequences of in-water behaviors.For exiting behavior recognition, we analyzed 12.5 h of camera with a total number of 337 "exiting" behavior. For "wandering" behavior recognition, we analyzed 20.7h of drones with a total number of 225 "wandering" behavior. Through this experiments, a total of 633 individual body trajectories were obtained.

Table 1 presents the statistics of behavior prediction results obtained from our model. The table provides a breakdown of the correct and incorrect predictions for different behavior categories, along with the corresponding accuracy. From the results, we can observe that the model performs well in predicting different behaviors. For the "entering" behavior category, the model achieved an accuracy of 88.24%, correctly predicting 90 instances out of a total of 102 samples. However, there were 12 incorrect predictions in this category.

Table 1. Prediction Results Statistics

Category	Correct Count	Incorrect Count	Accuracy
entering behavior	90	12	88.24%
exiting behavior	334	3	99.11%
wandering behavior	209	16	92.94%

The "exiting" behavior category displayed impressive performance, with an accuracy of 99.11%. The model accurately predicted 334 instances out of 337 samples, indicating a high level of precision in identifying this behavior. Only 3 incorrect predictions were observed in this category.

Regarding the "wandering" behavior category, the model achieved an accuracy of 92.94%. It correctly predicted 209 instances out of 225 samples, with 16 incorrect predictions. Overall, our model demonstrates strong performance across different behavior categories. The accuracy scores of 88.24%, 99.11%, and 92.94% for the "entering," "exiting," and "wandering" behaviors, respectively, showcase the effectiveness of our approach in accurately identifying and predicting these specific behaviors. Please note that these results are subject to variations based on dataset characteristics, model architecture, and training techniques.

The image segmentation accuracy of the dolphin body achieved an average precision of 92%. The entering behavior recognition model performed well under different postures and illumination conditions, our LSTM [11] model achieved high average precision (89.79%) on the entering behavior recognition scenario (Table 2).

Table 2. Model Performance Evaluation

Metrics	Score
Accuracy	94.98%
Recall	88.79%
F1-score	89.97%

5 Conclusion

In this paper, we have conducted three tasks: collection and annotation of the dataset, experiments with behavior recognition algorithms. Specifically, we used a two-stage annotation strategy to build our dolphin dataset. Finally, we conducted experiments to compare and select an optimal behavior recognition algorithm that balances efficiency and performance. With this algorithm, we facilitate computer vision to solve problems in the study of marine mammal behaviors with less human effort.

Acknowledgment. This research is supported by the Southern Marine Science and Engineering Guangdong Laboratory, Zhuhai (SML2021SP202). Additionally, it is also supported by the Shenzhen science and technology program (JCYJ202103241222 03009), the National Natural Science Foundation of China (No. 62232008), the Guang-Dong Basic and Applied Basic Research Foundation (2020A1515110199), the Guangdong HUST Industrial Technology Research Institute, the Guangdong Provincial Key Laboratory of Manufacturing Equipment Digitization (2020B1212 060014), the China's University-Industry Research Collaboration Fund for Innovation (2021ZYA03005), and Key Laboratory of MIIT for Intelligent Products Testing and Reliability 2023 Key Laboratory Open Project Fund (No. CEPREI2023-02).

References

1. Deng, J., Dong, W., Socher, R., Li, L.J., Li, K., Fei-Fei, L.: Imagenet: a large-scale hierarchical image database. In: 2009 IEEE Conference on Computer Vision and Pattern Recognition, pp. 248–255. IEEE (2009)
2. He, K., Zhang, X., Ren, S., Sun, J.: Deep residual learning for image recognition. In: Proceedings of the IEEE Conference on Computer Vision and Pattern Recognition, pp. 770–778 (2016)
3. Huang S L, Karczmarski L, C.J.: Demography and population trends of the largest population of indo-pacific humpback dolphins. Biological Conservation (2012)
4. Karczmarski, et al.: Socio-spatial dynamics of a coastal delphinid in a heavily anthropogenically impacted estuarine seascape (2019)
5. Kirillov, A., et al.: Segment anything. arXiv preprint arXiv:2304.02643 (2023)
6. Koonce, B., Koonce, B.: Efficientnet. Convolutional Neural Networks with Swift for Tensorflow: Image Recognition and Dataset Categorization, pp. 109–123 (2021)
7. Luo, W., Yang, W., Zhang, Y.: Convolutional neural network for detecting odontocete echolocation clicks. J. Acoustical Soc. America **145**(1), EL7-EL12 (2019)
8. Mathis, A., et al.: Deeplabcut: markerless pose estimation of user-defined body parts with deep learning. Nat. Neurosci. **21**(9), 1281–1289 (2018)
9. Miller, P.S.: A population viability analysis for the chinese white dolphin (sousa chinensis) in the pearl river estuary. IUCN/SSC Conservation Breeding Specialis t Group, Hong Kong (2016)
10. Mo, H., Zhang, Y., Su, C., Zheng, Y.: Research on the detection algorithm of dorsal fin of chinese white dolphin based on yolov4. In: 2021 IEEE International Conference on Power, Intelligent Computing and Systems (ICPICS), pp. 778–781. IEEE (2021)
11. Yu, Y., Si, X., Hu, C., Zhang, J.: A review of recurrent neural networks: Lstm cells and network architectures. Neural Comput. **31**(7), 1235–1270 (2019)
12. Zhang, H., et al.: Dino: Detr with improved denoising anchor boxes for end-to-end object detection. arXiv preprint arXiv:2203.03605 (2022)

M-GFCC: Audio Copy-Move Forgery Detection Algorithm Based on Fused Features of MFCC and GFCC

Dongyu Wang, Canghong Shi$^{(\boxtimes)}$, Junrong Li, Jiaxin Gan, Xianhua Niu, and Ling Xiong

Xihua University, Chengdu 610039, China
canghongshi@163.com

Abstract. Audio copy-move forgery has seriously affected the authenticity of audio, and copy-move forgery detection and localization of audio has become an urgent problem. In this paper, we propose an audio copy-move forgery detection algorithm based on the fusion of mel frequency cepstrum coefficient and gammatone frequency cepstrum coefficient feature fusion. Firstly, the algorithm extracts the voiced speech segments in the audio using the algorithm of voice activity detection, and then extracts the MFCC and GFCC features from each voiced speech segment, and then the corresponding weights are assigned to the two features for weighted summation to get the fused features. Finally, the similarity between the fused features of each voiced speech segment is calculated by using the dynamic time warping algorithm, and the part with the DTW distance less than the threshold is determined to be copy-move forgery. The experiments on publicly available replicated mobile forgery databases show that the algorithm not only enables precise localization of tampered audio, but also has high robustness.

Keywords: Copy-move forgery detection · Mel frequency cepstral coefficients · Gammatone frequency cepstral coefficients · Dynamic time warping · robustness

1 Introduction

With the advancement of digital multimedia technology, digital audio has gained widespread usage in our daily lives. In scenarios where these audio recordings are needed as evidence in court or other contexts, the authenticity of these audio files becomes crucial. As a result, digital audio tampering detection technology has emerged as a prominent area of interest, particularly for passive forensic purposes. One common form of audio tampering is audio copy-move forgery, where users copy segments of an audio clip and paste them elsewhere within the same audio file. For instance, by copying the word "not" from a particular segment of speech and pasting it into another part of the same audio, the original content "He is present." could be altered to "He is not present," thereby completely changing

H. Jin et al. (Eds.): IAIC 2023, CCIS 2058, pp. 220–234, 2024.
https://doi.org/10.1007/978-981-97-1277-9_17

the meaning of the original audio content. Figure 1 illustrates an audio file that has undergone copy-move forgery. The two red rectangular boxes indicate the copy-move forgery sections. Detecting such manipulations solely through human auditory assessment is challenging, resulting in lowered detection accuracy. In most cases, users tend to add common post-processing operations to audio tampering, such as adding noise or re-compressing the audio, thus making audio copy-move forgery detection operations more difficult.

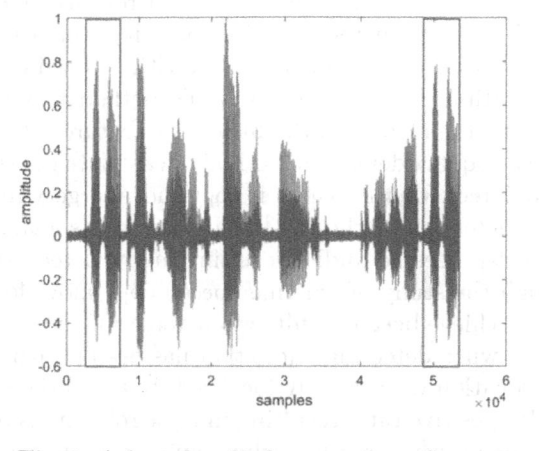

Fig. 1. A forged audio by copy-move forgery.

In recent years, several experts have adopted various methods to detect copy-move forgery in audio signals. Xie et al. [1] proposed a voice copy-move forgery detection algorithm that incorporates multiple features. Firstly, they employed a dual-threshold method to extract voiced segments from the audio files. Subsequently, they utilized the C4.5 decision tree to fuse features including gammatone cepstral coefficients feature, mel frequency cepstral coefficients feature, pitch feature, and discrete Fourier transform coefficients feature, enabling audio forgery detection. However, this approach exhibits high time complexity, which is not conducive to batch processing for efficient detection. Liu et al. [2] proposed a rapid audio forgery detection algorithm for copy-move. This method involves segmenting audio clips into phonemes, followed by extracting features through discrete Fourier transform (DFT). Subsequently, sorting features for each audio segment, and finally, the Pearson correlation coefficient is employed to measure the similarity between adjacent audio segments. This algorithm effectively reduces the time complexity from $O(n^2)$ to $O(nlogn)$. However, when detecting post-processed audio, the algorithm exhibits relatively lower precision rates. Yan et al. [3] proposed an audio copy-move forgery detection algorithm based on pitch similarity. Initially, they extracted pitch features between various voiced segments and then employed Pearson correlation coefficient and mean square deviation to compare the pitch similarities among these segments. However, this method is difficult to accurately localize when detecting post-processed audio. In order to enhance algorithm robustness, Yan et al. [4] optimized their approach and introduced a voice copy-move forgery detection algorithm based on pitch and formant. Nonetheless, this method exhibits limited robustness against post-processing operations that can alter pitch. Ustubioglu et al. [5] proposed an audio copy-move forgery detection algorithm based on pitch and modified discrete cosine transform. After utilizing a pitch tracking algorithm to extract voiced segments, MDCT coefficients are extracted for each audio segment. Finally, the algorithm calculates the similarity between speech segments using the Euclidean distance. However, this algorithm exhibits limited robustness in detecting tampering in audio that has undergone post-processing operations. Imran et al. [6]proposed a copy-move forgery detection algorithm based on 1D local binary patterns. The algorithm first utilizes the VAD method to segment the voiced and muted segments, then extracts the histogram features of each syllable, and finally uses the mean square deviation method to calculate the similarity between the histograms to detect whether there is copy-move forgery in the audio. However, the endpoint detection method of the algorithm has a large detection error when detecting post-processed audio, resulting in low detection accuracy of the algorithm. Through the study of existing speech copy-move forgery techniques, these methods can achieve better results when detecting audio that has not been post-processed, but when detecting audio that has been post-processed, not only is the tampering location inaccurately localized, but also these algorithms tend to have a high false positive rate, resulting in poor robustness of the algorithms. Therefore, in this paper we propose a robust audio copy-move forgery detection algorithm based on the fusion of MFCC features and GFCC features. The experiments indicate that our proposed audio copy-move forgery detection algorithm can accurately pinpoint the tampering location and exhibits a higher detection accuracy compared to state-of-the-art algorithms.

In later sections, Sect. 2 introduces the basics, which mainly includes the extraction of MFCC speech features and GFCC speech features. Section 3 mainly proposes this audio copy-move forgery detection algorithm. Section 4 gives the experimental results as well as comparison experiments. Finally, Sect. 5 summarizes the entire text.

2   Theory Basis

2.1   MFCC Feature Extraction

Mel frequency cepstral coefficient (MFCC) is a very important feature extraction technique widely used in the field of speech signal processing and speech recognition [7,8]. The MFCC feature can simulate the auditory perception of the human ear in a way that captures the basic acoustic properties of speech, with a high level of anti-interference, and still has a high level of recognition when recognizing audio that has undergone post-processing. The MFCC feature extraction flowchart is shown in Fig. 2. In the process of MFCC feature extraction, firstly, the speech x(n) is pre-processed as $x_i(m)$, which mainly includes pre-emphasis, frame blocking, and adding hamming window, and then the fast

Fourier transform is performed on each frame of the signal after adding the window to get the spectral energy $E_i(k)$, and then the Mel-filter bank is passed to get the mel spectral energy $H(i, m)$, and finally, the MFCC features are obtained by performing the discrete cosine transform. The MFCC features are shown in Eq. (1) [9].

$$mfcc(i,n) = \sum_{m=1}^{M} \log[H(i,m)] \cdot \cos\left[\frac{\pi \cdot n \cdot (2m-1)}{2M}\right] \tag{1}$$

where M denotes the number of Mel filters, i denotes the i-th frame of the speech signal, and n denotes the n-th column of the i-th frame of the signal.

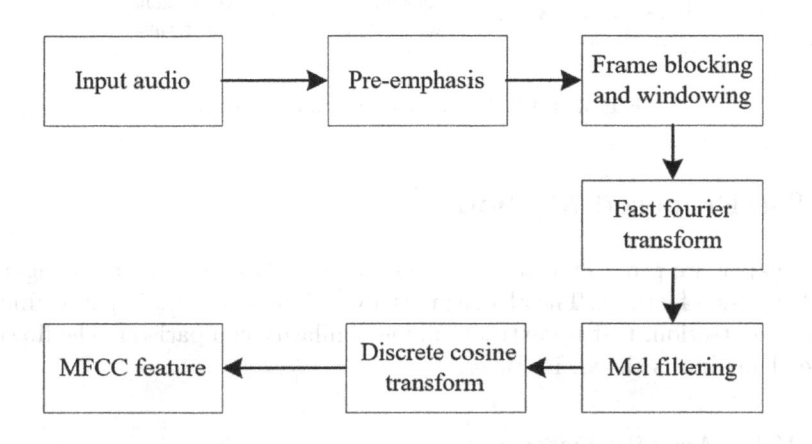

Fig. 2. MFCC feature extraction flowchart.

2.2   GFCC Feature Extraction

The gammatone frequency cepstral coefficients (GFCC) feature extraction process is similar to that of MFCC, but instead of using the triangular filter bank used in MFCC, we replace it with a set of gammatone filters [10]. The GFCC feature extraction flowchart is shown in Fig. 3. The gammatone filter bank consists of M filters with different central frequencies, as shown in the time-domain expression in Eq. (2) [11].

$$g_i(t) = b_i^n t^{n-1} e^{-2\pi b_i t} \cos(2\pi f_i + \varphi) U(t), t \geq 0, i = 1, 2, \cdots, M \tag{2}$$

After passing through the gammatone filter banks, we proceed with the discrete cosine transform, ultimately generating the GFCC feature parameters, as shown in Eq. (3).

$$gfcc(m) = \sqrt{\frac{2}{M}} \sum_{i=0}^{M-1} Y(i) \cos\left[\frac{\pi m(2i-1)}{2M}\right] \tag{3}$$

where Y(i) denotes the signal after the i-th filter, M represents the number of filters, and m denotes the m-th dimension of the GFCC feature.

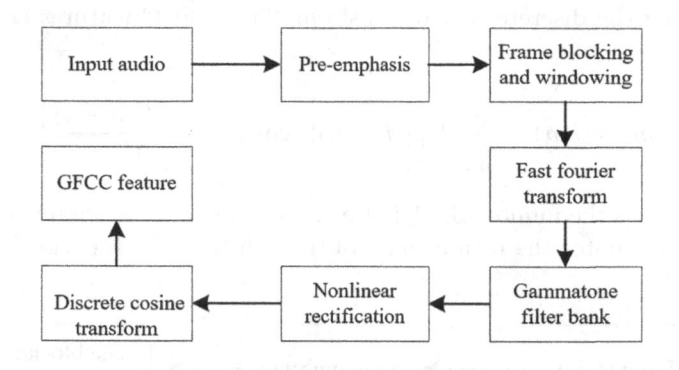

Fig. 3. GFCC feature extraction flowchart.

3   The Proposed Method

In this paper, we propose a robust audio copy-move forgery detection algorithm based on fusion features. The algorithm is divided into three main parts which are endpoint detection, feature extraction and similarity comparison. The flowchart of the algorithm is shown in Fig. 4.

3.1   Voice Activity Detection

In this paper, we use an endpoint detection algorithm based on short-term energy and spectral spreading to achieve accurate endpoint detection [12]. In speech signals, voiced speech segments are often characterized by higher energy levels, whereas non-speech segments typically exhibit lower energy levels. On the other hand, spectral spread feature serves as a crucial indicator describing the distribution of spectral energy. The introduced approach combines the short-term energy feature and spectral spread feature to comprehensively consider the energy and spectral distribution of the signal. We are able to accurately identify the positions of speech segments, thus achieving precise endpoint detection. Assuming the speech signal is denoted as x(n), two feature sequences, namely short-time energy and spectral spread, are extracted from the speech signal. First, the speech signal x(n) is segmented into frames and then subjected to windowing to obtain $y_i(n)$. The short-time energy is defined as shown in Eq. (4).

$$E(i) = \sum_{n=0}^{L-1} y_i^2(n) \tag{4}$$

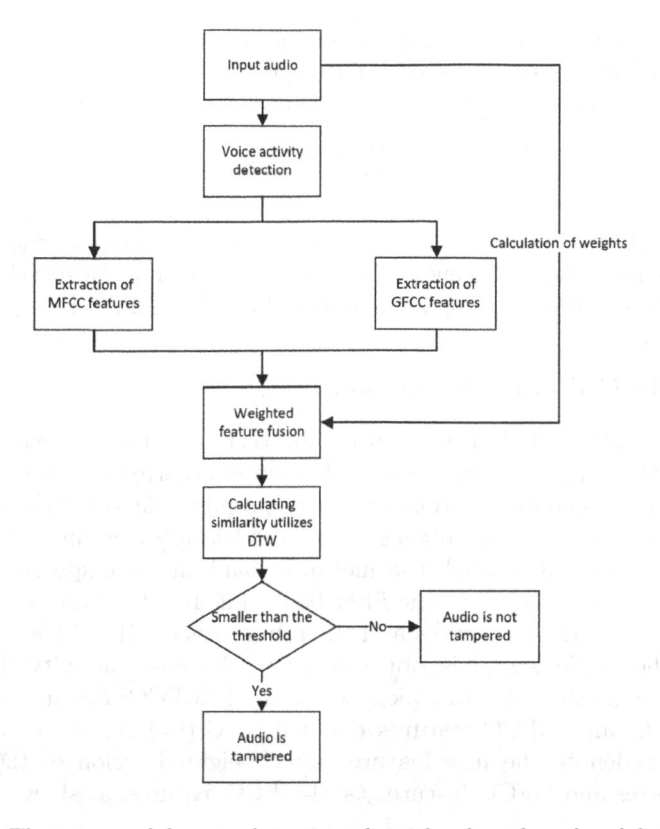

Fig. 4. The proposed forgery detection algorithm based on fused features.

where E(i) denotes the short-term energy of the signal of the i-th frame, $y_i(n)$ denotes the value of the signal of the i-th frame. L denotes the length of the frame, and fn denotes the total number of frames obtained after segmentation [13,14]. The spectral spread is defined as shown in Eq. (5).

$$spread = \sqrt{\frac{\sum_{k=b_1}^{b_2} (f_k - \mu_1)^2 s_k}{\sum_{k=b_1}^{b_2} s_k}} \tag{5}$$

where b_1 and b_2 represent the band edges, f_k is the frequency, s_k is the spectral value. μ_1 is the spectral centroid, the centroid is defined as shown in Eq. (6).

$$centroid = \frac{\sum_{k=b_1}^{b_2} f_k s_k}{\sum_{k=b_1}^{b_2} s_k} \tag{6}$$

After separately extracting the two feature sequences of short-term energy and spectral spread, calculate histograms for each feature sequence. Then, perform smoothing on the histograms and extract local maxima. Calculate the threshold T using Eq. (7). As a result, we can obtain the threshold T_1 for short-term energy feature and T_2 for spectral spread feature. Finally, we merge consecutive frames where both features exceed their respective thresholds. These continuous frame segments constitute the extracted speech segments.

$$T = \frac{W \times M_1 + M_2}{W + 1} \tag{7}$$

where M_1 and M_2 represent the first and second local maxima, respectively. W stands for weight. Using this endpoint detection algorithm the speech signal x(n) is segmented into k voiced segments, denoted as $S_i(i=1,2,...,k)$.

3.2   M-GFCC Feature Extraction

Both MFCC and GFCC features may exhibit certain deficiencies when used individually, as they might not fully reflect the comprehensive nature of speech information [15–17]. Therefore, in this article, we propose a feature fusion method to leverage their respective advantages. The fused features contain both the speech information extracted through the mel filter bank and the speech information extracted through the gammatone filter bank. The fused features emphasize the high accuracy of MFCC features and the robustness of GFCC features.

When the audio signal is obtained after the voice activity detection of the voiced segment $S_i(i=1,2,...,k)$, we extract MFCC features denoted as $M_i(i=1,2,...,k)$ and GFCC features denoted as $G_i(i=1,2,...,k)$ for each voiced segment. We denote the new features after weighted fusion of the extracted MFCC features and GFCC features as M-GFCC features, as shown in Eq. (8).

$$M - GFCC_i = \omega_{MFCC} \times M_i + \omega_{GFCC} \times G_i \tag{8}$$

$$\omega_{MFCC} + \omega_{GFCC} = 1 \tag{9}$$

where ω_{MFCC} denotes the weight corresponding to the MFCC feature and ω_{GFCC} denotes the weight corresponding to the GFCC feature.

In this article, we determine the weights for ω_{MFCC} and ω_{GFCC} as follows. Different combinations of weights for ω_{MFCC} and ω_{GFCC} will affect the robustness of the fused features. When we extract the audio features after post-processing, the higher robustness of the features will effectively increase the accuracy of the algorithm in detection. Firstly, the sum of the weights for ω_{MFCC} and ω_{GFCC} is set to 1, as shown in Eq. (9). Subsequently, we assign different weights to ω_{MFCC} and ω_{GFCC} separately and conduct a batch of experiments. We calculate the dynamic time warping(DTW) distance between voiced segments for different corresponding weight combinations. Since a smaller DTW value indicates higher similarity between voiced segments, the values of ω_{MFCC} and ω_{GFCC} that correspond to the minimum DTW are considered the optimal weights. For this reason we randomly selected audio files for batch experiments, and the results of the experiments are shown in Fig. 5, where the horizontal coordinates indicate the different combinations of weights of ω_{MFCC} and ω_{GFCC}, and the vertical coordinates indicate the percentage of the number of audios when the DTW achieves a very small value. It can be clearly seen from Fig. 5 that when the ω_{MFCC} weights are set to 0.3 and ω_{GFCC} weights are set to 0.7, the probability of DTW obtaining minimum values is the highest, so it is set as the final weight combination in this paper. Figure 6 shows the copy-move forgery audio signal without post-processing, and Fig. 7 shows the M-GFCC features extracted from the audio signal, we can observe that different audible segments have different M-GFCC features, and if they are copy-move forgery, they have very similar M-GFCC features, Therefore, we can determine whether the audio has been copy-move forgery by comparing the feature similarity. Figure 8 shows the audio signal after adding 10db gaussian white noise, and Fig. 9 shows the M-GFCC features extracted from the post-processed audio, which can be very similar to the copy-move forgery segments, and thus the features have high robustness.

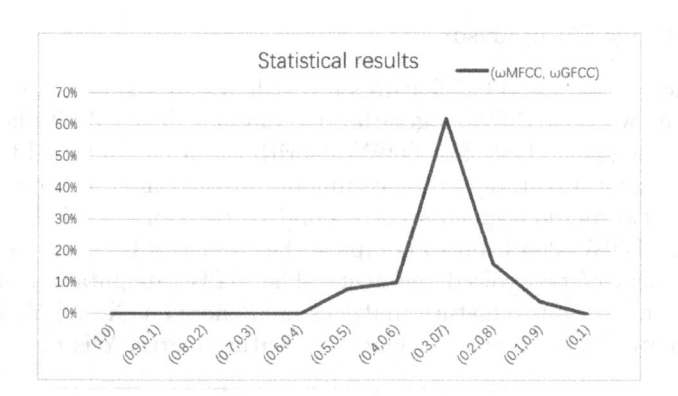

Fig. 5. Statistical results of different weight combinations.

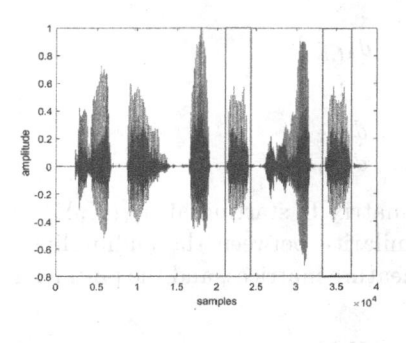

Fig. 6. An original copy-move audio.

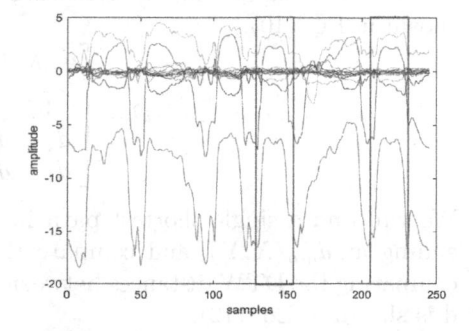

Fig. 7. M-GFCC feature of the original audio.

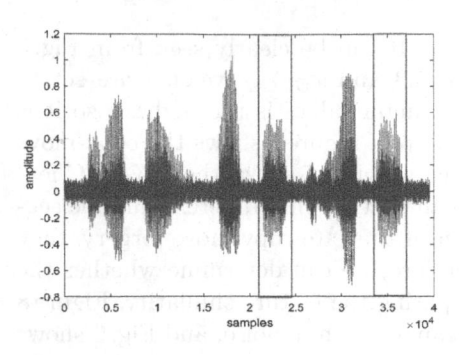

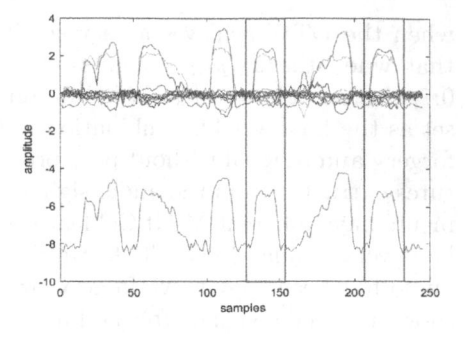

Fig. 8. A copy-move audio with 10 dB Gaussian noise.

Fig. 9. M-GFCC feature of the audio with 10 dB Gaussian noise.

3.3   Similarity Comparison

After extracting the M-GFCC features for each voiced segment, we employ the dynamic time warping(DTW) algorithm to compute the similarity between different voiced segments [18]. The DTW algorithm is a method used to calculate similarity between two time series, commonly utilized in speech recognition to determine if two audio segments correspond to the same word. In this study, we apply the DTW algorithm to compute the similarity between the M-GFCC feature matrices of two voiced segments. The DTW computation proceeds as follows: assuming the two feature matrices are denoted as X and Y. The length of feature matrix X is m, and the length of feature matrix Y is n.

$$dmn(X,Y) = \sqrt{\sum_{k=1}^{K} (x_{k,m} - y_{k,n}) \times (x_{k,m} - y_{k,n})} \qquad (10)$$

$d_{mn}(X,Y)$ denotes the distance between the m-th sample in the X-feature matrix and the n-th sample in the Y-feature matrix, and $d_{mn}(X,Y)$ is calculated as shown in Eq. (10).

$$D = \begin{bmatrix} d_{1,N} & d_{2,N} & \dots & d_{M,N} \\ \vdots & \vdots & \ddots & \vdots \\ d_{1,2} & d_{2,2} & \dots & d_{M,2} \\ d_{1,1} & d_{2,1} & \dots & d_{M,1} \end{bmatrix} \qquad (11)$$

We will find a single shortest path in the matrix D starting at $d_{11}(X,Y)$ and ending at $d_{mn}(X,Y)$, and compare the similarity between the audio clips by comparing the DTW distances between the feature matrices, and the path length d is shown in Eq. (12).

$$d = \sum_{m \in ix, n \in iy} d_{mn}(X,Y) \qquad (12)$$

4   Experimental Results

In this section, we focus on describing the results of our experiments on the copy-move forgery databases we have generated, as well as comparing them to three state-of-the-art algorithms.

4.1   Performance Metrics

To evaluate the experimental performance of the algorithm, we used two performance metrics: Precision and Recall.

Precision refers to the percentage of correctly detected tampered audio instances among the total number of detected tampered audio instances. Precision is defined by Eq. (13):

$$precision = \frac{TP}{TP + FP} \tag{13}$$

Recall refers to the percentage of correctly detected tampered audio instances among all tampered audio instances. Recall is defined by Eq. (14):

$$recall = \frac{TP}{TP + FN} \tag{14}$$

where TP represents the number of correctly detected tampered audio samples, FP represents the number of wrongly detected untampered audio samples, and FN represents the number of undetected tampered audio samples.

4.2   Copy-Move Forgery Dataset

In this article, the manipulated speech database we used was generated from the TIMIT database [19]. Each audio in TIMIT is approximately 4 s long, with an audio sampling rate of 8kHz. For the manipulated database, we randomly selected voiced segments from each audio file and pasted them into other parts of the same audio file. In the copy-move forgery database, there were 368 forgery audio files generated based on pitch, and 148 forgery audio files generated based on voice activity detection. We randomly selected 125 audio files from each of the pitch-based and vad-based databases to create a mixed database. Furthermore, we applied some common post-processing operations to the copy-move forgery audio, such as adding noise, median filtering, and MP3 compression [20]. In the end, we obtained a total of 1500 forgery audio files in the mixed database for experimentation. The configuration environment used in this experiment is as follows: Intel(R) Core(TM) i5-12400F 2.50 GHz, 32 GB DDR4 RAM, Windows 10 operating system, and MATLAB 2023a software as the experimental platform. As a result, the copy-move forgery audio files are obtained https://drive.google.com/file/d/1i9TmYy_tZw6PUBrVN8oO7KVAJqe14r93/view?usp=sharing [5].

4.3   Threshold Selection

In this section, we will explain the setting of the DTW threshold, and when the DTW values between the speech segments we are calculating, it is obvious to find that the DTW values between the duplicated speech segments will be much smaller than the distance between the non-duplicated speech segments. We count the DTW values of the audio segments in the copy-move forgery speech database, and the statistical results are shown in Table 1. When the DTW threshold we set is 20, the proportion of tampered segments is 98.80% and the proportion of non-tampered segments is 0.80%, and the detection accuracy is highest under this threshold.

Table 1. Statistical results of DTW

DTW	0–10	0–20	0–30
Duplicated	85.20%	98.80%	99.60%
Normal	0%	0.80%	2.40%

4.4   A Example of Detection

In this subsection, we will demonstrate the detection process by selecting an audio from the copy-move forgery database with 20db Gaussian white noise added as an example. Firstly, we extract the voiced segments in the audio file by using the speech activity detection algorithm, and the detection results of the turbid segments are shown in Fig. 10. We can observe that this endpoint detection algorithm can accurately divide the selected audio file into 8 voiced segments. The DTW values between the feature matrices of the individual voiced segments are then calculated. It is evident from Table 2 that the DTW value between voiced segment 2 and voiced segment 5 is 4.780, and between voiced segment 3 and voiced segment 6, the DTW value is 3.624. These two DTW values are significantly smaller than the DTW values between other voiced segments. If the DTW value between two voiced segments is smaller than the threshold we set, we will consider it as a duplicated speech segment. Therefore, it can be seen that our extracted speech copy-move forgery detection algorithm can not only detect single instances of copy-move forgery but can also detect multiple instances of copy-move forgery.

4.5   Performance Evaluation

The precision and recall of the mixed audio copy-move forgery database generated by pitch and vad using this algorithm is shown in Table 3. When detecting unprocessed forged audio, the algorithm achieves a precision of 98.80% and a recall of 98.40%. When detecting audio with the addition of 30dB Gaussian white noise, the algorithm achieves a precision of 97.98% and a recall of 97.20%. When detecting audio with the addition of 20dB Gaussian white noise, the algorithm achieves a precision of 96.37% and a recall of 95.60%. When detecting audio processed with

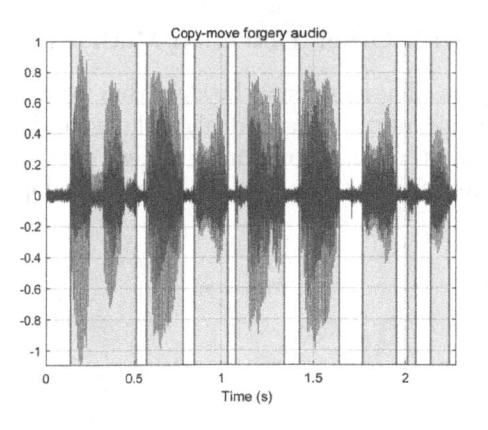

Fig. 10. A forged audio by copy-move forgery.

Table 2. DTW values between voiced segments.

Segments	1	2	3	4	5	6	7	8
1	0	73.639	72.691	100.909	73.236	70.362	144.524	86.637
2	73.639	0	37.957	40.117	**4.780**	37.850	102.435	44.147
3	72.691	37.957	0	47.947	37.769	**3.624**	69.557	26.801
4	100.909	40.117	47.947	0	39.817	47.750	114.602	63.370
5	73.236	**4.780**	37.769	39.817	0	37.614	102.225	44.271
6	70.362	37.850	**3.624**	47.750	37.614	0	69.422	26.015
7	144.524	102.435	69.557	114.602	102.225	69.422	0	31.102
8	86.637	44.147	26.801	63.370	44.271	26.015	31.102	0

median filtering post-processing, the algorithm achieves a precision of 98.79% and a recall of 98.00%. When detecting audio compressed with a 32kbps MP3 compression rate, the algorithm achieves a precision of 95.90% and a recall of 93.60%. When detecting audio compressed with a 64kbps MP3 compression rate, the algorithm achieves a precision of 97.52% and a recall of 94.40%.

Table 3. Detection results of the proposed algorithm on mixed database.

Method	The proposed method	
Type of attack	precision	Recall
No attack	**98.80%**	**98.40%**
Adding 30dB Gaussian noise	**97.98%**	**97.20%**
Adding 20dB Gaussian noise	**96.37%**	**95.60%**
Medianfiltered	**98.79%**	**98.00%**
32-bit compressed	**95.90%**	**93.60%**
64-bit compressed	**97.52%**	**94.40%**

4.6   Comparison with Other Detection Methods

In order to validate the effectiveness of the algorithm for detection, we conducted comparisons with some more advanced algorithms in the same experimental environment and using the same copy-move forgery dataset as described above. These algorithms are referred to as Yan [4], Ustubioglu [5], and Imran [6].

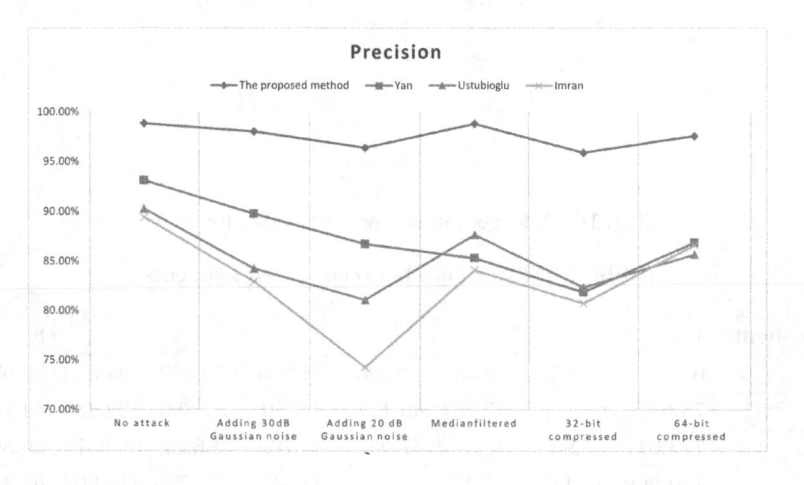

Fig. 11. Results of precision detection by respective methods on mixed database.

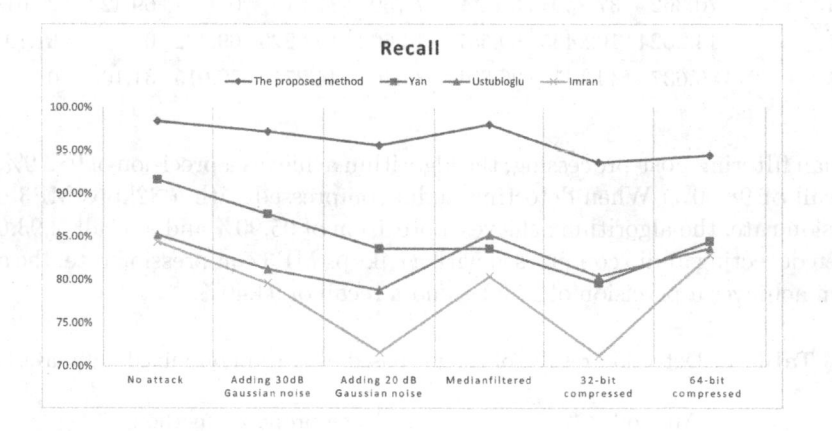

Fig. 12. Results of recall detection by respective methods on mixed database.

As can be seen in Fig. 11 and Fig. 12, we proposed algorithm significantly outperforms the other three algorithms in terms of precision and recall on mixed database.

5   Conclusions

In this paper, we propose a audio copy-move forgery detection algorithm based on dual feature fusion of MFCC and GFCC. By conducting experiments on the mixed audio copy-move forgery database generated from the TIMIT database, the experimental results show that the algorithm not only can effectively pinpoint the location of duplicated segments, but also has high accuracy in detecting post-processed audio files. We also compare the algorithm with the more advanced literature in the field, and the algorithm also has a high precision and recall rate. However, the algorithm also has some shortcomings, if when we use the speech activity detection algorithm to detect a speech segment that is extremely short, the DTW of that segment with other segments may also be less than the threshold value, resulting in a certain false detection rate. In our future work, we will continue to improve the algorithm to solve this problem.

Acknowledgements. This work was supported by the Natural Science Foundation of Sichuan Pvovince program (Grant No.2023NSFSC0470), National College Students' innovation and entrepreneurship training program (No. 202310623016), Provincial College Student' innovation and entrepreneurship training program (No. S202210623083), and the National Natural Science Foundation of China (NSFC) program (No.62171387, No.62202390).

References

1. Xie, Z., Wei, L., Liu, X., Xue, Y., Yeung, Y.: Copy-move detection of digital audio based on multi-feature decision. J. Inform. Sec. Appli. **43**, 37–46 (2018)
2. Liu, Z., Lu, W.: Fast copy-move detection of digital audio. In: 2017 IEEE Second International Conference on data Science in Cyberspace (DSC), pp. 625–629. IEEE (2017) -
3. Yan, Q., Yang, R., Huang, J.: Copy-move detection of audio recording with pitch similarity. In: 2015 IEEE International Conference on Acoustics, Speech and Signal Processing (ICASSP), pp. 1782–1786. IEEE (2015)
4. Yan, Q., Yang, R., Huang, J.: Robust copy-move detection of speech recording using similarities of pitch and formant. IEEE Trans. Inf. Forensics Secur. **14**(9), 2331–2341 (2019)
5. Ustubioglu, B., Küçükuğurlu, B., Ulutas, G.: Robust copy-move detection in digital audio forensics based on pitch and modified discrete cosine transform. Multimedia Tools Appli. **81**(19), 27149–27185 (2022)
6. Imran, M., Ali, Z., Bakhsh, S.T., Akram, S.: Blind detection of copy-move forgery in digital audio forensics. IEEE Access **5**, 12843–12855 (2017)
7. Qi, J., Wang, D., Jiang, Y., Liu, R.: Auditory features based on gammatone filters for robust speech recognition. In: 2013 IEEE International Symposium on Circuits and Systems (ISCAS), pp. 305–308. IEEE (2013)
8. Li, Z., Gao, Y.: Acoustic feature extraction method for robust speaker identification. Multimedia Tools Appli. **75**, 7391–7406 (2016)
9. Zheng, C., Jia, N.: Research on musical sentiment classification model based on joint representation structure. J. Phys: Conf. Ser. **1237**, 022086 (2019)

234 D. Wang et al.

10. Shi, X., Yang, H., Zhou, P.: Robust speaker recognition based on improved gfcc. In: 2016 2nd IEEE International Conference on Computer and Communications (ICCC), pp. 1927–1931 (2016)
11. Cao, Y., Yan, J., Sun, K., Luo, X.: Hydroacoustic target detection based on improved gfcc and lightweight neural network. In: 2023 42nd Chinese Control Conference (CCC), pp. 6239–6243. IEEE (2023)
12. Giannakopoulos, T.: A method for silence removal and segmentation of speech signals, implemented in matlab, vol. 2. University of Athens, Athens (2009)
13. Gao, H., Xue, L., Zhang, C.: Vehicle class recognition based on vehicle's sound signal. In: ICTIS 2013: Improving Multimodal Transportation Systems-Information, Safety, and Integration, pp. 898–903 (2013)
14. Liu, Y., Ge, H., Sun, L., Hou, Y.: Complementary attention-driven contrastive learning with hard-sample exploring for unsupervised domain adaptive person re-id. IEEE Trans. Circuits Syst. Video Technol. **33**(1), 326–341 (2022)
15. Patni, H., Jagtap, A., Bhoyar, V., Gupta, A.: Speech emotion recognition using mfcc, gfcc, chromagram and rmse features. In: 2021 8th International conference on signal processing and integrated networks (SPIN), pp. 892–897. IEEE (2021)
16. Zhao, X., Wang, D.: Analyzing noise robustness of mfcc and gfcc features in speaker identification. In: 2013 IEEE International Conference on Acoustics, Speech and Signal Processing, pp. 7204–7208. IEEE (2013)
17. Tazi, E.B.: A robust speaker identification system based on the combination of gfcc and mfcc methods. In: 2016 5th International Conference on Multimedia Computing and Systems (ICMCS), pp. 54–58 IEEE (2016)
18. Sakoe, H., Chiba, S.: Dynamic programming algorithm optimization for spoken word recognition. IEEE Trans. Acoust. Speech Signal Process. **26**(1), 43–49 (1978)
19. Garofolo, J.S., Lamel, L.F., Fisher, W.M., Fiscus, J.G., Pallett, D.S.: Darpa timit acoustic-phonetic continous speech corpus cd-rom. nist speech disc 1–1.1. NASA STI/Recon technical report n, vol. 93, p. 27403 (1993)
20. Ustubioglu, A., Ustubioglu, B., Ulutas, G.: Mel spectrogram-based audio forgery detection using CNN. SIViP **17**(5), 2211–2219 (2023)

A Survey of Homogeneous and Heterogeneous Multi-source Information Fusion Based on Rough Set Theory

Haojun Liu[1,3], Xiangyan Tang[1,3(✉)], Taixing Xu[1], and Ji He[2]

[1] School of Computer Science and Technology, Hainan University,
Haikou 570228, China
tangxy36@163.com
[2] School of Engineering and Information Science, University of Wollongong,
Wollongong 2500, Australia
[3] Hainan Blockchain Technology Engineering Research Center, Haikou 570228, China

Abstract. Multi-source information fusion (MSIF) can be defined as the process of automatically analyzing and synthesizing information and data from multiple sensors or sources based on certain standards so as to achieve the required decisions and estimates, and it includes two types of information, homogeneous information and heterogeneous information. MSIF is also referred as multi-sensor information fusion. Rough set theory (RST) provides an effective method to process uncertain, inaccurate, or incomplete data. Therefore, many homogeneous and heterogeneous MSIF approaches based on RST have been put forward. In this paper, we summarize the homogeneous and heterogeneous MSIF based RST. Firstly, we introduce the background knowledge of rough set theory and multi-source information fusion. Secondly, we classify the existing homogeneous and heterogeneous MSIF models based on RST. Then, we discuss these MSIF models and summarize their application scenarios. At the end of this paper, we propose the challenges and future trends of homogeneous and heterogeneous MSIF based on RST from the perspectives of data processing and privacy security.

Keywords: Multi-source information fusion · Homogeneous and heterogeneous MSIF · Rough set theory

1 Introduction

Since the era of data explosion, we need to collect, process, and store a large amount of data in a more efficient way [2]. The characteristics of these multi-source, multi-mode, and multi-domain provide us with a comprehensive understanding of objects [37]. It is due to these characteristics of data that multi-source information fusion has been applied. MSIF refers to the technology of combining

H. Jin et al. (Eds.): IAIC 2023, CCIS 2058, pp. 235–246, 2024.
https://doi.org/10.1007/978-981-97-1277-9_18

multiple information sources in space or time according to a specific standard in order to achieve a consistent interpretation or description of the tested objects, and then ultimately, achieve greater performance in the information systems. Meanwhile, it also can be seen as a formal framework, which involves synthesizing information from different sources with the help of mathematical approaches and technical tools. Data is produced by everything around us and transmitted through various networks, sensors, and mobile devices [52]. Under the big data environment, acquisition of data is not a limitation to a single data source any more, but mostly obtained from multiple data sources, with various types of structured information hidden among different data sources. Because the core theory of MSIF is to fully utilize multi-information sources, we can make a combination of multi-information sources under the restrictions of redundancies in space or time, and even complementary information. In comparison to a single data source, MSIF is superior in detectability, reducing the degree of ambiguity in reasoning, enhancing system fault tolerance, enhancing interpretation and dynamic monitoring capabilities, effectively improving the utilization rate of remote sensing image data and so on. The rough set theory proposed by Pawlak [30], provides an valid method to process uncertain, inaccurate, or incomplete data, and its principle is to approximate and characterize imprecise or uncertain targets with given information or knowledge. Currently, RST has been successfully applied in plentiful fields such as decision analysis [29], machine learning [11], data mining [5] and so on. Due to the characteristics of rough set theory combing the data processing through MSIF, the interest in MSIF based on rough set theory has dramatically increased, leading a rising amount of technologies and approaches. For instance, Khan and Banerjee [21] presented a conception of multiple source approximation systems on the basis of Pawlak approximation space, and it laid the foundation to study MSIF in rough set theory. The major contributions of the survey are outlined as below:

- We summarize the existing MSIF models based on RST from the perspectives of homogeneous information and heterogeneous information.
- We discuss the advantages and existing issues of homogeneous and heterogeneous MSIF models based on RST. And we list a table of the application of these models.
- We analyze the challenges of the existing homogeneous and heterogeneous MSIF models based on RST, and put forward potential future trends.

The remaining sections of the paper is organized below. In Sect. 2, we will introduce the background knowledge of rough set theory and multi-source information fusion. In Sect. 3, we classify and summarize the homogeneous and heterogeneous MSIF models based on RST, and discuss the advantages and existing issues of these models. In Sect. 4, we analyze the challenges faced by MSIF based on RST and propose future trends. A summary of the entire paper is presented in Section VI.

2   Background

2.1   Rough Set Theory

Rough set theory, originating from a model with simple information, is another mathematical tool for processing uncertainty, following probability theory, fuzzy sets, and evidence theory. The principle idea of RST is to derive the decision or classification rules of the problem through knowledge reduction while keeping the classification performance at the same stage. Its basic idea can be simply seperated into two sections. One is to establish regulations and conceptions via the categories of relational databases. Another is to acquire knowledge with the help of classifications in target approximation and equivalence relation [55]. During the past several decades, massive researchers from all corners of the world have shown great interest in RST.

In RST, data with symbolic values is mainly studied in attribute selection, and it is defined as an information system. Normally, it can be expressed as a quadruple Q=(U,A,V,I). In the quadruple, U is a set of non-empty finite objects, also referred as the domain of discourse. A represents a set of non-empty finite attributes, $V_a \in A$, V_a represents the range of attribute A. $I : U \to V_a$ is an information function, $\forall x \in U, a \in A$, I(x,a) represents the value of x on the property a, then $I(x, a) \in V_a$. Given M represents a set of condition-attribute while N is a set of decision-attribute. If $A = M \cup N$ and $M \cap N = \emptyset$, then this information system can be referred as a decision system. For instance, U represents a set of non-empty finite objects, namely the domain of discourse. Then U corresponds to U={P1,P2,P3,P4,P5}, and except for the first column, other columns stand for attributes. Attributes are divided into two types, condition-attribute and decision-attribute respectively. In the instance, the last column is decision-attribute. For example, M_2 is the condition-attribute of myodynia, then the form of its representation is M_2={✓, ×, ✓, ✓, ×}. N_1 is the decision-attribute of epidemic, then the form of its representation is N_1={×, ×, ✓, ✓, ×}. It is a single-decision system if there is only one decision-attribute, while it is a multi-decision system if there are two or more than two decision-attribute. An example is shown in Table 1 below.

Table 1. A concept example for rough set theory.

Patient	Headache	Myodynia	Hyperthermia	Epidemic
P1	✓	✓	×	×
P2	✓	×	×	×
P3	✓	✓	✓	✓
P4	×	✓	✓	✓
P5	×	×	✓	×

2.2   Multi-source Information Fusion

Multi-source information fusion (MSIF) refers to the technology of combining multiple information sources in space or time according to a specific standard in order to achieve a consistent interpretation or description of the tested objects, and then ultimately, achieve greater performance in the information systems. As information comes from everywhere, two types of information are classified, homogeneous information and heterogeneous information. A single data source is only available to partial information segment of the target object, while multi-data sources can mirror the general information of the merged object in a perfect and accurate way. According to the layer of processing information sources, MSIF is generally seperated into three types, including data-level, feature-level and decision-level. In the process of MSIF, it is generally seperated into four phases, including collecting and organizing, processing, analyzing and decision-making, and conclusion output. Theoretically, data-level fusion can retain a large amount of the original information, and achieve the most precise fusion outcomes as much as possible by providing the most detailed information for the target.

Currently, almost all of the approaches in MSIF integrate into multifarious theories, such as probability theory [8], fuzzy set theory [3], possibility theory [54], D-S theory [16], etc. Among the major challenges in MSIF, uncertainty issue, involving uncertainty both in data itself and fusion process is an urgent one [53]. Furthermore, there are the problems of data association, the precision of the association, calibration and matching of the design data fusion process, and data inconsistency, which involves the inconsistency of sensor frequencies, the inconsistency of sensor dimensions and the difference of data forms in the data fusion process [42].

3   Homogeneous and Heterogeneous MSIF in RST

In this section, we will discuss the homogeneous and heterogeneous MSIF models based on RST respectively.(see Fig. 1)

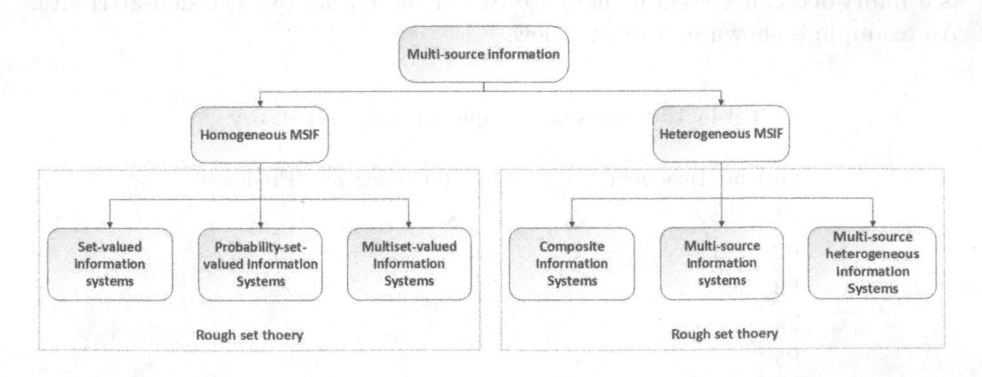

Fig. 1. Classification of homogeneous and heterogeneous MSIF models in RST.

3.1   Homogeneous MSIF

Set-Valued Information Systems (SvISs). A SvIS stands for a major universal model of a single value information system. Due to the lack of information and diverse information sources, set-valued data usually arise in these systems, so it is necessary to set up a mechanism to deal with these data, so as to achieve multi-source information fusion efficiently. Yao et al. specifically unveiled SvIS and proposed certain operations of set-valued data on basis of sets [44]. Wei et al. proposed two various types of fuzzy-rough set patterns on account of fuzzy correspondence class and fuzzy correspondence degree respectively [39]. William proposed to represent the eigenvectors of eigenvalues through as a set of strings, and proved that many application problems in the real world are highly efficient to be expressed by set-valued characteristics [9]. Guan et al. established tolerance relation, adopting the principle of optimal decision from a SvIS [13]. Qian et al. proposed two dominance relations on account of rough set theory for SvISs [31]. Zhang et al. presented the rejuvenation of rough-set approximations for relation matrices in a SvIS using incremental techniques [49]. Chen and Li investigated major rules of incremental rejuvenation of approximation while adding attributes and objects into SvISs [7]. Dai and Tian presented an attribute simplicity method in SvISs according to discriminative-ability relation [10]. Lang et al. put forward an incrementing technique to attribute simplicity in SvISs for the reason to attest the attribute simplicity in a changing SvIS [22]. Bao and Yang studied an attribute simplicity approach for SvISs, which takes attributes with preferred ordered fields and fuzzy decision attributes into account [4]. In view of the changing of criterion values in SvISs that relies on knowledge rejuvenation, Luo et al. came up with the updating features of approximate lasting preservation in SvISs [28]. Zhuang et al. put forward a multi-granularity dominance relation pattern on the basis of SvIS and applied it to multi-source information systems [58]. Zhang and Yang put forward a universal model of feature selection and approximation inference for massive-scale SvISs so as to integrate measurable and dominance-based rough sets [46].

Probability Set-Valued Information Systems (PSvISs). Information systems with discrete probability distributions are proposed as probability set-valued information systems. Huang et al. investigated the concept of PSvISs for the first time and proposed a tolerance-based mutable accuracy rough set framework [18]. In order to form a frame for granularity computing and comprehend the uncertainty nature, Xie et al. investigated the information structures and uncertainty measures in a PSvIS [41]. Li et al. came up with incrementing strategies for computing approximations of feature relations in a fragmentary information system [23]. Luo et al. proposed a incrementing method based on matrices for approximations rejuvenation in PSvISs [27]. Yang et al. developed incrementing approaches to maintain a rough approximation of multiple particles under dynamic granulation [43]. Wang et al. studied the incrementing mechanisms of attribution reduction in view of three diverse types of information entropy [38]. Shu et al. investigated incrementing methods of computing the positive fields and tolerance grades to achieve attribute simplicity on the basis of positive fields [34].

Multiset-Valued Information Systems (MvISs). A MvIS is referred to a information system with multiple sets of information values, a database that depicts the relations between an object and an attribute. MvISs represent the outcome of the fusion of multi-classification information systems, and this model helps to process the missing values in the data sets. Zhao and Hu [57] proposed a three-way decision approach combing rough sets of decision theory in MvISs. At present, research on MvISs are mostly related to mathematical properties [12,33,36]. Therefore, a great amount of work in this field still remain further investigation.

3.2 Heterogeneous MSIF

Composite Information Systems (CISs). Since a number of diverse kinds of data exist in information systems, Zhang et al. took the lead to come up with a unified information framework, which is named after a CIS, so as to process data like categorical data, missing data and everything of the sort, making the information systems more diverse and elaborate. Meanwhile, CISs are actually subjected to the normal information systems, developing on the basis of these systems [47]. Hu et al. developed an effectual hybrid attribute simplicity algorithm on account of a universal fuzzy rough model for the first time so as to process heterogeneous data, but they only considered two kinds of data features, namely numerical features and categorical features [15]. As CISs keep making great progress in these years, they have become an significant research field in numerous application scenarios. For instance, Chen et al. put forward an attribute simplicity approach for a decision-information system with symbolic and practical-value conditional attributes combing with the typical and the fuzzy rough set models [6]. Zhang et al. put forward matrix-based approaches for computations of the approximations, positive, boundary and negative regions in CISs, which plays a significant role for feature selection and knowledge [48]. Moreover, to accelerate the rough set approximation calculation process, they innovatively proposed a boolean matrix representation of upper and lower approximations in the CISs, and devised a parallel method for computations of approximations operators on the basis of matrices, then applied it on Multi-GPU. [50]. As it is possible to define fuzzy RST for diverse kinds of attributes to estimate the similarity among objects, Zhang et al. developed an information entropy approach for feature selection of hybrid data sets [56]. Additionally, a number of researchers have investigated the nature and structures of mixed information systems by deeper research of CISs [45]. Huang et al. enlarged the CISs models by integrating multiple kinds of attributes and multiple compound relations from diverse data sources [19].

Multi-source Information Systems (MsISs). In the same domain, information from different sources exists at each time point, and with the change of temporal information, a single information system appears, which is called a MsIS. Huang et al. [20] put forward two novel matrix operators to construct rough approximate representations based on matrices. Then, with the variation of objects, attributes and values in MsISs, location information among some elements is transmitted in each composite relation matrix by using the previously accumulated matrix information, and the approximate maintenance increment mechanism based on matrices is proposed.

Multi-source Heterogeneous Information Systems (MsHISs). A multi-source heterogeneous information system indicates an integrated system consisted of components from many various sources, which includes hybrid data (structured and unstructured) and discrete data (data distribute from diverse systems or platforms). Wilson et al. proposed three novel heterogeneous distance functions, which is to cope applications consisting of nominal and continuous properties at the beginning [40]. Liu et al. investigated the acquisition of heterogeneous and spatial data, and it has been a significant issue in smart city information platform [25]. Zhang et al. investigated approaches of big data fusion and heterogeneous data, focusing on applying deep learning approaches in fusion with multiple source heterogeneous data [51]. To eliminate data redundancy and extend the network lifetime, Liu put forward multiple source heterogeneous data fusion in view of cognitive semantics [26]. Liu et al. made a summary of methodologies of multiple source heterogeneous big data fusion on the basis of deep learning, classifying them into three types [24]. Sun et al. [35] presented a multi-granularity fuzzy rough set methodology in view of a MsHIS to solve multi-attribute group decision making issues. Hadi et al. [1] considered a multi-sensor network fusion model for 3D data registration using inertial planes, underlying geometric relation and transform model uncertainty. Quellec et al. [32] a novel content-based heterogeneous information retrieval framework is proposed, which aims to retrieve potentially incomplete documents from a database so as to depict every single image in the document through its information fusion. Gang et al. [14] came up with a novel model that used relevant meta-information to describe mobile objects, and proposed a map-matching algorithm called IF-Match that can handle many ambiguous cases where existing methods do not match correctly. Huang et al. [17] put forward an evolving optimal fuzzy system with information fusion of heterogeneous distributed computing and polar-space dynamic model for online motion control of Swedish redundant robots.

3.3 Discussion and Application of the Homogeneous and Heterogeneous MSIF in RST

The existing homogeneous and heterogeneous MSIF based on RST has its advantages and limitations. In homogeneous MSIF, the uncertainty and quality of source information can be better preserved. Moreover, due to the integration of

RST, the effect of feature selection has been improved, which helps to select the most representative feature from multiple sources so as to reduce redundant information and enhance the efficiency and precision of information fusion. Furthermore, it can provide an intuitive perspective to analyze and interpret the results of information fusion. However, as data keep increasing rapidly, it may require more computing resources and time to deal with large-scale fusion problems, and whether the format and structure of the data are consistent is also an urgent research filed. For heterogeneous MSIF, it can integrate multiple categories of information sources from multiple fields to facilitate more comprehensive decision making and analysis. Besides, due to the flexibility of heterogeneous MSIF approaches, it can adapt to the differences of various data sources, and it is better in dealing with complex multi-source data. However, because heterogeneous information fusion eventually needs to form into a standardized expression, semantic consistency will lead to increasing difficulty of its fusion. Table 2 summarizes the existing applications of homogeneous and heterogeneous MSIF based on RST.

Table 2. Applications in MSIF models.

MSIF models	MSIF models in RST	Applications
Homogeneous MSIF	SvISs	Incrementing fusion
	SvISs	attribute simplicity
	SvISs	Uncertainty measuring
	SvISs	Optimization decision regulations
	PSvISs	Incrementing fusion
	PSvISs	Uncertainty measuring
	MvISs	Three-way decision
Heterogeneous MSIF	CISs	Incrementing fusion
	MsISs	Uncertainty measuring
	MsISs	Optimization decision regulations
	MsISs	Approximation space fusion
	MsHISs	Agent evaluation
	MsHISs	Group decision making

4   Challenges and Future Trends

As the introduction and discussion can be clearly seen in the above sections, we understand the latest approaches and applications on homogeneous and heterogeneous MSIF based on RST. However, some challenges still remain unsolved.

4.1   Data Processing

With the continuous development of the era of big data, the amount of data has increased dramatically, which has become a huge challenge for MSIF. One of the biggest challenges in MSIF is to handle the difference in data quality from different data sources, we may solve the noise, missing values and error values of data sources with the help of data enhancement and data completion methods to improve the accuracy and reliability of fusion results. At the same time, the data quantity and data scale keep increasing, so as the computational complexity of multi-source information fusion. Therefore, we can explore the combination of deep learning methods and rough set so as to improve the efficiency and accuracy of MSIF.

4.2 Privacy and Security

As the application of information fusion in various fields increases, privacy and security issues have become increasingly crucial. Research in the future needs to consider how to protect data privacy and security in information fusion, such as integrating the immutable feature of blockchain technology to ensure data security, or combing privacy protection feature in federal learning to information fusion, so as to achieve the purpose of privacy protection.

5 Conclusions

In this survey, we have retrospected the state-of-the-art research work of homogeneous and heterogeneous multi-source information fusion based on rough set theory. First, we have introduced the background knowledge of MSIF and RST. Second, we have summarized the latest MSIF based on RST from the perspectives of Homogeneous information and heterogeneous information. Then, we have discussed these MSIF models and summarized their application scenarios. Finally, we have proposed the challenges and potential future trends of homogeneous and heterogeneous MSIF based on RST from the perspectives of data processing and privacy security. We hope that with the help of this work, further research of homogeneous and heterogeneous MSIF based on RST can be carried on.

Acknowledgements. This work was supported by Hainan Provincial Natural Science Foundation of China (Grant No. 620MS021), National Natural Science Foundation of China (NSFC) (Grant No. 62162022,62162024), the Major science and technology project of Hainan Province (Grant No. ZDKJ2020012), Youth Foundation Project of Hainan Natural Science Foundation(621QN211).

References

1. Aliakbarpour, H., Prasath, V.S., Palaniappan, K., Seetharaman, G., Dias, J.: Heterogeneous multi-view information fusion: review of 3-D reconstruction methods and a new registration with uncertainty modeling. IEEE Access **4**, 8264–8285 (2016)

2. Alsolbi, I., Shavaki, F.H., Agarwal, R., Bharathy, G.K., Prakash, S., Prasad, M.: Big data optimisation and management in supply chain management: a systematic literature review. Artif. Intell. Rev. **56**(3), 1–32 (2023)
3. de Andrés-Sánchez, J.: A systematic review of the interactions of fuzzy set theory and option pricing. Expert Syst. Appl. **223**, 119868 (2023)
4. Bao, Z., Yang, S.: Attribute reduction for set-valued ordered fuzzy decision system. In: 2014 Sixth International Conference on Intelligent Human-Machine Systems and Cybernetics. vol. 2, pp. 96–99. IEEE (2014)
5. Chan, C.C.: A rough set approach to attribute generalization in data mining. Inf. Sci. **107**(1–4), 169–176 (1998)
6. Chen, D., Yang, Y.: Attribute reduction for heterogeneous data based on the combination of classical and fuzzy rough set models. IEEE Trans. Fuzzy Syst. **22**(5), 1325–1334 (2013)
7. Chen, H., Li, T., Tian, H.: Approaches for updating approximations in set-valued information systems while objects and attributes vary with time. Rough Sets and Intelligent Systems-Professor Zdzisław Pawlak in Memoriam **1**, 229–248 (2013)
8. Cheng, H., Zhao, J., Fu, M.: Research on the method of multi-source information fusion based on Bayesian theory. In: 2018 IEEE 3rd Advanced Information Technology, Electronic and Automation Control Conference (IAEAC), pp. 1760–1763. IEEE (2018)
9. Cohen, W.W.: Learning trees and rules with set-valued features. In: AAAI/IAAI, Vol. 1. pp. 709–716 (1996)
10. Dai, J., Tian, H.: Fuzzy rough set model for set-valued data. Fuzzy Sets Syst. **229**, 54–68 (2013)
11. Gangadhari, R.K., Khanzode, V., Murthy, S.: Application of rough set theory and machine learning algorithms in predicting accident outcomes in the Indian petroleum industry. Concurrency Comput. Pract. Experience **34**(26), e7277 (2022)
12. Girish, K., John, S.J.: Multiset topologies induced by multiset relations. Inf. Sci. **188**, 298–313 (2012)
13. Guan, Y.Y., Wang, H.K.: Set-valued information systems. Inf. Sci. **176**(17), 2507–2525 (2006)
14. Hu, G., Shao, J., Liu, F., Wang, Y., Shen, H.T.: IF-matching: towards accurate map-matching with information fusion. IEEE Trans. Knowl. Data Eng. **29**(1), 114–127 (2016)
15. Hu, Q., Xie, Z., Yu, D.: Hybrid attribute reduction based on a novel fuzzy-rough model and information granulation. Pattern Recogn. **40**(12), 3509–3521 (2007)
16. Hua, Z., Jing, X.: An improved belief Hellinger divergence for Dempster-Shafer theory and its application in multi-source information fusion. Appl. Intell. **53**(1), 1–20 (2023)
17. Huang, H.C.: An evolutionary optimal fuzzy system with information fusion of heterogeneous distributed computing and polar-space dynamic model for online motion control of Swedish redundant robots. IEEE Trans. Industr. Electron. **64**(2), 1743–1750 (2016)
18. Huang, Y., Li, T., Luo, C., Fujita, H., Horng, S.J.: Dynamic variable precision rough set approach for probabilistic set-valued information systems. Knowl.-Based Syst. **122**, 131–147 (2017)
19. Huang, Y., Li, T., Luo, C., Fujita, H., Horng, S.J., Wang, B.: Dynamic maintenance of rough approximations in multi-source hybrid information systems. Inf. Sci. **530**, 108–127 (2020)
20. Huang, Y., et al.: A novel border-rich Prussian blue synthetized by inhibitor control as cathode for sodium ion batteries. Nano Energy **39**, 273–283 (2017)

21. Khan, M.A., Banerjee, M.: Formal reasoning with rough sets in multiple-source approximation systems. Int. J. Approximate Reasoning **49**(2), 466–477 (2008)
22. Lang, G., Li, Q., Yang, T.: An incremental approach to attribute reduction of dynamic set-valued information systems. Int. J. Mach. Learn. Cybern. **5**, 775–788 (2014)
23. Li, T., Ruan, D., Geert, W., Song, J., Xu, Y.: A rough sets based characteristic relation approach for dynamic attribute generalization in data mining. Knowl.-Based Syst. **20**(5), 485–494 (2007)
24. Liu, J., Li, T., Xie, P., Du, S., Teng, F., Yang, X.: Urban big data fusion based on deep learning: an overview. Inf. Fusion **53**, 123–133 (2020)
25. Liu, S., Peng, L., Chi, T., Wang, X.: Research on multi-source heterogeneous data collection for the smart city public information platform. In: 2016 IEEE International Geoscience and Remote Sensing Symposium (IGARSS), pp. 623–626. IEEE (2016)
26. Liu, Y.: Multi-source heterogeneous data fusion based on perceptual semantics in narrow-band internet of things. Pers. Ubiquit. Comput. **23**(3–4), 413–420 (2019)
27. Luo, C., Li, T., Chen, H.: Dynamic maintenance of approximations in set-valued ordered decision systems under the attribute generalization. Inf. Sci. **257**, 210–228 (2014)
28. Luo, C., Li, T., Chen, H., Lu, L.: Fast algorithms for computing rough approximations in set-valued decision systems while updating criteria values. Inf. Sci. **299**, 221–242 (2015)
29. Pawlak, Z., Sowinski, R.: Rough set approach to multi-attribute decision analysis. Eur. J. Oper. Res. **72**(3), 443–459 (1994)
30. Pawlak, Z.: Drawing conclusions from data-the rough set way. Int. J. Intell. Syst. **16**(1), 3–11 (2001)
31. Qian, Y., Dang, C., Liang, J., Tang, D.: Set-valued ordered information systems. Inf. Sci. **179**(16), 2809–2832 (2009)
32. Quellec, G., Lamard, M., Cazuguel, G., Roux, C., Cochener, B.: Case retrieval in medical databases by fusing heterogeneous information. IEEE Trans. Med. Imaging **30**(1), 108–118 (2010)
33. Riesgo, Á., Alonso, P., Díaz, I., Montes, S.: Basic operations for fuzzy multisets. Int. J. Approximate Reasoning **101**, 107–118 (2018)
34. Shu, W., Shen, H.: Updating attribute reduction in incomplete decision systems with the variation of attribute set. Int. J. Approximate Reasoning **55**(3), 867–884 (2014)
35. Sun, B., Ma, W., Chen, X., Li, X.: Heterogeneous multigranulation fuzzy rough set-based multiple attribute group decision making with heterogeneous preference information. Comput. Ind. Eng. **122**, 24–38 (2018)
36. Syropoulos, A.: Mathematics of multisets. In: Calude, C.S., PǍun, G., Rozenberg, G., Salomaa, A. (eds.) WMC 2000. LNCS, vol. 2235, pp. 347–358. Springer, Heidelberg (2001). https://doi.org/10.1007/3-540-45523-X_17
37. Tang, Q., Liang, J., Zhu, F.: A comparative review on multi-modal sensors fusion based on deep learning. Sig. Process. **213**, 109165 (2023)
38. Wang, F., Liang, J., Qian, Y.: Attribute reduction: a dimension incremental strategy. Knowl.-Based Syst. **39**, 95–108 (2013)
39. Wei, W., Cui, J., Liang, J., Wang, J.: Fuzzy rough approximations for set-valued data. Inf. Sci. **360**, 181–201 (2016)
40. Wilson, D.R., Martinez, T.R.: Improved heterogeneous distance functions. J. Artif. Intell. Res. **6**, 1–34 (1997)

41. Xie, X., Li, Z., Zhang, P., Zhang, G.: Information structures and uncertainty measures in an incomplete probabilistic set-valued information system. IEEE Access **7**, 27501–27514 (2019)
42. Yang, F., Zhang, P.: MSIF: multi-source information fusion based on information sets. J. Intell. Fuzzy Syst. (Preprint), 1–10 (2023)
43. Yang, X., Qi, Y., Yu, H., Song, X., Yang, J.: Updating multigranulation rough approximations with increasing of granular structures. Knowl.-Based Syst. **64**, 59–69 (2014)
44. Yao, Y., Noroozi, N.: A unified model for set-based computations. In: Soft Computing: 3rd International Workshop on Rough Sets and Soft Computing, pp. 252–255. Citeseer (1994)
45. Zeng, J., Li, Z., Zhang, P., Wang, P.: Information structures and uncertainty measures in a hybrid information system: Gaussian kernel method. Int. J. Fuzzy Syst. **22**(1), 212–231 (2020)
46. Zhang, H.Y., Yang, S.Y.: Feature selection and approximate reasoning of large-scale set-valued decision tables based on α-dominance-based quantitative rough sets. Inf. Sci. **378**, 328–347 (2017)
47. Zhang, J., Li, T., Chen, H.: Composite rough sets. In: Lei, J., Wang, F.L., Deng, H., Miao, D. (eds.) AICI 2012. LNCS (LNAI), vol. 7530, pp. 150–159. Springer, Heidelberg (2012). https://doi.org/10.1007/978-3-642-33478-8_20
48. Zhang, J., Li, T., Chen, H.: Composite rough sets for dynamic data mining. Inf. Sci. **257**, 81–100 (2014)
49. Zhang, J., Li, T., Ruan, D., Liu, D.: Rough sets based matrix approaches with dynamic attribute variation in set-valued information systems. Int. J. Approximate Reasoning **53**(4), 620–635 (2012)
50. Zhang, J., Zhu, Y., Pan, Y., Li, T.: Efficient parallel Boolean matrix based algorithms for computing composite rough set approximations. Inf. Sci. **329**, 287–302 (2016)
51. Zhang, L., Xie, Y., Xidao, L., Zhang, X.: Multi-source heterogeneous data fusion. In: 2018 International Conference on Artificial Intelligence and Big Data (ICAIBD), pp. 47–51. IEEE (2018)
52. Zhang, P., et al.: Multi-source information fusion based on rough set theory: a review. Inf. Fusion **68**, 85–117 (2021)
53. Zhang, P., Li, T., Wang, G., Wang, D., Lai, P., Zhang, F.: A multi-source information fusion model for outlier detection. Inf. Fusion **93**, 192–208 (2023)
54. Zhang, P., et al.: A possibilistic information fusion-based unsupervised feature selection method using information quality measures. IEEE Trans. Fuzzy Syst. **31**(9), 2975–2988 (2023)
55. Zhang, Q., Xie, Q., Wang, G.: A survey on rough set theory and its applications. CAAI Trans. Intell. Technol. **1**(4), 323–333 (2016)
56. Zhang, X., Mei, C., Chen, D., Li, J.: Feature selection in mixed data: a method using a novel fuzzy rough set-based information entropy. Pattern Recogn. **56**, 1–15 (2016)
57. Zhao, X.R., Hu, B.Q.: Three-way decisions with decision-theoretic rough sets in multiset-valued information tables. Inf. Sci. **507**, 684–699 (2020)
58. Zhuang, Y., Liu, W., Fan, M., Li, J.: Multi-granulation dominance relation and information fusion based on set-valued information system. PR & AI **28**(8), 742–749 (2015)

Formula Maintenance of Single Material Tobacco Compatibility Based on Co-formulation Analysis Method

Yuxuan Dong[1], Kaihu Hou[1], Yaqin Liu[2(✉)], Xiantao Ma[2], Ming Chen[2], Yunfa Gao[2], Jiming Dai[2], and Jie Long[3]

[1] Faculty of Mechanical and Electrical Engineering, Kunming University of Science and Technology, Kunming 650500, China
[2] Qilin Redrying Plant, Yunnan Leaf Tobacco Redrying Co., Ltd., Qujing 655000, China
2428097237@qq.com
[3] Yunnan Provincial Tobacco Quality Supervision and Inspection Station, Kunming 650106, China

Abstract. To explore the relationship between the compatibility of different grades of single-material tobacco from different producing areas and the formula of the cigarette leaf group, a maintenance method for flue-cured tobacco formula based on co-formulation analysis was proposed. First, historical formula data provided by a redrying plant were used to select high-frequency single-material tobaccos. Then, the two groups of high-frequency single-material cigarettes were organized into a co-formulation matrix. The correlation matrix was constructed based on the correlation between the two groups of high-frequency single-material cigarettes, and the compatibility score matrix of single-material cigarettes was established based on the similarity between the two groups. Eventually, cluster analysis was conducted on the compatibility score matrix to explore the compatibility between single-material tobaccos. The results revealed the existence of 25 high-frequency single-material cigarettes. The clustering analysis categorized these cigarettes into two groups. One group showed closer relationships, indicating a higher probability of these cigarettes appearing together in the same formula. The other group showed a lower probability of co-occurring in the same formula. Importantly, these clustering results accurately reflected actual production scenarios. This method effectively captures the compatibility between single cigarettes in the cigarette leaf group formula, offering valuable support for formula maintenance and enhancing the utilization rate of single cigarettes.

Keywords: Co-formulation matrix · Compatibility · Single-grade tobacco strip

1 Introduction

Tobacco leaf is a crucial component of the tobacco industry, serving as the primary raw material for Chinese cigarettes and the backbone of industry growth. As the primary focus of research and usage of raw materials in industrial enterprises, the formulation of the cigarette leaf group directly impacts the utilization rate of tobacco leaves, as well as the quality and cost of cigarette products. Formula threshing involves mixing raw tobacco of varying grades, producing areas, and parts during the redrying process according to specific ratios. Formula design and maintenance are vital for successful formula threshing. Compatibility is a key characteristic of tobacco leaves, and the investigation of tobacco compatibility is necessary for determining the formula module [1]. Meanwhile, the unresolved long-term imbalances between tobacco supply and demand further exacerbate this issue. As a unique agricultural product, tobacco is greatly influenced by various factors, leading to a reduction in single-material tobacco production capacity and subsequent fluctuations in the quality of flue-cured tobacco formulations [2]. Therefore, enhancing the utilization rate of flue-cured tobacco and improving the stability of flue-cured tobacco formulas have become pressing concerns for tobacco industry enterprises.

For a considerable period of time, numerous scholars have engaged in studying the preservation of flue-cured tobacco formula from various perspectives. The cigarette leaf group formula is created by blending individual cigarettes with different origins and grades according to a specific formula. To ensure the consistency in cigarette taste, it is imperative to periodically maintain the formula of the cured tobacco leaf group [3]. The quality of individual tobacco materials directly correlates with the overall quality of cigarettes. The sensory quality of individual tobacco materials is assessed based on their customary chemical components, enabling the acquisition of substitute tobacco materials that can serve the purpose of formula maintenance [4–7]. Relying solely on the chemical composition for assessing the similarity between various grades of individual cigarettes in the leaf group formula is a rather limited approach. Consequently, this study utilized the single-material tobacco ratio, sensory evaluation, and near-infrared spectroscopy to assess and analyze the interchangeable tobacco materials within the leaf group formula, as well as the formula itself before and after replacement. The aim of these analyses was to maintain the integrity of the cigarette leaf group formula [8–10].

Although some explorations have been made on the maintenance of cigarette leaf group formula at this stage, the existing research basically does not consider too much the compatibility between single-material strips of origin and grade. Association rules were used to directly mine the replacement rules of single-material tobacco to assist the maintenance of flue-cured tobacco formula and ensure the compatibility of single-material tobacco [11]. However, this study only considered the compatibility between non-main-material single-material tobacco. For the evaluation of the compatibility between single cigarettes, the method of sensory quality evaluation of single cigarettes was used [12–15], but these studies lacked the analysis and application of historical data. Therefore, it is necessary to develop a method for the maintenance of flue-cured tobacco formula based on the compatibility of single tobacco by using historical formula data.

The frequency of single cigarette use reflects the number of times a single cigarette is utilized in the brand formula, which provides a measurement of the impact on the brand when it is absent [16]. This paper presents an intelligent decision-making method for cultivating flue-cured tobacco formula maintenance, particularly focusing on analyzing the relationship between tobacco leaf formula and historical data. Our objective is to explore the compatibility patterns among different single-material tobacco based on their frequency and co-occurrence frequency in various flue-cured tobacco formulas. Firstly, we select high-frequency words through co-formulation analysis. Secondly, we construct co-formulation matrix and compatibility score matrix for single-material tobacco, followed by cluster analysis. Finally, we conduct a visualization analysis of high-frequency single-material tobacco in flue-cured tobacco formula. Our aim is to offer intelligent decision-making maintenance methods for flue-cured tobacco formula, along with theoretical support for enhancing the utilization rate and ensuring the stability of single-material tobacco in the flue-cured tobacco formula.

2 Methods and Framework

2.1 Co-formulation Analysis

Co-term Analysis, a content analysis method proposed by French bibliometrics in the mid-70s of the 20th century [17], utilizes word frequency analysis to extract keywords or subject words from relevant literature databases. Typically, these are high-frequency words that appear above a certain threshold and can represent the research topic or direction of a particular subject. Next, the number of simultaneous occurrences of these high-frequency words in the same article is counted in pairs, resulting in a co-word matrix. Subsequently, analysis is conducted based on this co-word matrix [18]. Through the application of bibliometric methods such as co-word matrix analysis and concept word co-occurrence clustering analysis, researchers have objectively revealed the research trends in various fields, including education, medical nursing, and aviation, among others [19–24]. This analysis allows for the identification of relationships between high-frequency words, shedding light on their underlying meanings. Such insights serve as a valuable reference for further research.

Based on the concept of co-word analysis, the proposed method is known as co-formulation analysis. This method analyzes the content of formulas by examining the phenomenon in which certain ingredients appear together in the same formula. It then assesses the relationship between these components to conduct structural research on the formula. Key ingredients are extracted from multiple formulations using the Price formula, and their frequency must exceed a certain threshold. Subsequently, a co-occurrence matrix is constructed based on the number of times two ingredients appear in the same formula, enabling further research and analysis.

The co-formulation matrix serves as an indicator of the relationship strength between two individual cigarettes. A higher frequency of occurrence of two single cigarettes in the same formula implies a stronger correlation between them. The matrix is designed to show relationships without any relative symmetry, with the diagonal being the axis of symmetry. This matrix forms the foundation for analysis using visualization tools [25]. The number of rows and columns in the matrix corresponds to the number of high-frequency vital levels in the data model. The diagonal elements hold no significance and are replaced with 0. The correlation between each row and column is expressed in terms of specific levels and other levels [26].

A co-formulation matrix with n key levels is defined as formula (1):

$$M = \begin{bmatrix} C(a_1, a_1) & C(a_1, a_2) \cdots C(a_1, a_n) \\ C(a_2, a_1) & C(a_2, a_2) \ \ldots \ C(a_2, a_n) \\ \vdots & \vdots & \vdots \\ C(a_n, a_1) & C(a_n, a_2) \ \cdots \ C(a_n, a_n) \end{bmatrix} \tag{1}$$

In formula (1), $C(a_i, a_j)$ represents the co-occurrence degree of the critical grades a_i and a_j, which can be calculated from the number of occurrences of a_i and a_j:

$$C(a_i, a_j) = f(a_i|a_j)/f(a_i) \tag{2}$$

The formula (2) calculates the relative co-occurrence degree of a_i relative to a_j, where: $f(a_i|a_j)$ represents the number of times that co-occur in the same formula in the formula; $f(a_i)$ denotes the number of occurrences in the formulation.

In order to eliminate the influence of frequency disparity and evaluate the compatibility of single tobacco, the compatibility score matrix of single tobacco was constructed based on the co-party matrix. The compatibility score matrix of single tobacco is the basis of cluster analysis, which can store the similarity between n objects. The expression form is an $n \times n$ dimensional matrix, where $d(i, j)$ is the quantitative representation of the dissimilarity between object i and object j, which is usually a non-negative value. When object i and object j are more similar, the value is closer to 0. The more different the two objects, the greater the value. When $d(i, j) = d(j, i)$, and $d(i, i)=0$, matrix (3) can be obtained.

$$\begin{bmatrix} 0 \\ d(2, 1) & 0 \\ d(3, 1) & d(3, 2) & 0 \\ \vdots & \ldots\ldots\ldots\ldots \\ d(n, 1) & d(n, 2) \cdots\cdots 0 \end{bmatrix} \tag{3}$$

2.2 Co-formulation Clustering

Using the Co-formulation analysis approach, cluster analysis is performed using a single cigarette compatibility score matrix. Cluster analysis is a statistical method used to categorize individuals based on their characteristics. It relies on the classification of data similarity and merges objects that are close to each other into the same category until all individuals are grouped [27]. In the cluster analysis of key levels, each high-frequency level is considered as an individual and then merged based on their distances. This process continues until all key levels are grouped together, eventually forming a cluster.

2.3 Multidimensional Scaling Analysis

The structure of the multi-dimensional scale analysis is highly intuitive. It utilizes the co-formulation matrix to visually depict the relationship among key ingredients in a formula, representing them in a low-dimensional space. The distance between the ingredients on the plane indicates the similarity between them. When the distance between key ingredients is shorter, it suggests a higher level of similarity in their properties.

The basic principle is as follows: First, given n d-dimensional sample data, such as matrix (4):

$$
\begin{bmatrix}
x_{11} & x_{12} & \cdots & x_{1d} \\
x_{21} & x_{22} & \cdots & x_{2d} \\
& \vdots & \ddots & \vdots \\
x_{n1} & x_{n2} & \cdots & x_{nd}
\end{bmatrix}
\tag{4}
$$

The Euclidean distance between these n samples can be expressed as a matrix (5).

$$
\begin{bmatrix}
\delta_{11} & \delta_{12} & \cdots & \delta_{1n} \\
\delta_{21} & \delta_{22} & \cdots & \delta_{2n} \\
& \vdots & \ddots & \vdots \\
\delta_{n1} & \delta_{n2} & \cdots & \delta_{nn}
\end{bmatrix}
\tag{5}
$$

Here, $\sigma_{ij}(X) = \delta_{ij}$ is the Euclidean distance between sample i and sample j, which has:

$$
\delta_{ij} = \|X_i - X_j\| = \sqrt{\sum_{s=1}^{d}(x_{is} - x_{js})^2}
\tag{6}
$$

Assume that these n samples are mapped into a new pp-dimensional space, and the matrix form is shown in matrix (7):

$$
Y =
\begin{bmatrix}
y_{11} & y_{12} & \cdots & y_{1p} \\
y_{21} & y_{22} & \cdots & y_{2p} \\
& \vdots & \ddots & \vdots \\
y_{n1} & y_{n2} & \cdots & y_{np}
\end{bmatrix}
\tag{7}
$$

where $p < \, < d$. Generally, p takes 2 or 3. In general, the Metric MDS is to minimize the following objective function:

$$
\sigma(X) = \sum_{i=1}^{n}\sum_{j=1}^{n}(\delta_{ij} - \|y_i - y_j\|)^2
\tag{8}
$$

Euclidean distance can be arbitrarily rotated and transformed because these transformations do not change the distance between samples. Therefore, for Y, the solution is not unique.

3   Data Sources and Process

3.1   Data Sources

This paper utilizes the formula data from Yunnan Kirin Redrying Plant in the years 2019 and 2020 as the primary data source. The dataset includes a total of 42 formula data, consisting of 252 individual tobacco grades, which are referred to as the key grades in this study.

3.2   Research Tools

The main research tools utilized in this study are Python3.6 and SPSS19.0. Python3.6 was employed to construct the Co-formulation matrix, while SPSS19.0 was utilized for subsequent data analysis.

3.3   Research Process

Data pre-processing. In the composition of the tobacco leaf blend, each individual tobacco material has a different proportion. It includes the primary material tobacco which consists of high-quality tobacco leaves with exceptional aroma and is used in large quantities. This primary material tobacco is crucial in determining the cigarette style and providing aroma. The excipient tobacco, on the other hand, plays a coordinating and enhancing role in the aroma. It is generally used in small proportions and often consists of different flavorful or different types of tobacco leaves compared to the primary material tobacco. Additionally, the harmonized tobacco is included in the cigarette leaf blend formula to regulate the strength and smoke concentration. It plays a role in achieving a balanced formula. Lastly, the filled tobacco has relatively neutral tobacco leaf quality. It may not have exceptional aroma quality or obvious quality defects, but it is preferred for its better filling capabilities and cost reduction. In accordance with the ABC classification method, all the different types of single-material tobacco in the blend formula are divided. The single-material tobacco with a proportion ranging from 65% to 100% is classified as Class A or main material tobacco. The single-material tobacco with a proportion ranging from 25% to 65% is classified as Class B or auxiliary material tobacco. The single-material tobacco with a proportion ranging from 5% to 15% is classified as Class C or harmonized tobacco. Any single-material tobacco with a proportion less than 5% is classified as Class D or filled tobacco.

As tobacco is a unique agricultural product, the formulation of the tobacco leaf blend also takes into account expert evaluations and recommendations in actual production. Based on previous data and expert suggestions, it is concluded that the inclusion of D-type single cigarettes will not impact the stability and flavor of the blend. Thus, to ensure accuracy in determining the high-frequency words, the D-type single cigarettes were excluded from consideration. Consequently, the original set of 252 key elements was reduced to 131.

The data consists of eight different places of origin, namely A, B, C, D, E, F, G, and H. Each place of origin is further differentiated by different grades, denoted by numbers. For instance, "Zhanyi C3F-C03" is represented as A19, while "Xiangyun B1F-B01" is indicated as D1.

Determination of High-Frequency Words. Through data preprocessing, this paper obtained a total of 131 significant cigarette grades from a pool of 42 formulations. The selection of high-frequency keywords was based on a threshold determined by the citation frequency. The specific calculation of the high-frequency threshold is outlined in formula (9), which is derived from the Price formula.

$$M = 0.749 \cdot \sqrt{N_{max}} \qquad ((9))$$

Among them, M is the high-frequency threshold, and N_{max} is the highest frequency of interval single material smoke use. In order to establish the high-frequency threshold, a total of 42 formula information was analyzed. The highest frequency of usage was found in the case of the single material tobacco 'A19', which occurred a total of 11 times. That is, N_{max} was 11. The high-frequency threshold was calculated based on the Price formula. Following the research requirements, keywords with a word frequency of 3 or higher were selected as high-frequency key grade single material tobacco. As a result, a total of 25 high-frequency key grade single material tobacco were identified. For Co-formulation analysis, Python 3.6 was utilized to construct a Co-formulation matrix. Cluster analysis and multidimensional scaling analysis were conducted using SPSS 19.0.

4 Results and Analysis

4.1 Confirmation and Statistical Analysis of High-Frequency Key Level

The refinement of the formula is represented by the key grade, which encapsulates the core aspects of the formula. Consequently, the high-frequency key grade serves as an indicator of the formula's research focal points. Following data preprocessing, a comprehensive count was conducted, resulting in a total of 131 formula grades, with a collective frequency of 247 occurrences. Among these, 25 high-frequency key grade single cigarettes (frequency ≥ 3) were identified, corresponding to a total frequency of 103 occurrences. This accounts for 41.70% of the overall frequency of the grade, thereby satisfying the requirements outlined in Table 1.

Table 1. 25 high-frequency key-level statistical tables.

Number	level	Frequency	Number	level	Frequency	Number	level	Frequency	Number	level	Frequency
1	A19	11	8	A13	4	14	B13	3	20	D18	3
2	D19	7	9	A22	4	15	B17	3	21	D22	3
3	E21	6	10	A25	4	16	C1	3	22	F1	3
4	A14	5	11	D13	4	17	C12	3	23	F13	3
5	B18	5	12	E18	4	18	C20	3	24	F18	3
6	C17	5	13	B1	3	19	D14	3	25	F21	3
7	E13	5									

Table 1 reveals that the prominent tobacco formulas predominantly utilize the following key grades: A19, D19, E21, A14, and B18. Nonetheless, it is crucial to conduct additional data mining and analysis in order to comprehend the interconnections and internal dynamics among these critical grades of tobacco.

4.2　Constructing the Co-formulation Matrix

Using Python 3.6 software, a Co-formulation matrix of size 25 × 25 was created. Table 2 illustrates that the diagonal values are all set to 0, while the values in each position indicate the count of high-frequency key-level single-grade cigarette pairs.

Table 2. 25 high-frequency key grade co-formulation matrix (part).

	A19	D19	E21	E13	A14	B18	C17	A22
A19	0	4	1	2	2	3	3	2
D19	4	0	0	0	1	1	1	1
E21	1	0	0	0	0	0	0	1
E13	2	0	0	0	1	1	2	1
A14	2	1	0	1	0	3	0	0
B18	3	1	0	1	3	0	0	0
C17	3	1	0	2	0	0	0	0
A22	2	1	1	1	0	0	0	0

4.3　Construction of Single Material Tobacco Compatibility Score Matrix

To better display the covariant relationship between high-frequency key grades of single-grade cigarettes, the generated Co-formulation matrix is imported into SPSS19.0, and the binary Ochiai $= (C_{ij}/C_i \times C_j)$ is used to convert the Co-formulation matrix into a correlation matrix. By subtracting all the correlation matrix values from 1, a single cigarette compatibility score matrix of size 25 × 25 was ultimately obtained, as presented in Table 3. In this single cigarette compatibility score matrix, a lower value implies a closer relationship between the two high-frequency key grades. In other words, the two single cigarettes are frequently employed in the same formulation. The diagonal value of 0 indicates the relationship between the key grade and itself.

4.4　Cluster Analysis of High-Frequency Key Grade Single Material Tobacco

Cluster analysis is an exploratory analysis technique used to automatically classify sample data based on the relationships between variables. In the context of key grades, cluster analysis seeks to identify clusters that exhibit a high frequency of pairs of key grades appearing together in the same formula. In this study, system clustering and clustering analysis of key grades were performed using SPSS software. Figure 2 illustrates the results of the cluster analysis, where the 25 key grades are categorized into two groups. The vertical axis represents high-frequency key grades, while the horizontal axis (ranging from 0 to 25) reflects the distance between the key grades on the left. The clustering of two key grades together indicates their close relationship. The single-tobacco categorized as type one had a lower value in the compatibility score matrix, indicating a

Table 3. Single cigarette compatibility score matrix (part).

	A19	D19	E21	E13	A14	B18	C17	A22
A19	0	0.45	0.833	0.429	0.411	0.497	0.519	0.682
D19	0.45	0	0.653	0.32	0.526	0.345	0.423	0.42
E21	0.833	0.653	0	0.702	0.896	0.74	0.654	0.428
E13	0.429	0.32	0.702	0	0.419	0.458	0.484	0.645
A14	0.411	0.526	0.896	0.419	0	0.405	0.61	0.703
B18	0.497	0.345	0.74	0.458	0.405	0	0.61	0.455
C17	0.519	0.423	0.654	0.484	0.61	0.61	0	0.45
A22	0.682	0.42	0.428	0.645	0.703	0.455	0.45	0

closer relationship among them. This suggests a higher likelihood of these single-tobacco appearing together in the same formula. On the other hand, the single-tobacco categorized as type two had higher values in the compatibility score matrix, but the relationship among them was not as significant as that of type one. This implies a lower probability of simultaneous occurrence among these single-tobacco in the same formula, meaning that the single-tobacco used in the same formula exhibits higher compatibility (Fig. 1).

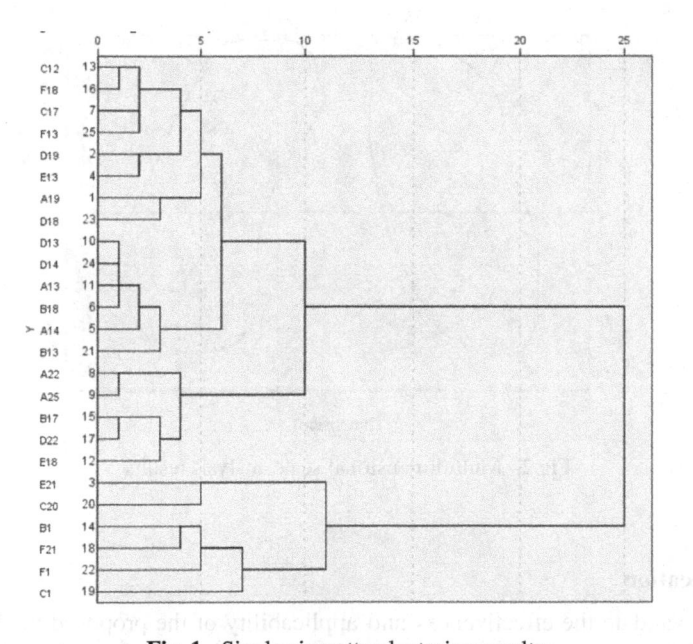

Fig. 1. Single cigarette clustering results

4.5 Multidimensional Scaling Analysis

Cluster analysis is a data exploration technique used to identify patterns and relationships within a dataset. In the context of this study, cluster analysis was employed to examine the interrelationships between key grades of single cigarettes [28]. By analyzing the single cigarette compatibility score matrix of these key grades, a multidimensional scale analysis was executed using the SPSS software. This analysis aimed to reduce the dimensionality of the data and calculate the Euclidean distance model. Subsequently, a visualization map was generated to visually depict the cluster groups and the similarity between key grades of single cigarettes. This visual representation provides an intuitive and comprehensive understanding of the relationships among the key grades within the dataset. The results are presented in Fig. 3. In this figure, dimension 1 represents centripetal degree, which signifies the extent of mutual influence among different fields. Dimension 2, on the other hand, represents density, which reflects the level of internal correlation within each field. From Fig. 3, it is evident that the clustering results of key grade single material tobacco align with those obtained from the analysis in Fig. 2. The clustering can be broadly categorized into two types, demonstrating a pattern of concentrated distribution. By examining the position of each group on the coordinates, it is observed that the proximity between the two groups can also be ascertained. Given the substantial distance between the two groups and the limited connections present, it can be inferred that the interactions and mutual influence between the two groups are relatively low, consequently impacting their compatibility.

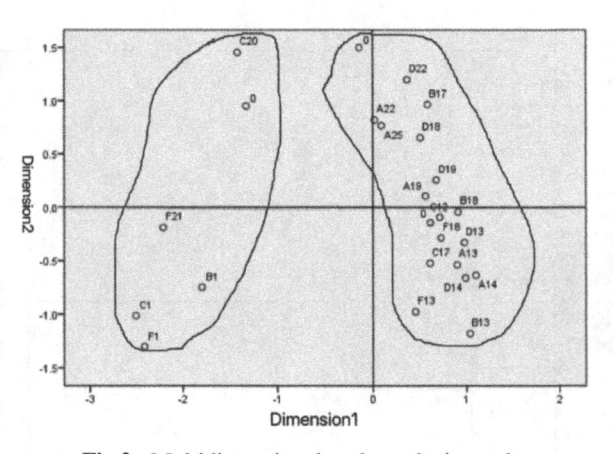

Fig.2. Multidimensional scale analysis results

4.6 Verification

In order to validate the effectiveness and applicability of the proposed method in this paper, the following analysis and verification activities have been conducted:

Comparative Verification of Association Rules. The association rules were used to analyze the formula data of the key grade single cigarette. The obtained results are presented below:

Table 4. Examples of strong association rules

antecedents	consequents	support	confidence	lift
frozenset({'D19'})	frozenset({'A19'})	0.1379	0.5714	1.5065
frozenset({'D22'})	frozenset({'D19'})	0.1034	1	4.1429

Table 5. Examples of weak association rules

antecedents	consequents	support	confidence	lift
frozenset({'E18'})	frozenset({'D19'})	0.1379	0.2414	0.069
frozenset({'D13'})	frozenset({'D19'})	0.1379	0.2414	0.069
frozenset({'D13'})	frozenset({'B18'})	0.1379	0.1724	0.069
frozenset({'A22'})	frozenset({'A19'})	0.1379	0.3793	0.069
frozenset({'A25'})	frozenset({'A19'})	0.1379	0.3793	0.069
frozenset({'D13'})	frozenset({'A14'})	0.1379	0.1724	0.069
frozenset({'A13'})	frozenset({'D14 '})	0.1379	0.1034	0.069
frozenset({'A13'})	frozenset({'D13'})	0.1379	0.1379	0.069
frozenset({'D13'})	frozenset({'A13'})	0.1379	0.1379	0.069
frozenset({'E18 '})	frozenset({'D22', 'D19'})	0.1379	0.1034	0.069

Based on the analysis of association rules, the results, depicted in Table 4 and Table 5, demonstrate that a confidence level above 0.1034 indicates a strong association rule. This signifies a higher likelihood of the key grade single cigarettes appearing in the same formula. On the other hand, a confidence level of 0.069 represents a weak association rule with a low probability of the key grade single cigarettes appearing in the same formula. A comparison between the association rule results, clustering results, and multidimensional scale map indicates consistency among the findings. This illustrates the practicality and effectiveness of employing the co-formulation matrix for cluster analysis to examine the compatibility between single cigarettes in the cigarette formula.

Verification of Actual Formula Data. To validate the efficacy of the proposed method and ensure its consistency with real-world performance, we selected the 2021 formula data of Qilin Redrying Plant in Yunnan Province for verification. This included 26 formula data comprising 201 single-feed tobacco grades. Prior to analysis, the data underwent preprocessing, and class D single-feed tobacco was excluded from the dataset. The single-feed tobacco grades originated from 9 different regions, denoted as A, B, C, D, E, F, G, H, and I. Each grade was assigned a specific number to distinguish it. We selected a total of 36 key single-feed smoke grades, with a frequency greater than or equal to 2, based on the Price formula. Subsequently, common square analysis and cluster analysis were performed. The obtained results are presented Fig. 3.

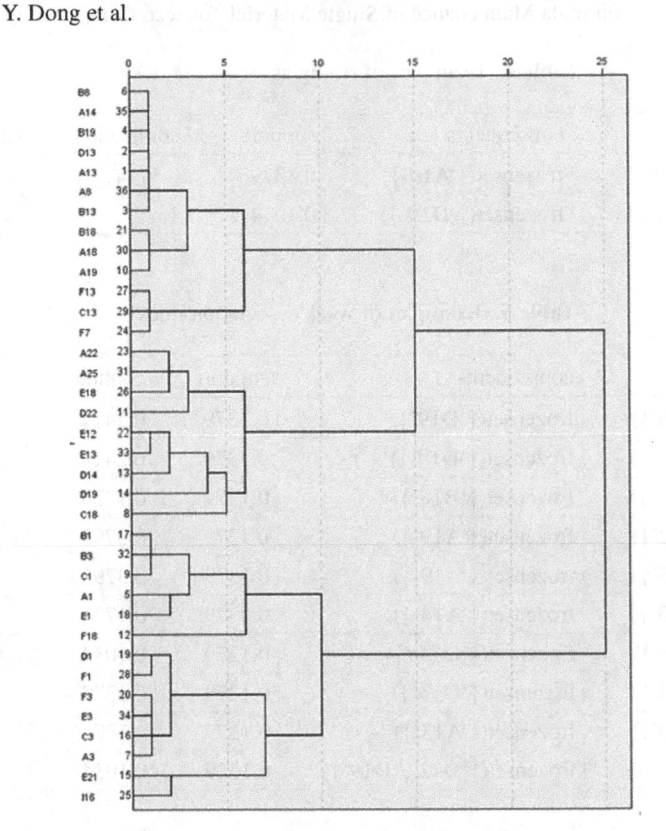

Fig. 3. Single cigarette clustering verification results

From Fig. 3, it can be observed that the 36 key grades can be divided into two distinct categories. One category exhibits a lower single cigarette compatibility score matrix, suggesting a closer relationship among these key grades. This means that there is a higher probability of single-material smoke appearing in the same formula. On the other hand, the other category demonstrates the opposite trend. The clustering results of key grade single material tobacco from 2021 were compared with those from 2019 and 2020. The results from 2021 generally followed the expected pattern, with only a small number of abnormal values. These anomalies can be attributed to factors such as tobacco leaf yield and human factors. Overall, this method can provide valuable guidance for tobacco leaf formulation and preparation.

5 Discussions

Through historical data mining, this study aims to explore the compatibility relationship between the origins and grades of single-material tobacco. By maintaining the flue-cured tobacco formula, it is possible to enhance the utilization rate of flue-cured tobacco in actual tobacco production. This, in turn, helps to maintain the stability of cigarette taste and quality, while ensuring the relevant interests of industrial enterprises. Wang Nan et al. [29] utilized non-negative matrix factorization in order to uncover the compatibility rules between individual cigarettes, thereby effectively preserving the sensory quality of cigarettes. Similarly, Feng Runze et al. [30] determined the co-occurrence frequency of individual cigarettes and calculated the mutual information between them. This enabled them to identify the compatibility between single cigarettes, addressing the challenge of multiple single cigarette replacements in practical production and ultimately ensuring the stability of the cigarette formula. The research objective of this paper aligns with theirs.

6 Conclusion

This paper aims to improve the utilization rate of tobacco leaves, ensure the quality of cigarette products, and reduce costs in the tobacco industry by considering the historical data of tobacco leaf formula. To achieve this, a maintenance method of tobacco leaf formula based on co-formulation analysis is proposed. The study systematically analyzes the historical formula data of Kirin Redrying Plant in Yunnan Province in 2019 and 2020. The frequency of use of single-material tobacco in each formula is calculated, and high-frequency single-material tobacco is extracted. Firstly, a co-formulation matrix and a single-material tobacco compatibility score matrix are constructed. Then, cluster analysis is performed to identify the close relationship and compatibility patterns between single-material tobacco and the key single-material tobacco used in the flue-cured tobacco formula. This provides a theoretical basis for formulation and maintenance. This method overcomes the limitations of traditional tobacco leaf formula in utilizing historical data for mining value. It systematizes the use of historical data and data mining, thereby making it convenient for tobacco enterprises to apply and accumulate knowledge and experience. The proposed maintenance method based on co-party analysis allows for a deeper understanding of the use of single tobacco in the flue-cured tobacco formula. This provides a theoretical basis for formula maintenance, reduces the workload of formula maintenance, and introduces new ideas for formula research. Overall, this study contributes to improving tobacco leaf formula maintenance and offers practical implications for the tobacco industry.

Fund Project. China National Tobacco Corporation Yunnan Tobacco Company Science and Technology Plan Key Project "Research on Basic Conditions for Intelligent Grading of Tobacco Leaves" (2020530000241003); Key Project of Science and Technology Plan of Yunnan Tobacco Company of China National Tobacco Corporation "Establishment and Application of Artificial Intelligence Grading Model Based on Basic Samples of Flue-cured Tobacco" (2021530000241012); Kunming University of Science and Technology 2022 Student Extracurricular Academic Science and Technology Innovation Fund Project "Xinxin 'Agriculture' - Modular Intelligent Classifier for Agricultural Products" (2022KJ117).

References

1. Wang, J., Shu, J., Li, W.: Evaluation of compatibility of characteristic flue-cured tobacco variety CB-1 in Fujian Province. J. Anhui Agric. Univ. **49**(02), 352–358 (2022)

2. Li, Q., et al.: Tobacco leaf substitution method based on tobacco pyrolysis difference analysis. Tobacco Sci. Technol. **51**(08), 76–84 (2018)
3. Li, S., et al.: Tobacco leaf replacement and cigarette formula maintenance based on near-infrared spectroscopy similarity. Tobacco Sci. Technol. **53**(02), 88–93 (2020)
4. Yang, C., et al.: Chemical composition analysis of cigarette leaf group and its relationship with formula maintenance. Tobacco Sci. Technol. **52**(04), 76–84 (2019)
5. Gao, Y., Zhang, Y., Huang, P.: Application of cluster analysis in cigarette formulation. China Agric. Bull. **02**, 103–106 (2006)
6. Yan, K., Wang, J., Qu, J., Li, X.: Correlation analysis between smoking quality and main physical and chemical indexes of flue-cured tobacco in Henan. Tobacco Sci. Technol. **01**, 5–9 (2001)
7. Wang, Y., et al.: Availability of chemical components in Qujing tobacco leaves and its effect on sensory smoking quality. Tobacco Sci. Technol. **11**, 67–73 (2014)
8. Mi, J., et al.: Upper bound analysis of tobacco blend proportion based on near infrared spectral projection and monte carlo method. Spectro. Spectral Anal. **31**(04), 915–919 (2011)
9. Yang, Q., et al.: Application and analysis of sensory quality model of tobacco leaf ratio in leaf group formula design. J. Food Biotechnol. **36**(07), 756–761 (2017)
10. Qin, Y., Zhang, M., Yang, N., Shan, Q.: Kernel mapping and Rank-Order distance based locality preserving projection similarity measure. Spectro. Spectral Anal. **41**(10), 3117–3122 (2021)
11. Luo, X., Wang, N., Zhang, Z., Zhao, L., Tang, J., Tang, J.: Application of association rule mining method in intelligent cigarette formula maintenance. Chin. J. Tobacco **24**(03), 21–29 (2018)
12. Wang, J., Du, Y., Li, X.: Evaluation and analysis on tobacco leaf compatibility. Chin. Tobacco Sci. **31**(01), 65–69 (2010)
13. Zhang, Q., Ma, J., Dong, G.: Compatibility of introduced American flue-cured tobacco cultivars NC102 and NC297. Chin. Tobacco Sci. **17**(06), 19–26 (2011)
14. Li, W., Jiang, C., Tang, Y.: Analysis on compatibility of peculiar tobacco leavies from different producing areas. Guizhou Agric. Sci. **05**, 53–55 (2008)
15. Wang, J., Yuan, H., Li, X.: Assessment method and rule of tobacco leaf compatibility. Tobacco Sci. Technol. **06**, 8–11 (2007)
16. Wu, Y.: Cigarette formulation rule extraction and formulation maintenance based on Bayesian network. Northeastern University (2017)
17. Feng, L., Cold, F.: Advances in co-word analysis. Chin. J. Library Sci. **02**, 88–92 (2006)
18. Zhang, Q., Ma, F.: Foreign knowledge management research paradigm-based on co-word analysis. J. Manag. Sci **06**, 65–75 (2007)
19. Ni, B., He, M., Cao, B.: Status quo and research trends of neurosurgical departments in China: bibliometric and scientometric analyses (Preprint). (2020)
20. Wang, Z., Deng, Z., Wu, X.: Status quo of professionalpatient relations in the internet era: bibliometric and co-word analyses. Int. J. Environ. Res. Public Health (7), (2019)
21. Li, X., Ma, L., Liu, Y.: Mapping theme trends and knowledge structures of dignity in nursing: a quantitative and co-word biclustering analysis. J. Adv. Nurs. **78**(7), 1980–1989 (2022)
22. Wang, L., Yang, G., Jiang, X.: Research hotspots and trends in nursing education from 2014 to 2020: a co-word analysis based on keywords. J. Adv. Nurs. **78**(3), 787–798 (2022)
23. Deng, S., Wang, W., Zhang, Y., Zhang, C., Tu, S.: Case study of aero-engine system failure based on co-word analysis. J. East China Univ. Sci. Technol. (Nat. Sci. Edn.) **47**(05), 635–646 (2021)
24. Shi, B., Wei, W., Qin, X.: Mapping theme trends and knowledge structure on adipose-derived stem cells: a bibliometric analysis from 2003 to 2017. Regen. Med. **14**(1), 33–48 (2019)

25. Bai, Y., Luo, J., Xu, Z., Lin, L.: Hotspot analysis of cross-border e-commerce teaching research-knowledge map based on co-word matrix. Laboratory Res. Explor. **39**(10), 264–269 (2020)
26. He, M., Wu, X., Chang, M., Ren, W.: Distributed collaborative filtering recommendation based on user co-occurrence matrix multiplier. Comput. Sci. **43**(S2), 428–435 (2016)
27. Li, C., Zhi, l., Ji, X., Wang, X.: Research progress of information science in china-statistical analysis based on keywords of journal papers. Library Inf. Work **54**(24), 31–36 (2010)
28. Jiang, H., He, S., Wang, T.: Research hotspots and development trends of college students ' network moral behavior-knowledge map analysis based on co-word matrix. J. Qinghai Normal Univ. (Soc. Sci. Ed.) **43**(01), 31–38 (2021)
29. Wang, N., Luo, X., Zhang, Z., Yu, Y., Tang, J., Tang, J.: Cigarette formula maintenance method based on non-negative matrix factorization. Tobacco Technol. **52**(08), 67–76 (2019)
30. Feng, R., Luo, X., Zhang, Z., Tang, J., Jordanna.: Cigarette formula maintenance method based on mutual information of single cigarette. Tobacco Sci. Technol. **54**(03), 65–71 (2021)

Multi-physical Field Collaborative Simulation Optimization Technology and Reliability Analysis of Power Amplifiers

Zhenbing Li[1], Junjie Huang[1], Shilin Jia[1], Jinrong Zhang[1], Jialong Fu[1], Xiaochuan Fang[2], Gang Li[3], and Guangjun Wen[1(✉)]

[1] School of Information and Communication Engineering, University of Electronic Science and Technology of China, Chengdu, Sichuan, People's Republic of China
wgj@uestc.ed
[2] Antennas and Electromagnetics Research Group, School of Electronic Engineering and Computer Science, Queen Mary University of London, London E1 4NS, UK
[3] School of Information and Software Engineering, University of Electronic Science and Technology of China, Chengdu, Sichuan, People's Republic of China

Abstract. As the core RF front-end device of wireless communication systems, the performance of millimeter wave power amplifier chips determines the communication distance/quality of the system. In order to improve the communication distance of the system, power amplifiers are often required to operate in a saturated state, which can cause severe warping and deformation of PA chips due to high temperature accumulation. This seriously affects the performance of RF front-end, and has become a key issue that needs to be urgently solved in the industry. This paper takes MATLAB as the core to build a Circuit-Electromagnetic-Thermal-Force Multiple Physical Field collaborative simulation and optimization platform. In HFSS, using vbs script to achieve fully parameterized modeling and full band circuit electromagnetic joint simulation. In ANSYS Workbench, thermodynamic simulation is implemented using SCDM and Mechanical in IronPython environment. In this paper, a PA chip based on GaN process is simulated by using the Multi-Physical Field collaborative simulation and optimization platform. Aiming at its high temperature agglomeration problem, the number and layout of transistor count are optimized, and the maximum temperature is successfully reduced by about 70 °C. In order to compensate for the circuit performance degradation caused by layout, this paper built an automatic optimization platform to optimize it, and the gain increased by about 8.4 dB compared with that before optimization, which proved the availability, reliability and progressiveness of the multi-physical field collaborative simulation optimization platform.

Keywords: Circuit-Electromagnetic-Thermal-Force Multiple Physical Field · Millimeter Wave Power Amplifier Chip · Co-Simulation Design · Reliability Analysis

H. Jin et al. (Eds.): IAIC 2023, CCIS 2058, pp. 262–276, 2024.
https://doi.org/10.1007/978-981-97-1277-9_20

1 Introduction

With the rapid development of wireless communication systems, especially the popularity of 5G communication, the demand for output power of RF front-end circuits is becoming increasingly high, and the performance of RF front-end circuits is mainly determined by power amplifiers (PA). With the continuous innovation of advanced packaging technology and semiconductor materials and processes, although the packaging size of components continues to decrease, the output power value of GaN based power amplifier products continues to increase, further confirming the speculation of the "post Moore era" [1]. However, high integration miniaturized power amplifier products are prone to warping and deformation due to high temperature accumulation, which can lead to the failure of the power amplifier in the specified working modes [2, 3]. In order to improve the processing yield of power amplifier products, developing Multiphysics simulation collaborative simulation and optimization platform technology integrating circuit, electromagnetic, thermal and force has become a current research hotspot [4, 5]. The typical Multiphysics simulation technology involves model construction [6, 7], calling of interface functions, multi-objective collaborative simulation [8, 9], data flow transmission and iterative optimization [10]. Through this technology, better reliability analysis and optimization methods can be provided for the design of target circuits.

At present, many can already simulate non-single physical quantities. For example, ADS can simulate circuits and electromagnetic characteristics [11], HFSS can perform complete electromagnetic characteristics simulation and partial circuit characteristics simulation [12], and ANSYSWorkbench can also perform thermal and mechanical simulations [13]. However, there are still many problems using a single platform for simulation, such as different focuses and inconsistent simulation and optimization algorithms. Currently, designers often prefer to use the most reliable platform to provide reliable data. If the most reliable software in various fields can be called based on a certain platform to perform collaborative simulation and optimization on the target circuit, the final circuit obtained will inevitably be recognized by the industry, with guaranteed reliability and greatly improved design efficiency.

In response to the above requirements, this article proposes a reliability analysis method and simulation optimization platform based on MATLAB to achieve collaborative simulation optimization of multiple physical quantities. By programmatically using ADS, HFSS, and ANSYSWorkbench interfaces in MATLAB, parameterized modeling and rapid simulation of the target circuit can be achieved using three software, ADS builds a circuit electromagnetic coordination simulation optimization platform containing source devices based on TADSInterface, and it performs power characteristic simulation, and calculates thermal dissipation power; HFSS is responsible for programming parameter modeling based on vbs and full band electromagnetic simulation of passive circuits; ANSYSWorkbench is responsible for loading thermal loads and boundary conditions on thermodynamic models, and performing temperature and thermal force simulations. To verify the reliability of the platform, this article takes a GaN power amplifier chip [14] operating at 22 GHz ~ 27 GHz as the test object, and completes the establishment, automated assignment, and simulation of a GaN PA parameterized full electromagnetic simulation model based on vbs script language; A GaN PA parameterized thermodynamic simulation model written in IronPython language environment and used Mechanical for assignment and simulation. Through multi-physical field collaborative simulation of the power amplifier, it was found that its maximum temperature is as high as 206.6 °C, which is close to the maximum operating temperature of GaN HEMT, severely affecting its circuit performance and posing a risk of chip burnout. To solve the problem of high-temperature agglomeration, the idea of homogenizing the temperature field distribution of the chip was adopted, and the number and layout of transistors were optimized, successfully reducing the maximum temperature by about 70 °C. At the same time, in order to compensate for the circuit performance degradation caused by layout changes, this article has built an automatic optimization platform to optimize it.

2 The Process of Building a Multi-physical Field Collaborative Simulation and Optimization Platform

2.1 Basic Concepts

The design idea of the Circuit-Electromagnetic-Thermal-Force multi-physical field collaborative simulation optimization platform proposed in this article is to use MATLAB as the main control platform, and then use external interfaces to call ADS, HFSS, ANSYSWorkench for circuit, electromagnetic, thermal, and stress deformation simulation. Among them, HFSS (Full EM) uses finite element method and efficient automatic mesh generation algorithm for full electromagnetic simulation of three-dimensional structures, ADS (Circuit) uses active device SPICE model and SnP file suitable for all frequency bands for circuit structure simulation and optimization based on genetic algorithm, and ANSYSWorkbench calculates the thermal and mechanical stress distribution of the target circuit based on solid finite element method. The architecture of the multi-physics collaborative simulation platform based on MATLAB, ADS, HFSS, and ANSYS is shown in Fig. 1.

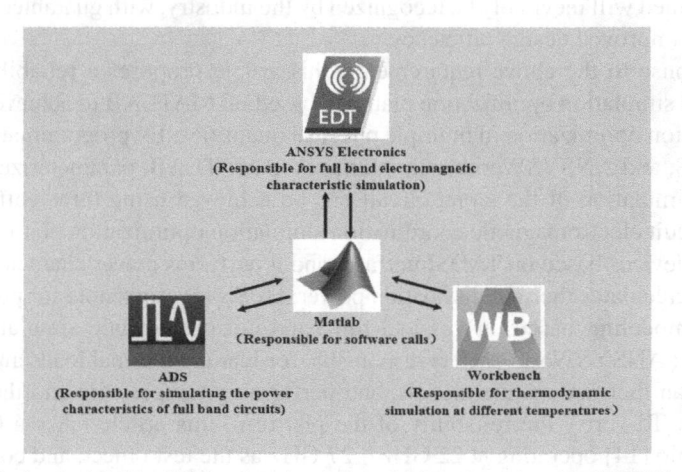

Fig. 1. Multi-physical field collaborative simulation platform based on ADS, HFSS, and ANSYS

2.2 Design of a Multi-physical Field Collaborative Simulation and Optimization Platform

In order to achieve collaborative simulation and optimization of multiple physical fields, it is necessary to achieve data transfer between various EDA software, generating and transmitting data streams based on MATLAB, that is, the cyclic iteration of physical parameters, which is the core operation of the multi-physical field collaborative simulation platform. Its main processes include:

1. ADS-HFSS integration: Import the SnP generated by the full band automated electromagnetic simulation of the target circuit into ADS for RF power characteristic simulation, and optimize the RF device parameters through MATLAB to improve the electromagnetic circuit performance;
2. ADS-ANSYS Workbench integration: After conducting thermal dissipation power P_{diss} simulation on the target circuit by ADS, the data is transmitted to ANSYS-Workbench for temperature and stress deformation simulation. For power amplifier chips, the thermal dissipation power is determined by the combination of DC power consumption and input/output power, as shown in the following equation:

$$P_{diss} = P_{dc} - (P_{out} - P_{in}) \tag{1}$$

To achieve MATLAB's control of HFSS, ADS, and ANSYS Workbench, it is necessary to implement three sub interfaces: MATLAB-ADS, MATLAB-HFSS, and MATLAB-ANSYS. The functional block diagram of the implementation is shown in Fig. 2. The functions that each sub platform interface needs to be completed include establishing simulation models, modifying model parameters, executing simulations, post-processing results, and data transmission.

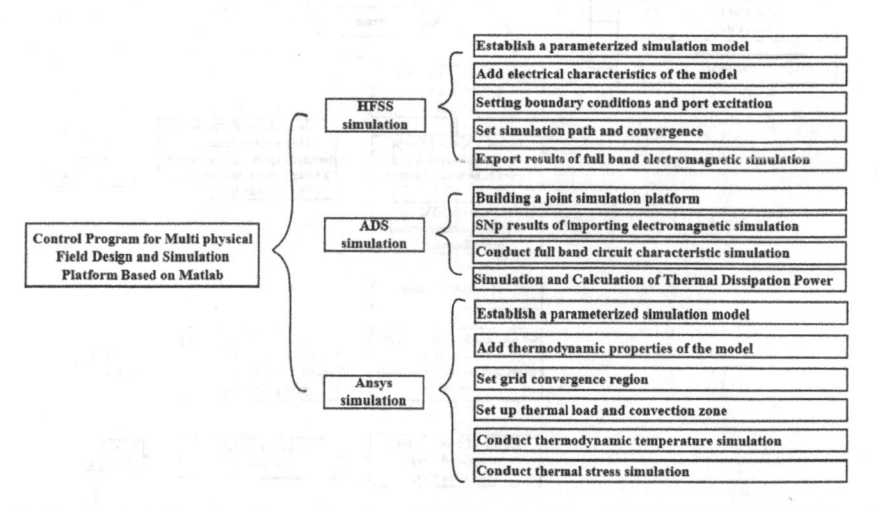

Fig. 2. Implementation functions of various interfaces in the multi-physical field collaborative simulation platform

The process of multi-physical field full wave electromagnetic simulation is shown in Fig. 3, which mainly includes parameterized modeling, material allocation, excitation port setting, boundary condition setting, and solution setting; The multi-physical field thermodynamic simulation process is shown in Fig. 4, which mainly includes parameterized modeling, material thermodynamic attribute settings, boundary condition settings, initialization settings, solution settings, and data transfer between modules.

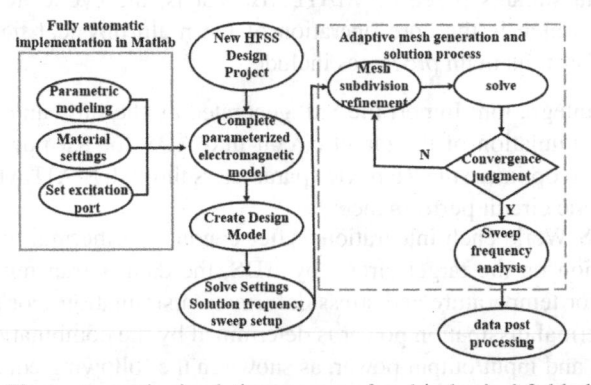

Fig. 3. Electromagnetic simulation process of multi-physical field platform

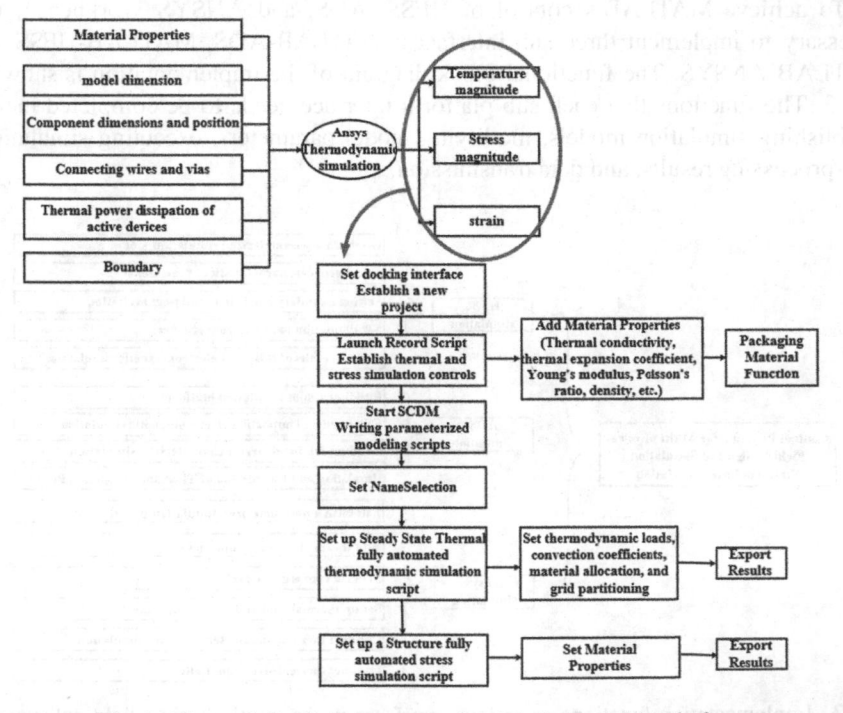

Fig. 4. Thermal simulation process of multi-physical field platform

Parameterized Electromagnetic Model and Automation Simulation Based on vbs. MATLAB-HFSS establishes data transmission through an interface, controls

HFSS to parameterize electromagnetic structure modeling, set simulation conditions, and output formats in a vbs script using MATLAB, and ultimately achieves programmatic design of full wave electromagnetic characteristics simulation of passive circuits in the circuit model. The pre-processing of electromagnetic simulation relies on the MATLAB-HFSS-API toolkit to refine and smooth complex models. The API toolkit provides various encapsulated and integrated interface functions for designers to facilitate secondary development, mainly including the following five parts.

1. 3D modeler: Contains the creation of various geometric entities (boxes, polyhedra, cylinders), modification of geometric entity attributes, stretching, displacement, cutting, merging, and other operations.
2. Analysis: Contains settings for excitation ports, settings for solving scans, and insertion of specified solving methods.
3. Boundary: Select the port type (lumped port, wave port, ring port) and boundary conditions (radiation boundary condition, ideal conductor boundary condition).
4. Contrib: Add and set parameter variables, define the export method and path of the results, obtain SnP files and various simulation result data.
5. General: Add material properties, set the docking data flow method for MATLAB-HFSS, create and save projects, and insert new engineering designs.

MATLAB API provides numerous modeling and simulation function tools, such as using HfssBox in the 3dmoder folder to draw rectangular entities and construct various RF devices and microstrip lines when modeling power amplifier chips; HfssMove to move entities and adjust device spacing; HfssCylinder draws cylinders to construct through holes, blind holes, etc.; HfssRectangle to draw the plane where the aggregated port Port is located, etc. Fully automated modeling based on vbs script development can break away from traditional recording/manual modeling methods, and programmatic modeling can be achieved by utilizing the relevant functions in the MATLAB API toolkit shown above. Compared with traditional manual modeling, using MATLAB for programmatic modeling not only improves the accuracy of modeling, but also makes parameter modification of devices more efficient and convenient for later secondary development and optimization of models, especially for complex multi-layer models such as power amplifier chips. During modeling, it is not necessary to use HFSS to find the specific location of the model and perform repetitive operations. The model can be reconstructed directly through a few lines of commands.

After modeling, the material allocation port excitation and radiation boundary conditions are set using the hfssAssignMaterial function, hfssAssignLumpedPort function, and hfssAssignRadiation function, respectively. Then, the hfssInsertSolution function and hfssInterpolatingSweep function are used to set the grid division conditions and sweep settings. Finally, the hfssSolveSetup function is used to control the HFSS for electromagnetic simulation, and the hfssExportNetworkSNPFile function is used to export the SnP file.

Then, the SnP file is transferred to the ADS joint simulation platform through MATLAB to simulate the RF power characteristics, and the corresponding device parameters are modified by reading the netlist.log file, continuously conducting circuit electromagnetic joint simulation. On this basis, the MATLAB loop can be used to repeatedly iterate variables and set the objective index as the optimization objective to achieve automated circuit electromagnetic optimization and find the global optimal solution.

Modeling and Mechanical Simulation of SCDM Based on Python

To program the ANSYSWorkbench thermodynamic simulation process, it is necessary to first use the SCDM platform to construct a parameterized thermodynamic 3D model in the IronPython language environment. The input parameters include material properties, model size, component size position, connection line and via size position, output power and efficiency of the source device, simulation boundary conditions, etc. SpaceClaim Direct Modeler (SCDM) is an industry-leading direct modeling tool of rapid conceptual design and geometric operations. It is equipped with pre-processing tools based on CAE analysis, which can achieve model repair, preparation, and optimization.

Generally, the five code blocks for constructing the target circuit model based on SCDM programming are shown in Fig. 5.

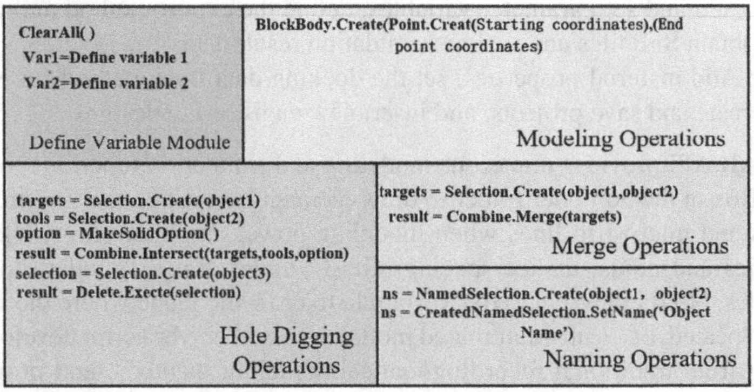

ClearAll( ) Var1=Define variable 1 Var2=Define variable 2 Define Variable Module	BlockBody.Create(Point.Creat(Starting coordinates),(End point coordinates) Modeling Operations	
targets = Selection.Create(object1) tools = Selection.Create(object2) option = MakeSolidOption( ) result = Combine.Intersect((targets,tools,option) selection = Selection.Create(object3) result = Delete.Execte(selection) Hole Digging Operations	targets = Selection.Create(object1,object2) result = Combine.Merge(targets) Merge Operations	
	ns = NamedSelection.Create(object1, object2) ns = CreatedNamedSelection.SetName('Object Name') Naming Operations	

Fig. 5. Schematic code for each module of the parameterized PA thermodynamic model

After completing the construction of the thermodynamic simulation model, the Iron-Python language script is written using Mechanical Scripting on the ANSYS platform to simulate the temperature cloud map and stress deformation. The output results include the temperature magnitude, stress distribution, and shape variables of each node in the model, and post-processing is performed on the output results to analyze whether the maximum temperature exceeds the rated range of the component or the maximum temperature that the material can withstand, whether the maximum shape variable exceeds the maximum allowable threshold of the component material, and whether the deformation at the solder joint will affect the electrical connection.

To facilitate programmatic ANSYS thermodynamic simulation, it is necessary to create datasets for independent temperature and stress simulations and establish data flow directions. Therefore, the pre-simulation processing includes: material attribute settings, global grid partitioning, internal thermal load distribution, natural convection coefficient settings, and insertion of simulation controls; The post-processing includes adjusting the size of the results, exporting temperature cloud maps and temperature data at each coordinate point, and transferring the thermal simulation results to the deformation stress simulation control, as shown in Fig. 6.

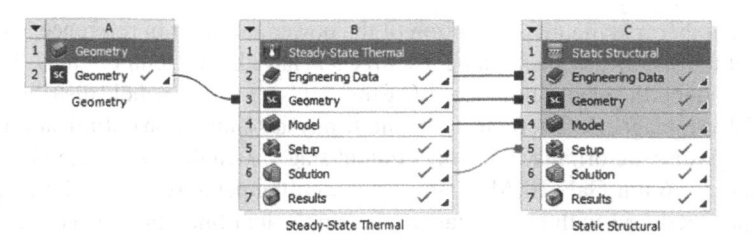

Fig. 6. Thermodynamic simulation data flow transfer method created based on IronPython

3 Analysis and Verification Example of Multi-physical Field Simulation Optimization Platform

To validate the multi-physical field collaborative simulation optimization platform proposed in this paper, a 1 W GaN power amplifier chip operating at 22 GHz–27 GHz [14] was used as the test object to analyze its reliability and provide an optimization plan. The model of GaN power amplifier is based on the ADS layout and Momentum3D schematic diagram in the text, and the final programmed automatic construction of HFSS full band simulation model is shown in Fig. 7.

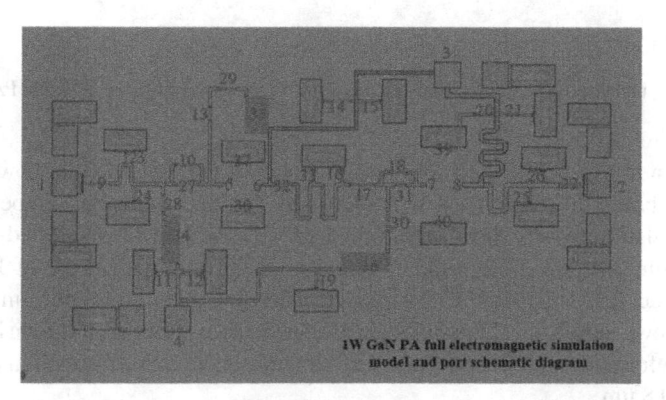

Fig. 7. GaN PA HFSS full wave electromagnetic simulation model and port setting diagram created based on vbs script

After completing electromagnetic simulation, ADS RF characteristics are simulated through data flow on multi-physical field platforms. Compared with the DC characteristics, the DC characteristics of GaN power amplifier full band circuit electromagnetic joint simulation are consistent with the electromagnetic simulation results of ADS Momentum circuit provided in reference [14], and the RF performance is also highly consistent.

3.1 Optimization of Transistor Layout Based on Thermodynamic Simulation

On the basis of the programmed GaN power amplifier chip HFSS model, improving the thermodynamic model requires the introduction of HEMT transistors. According to independent thermodynamic simulation of the power amplifier in reference [14], it was found that transistors are the main heat source and have higher temperatures. Therefore, this article adopts the idea of homogenizing the chip's temperature field distribution in the Circuit-Electromagnetic-Thermal-Force multi-physics simulation optimization method. By decomposing the original PA chip's output stage, a single 8×46 um GaN HEMT, into two 4×46 um GaN HEMTs, the maximum temperature within the chip can be significantly reduced without increasing the chip's area limitation. This improvement enhances the chip's reliability and lifespan. The parameterized transistor model and ground hole distribution are illustrated in Fig. 8.

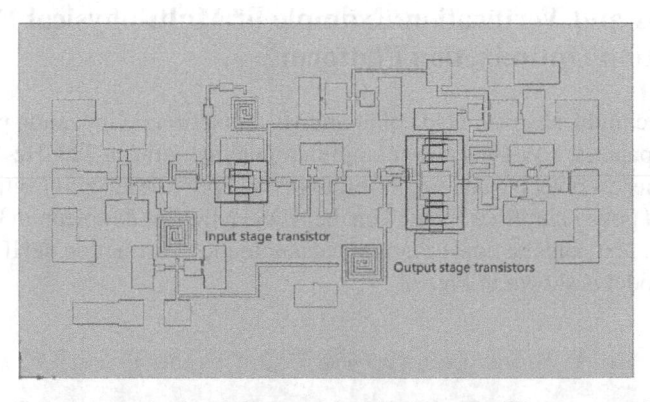

Fig. 8. Thermodynamic simulation model of GaN based millimeter wave PA chip

As shown in Fig. 9, the thermodynamic model of the SCDM GaN power amplifier chip written based on IronPython was thermally simulated at room temperature of 25 °C. The simulation results show that compared with the HEMT integrated in reference [14], the optimized power amplifier chip has a maximum temperature of 136.47 °C, a temperature decrease of 70.13 °C. Moreover, according to the stress deformation results of the GaN power amplifier chip after steady-state thermal simulation shown in Fig. 10, its maximum deformation is concentrated near the transistor, with a maximum deformation variable of 0.8 um.

3.2 Building an Optimization Platform for PA Electromagnetic Characteristics Based on the New Layout

Due to the influence of optimized temperature and stress results, the layout and via-hole layout of transistors are different from those in reference [14]. Therefore, it is necessary to re-optimize the modified GaN power amplifier chip based on thermodynamic models through circuit electromagnetic multi-physical field iteration to achieve a RF power amplifier chip design that meets the Circuit-Electromagnetic-Thermal-Force standard at the same time.

To achieve programmatic circuit optimization, MATLAB is used as the optimization platform, and the netlist.log file and dataset.ds file are accessed through interface functions (netlist.log is used to access ADS engineering variables; dataset.ds is used to store

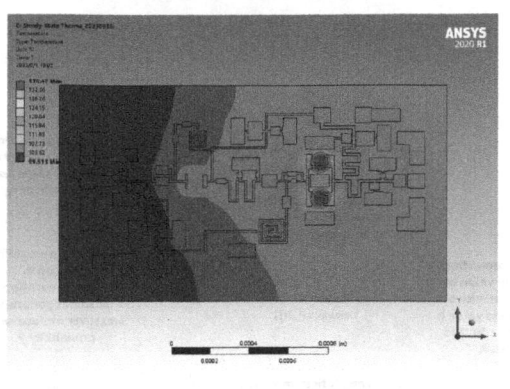

Fig. 9. Temperature cloud map for simulating the thermal effect of 1 W GaN PA based on multiple physical fields

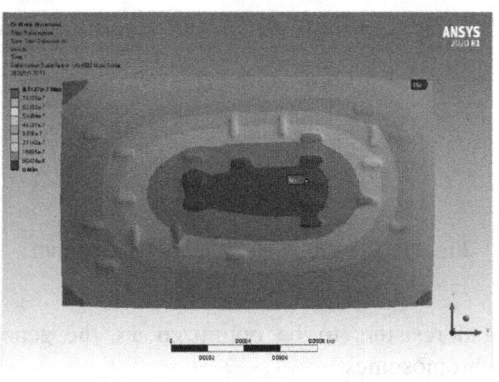

Fig. 10. Simulation results of stress and deformation in 1 W GaN PA based on multiple physical fields

simulation result data). The optimization effect is achieved by iteratively modifying the variable values in the netlist.log file through optimization functions. In order to restore the RF performance of PA, iterative optimization is focused on the matching circuit and negative feedback circuit of GaN PA chip, with the relevant capacitance and resistance of this part of the circuit set as optimization variables.

This method chosen in this article is based on genetic algorithm optimization, which simulates the natural selection law of survival and survival in biological populations. Through the iterative process of selection, crossover, mutation, and reselection, the most suitable individual for the environment is gradually selected, which is to obtain the minimum value of the fitness function. The optimization process is shown in the figure below:

Taking S parameter optimization as an example, the optimization objective is set to:

$$S_{11} < -18 @ 23.8\text{GHz} < \text{Freq} < 24.2\text{GHz}$$

$$S_{21} > 19.8 @ 22\text{GHz} < \text{Freq} < 26\text{GHz}$$

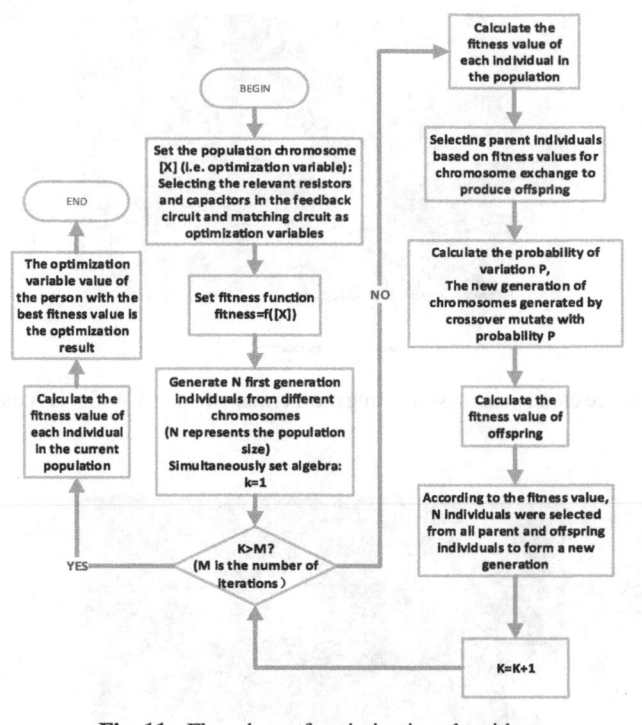

Fig. 11. Flowchart of optimization algorithm

The capacitors and resistors to be optimized are the genes of the optimized population, forming chromosomes:

$$X = [R^0, R^1, R^2, R^3, R^4, C^0, C^1, C^2, C^3, C^4, C^5, C^7]$$

X serves as the basis for determining individual fitness. Set fitness functions based on optimization objectives: fitness $= f(X)$. The genetic algorithm iteratively searches for the minimum fitness value and corresponding chromosome X. Modify the R0 ~ R4, C0 ~ C7 values in the netlist. Log file in the project using MATLAB and ADS_ The RunSimulation () function in the API controls ADS for simulation, and reads the dataset. ds file after simulation to obtain the S parameters at each frequency point of the modified circuit. If the S parameter of the individual meets or approaches the optimization goal, the fitness value will be smaller, and vice versa, the fitness value will be larger. Based on this idea, the fitness function is set as Eqs. (2):

$$f = \sum_{Freq=23.8*10^9}^{Freq=24.2*10^9} [max\{S_{11}(Freq) + 18, 0\} + \varepsilon(S_{11}(Freq) + 18)]$$

$$+ \sum_{Freq=23.8*10^9}^{Freq=24.2*10^9} [max\{19.8 - S_{21}(Freq), 0\} + \varepsilon(S_{21}(Freq) - 19.8)] \qquad (2)$$

$$+50 * max\{S_{11}(24 * 10^9) + 18, 0\} + 50 * max\{S_{21}(24 * 10^9) - 19, 0\}$$

After setting the optimization variable range, population size, and iteration count, the optimization process begins. Based on the fitness function values of the parent individuals, individualsare chosen for chromosome crossover and mutation to generate a new generation, thus iteratively optimizing the solution.

In terms of code implementation, the built-in genetic optimization function ga () provided by MATLAB is employed. Its function expression is as follows: [x_best, fval] = ga (fitness, nvars, A, b, Aeq, beq, lb, ub, nonlcon, options).

The pseudo code of this optimization algorithm is as follows (Table 1):

Table 1. Pseudocode of Genetic Algorithm

Genetic optimization function

```
1  Function  fitness=f(X)
/*Where X=[R⁰,R¹,R²,R³,R⁴,C⁰,C¹,C²,C³,C⁴,C⁵,C⁶].That is, the
resistors and capacitors related to the matching circuit
and feedback circuit. This function is used to solve for
the fitness value for each simulation*/
2  [S₁₁,S₂₁,S₁₂,S₂₂]=ADS_MTALB(X);
/*The function ADS_MTALB() is used to modify variable
values,simulate and return simulation results.Modify the
netlist.log file in the ADS project file through TADSIn-
terface.ChangeParameter(), and modify the value of the
optimization variable recorded in it to X.It then runs
the simulation using the TADSInterface (). RunSimulation
() function and reads the dataset.ds file to obtain the
S parameters for the simulated frequency range.*/
3
```

$$f = \sum_{Freq=23.8*10^9}^{Freq=24.2*10^9} [max\{S_{11}(Freq) + 18,0\} + \varepsilon(S_{11}(Freq) + 18)]$$
$$+ \sum_{Freq=23.8*10^9}^{Freq=24.2*10^9} [max\{19.8 - S_{21}(Freq),0\} + \varepsilon(S_{21}(Freq) - 19.8)]$$
$$+50 * max\{S_{11}(24 * 10^9) + 18,0\} + 50 * max\{S_{21}(24 * 10^9) - 19,0\}$$

```
/* Set optimization goals:
S₁₁>-18@23.8GHz<Freq<24.2GHz,S₂₁>19.8@22GHz<Freq<26GHz*/
4  END;
5  A,b,Aeq,beq,nonlcon=[ ];
6  nvars=13;
/*Set equality and inequality constraints between varia-
bles and the number of variables*/
7  options=gaoptimset('PopulationSize', 30, 'Genera-
tions', 50);
/* Set the number of populations and iterations*/
8  lb=[R0_min,R1_min,…,C6_min];
9  ub=[R0_max,R1_max,…,C6_max];
/*Set the upper and lower limits of optimization varia-
bles*/
10
[x_best,fval]=ga(fitness,nvars,A,b,Aeq,beq,lb,ub,nonlcon
,options);
/*Optimizing circuit S parameters through the ga() func-
tion*/
```

The performance of the chip optimized by MATLAB controlled ADS is shown in Fig. 12. The S21 at the central operating frequency point has increased by about 8.4 dB compared to before optimization, which can achieve the optimization goal.

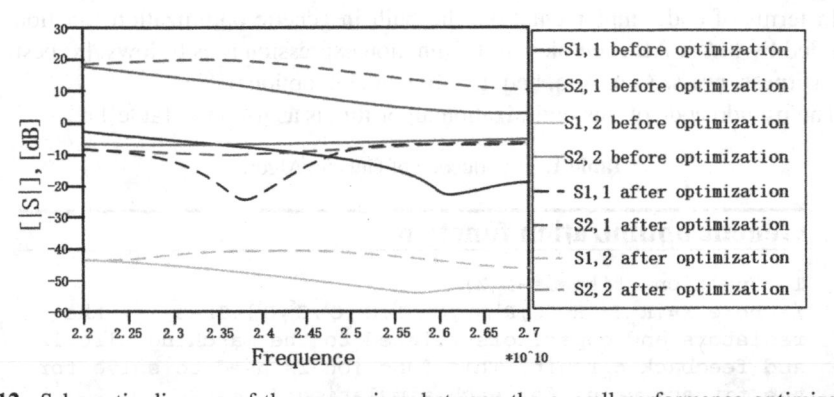

Fig. 12. Schematic diagram of the comparison between the overall performance optimization results of PA with optimized heat dissipation layout and before optimization

4 Summarize

In conclusion, the final reliability design recommendations for GaN millimeter-wave PA chip achieved through Circuit-Electromagnetic-Thermal-Force multi-physical field simulation optimization are as follows:

Firstly, due to the fact that the GaN millimeter wave PA chip currently uses a four finger integrated HEMT transistor with OMMIC process, thermodynamic simulations have found that its temperature is close to 200 °C, which has a significant impact on the reliability between the core and channels. Therefore, through verification, this article proposes an optimization method that changes the output stage transistor to two transistors with increased gate length, and the distance between each transistor can be adjusted to effectively reduce the impact of temperature.

Secondly, in this article a fully position parameterized VIA hole layout for GaN millimeter wave PA chips has been preliminarily designed. The secondary development of the chip can refer to the current heat source distribution and stress deformation, and adjust the size and layout of VIA.

Thirdly, this article is a multi-physical field simulation and reliability analysis based on the unpacked layout (bare sheet) of 1 W GaN PA. Based on the analysis results of the bare sheet, some reference can be provided for the packaging of 1 W GaN PA. For example, the substrate needs to choose materials with good heat dissipation (it is recommended to have a thermal conductivity greater than common FR4 or Rogers), and it is also recommended to consider whether to add heat dissipation fins or use wind tunnel effects to improve forced convection efficiency during packaging, Accelerate convection or conduction as much as possible, so that the maximum temperature of Die does not exceed 120 °C.

5 Conclusions

This article develops a Circuit Electrical Thermal Force multi-physical field joint simulation and optimization platform based on MATLAB, including programmatic modeling and full band circuit electromagnetic collaborative simulation, an optimal via-hole layout thermodynamic model, temperature and stress deformation simulation written in IronPython language, to guide the modification of circuit layout and achieve full process simulation analysis of circuit, electromagnetic, thermal, and stress. At the same time, in order to verify the reliability of the multi-physical field simulation proposed in this article, a GaN based power amplifier chip operating at 22 GHz–27 GHz was optimized through simulation. The simulation results showed that the layout size was increased by 2.36%, and the temperature decreased by 40%. Through optimization, the gain after modifying the layout was increased by 8.4 dB, proving the availability and reliability of the proposed multi-physical field platform. Modification suggestions were proposed for the later optimization of the chip, providing guidance for the design of third-generation compound semiconductor GaN power amplifier chips of the same type.

References

1. Liu, F., et al.: Innovative Sub-5- μ m microvias by picosecond UV laser for post-moore packaging interconnects. IEEE Trans. Components, Packaging Manuf. Technol. **9**(10), 2016–2023 (2019)
2. Meneghini, M., et al.: Power GaN HEMT degradation: from time-dependent breakdown to hot-electron effects. In: 2018 IEEE International Electron Devices Meeting (IEDM), pp. 30.5.1–30.5.4. IEEE, San Francisco (2018)
3. Mahrokh, M., Yu, H., Guo, Y.: Thermal modeling of GaN HEMT devices with diamond heat-spreader. IEEE J. Electron. Devices Soc. **8**, 986–991 (2020)
4. Duque, J.L., et al.: Multi-physics analysis of human exposition to electromagnetic fields by 5G systems. In: 2022 IEEE USNC-URSI Radio Science Meeting (Joint with AP-S Symposium), pp. 27–28. IEEE, Denver (2022)
5. Liu, X., Fan, M., Hu, Y., Li, H., Liu, F., Kang, J.: Simulation methods of multi-physics effects in nano-scale CMOS. In: 2022 International Electron Devices Meeting, pp. 15.4.1–15.4.4. IEEE, San Francisco (2022)
6. Na, W., Zhang, W., Yan, S., Feng, F., Zhang, W., Zhang, Y.: Automated neural network-based multiphysics parametric modeling of microwave components. IEEE Access **7**, 141153–141160 (2019)
7. Rivière, N., Stokmaier, M., Goss, J.: An innovative multi-objective optimization approach for the multiphysics design of electrical machines. In: 2020 IEEE Transportation Electrification Conference & Expo (ITEC), pp. 691–696. IEEE, Chicago, IL, USA (2020)
8. Blair, C., López Ruiz, S., Morales, M.: A MultiPhysics simulation vision from antenna element design to systems link analysis. In: 2019 International Conference on Electromagnetics in Advanced Applications, pp. 1420–1422. IEEE, Granada, Spain (2019)
9. Jia, Z.L., Xue, Z.S., Liu, X.Y., Zhang, H.H.: Electromagnetic-circuital-thermal multiphysics simulation of microwave amplifier. In: 2022 Photonics & Electromagnetics Research Symposium, pp. 1155–1158. IEEE, Hangzhou (2022)
10. Guerrieri, S.D., Ramella, C., Catoggio, E., Bonani, F.: Variability-aware MMIC design through multiphysics modelling. In: 2022 IEEE MTT-S International Conference on Numerical Electromagnetic and Multiphysics Modeling and Optimization, pp. 1–4. IEEE, Limoges, France (2022)

11. Bello, H., Oyeleke, O., Usman, A.D., Bello, T., Muhammad, I., Zakariyya. O.S.: Modelling and realization of a compact CPW transmission lines using 3D mmics technology in ADS momentum. In: 2019 15th International Conference on Electronics, Computer and Computation (ICECCO), pp. 1–7. IEEE, Abuja, Nigeria (2019)
12. Niu, Z., Zhang, B., Fan, Y.: A 220GHz miniaturized integrated front end based on solid-state circuits. In: 2021 International Conference on Microwave and Millimeter Wave Technology, pp. 1–3. IEEE, Nanjing, China (2021)
13. Zhongwei, L., Zhaiqi.: Thermal stress analysis and optimization design of high temperature and high pressure valve based on the workbench. In: 2015 8th International Conference on Intelligent Computation Technology and Automation, pp. 1055–1058. IEEE, Nanchang, China (2015)
14. Lin, P., Jianqiang, C., Zhihao, Z., Yang, H., Tong, W., Gary, Z.: Design of broadband high-gain GaN MMIC power amplifier based on reactive/resistive matching and feedback technique. IEICE Electron. Express **18**(19), 1–6 (2021)

Research and Design of Hydrological Data Visualization Based on Digital Twin

YuDan Zhao[1], Wu Zeng[2(✉)], Ying Ni[1], Peng Xia[1], and RuoChen Tan[3]

[1] School of Electrical and Electronic Engineering, Wuhan Polytechnic University,
Hubei 430023, China
[2] School of Mathematics and Computer Science, Wuhan Polytechnic University, Hubei 430023,
China
zengwu@whpu.edu.cn
[3] Computer Science and Engineering, University of California, San Diego, La Jolla , CA 92093,
USA

Abstract. In the context of digital transformation, cities and enterprises are striving to build a digital industrial chain, cultivate a digital ecosystem, and support high-quality economic development. Therefore, the use of visualization technology to assist decision-makers in rational planning has become a hot spot. Taking wuhan city as an example, combined with 3D modeling technology, it is aimed at smart cities and based on digital twins to create multiple scenarios for hydrological data application services and improve hydrological information services. First, we collected the data released by the china hydrology and water resources station; then, we visualized the hydrological data of the yangtze river Hankou station by using methods such as view juxtaposition and 3D interaction; after that, we constructed a 3D scene based on the real scene of the yangtze river Hankou basin, and used algorithms to the water body model is optimized; finally, the interaction between data and scenes is designed, various functions are realized by using high-level programming language design, and the water level changes in the flood season are simulated to help analyze and understand data more clearly, and assist decision makers in making decisions.

Keywords: Data Visualization · Hydrological Data · Digital Twin

1 Introduction

Under the tide of the global digital economy, carrying out digital transformation has become an inevitable choice to adapt to the digital economy and seek survival and development. Therefore, the application of data visualization technology has become the focus of researchers. In 2020, China experienced the worst flood situation since 1998 [1], and the Yangtze River basin flood occurred. The national hydrological department applied data visualization technology to closely monitor the water and rain situation, effectively achieving flood control, drought relief and disaster reduction.

Visualization [2] is seen as building a mental model or image of something in the human mind, drawing conclusions through information enlightenment, analyzing data

H. Jin et al. (Eds.): IAIC 2023, CCIS 2058, pp. 277–289, 2024.
https://doi.org/10.1007/978-981-97-1277-9_21

efficiently, and making decisions thousands of miles away. Peng [3] and others aimed at the low utilization rate of existing water resource management methods, based on big data, cloud computing and other technologies, designed and implemented a water resource management system, using various graphics such as histograms and line graphs from different dimensions, the visual display of water resources data realizes the refined management of water resources. The father of digital twins, Michael Grieves [4] first established the concept of digital twins in his speech on product life cycle and management in 2002. In 2019, Italian scholar Macchione [5] proposed the use of 3D visualization products to represent flood events in a 3D environment to balance risk communication and actual needs. Taking the Qiantang River Basin as the research object, Zhao [6] built a multi-scale spatial geographic information model of the watershed through key technologies such as air-ground integrated data acquisition, GIS and BIM 3D modeling, and multi-source heterogeneous model aggregation and integration. With the model as the carrier, the integration and sharing of watershed information resources is realized, and a new technical idea is provided for the intelligent management of watersheds.

It can be seen that the use of digital twin technology to study hydrological data visualization [7] is an emerging field, but its theoretical research has relatively mature methods. The traditional method uses a variety of 2D charts [8] with different functional charts to represent water resources data, but it has a strong sense of separation from reality; while the BIM model is highly sophisticated and professional, and is relatively friendly to professional researchers, but it is slightly redundant for non-professionals such as planning resource managers.

This study combines the ECharts [9] two-dimensional chart library to visualize the hydrological data of wuhan Hankou station, uses the Unity engine to construct a three-dimensional scene of the Hankou river basin in the yangtze river, optimizes the water body model, and reasonably designs the interaction between the data and the scene. Connect numbers to the real world, helping to dissect and understand data more clearly.

2   Data Collection and Processing

The data used in the research can be divided into three categories, namely water level data, weather data, and three-dimensional data, as shown in Table 1. Among them, the water level data comes from the open source information of the official homepage of yangtze river water conservancy [10]; the weather data comes from the historical data of weather homepage [11] and zhenqi homepage [12], in which the daily AQI data and PM2.5 concentration data are based on the environmental protection, the average result obtained from the hourly data calculation of the terminus has high data rigor; the city map data in the 3D data comes from openstreetmap [13], and the rest of the model resources are provided by the well-known open source 3D model homepage archive3D [14].

For water level data and weather data, we manually obtain the required data from the website and store it in excel format. First, we filter the data granularity and dimension, and then fill in the missing values with zeros. Combined with visual analysis requirements, the data is pre-processed. Processing, and finally convert the data into.*json* format for storage, which is convenient for inputting into the visualization system.

Table 1. Data collection information form.

Data category	Data content	Data description	Data Format
Water level data	Flow	The unit is m^3/s	.excel
	Water level	The unit is m (1985 National Elevation Datum)	
	Water temperature	Unit is °C	
Weather data	Maximum temperature	Unit is °C	.excel
	Lowest temperature	Unit is °C	
	Weather	Divided into sixteen types: sunny, cloudy, sunny to cloudy, fog, sleet, thunderstorm, light rain, heavy rain, moderate rain, heavy rain, heavy snow, light snow, moderate snow, rain, hail, cloudy	
	Wind direction	Including wind direction and grade	
	Air quality	Divided into the following eight categories: AQI, $PM_{2.5}$, PM_{10}, CO, SO_2, NO_2, O_3, quality level The unit of each index is $\mu g/m^3$ (CO is mg/m^3) The quality level is divided into five categories: lightly polluted, moderately polluted, heavily polluted, good, and excellent	
3D data	City map data	Points of interest, road network, buildings vector	.osm
	Model resource	Large buildings, skyboxes, water models	.obj

For the overall city data in 3D scenes, we manually obtain the required range of .*osm* format data from the open source map website, first use QGIS software to preprocess the data such as layer splitting, delete redundant roads and buildings, etc. The processed data files are exported to .*ship* format, and then Blender software is used to process the .*ship* file for extrusion and other operations. Finally, the data is converted to .*obj* format, which is imported into the Unity development tool to build the required 3D visual scene.

3 Visual Interaction of Hydrological Data

3.1 Time Series Dimension

In general, the standard display is used as a carrier for horizontal and vertical pixel arrangement, which is more suitable for the representation of 2D information. Research by Colin Ware [15] shows that most of the visual information comes from two-dimensional space.

In the 2D chart part of data representation, weather data is collected using an overall data visualization board, as shown in Fig. 1, using a variety of visual charts, each individual chart can be enlarged, link view. Animation interactions such as transitions, zooming, and filtering are highly personalized, allowing details to be presented on demand.

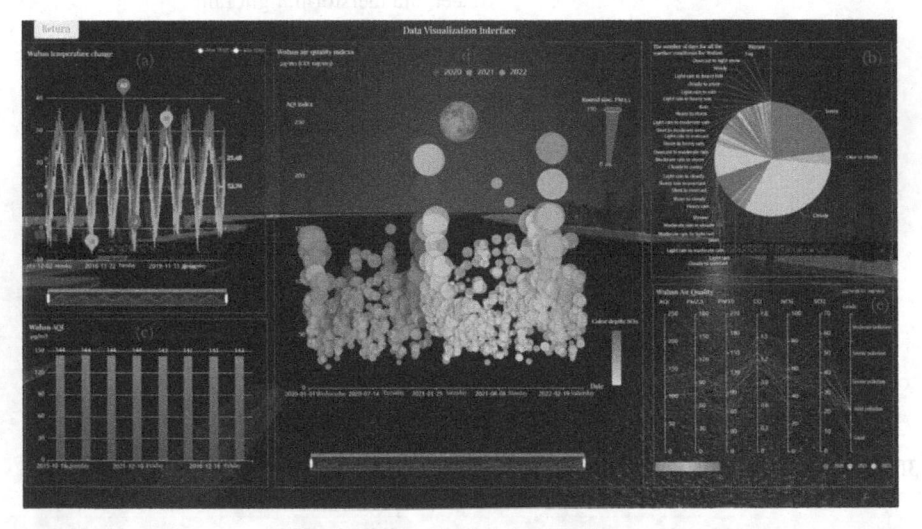

Fig. 1. Data visualization interface

We can analyze the meaning conveyed by the chart from multiple angles through different interactive methods.

For example, after zooming in on (d) in the data board, as shown in Fig. 2, use the scatter diagram to represent the various indexes of Wuhan's air quality. The three colors of red, yellow and green correspond to the three years of 2020, 2021 and 2022, and the size of the circle represents PM2.5, the degree of brightness indicates SO2, the vertical axis indicates AQI, and the horizontal axis indicates the date. You can use the zoom axis below to view any range. Click any dot to display the seven index parameter values and quality levels corresponding to the corresponding date, which is convenient for users to further analyze. In addition, we can see that in December and January every year, the various indexes increase significantly, and the air quality drops significantly.

As another example, after zooming in on (a) in the data dashboard, the line chart of historical temperature changes in Fig. 3 shows the daily temperature changes in Wuhan from 2013 to 2022. Blue represents the highest temperature and green represents the

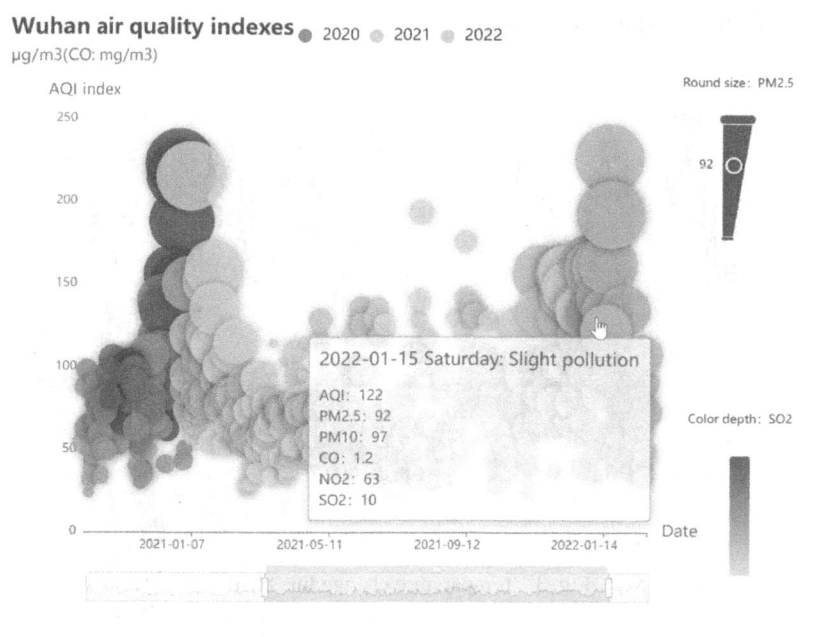

Fig. 2. Air quality indexes and quality grades

lowest temperature. From the figure, we can see obvious seasonal characteristics. In recent years, the highest temperature in Wuhan has reached 40 °C, the lowest temperature has been −2 °C, the highest temperature has averaged 21.46 °C, and the lowest temperature has averaged 12.74 °C. You can view the average of the highest and lowest temperatures in any range by manipulating the zoom bar below.

For another example, after zooming in on (b) in the data dashboard, as shown in the pie chart in Fig. 4(1), you can use the selection box on the upper left to select a different time period for the next step. The number of days for various weather conditions in Wuhan during this period It can be seen from Fig. 4(1) that in January-December 2020, the top three weather conditions with the largest proportion are cloudy, sunny and cloudy to sunny. You can also choose weather conditions or wind direction conditions for targeted analysis, as shown in Fig. 4(2), it can be seen that the number of days with northeast wind, north wind, and east wind during this period occupies the top three wind direction conditions.

The many 2D charts shown above can meet the basic needs of analysis, but if matched with corresponding auxiliary materials such as 3D scenes, sound effects, and videos, the data display can be more intuitive and easy to understand. For example, the simulated display of cloudy and sunny weather, the transition between morning and evening, and the change of winter snow and summer rain, supplemented by corresponding sound effects, is more friendly and immersive to ordinary people such as ordinary people who are concerned about the history and current situation of wuhan and the yangtze river basin.

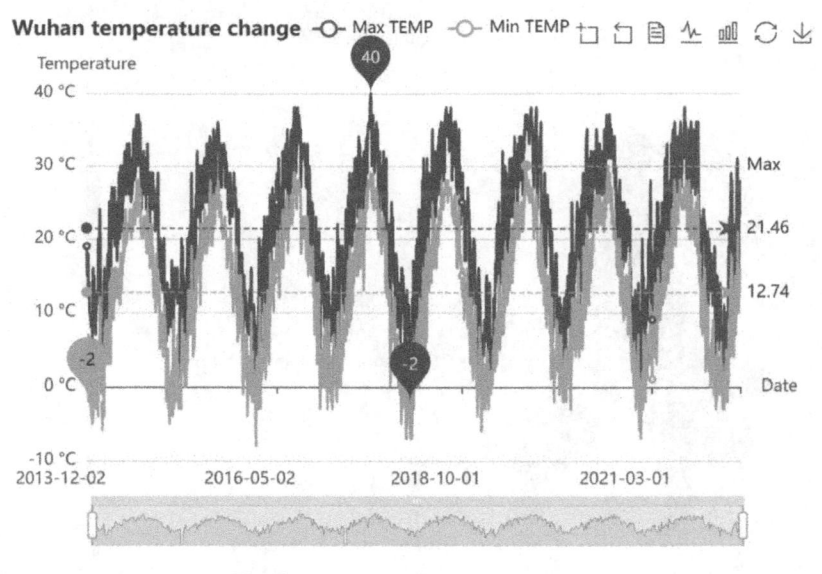

Fig. 3. Temperature change line chart

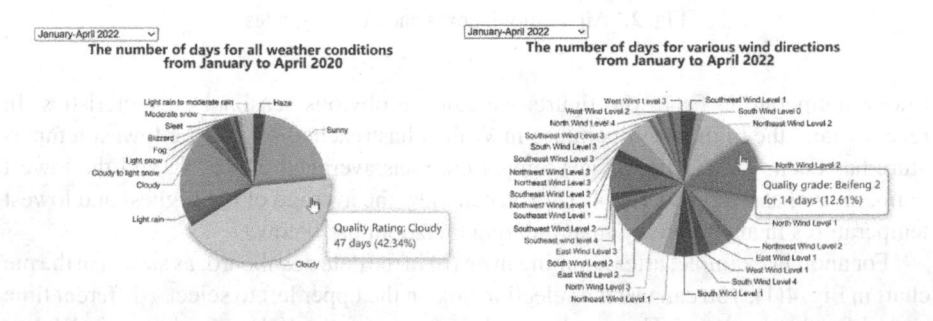

Fig. 4. (1) Days of various weather conditions Figure (2) Days of wind direction

3.2 Spatio-Temporal Visual Interaction

Build a 3D visualization scene based on the real scene, as shown in Fig. 5, which shows the scene of snowing in the daytime and raining in the daytime, snowing in the night and raining in the night, and the double sensory stimulation of vision and hearing achieves an immersive experience. On this basis, the numerical changes of water data are combined with the changes of water data in 3D scenes to break the barriers of dimensions and map numbers to reality.

Next, Fig. 6 is an example of the interaction between a 3D scene and a 2D chart. It can display the water level trend graph of Hankou station in 1998 and 2020. The upper left corner area is the query box. Here, you can choose to query the water level information of Hankou station on a certain day and time. After clicking on the water level, the water body model in the 3D scene will also be presented corresponding height changes.

Fig. 5. Different morning and evening rain and snow scenes

Fig. 6. Example of interaction between 3D scene and 2D chart

Figure 7 is the real scene of the Yangtze River Basin, and Fig. 8 is the simulated scene. It can be seen that the simulated scene has a high degree of reality, but there is still room for improvement. The water rendering effect occupies the largest field of view in the scene. At the same time, the water model also participates in the simulation of the water level and the change simulation of the weather and precipitation. In the future, it can participate in more interactive designs, such as the sediment content in the precipitation. How many and so on. For better results, we have applied the real-time water body algorithm to improve the water body model.

Fig. 7. The real scene in the Yangtze River Basin

Fig. 8. The simulated scene

4   Real-Time Water Body Simulation Algorithm

The effect of water body simulation [16] not only refers to the similarity in shape, but also has a corresponding sense of reality in terms of physical characteristics. According to different standards, it can be divided into multiple parts. Among them, the effect of water surface fluctuation and water body color effect attract the most attention. Empirical simulation [17] and other methods are used to express the wave form of water bodies, which can achieve very realistic effects, and rendering techniques such as environment mapping and ray tracing [18] are used to realize the effects of refraction, reflection and caustics of water bodies on light.

4.1   Simulation of Water Surface Fluctuation Based on Empiricism

Based on empiricism, the wave effect of the water surface can be realized by superposition of sinusoidal ripples. The method of using the classic FFT algorithm [19] to generate water surface fluctuations can achieve very satisfactory realistic effects. For example, the offline FFT baking water surface technology used in Assassin's Creed 3 [20], the ripple effect of the water surface is very real, so it has always been more popular.[21] was the first to use the FFT algorithm to generate the water surface, but Tessendorf [22] published a paper on Siggraph on the simulation of the ocean water surface that is more far-reaching. This method changes and calculates each frame in the frequency domain, it is then expressed in the space domain by fast Fourier transform.

In the one-dimensional case, n sine waves are superimposed to obtain a composite wave effect, such as formula (1):

$$h = \sum_{i=1}^{n} h_{0i}\cos(\omega_i \cdot x - \omega_{ti} + \varphi_i) \tag{1}$$

Through Euler's formula, rewrite the above formula into Fourier transform form, such as formula (2):

$$h = re\left(\sum_{i=1}^{n} h_{0i}\cos(\omega_i \cdot x - \omega_{ti} + \varphi_i) + i * \sin((\omega_i \cdot x - \omega_{ti} + \varphi_i))\right) \tag{2}$$

That is to get the formula (3):

$$h = re \sum_{i=1}^{n} h_{0i}e^{(-i\omega_{ti}+i\varphi_i)}e^{i\omega_i x} \tag{3}$$

Based on the same derivation, we can also easily obtain the Fourier transformation form in the two- dimensional case:

$$h(x, t) = \sum_{k} \tilde{h}(k, t)e^{ikx} \tag{4}$$

$$\tilde{h}(k, t) = \tilde{h}_0(k)e^{i\omega(k)t} + \tilde{h}_0^*(-k)e^{-i\omega(k)t} \tag{5}$$

Among them, x is the coordinate vector (x, y) on the two-dimensional plane, and k it is the frequency domain vector (kx, ky). In order to construct the water surface, we only need to obtain t the height value of each (x, x) coordinate point at any time. We can first obtain the height value in the frequency domain, and then obtain the corresponding height value in the space domain through the inverse Fourier transformation according to (4) and (5).

The equal ω constant comes from the ocean wave spectrum and statistical laws. Tessendorf gives the calculation methods of these constants in his paper, such as formula (6) and formula (7).

$$\tilde{h}_0(k) = \frac{1}{\sqrt{2}}(\xi_r + i\xi_i)\sqrt{P_h(k)} \tag{6}$$

$$P_h(k) = A\frac{\exp(-1/(kL)^2)}{k^4}\left|\hat{k} \cdot \omega\right|^2 \tag{7}$$

Among them $L = V^2/g$, represents V the maximum wavelength under the influence of wind speed; A is a constant, reflecting the influence of global factors on the waveform; k is the wave number, $k = 2\pi/\lambda$; ξ_r and ξ_i is a random number generated by a Gaussian function.

4.2 Real-Time Water Algorithm Water Color

The water color we observe is the effect of sky light and ambient light after reflection, refraction, and scattering reactions [23]. When the light finally reaches the human eye, it will be affected by the water surface and environmental conditions.

We know that sunlight and sky light are mixed in the incident light, and the sky light itself is caused by the atmospheric scattering effect of sunlight [24], and its intensity value is complicated to calculate based on the physical light path. Among them is the texture sampling [25] function, which is used to complete two-dimensional texture sampling; it is the color gradient texture of the incident light on the water body. According to the actual scene, the pure white light (R = 255, G = 255, B = 255) is the starting point of the texture. Refer to the gradient texture generated by the actual environment as shown in Fig. 9; it is the mixing coefficient, ranging from [0,1]. When the coefficient changes from small to large, the color value of the incident light on the water body gradually changes from gradient The leftmost end of the texture is offset to the rightmost end, that is, the transition from the color of the original sun light to the color of the sky light.

Fig. 9. Gradient texture

Figure 10(1) shows the effect when the FFT algorithm and color texture sampling are not applied. The final simulation effect is shown in Fig. 10(2), although there is still a certain gap with the real river water effect shown in Fig. 10(3). But compared with Fig. 10(1), there has been some progress.

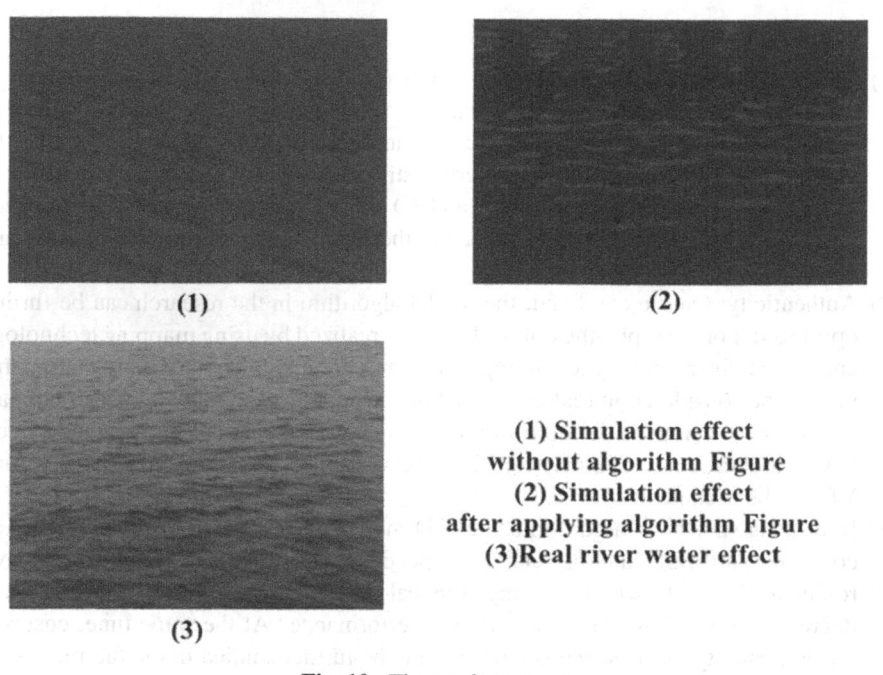

(1) Simulation effect
without algorithm Figure
(2) Simulation effect
after applying algorithm Figure
(3)Real river water effect

Fig. 10. The result contrast

5 Conclusion

Under the background of the construction of smart cities [26], we have developed a multi-bit exploratory visual interactive system by organically combining graphics rendering technology and big data visual analysis technology, which provides a new way to assist decision-making, interactive analysis The correlation between the weather conditions in Wuhan and the water level and flow of Hankou station not only enables people to have a deeper understanding of the tide itself, but also provides a basis for government departments to formulate policies for preventing and controlling floods at Hankou station.

The difficulties and innovations of this study are as follows:

(1) Applying digital twin and data visual analysis technology to construct high-precision digital models, build a water resource utilization planning and management platform, and design and achieve better interaction between data and models in terms of technology and interaction effects are extremely challenging tasks.
(2) The 3D water body model has been improved to improve the accuracy of the 3D scene. Based on empiricism, the FFT algorithm is used to improve the accuracy of 3D water body model presentation and connect the data with the real world.
(3) Apply digital twin technology to hydrological data management, design a more friendly interactive graphical interface, improve the intuitiveness of data, study and analyze the data of water resources utilization change at Hankou station in wuhan city, summarize the enlightenment brought by the information behind the historical data, and assist decision-making.

Future job outlook:

(1) Data analysis and UI design: In the research, the use of mathematical ideas to analyze the data is less, and the real-time data has a large delay. Next, the API interface is used to access real-time data and add auxiliary analysis, which will make the system more perfect. In addition, there is still a lot of room for improvement in the association design between numbers and 3D scenes, such as the association between wind direction and water in the data, and the interaction between precipitation and water.

(2) Authenticity: On the one hand, the model algorithm in the research can be further optimized. For example, the color of water is realized by using mapping technology and optical fiber tracking technology in the research, but there are more factors that need to be considered in reality; On the one hand, although multi-dimensional data visualization can provide more intuitive visual effects and auxiliary sound effects have a higher sense of immersion, 2D screen displays are still used to carry it, and VR can be tried next.

(3) Rendering speed: A large number of 3D models have very high requirements on computer hardware, and the rendering speed is greatly affected by the device environment. Generally, when pursuing high real-time effects, it often sacrifices a certain degree of authenticity. How to improve performance? At the same time, ensuring the authenticity of the scene is a problem to be further studied in the future.

References

1. Feng, T.C., Wu, Q.Y.: Flood disasters in the Yangtze River Basin in 1998. People's Yangtze River **02**, 28–29 (1999)
2. Xia, J.Z., Li, J., Chen, S.M., et al.: A review of cross-research on visualization and artificial intelligence. Chinese Sci. Inf. Sci. **51**(11), 1777–1801 (2021)
3. Peng, H.Y., Zhou, J.H., Zhang, J.Y., et al.: Design and research of water resources management system based on big data. Softw. Eng.. Eng. **25**(03), 59–62 (2022)
4. Grieves, M.: Intelligent digital twins and the development and management of complex systems. Digital Twin 2 (2022)
5. Francesco, M., Pierfranco, C., Carmelina, C., et al.: Moving to 3-D flood hazard maps for enhancing risk communication. Environ. Modeling Softw. **111**, 510–522 (2019)
6. Zhao, X.Y., Mao, X.Y., Xu, H.Q., et al.: Construction and application of multi-scale spatial geographical information model of digital watershed——taking the qiantang river basin as an example. People's Yangtze River **52**(S2), 293–297 (2021)
7. Li, W., Wu, B.: Research on core data visualization in digital Twin City construction. Modern Inf. Technol. **6**(17), 128–131 (2022)
8. Zeng, W., Zhang, Y., Hu, K., et al.: Multi-dimensional visualization and simulation analysis of COVID-19 outbreak. Commun. Comput. Inf. Sci. **1423**, 541–543 (2021)
9. EChacrts Homepage. https://echarts.apache.org/zh/index.html. Accessed 17 May 2023
10. Yangtze River Water Conservancy Homepage. http://www.cjw.gov.cn/. Accessed 01 Feb 2023
11. Weather Homepage. https://lishi.tianqi.com/wuhan. Accessed 23 Apr 2022
12. Zhenqi Homepage. https://www.aqistudy.cn/historydata/. Accessed 23 Apr 2022
13. Openstreetmap Homepage. https://www.openstreetmap.org/. Accessed 17 May 2023
14. Archive3D Homepage. https://archive3d.net/. Accessed 17 May 2023

15. Colin Ware.: Information Visualization, 4th edn. (2019)
16. Wang, Y.: Research on realistic water surface generation and animation based on real data. Tianjin University of Technology (2021)
17. Guo, F.L., Chen, X.L., Liang, Y.J.: A review of water body simulation and rendering methods. Comput. Appl. **33**(S2), 224–228 (2013)
18. Zhu, G.X.: Research on the visual simulation method of nearshore waters based on the optical characteristics of water bodies. University of Electronic Science and Technology (2022)
19. Ken, P., David, F.: Pad: an alternative approach to the computer interface. Siggraph, 57–64 (1993)
20. Heer, J., Robertson G.: Animated transitions in statistical data graphics. IEEE Trans. Visualization Comput. Graph. **13**(6), 1240–1247 (2007)
21. Gary, A.M., Peter, A.W., John, F.M.: Fourier synthesis of ocean scenes. IEEE Comput. Graph. Appl., 16–23 (1987)
22. Jerry, T.: Simulating ocean water. Simulating Nature: Realistic and Interactive Techniques, Course Notes, Siggraph (1999)
23. Zhang, L.H.: Research on real-time simulation method of multi-scale water body. Yanshan University (2018)
24. Liu, W.M., Li, J.S., Lu, Z.W.: Research on real-time rendering technology of planetary atmospheric scattering effect based on GPU. J. Syst. Simul. **21**(S1), 103–105+109 (2009)
25. Wu, L., Chen, L.T., He, M.Y.: Interactive real-time water surface rendering. Comput. Appl. Res. (08), 2387–2389+2440 (2008)
26. Wang, H., Chen, X. W., Jia, F., et al.: Digital twin-supported smart city: Status, challenges and future research directions. Expert Syst. Appl., 217 (2023)

Design of Constraint FH-LFM DFRC Waveform

Rui Xu, Ruiming Wen, Gang Li, and Guangjun Wen[✉]

University of Electronic Science and Technology of China, Chengdu 611731, China
wgj@uestc.edu.cn

Abstract. An efficient spectrum sharing method is urgent for the increasing demand for frequency resources. Dual-functional radar-communication (DFRC) is expected to efficiently share the spectrum by minimizing performance deterioration. Existing DFRC waveforms based on radar waveform designs result in out-of-band leakage, high receiver complexity and high symbol error rate (SER). In this paper, we propose a DFRC waveform based on constrained frequency hopping (FH) and linear frequency modulation (C-FH-LFM). The FH sequence for the chips and positive/negative frequency modulation slope are adopted to embed the communication information, achieving a single frequency traversal within a single pulse. We give explicit rules of FH sequence to reduce the storage and processing of the matching signal and reduce the SER at receiver. Then, we derived the range of the time-bandwidth product and the upper limit of the distributable chip number, which is universal in modulation systems by positive/negative LFM slope. The simulation results show that, the proposed C-FH-LFM waveform reduces out-of-band power leakage of 12 dB compared with phase shift keying scheme, and provides 6 dB improvement of SNR under the same SER compared with unconstrained FH-LFM scheme.

Keywords: linear frequency modulation · dual-functional
radar-communication · frequency hopping · pulse compression

1 Introduction

The number of wireless communication equipment has shown an explosive growth trend with the 5G. The global communication industry has an increasingly urgent demand for wireless spectrum. The price of the existing authorized frequency band increases substantially, which requires the rapid development of additional spectrum resources [1]. To alleviate this contradiction, future communication systems will explore the feasibility of co-existing with other electronic devices in the same frequency band. The radar frequency band is regarded as the best candidate for this purpose [2]. For a long time, the development of radio frequency systems such as radar, communication and electronic warfare have been fragmented and independent, consuming tremendous spectrum and hardware resources and limiting the effectiveness of the combat platform. To serve a variety of civil or military emerging application scenarios and efficiently utilize spectrum resources, dual-functional radar-communication (DFRC) has attracted great attention

© The Author(s), under exclusive license to Springer Nature Singapore Pte Ltd. 2024
H. Jin et al. (Eds.): IAIC 2023, CCIS 2058, pp. 290–303, 2024.
https://doi.org/10.1007/978-981-97-1277-9_22

from academia and industry. According to the waveform type, DFRC can be mainly divided into two categories. The first category refers to the use of communication signal waveform to directly accomplish radar functions (detection or imaging) [3]. Orthogonal frequency division multiplexing (OFDM) is the representative technology. The second category is to embed communication information in the existing radar waveform according to some specific modulation modes to achieve shared communication [4]. Frequency hopping (FH) [5, 6] and Linear frequency modulation (LFM) are commonly used in this case, and the combination of the FH and LFM has become the forefront research direction.

Different from other existing frequency modulation schemes based on LFM [7, 8], the FH scheme within pulse utilizes the mapping of FH sequences to realize communication, and other modulation modes are adopted within each subpulse to increase the data rate [9-12]. The authors in [10] combine FH-LFM with phase shift keying (PSK), and adopt optimization algorithm to select waveforms with the optimal radar and communication performance. In [11, 12], FH is combined with PSK and multiple input multiple output (MIMO), and the optimization algorithm is also used for constraints. However, DFRC based on FH-LFM has some issues that need to be addressed to become more effective. Firstly, although PSK modulation has strong scalability, it has the inherent problem of out-of-band leakage, which is very dangerous for radar signals with high power. Secondly, same as other frequency modulation schemes, many existing works based on FH fail to achieve a single frequency traversal within a single pulse, resulting in spectrum reuses or voids. Because the actual working bandwidth affects the range resolution of radar, the reduction of radar spectrum utilization directly weakens the radar performance [1]. At the same time, although the waveform constraints obtained by the optimization algorithm achieve a certain degree of radar and communication performance enhancement, it sacrifices the waveform set that can be used to modulate data and hence reduces the data rate. Thirdly, the FH detection of multi-frequency-point greatly increases the communication detection error and the complexity of the matching processing at the communication receiver. Due to the randomness of the waveform by optimization algorithm, if each pulse transmits b bits, the communication receiver needs to restore 2^b kinds of corresponding matching waveforms for demodulation, which greatly increases the realization difficulty of the hardware at the receiver.

This paper focuses on the application of DFRC waveform design based on FH-LFM in the pulse radar system. Specifically, we propose a DFRC scheme based on constrained FH-LFM (C-FH-LFM). The pulse is divided into multiple chips, achieving a single traversal of available frequency points. The inter-chip FH sequence and the positive/negative frequency modulation slope in chip are used to embed the communication data. Then, we give explicit rules of FH rather than optimization algorithm. This allows the communication receiver to reduce the storage and processing of the matching signal and reduce the error detection. Meanwhile, the range of the time-bandwidth product and the upper limit of the distributable chip number are derived for C-FH-LFM, which is universal in positive/negative LFM modulation systems. Simulation results show that the proposed C-FH-LFM wave reduces out-of-band power leakage of 12 dB compared with PSK scheme, and reduces the communication symbol error rate (SER) compared with unconstrained FH-LFM scheme.

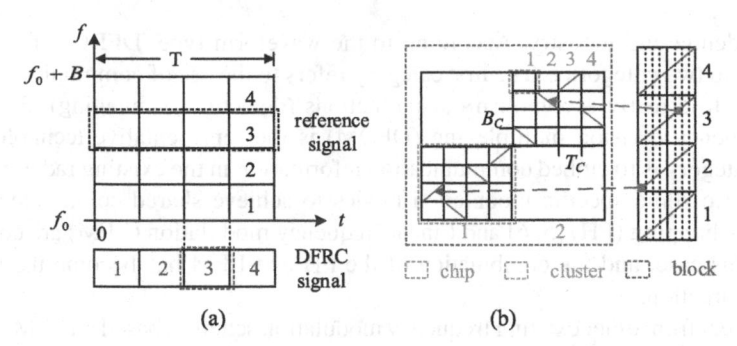

Fig. 1. Time and frequency (a) partition scheme; (b) modulation scheme.

The remainder of the paper is organized as follows. The DFRC waveform design is introduced in Sect. 2. The communication reception processing link design is provided in Sect. 3. Numerical simulation results are presented in Sect. 4. Conclusions and future research directions are finally provided in Sect. 5.

Notation: The bold capital letters represent vectors. $\lfloor . \rfloor$ represents the rounded down operation. $(\cdot)!$ stands for factorial operation. $(\cdot)^H$ represents the matrix transpose conjugate operation. % denotes remainder taking operations.

2 C-FH-LFM Waveform Design

The frequency of LFM within the pulse changes linearly with time. The waveform with increasing or decreasing frequency modulation slope is called positive/negative LFM. This section introduces the design principle of C-FH-LFM DFRC waveform and analyzes the radar and communication performance of waveform. With the time and frequency division and distribution of each radar pulse, the DFRC waveform realizes the single traversal of in-band frequency points. Communication data is embedded in FH sequences and LFM slopes.

The time-frequency framework of the DFRC waveform is shown as Fig. 1. We set T and B as the duration and bandwidth of the pulse, respectively. Each pulse is divided into N groups of reference signal in the frequency domain. The signal is also divided into N blocks in the time domain. We further subdivide each block into M clusters and each cluster into N chips. Thus, each pulse contains $Q = N^2M$ chips. Based on this model, the absolute position of the timing sequence of the k-th chip is expressed as

$$k = n_b[k]MN + m_c[k]N + n_c[k], k = 1, 2, ..., Q, \tag{1}$$

where n_b, m_c, n_c are the block, cluster and chip count of the chip, respectively. Any k is mapped to a corresponding combination of $[n_b, m_c, n_c]$. Therefore, $[k]$ is equivalent with $[n_b, m_c, n_c]$ in the following context. Within a pulse, we divide the modulated data into three parallel types. Firstly, we set $l_{PN}[k] = \pm 1$ as the chip symbol. The k-th chip selects positive/negative slope modulation according to $l_{PN}[k]$. Secondly, we set $l_{MS}[k] = 1, 2, \cdots, M$ as the cluster sequence. Every M clusters within a block form a cluster sequence and there are N cluster sequences. The $l_{MS}[k]$ value of N chips in the same cluster is the same. Thirdly, we set $l_{NS}[k] = 1, 2, \cdots, N$ as the chip sequence. Every N chips within a cluster form a chip sequence and there are MN chip sequences. $l_{MS}[k]$ and $l_{NS}[k]$ satisfy the following constraint.

$$l_{MS}[n_b, m_c, n_c] = l_{MS}[n_b, m_c, n_c'], \tag{2}$$

$$l_{MS}[n_b, m_c, n_c] \neq l_{MS}[n_b, m_c', n_c], \tag{3}$$

$$l_{NS}[n_b, m_c, n_c] \neq l_{NS}[n_b, m_c, n_c'], \tag{4}$$

$$\begin{aligned} l_{NS}[n_b, m_c, n_c] &\neq l_{NS}[n_b', m_c', n_c], \\ \text{if } l_{MS}[n_b, m_c, n_c] &= l_{MS}[n_b', m_c', n_c], \end{aligned} \tag{5}$$

where n_b', m_c', n_c' indicate different values from n_b, m_c, n_c respectively. (3) and (4) illustrate the non-repetition characteristics of single sequence for $l_{MS}[k]$ and $l_{NS}[k]$ respectively. (5) explains that if two clusters at different locations have the same cluster sequence $l_{MS}[k]$, the chip sequences $l_{NS}[k]$ within them are completely different. The corresponding center frequency of chip is defined by $l_{MS}[k]$ and $l_{NS}[k]$ as

$$\hat{f}_C[k] = f_0 + B_C[(l_{NS}[k] - 1)MN + (l_{MS}[k] - 1)N + n_c] - B_C/2, \tag{6}$$

where f_0 is the lowest working frequency of DFRC signal; $B_C = B/Q$ is the bandwidth of a single chip, and the tail term $B_C/2$ is the amendment of the chip center frequency. We set $l[k] = l_{PN}[k]((l_{NS}[k] - 1)M + l_{MS}[k])$ as the value of each chip and $l[k] = \pm 1, \pm 2, \cdots, \pm MN$. As shown in Fig. 1(a) and Fig. 1(b), we set parameters as $N = 4$, $M = 2$. The chip symbol $l_{PN}[k]$ of the 2 clusters in the 3nd block is $[1 -1 -1 1]$, $[-1 - 1 1 -1]$. The cluster sequence $l_{MS}[k]$ of the chips in the 3nd block is $[1 1 1 1]$, $[2 2 2 2]$. The chip sequence $l_{NS}[k]$ of the chips in the 2 clusters and the 3nd block is $[2 1 3 4]$, $[1 3 4 2]$. For the specific chip value, $l[3, 1, 1] = ((2 - 1)M + 1) = 3$ (red dotted arrow) and $l[3, 2, 2] = -((3 - 1)M + 2) = -6$ $[3, 2, 2] = -6$ (blue dotted arrow). Since chips, clusters and blocks are the time domain divisions of pulse, $l[k]$ can be represented by the cluster vector, block vector and pulse vector as

$$\begin{cases} M_c[n_b, n_c] = [l[n_b, m_c, 1], l[n_b, m_c, 2], \cdots, l[n_b, m_c, N]] \\ N_b[n_b] = [M_c[n_b, 1], M_c[n_b, 2], \cdots, M_c[n_b, M]] \\ D = [N_b[1], N_b[2], \cdots, N_b[N]] \end{cases}. \tag{7}$$

The C-FH-LFM pulse with Q chips and D is shown as

$$s(t, D) = \sum_{k=1}^{Q} u(t/T_C - k + 1)\hat{s}(k, t - (k - 1)T_C, D), \tag{8}$$

where $u(x) = 1$, $0 \leq x \leq 1$ is the rectangular window function; $T_C = T/Q$ is the period of each chip. $\hat{s}(k, t, D)$ is waveform of the k-th chip, which is shown as

$$\hat{s}(k, t, D) = e^{j\pi(2\hat{f}_C[k]t + l_{PN}[k]\mu(t - (k - 1/2)T_C)^2) + j\theta[k]}. \tag{9}$$

In (9), μ is the frequency modulation slope. The continuous phase frequency shift keying (CPFSK) technology is utilized to eliminate the phase jump between chips to reduce energy leakage out of band. $\theta[k]$ is the phase compensation corresponding to the k-th chip to maintain the phase continuity between the chips, whose value is the difference between the end phase of the previous chip and the reference starting phase of the current chip.

$$R_{DATA} = PRF \cdot \left(Q + \left\lfloor \log_2 \left((M!)^N \prod_1^N (N!)^M \right) \right\rfloor \right). \tag{10}$$

Specially, if $N = 1$, $M = Q$, C-FH-LFM turns into unconstrained sorted FH-LFM. In that case, the data rates becomes $PRF \cdot (Q + \lfloor \log_2(Q!) \rfloor)$. If $M = 1$ and $N = 1$, the waveform turns into unmodulated LFM. In Table 1, we compare the designed waveforms with the communication data rates of the other 4 existing schemes. The data rates of all schemes are affected by PRF and the number of subpulses Q. In the multi-antenna system, the data rate is also affected by the number of antennas A_t and optional atoms F. On the other hand, J means the number of communication demodulation judgment for frequency points. The higher the value J, The higher the demodulation link complexity and the SER [4] (referring to FSK). It can be seen that, J of the proposed scheme is $1/MN$ of the other schemes. The principle analyze is given in Sect. 3.

Table 2 shows the function comparison of multiple variables in each scheme. The symbol D_{MO}, O_C and O_R respectively represent the corresponding parameters used for data modulation, communication performance optimization and radar performance optimization. It should note that if the variable is implementing multiple functions, the corresponding multiple functions will be weakened. For example, if the phase parameter of the chips is involved in radar performance optimization such as out-of-band leakage suppression or sidelobe suppression, the available amount of modulation data by PSK will be reduced, i.e. the communication rate will be much lower than the corresponding value in Table 1. The last line of Table 2 indicates whether the scheme provides a constraint on the traversal of the available frequency points, which is related to the spectrum utilization of the waveform and further the radar range resolution. Meanwhile, compared with non-traversal sequences, the calculation complexity of frequency judgment for traversal FH sequences in communication receiver can be reduced from the power (Q^Q) to the factorial ($Q!$).

Based on Table 1 and Table 2, the characteristics of proposed C-FH-LFM are as follows. Firstly, The C-FH-LFM scheme reduces the judgment number, and further.

the communication receiver link complexity and SER. Then, the phase parameter is fully available for out-of-band leakage suppression. Moreover, The C-FH-LFM scheme realizes the full utilization of in-band resources by the traversal of available frequency points.

Table 1. Performance comparison of multiple waveforms

Scheme	Data Rate/bps	J
FH PSK [13]	$PRF \cdot M_t Q N_{psk}$	Q
FH [14]	$PRF \cdot Q \left\lfloor \log_2 \binom{F}{M_t} \right\rfloor$	Q
C-HIM [11]	$PRF \cdot Q \left(M_t + \left\lfloor \log_2 \binom{F}{M_t} M_t! \right\rfloor \right)$	Q
FH-FMCW [10]	$PRF \cdot \left(N_{psk} Q + \lfloor \log_2 Q! \rfloor \right)$	Q
Proposed	R_{DATA}	$\frac{Q}{MN}$

Table 2. Function comparison of multiple variables

	FH-FMCW [10]	C-HIM [11]	FH PSK [13]	FH selection [14]	Proposed
FH sequence index	D_{MO}, O_C, O_R	D_{MO}	D_{MO}	D_{MO}, O_R	D_{MO}
Phase	D_{MO}, O_C	D_{MO}, O_C	D_{MO}	O_R	O_R
MIMO index	D_{MO}, O_R	D_{MO}	D_{MO}, O_R	D_{MO}	No
Traversal of B	No	Yes	No	No	Yes

3 Communication Reception Processing Link Design for C-FH-LFM Waveform

3.1 Communication Reception Processing Link

We consider single-user communication transmissions in this section and assume that the transmitted signals transmit through line-of-sight (LoS) channels. In addition, we assume that the communication receiver adopts a uniform linear array (ULA) with A_r antennas. Thus, the communication signal received is expressed as:

$$y(k, t, \mathbf{D}) = \mathbf{w}^H \mathbf{a}(\theta) G_{PL} \sum_{k=1}^{Q} u(t/T_C - k + 1) h[k] \hat{s}(k, t - (k-1)T_C, \mathbf{D}) + z(k, t),$$

(11)

where $h[k]$ is the Rician fading coefficient of the k-th chip, G_{PL} is the path loss attenuation coefficient, and $z(k, t)$ additive Gaussian noise; θ is the arrival angle of the received signal, and $\mathbf{a}(\theta)$ is the corresponding steering vector; $\mathbf{w}$ is the antenna combination weight vector. We assume that the receiver has acquired the channel coefficient, and set the reference down-conversion signal of the k-th chip as:

$$s_{RE}(k, t) = u(t/T_c)e^{-j\pi(2t(f_0+B_C(k\%MN))+\mu(t-T_C/2)^2)}. \tag{12}$$

$s_{RE}(k, t)$ can be regarded as a cyclic signal with period of $T/MN = NT_C$, i.e., a cluster period. The effective band of $s_{RE}(k, t)$ is $[f_0, f_0 + NB_C]$. By mixing the reference signal with the received signal, and sampling with the sampling period T_S, the discrete output signal of a pulse is obtained as

$$\tilde{y}[k, i, \mathbf{D}] = \tilde{z}[k, i] + \mathbf{w}^H \mathbf{a}(\theta) \sum_{k=1}^{Q} u(i/I_S - k + 1)h[k]G_{PL}$$
$$\cdot e^{j\pi(2B_C iT_S N(l_{MS}[k]-1+l_{NS}[k]M)+(l_{PN}[k]-1)\mu(iT_S-T_C/2)^2)}, \tag{13}$$

where I_S is the total number of sampling in a chip and $i = 0, 1, \cdots, I_S - 1$ is the sampling count. $\tilde{z}[k, i]$ is the noise after down-conversion and sampling.

The waveform parameters and FH sequences in Fig. 2(a) are consistent with those in Fig. 1, where $N = 4$ and $M = 2$. Figure 2(b) shows the spectrum of $\tilde{y}[k, i, \mathbf{D}]$ with the period as cluster (NT_C). Since the $l_{MS}[k]$ of the 1st cluster is 1, its frequency offset in (b) is concentrated on $Ml_{NS}[k]NB_C$. The $l_{MS}[k]$ of the 2nd cluster is 2, and its frequency offset is concentrated on $(1 + Ml_{NS}[k])NB_C$. In both cases, the frequency offsets within the cluster are separated by MNB_C. It allows the estimating of $l_{NS}[k]$ to detect only N frequency points. We set $l'_{PN}[k], l'_{MS}[k], l'_{NS}[k]$ and $\mathbf{D}'$ as the estimates of $l_{PN}[k], l_{MS}[k], l_{NS}[k]$ and $\mathbf{D}$ respectively. The matching signal with estimation value $l'[k] = l'_{PN}[k]((l'_{NS}[k] - 1)M + l'_{MS}[k])$ for the k-th chip is shown as

$$x[k, i, l'] = u(i/I_S) e^{j\pi(2iT_S f'_{EO}[k]+(l'_{PN}[k]-1)\mu(iT_S-T_C/2)^2)}, \tag{14}$$

where $f'_{EO}[k]$ is the frequency offset estimation after the mixing of the received signal and the reference signal as

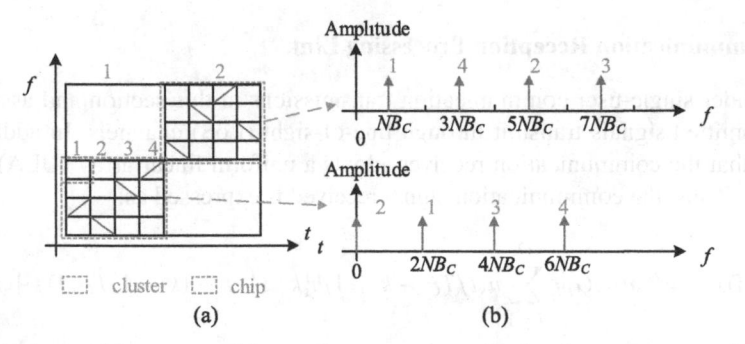

Fig. 2. (a) Time and frequency with waveform sequences; (b) the cluster spectrum of $\tilde{y}[k, i, \mathbf{D}]$ after down-conversion with reference signal.

$$f'_{EO}[k] = B_C[(l'_{NS}[k] - 1)MN + (l'_{MS}[k] - 1)N]. \tag{15}$$

We match $x[k, i, l']$ of different estimation values with $\tilde{y}[k, i, \mathbf{D}]$ in turn. Since $l'[k]$ is bounded by an interval and (2)-(5) provides the constraints, the detection of pulse vector $\mathbf{D}$ within a single pulse is achieved through the maximum likelihood (ML) detection as

$$\mathbf{D}' = \arg\min_{\mathbf{D}'} |\tilde{y}[k, i, \mathbf{D}] - \mathbf{w}^H \mathbf{a}(\theta)$$

$$\cdot \sum_{k=1}^{Q} u(t/T_C - k + 1)h[k]G_{PL}\hat{s}(k, t - (k-1)T_C, \mathbf{D}')|^2. \tag{16}$$

3.2 Demodulation Performance for FH

FH signals adopt orthogonal waveforms of different frequencies to reduce error in detection [10]. Its constraint is shown as

$$B_C T_C = p_{TB} = \eta, \ \eta \in \mathbb{N}_+, \tag{17}$$

where $\mathbb{N}_+$ represents the set of positive integers; p_{TB} is time-bandwidth product of the single chip. However, in application, the noise, interference, channel coefficient estimation error and other issues will make the output of unmatched signals not 0 after matched filter. These residual values deteriorates SER. Firstly, we discuss the unconstrained FH waveform scheme, and set the number of available frequency points for modulation as Q, which is consistent with the number of chips. $o(w_1, w_2)$ is the output of the matching filter at t = 0 for the received signal with frequency count w_1 and the matched signal with frequency count w_2, where $w_1, w_2 = 1, 2, \cdots, Q$ and $w_1 \neq w_2$. $o'(w_1)$ is the output of the matched filter for two signals with matched frequency count w_1. By Setting SNR_0 as the SNR in (13), the judgment value ratio and the corresponding error probability based on $o(w_1, w_2)$ and $o'(w_1)$ are

$$r_{FH}(w_1, w_2) = o(w_1, w_2)/o'(w_1), \tag{18}$$

$$P_{FE}(w_1, w_2) = \text{qfun}\left((1 - r_{FH}(w_1, w_2))\sqrt{SNR_0}\right), \tag{19}$$

respectively, where qfun($\cdot$) is Q function. After Q matched filters, Q outputs compare the value with each other. The error probability for the received signal with frequency count w_1 to estimate in unconstrained FH is

$$\overline{\Theta}(w_1) = 1 - \prod_{q=2}^{Q} (1 - P_{FE}(w_1, w_q)). \tag{20}$$

For the C-FH-LFM waveform with $M = 1$, $l_{MS}[k] = 1$ is fixed. Each chip only requires N times of matching and judgment for $l_{NS}[k]$, and the frequency interval is MNB_C. Thus, The error probability for the received signal with frequency count w_1 to estimate is

$$\Theta(w_1) = 1 - \prod_{d=2}^{N} (1 - P_{FE}(w_1, w_{(d-1)MN+1})) < \overline{\Theta}(w_1). \tag{21}$$

In this case, The rule of C-FH-LFM makes the matched filtering processing at the receiver only N/Q of the unconstrained scheme, and reduces the error rate. Combining (10) and (21), as N decreases and $M = Q/N^2$ increases, the above performance will be degraded, but the communication rate will be improved. And vice versa.

3.3 Upper Limit of Chip Number of Each Pulse

Considering that each chip needs to be demodulated independently, this sub-section analyzes the setting rules for time-bandwidth product p_{TB} of each chip, and then provides the upper limit of the dividable chip number within the pulse. The p_{TB} is expressed as

$$p_{TB} = f_C T_C = \mu T_C^2 = \mu(I_S T_S)^2, \tag{22}$$

where T_S is the sampling period, and I_S is the number of samples in a chip. According to the Nyquist sampling theorem, the signal must work in with frequency band less than half the sampling frequency.

$$\mu T_S \cdot Q \leq f_S/2 = 1/2T_S. \tag{23}$$

Then, we have $p_{TB} \leq I_S/2Q$. On the other hand, matched filtering generates a peak value v_{MA} at $t = 0$ with matched slope, and a lower amplitude v_{UN} with inverse slope in (14). These two amplitudes are expressed as

$$v_{MA} = \left\| \sum_{\varepsilon=1}^{I_s} e^{j\theta_{MA}} \right\|_{t=0} = I_S, \tag{24}$$

$$v_{UN} = \left\| \sum_{\varepsilon=1}^{I_s} e^{j2\pi(2\mu(\varepsilon T_S - t - T_C/2)^2 + \mu t^2) + j\theta_{UM}} \right\|_{t=0} = \left| \sum_{\varepsilon=1}^{I_s} e^{j4\pi\mu(\varepsilon T_S - T_C/2)^2} \right|, \tag{25}$$

where θ_{MA} and θ_{UM} are the residual phase of matched and unmatched case. We set the judgment input ratio as

$$r_V = v_{UN}/v_{MA} = \left| \sum_{\varepsilon=1}^{I_s} e^{j4\pi p_{TB}(\varepsilon/I_S - 1/2)^2} \right| / I_S \approx \left| \int_0^1 e^{j4\pi p_{TB}(g - 1/2)^2} dg \right|. \tag{26}$$

The parameter p_{TB} will affect the r_V and then the judgment performance of the signal. The smaller the r_V, the lower the SER. The curve for the change of r_V according to p_{TB} is shown in Fig. 3. With the increase of p_{TB}, r_V decreases overall, but the curve is irregular. When the p_{TB} is less than 3.6, the r_V increases sharply as the p_{TB} decreases. Considering the limitation of sampling rate in hardware, sampling points I_S in sub-pulse is limited. The p_{TB} value of the first minimum point of r_V is 3.6 as the reference lower limit in engineering. Therefore, p_{TB} needs satisfy

$$3.6 \leq p_{TB} \leq I_S/2Q. \tag{27}$$

Then, in view of $I_S = f_S T/Q$, we get the upper limit of dividable chips within a single pulse as

$$Q \leq \sqrt{f_S T/7.2}. \tag{28}$$

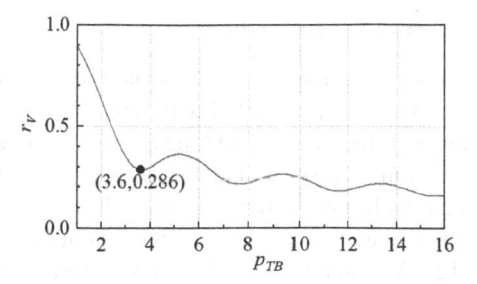

Fig. 3. r_V curve according to p_{TB}.

The conclusions of (27) and (28) are applicable to any scene where the positive/negative slope LFM waveform is adopted for communication, and also to the scene where this modulation mode is combined with other modulation modes.

4 Simulation Result

4.1 Simulation Setup

In this section, we compare the proposed C-FH-LFM waveform with existing works, in terms of waveform characteristic, radar performance and communication performance. MATLAB is used to evaluate radar and communication simulation scenarios with additive white noise, through Monte-Carlo simulations. Following existing works, the wireless channel between the radar and the communication receiver is considered to be the ideal Rician fading channel, and we set *SNR* as the SNR after the RF bandpass filter with B as bandwidth at the receiving end for radar or communication [17]. The parameters used in the simulation are listed in Table 3 [10, 12, 15].

Table 3. Parameter settings of C-FH-LFM waveform simulation

Parameter	Value
Modulation parameter M	2
Modulation parameter N	4
Chip modulation bandwidth B_C	1 MHz
Chip period T_C	7 us
Frequency modulation slope μ	0.143 MHz/us
Time-bandwidth product p_{TB}	7
LFM waveform width T	224 us
Sampling frequency f_S	64 MHz

4.2   C-FH-LFM Waveform Characteristic

Figure 4 shows the power spectral density (PSD) of the C-FH-LFM waveform and FH-LFM-2PSK waveform in [10]. [10] adopts FH within the chirp cycle and optimizes the radar and communication performance by selecting from the set of all waveforms. Under the condition of the same communication data rate as R_{DATA} in (10) and the same chip number as K, we select waveforms by the scheme in [10]. By statistical calculation, the out-of-band spectrum leakage, i.e. power summation out of $[-B/2, B/2]$ of C-FH-LFM is reduced up to 12 dB compared with FH-LFM-2PSK. [10] adopt PSK for data modulation, which produce large out-of-band leakage on the spectrum. In C-FH-LFM, the LFM slopes is used for data modulation, instead of phase. CPFSK is utilized to eliminate phase jumping between chips. In that case, the energy of the waveform can be concentrated within the working band, reducing interference for other systems as well as the risk of detection by adversaries.

4.3   Radar Performance

100,000 Monte Carlo simulations are carried out to verify the radar receiver operating characteristics of different waveform designs applied in same detection scenarios with target detection probability P_D as the metric. We assumed that the radar reflection scene obeyed Swerling-I [16]. The cell-averaging constant-false alarm rate (CA-CFAR) are adopted to conduct target detection. The original LFM represents the LFM waveform without data modulation. The FH-LFM-2PSK optimization scheme in [10] is consistent with the aforementioned. First, we set the false alarm rate $P_{FA} = 10^{-5}$ and provide the P_D under different SNRs in Fig. 5(a). Then, we set the reflecting signal from the target of interest with 5 dB SNR, and provide P_D with a varying P_{FA} in Fig. 5(b). As shown in the Fig. 5, The C-FH-LFM waveform proposed is slightly worse than the original LFM waveform with the same SNR and P_{FA}, and close to the scheme in [10]. The LFM characteristics are maintained, and the constraints provide the traversal of frequency points within a single pulse.

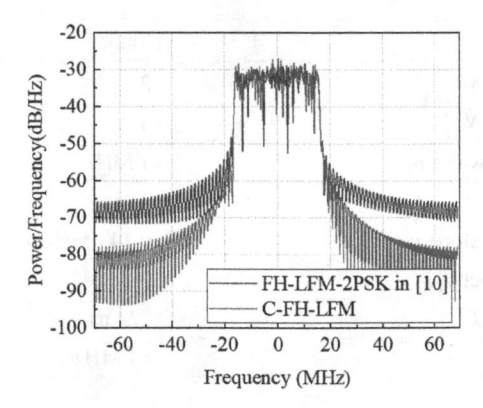

Fig. 4. Comparison of waveform PSD.

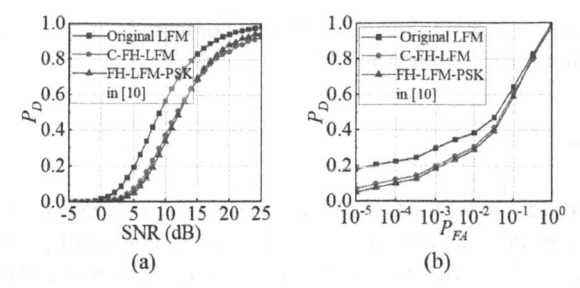

Fig. 5. Radar receiver operating characteristics of the C-FH-LFM original LFM and FH-LFM-PSK DFRC system for (a) P_D vs. SNR with $P_{FA} = 10^{-5}$; (b) P_D vs. P_{FA} with SNR = 5dB.

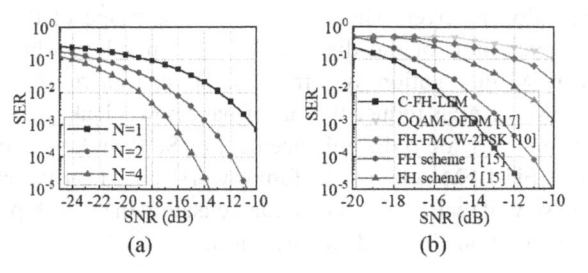

Fig. 6. SER comparison (a) for different parameter N of C-FH-LFM; (b) with other waveform schemes.

4.4 Communication Performance

Since data mapping is performed in LFM period, we adopt SER as the evaluation metric of communication performance. 100,000 Monte Carlo simulations are carried out to verify the communication performance of different waveform designs applied in the same communication scenarios with SER as the metric.

To show the FH demodulation performance improvement of proposed C-FH-LFM, the communication performances for LFM signal with different parameter N are simulated under the same bandwidth and pulse width. As N changes, we set $M = Q/(N^2)$, keeping Q constant. Figure 6(a) depicts the SER performance of C-FH-LFM waveform with different parameter N. In particular, if $N = 1$, the waveform is equivalent to unconstrained FH-LFM. The C-FH-LFM with $N = 4$ provides 6 dB improvement of SNR under same SER compared to the unconstrained FH-LFM ($N = 1$). Obviously, high N value can lower SER. However, the value of N is limited by Q, which is further limited by the derivation result in Sect. 3.

To compare with the communication performance of C-FH-LFM waveform, we simulate the communication performance of the waveform based on FH-FMCW-2PSK in [10], FH selecting in [15], and OQAM-OFDM in [17] under the same conditions of bandwidth and pulse width. Since some schemes cannot flexibly obtain arbitrary communication rates, we adopt waveform sets with similar communication rates for comparison. Figure 6(b) depicts the SER performance of C-FH-LFM waveform compared with those existing waveform schemes. We can observe that, the SER performance of C-FH-LFM waveform is better than that of other schemes. The C-FH-LFM scheme provides 2 dB improvement of SNR in same SER compared to the SER in [15] and 6 dB improvement compared to scheme in [10].

5 Conclusion

In this paper, we proposed a DFRC scheme based on C-FH-LFM waveform to improve spectrum utilization, and reduce out-of-band leakage, communication receiving link complexity, as well as SER. We utilized the inter-chip FH sequence and positive/negative frequency modulation slope in chip to embed the communication information, achieving a single frequency traversal within a single pulse. The explicit rules of FH sequence was provided to reduce the storage and processing of the matching signal and reduce the SER at communication receiver. Then, we derived the range of the time-bandwidth product and the upper limit of the distributable chip number, which is universal in positive/negative LFM modulation systems. The simulation results showed that, the proposed C-FH-LFM wave can reduce the out-of-band power leakage of 12 dB compared with PSK scheme and provide 6 dB enhancement of SNR under same SER compared with unconstrained FH-LFM scheme. In future work, based on the characteristics of waveform design, we will realize parameter compensation in each chip and optimize the radar performance based on C-FH-LFM waveform.

References

1. Liu, F., Masouros, C., Petropulu, A.P., Griffiths, H., Hanzo, L.: Joint radar and communication design: applications, state-of-the-art, and the road ahead. IEEE Trans. Commun.Commun. **68**(6), 3834–3862 (2020)
2. Griffiths, H., et al.: Radar spectrum engineering and management: technical and regulatory issues. Proc. IEEE **103**(1), 85–102 (2014)
3. Reichardt, L., Sturm, C., Grünhaupt, F., Zwick, T.: Demonstrating the use of the IEEE 802.11 p car-to-car communication standard for automotive radar. In: 6th European Conference on Antennas and Propagation (EUCAP), pp. 1576–1580. IEEE, Prague (2012)
4. Ma, D., Shlezinger, N., Huang, T., Liu, Y., Eldar, Y.C.: FRaC: FMCW-based joint radar-communications system via index modulation. IEEE J. Sel. Top. Signal Process. **15**(6), 1348–1364 (2021)
5. Gu, Y., Zhang, L., Zhou, Y., Zhang, Q.: Waveform design for integrated radar and communication system with orthogonal frequency modulation. Digital Signal Process. **83**(1), 129–138 (2018)
6. Wu, K., Guo, Y.J., Huang, X., Heath, R.W.: Accurate channel estimation for frequency-hop** dual-function radar communications. In: IEEE International Conference on Communications Workshops (ICC Workshops), pp. 1–6. IEEE, Dublin (2020)
7. Dou, Z.: Radar-communication integration based on MSK-LFM spread spectrum signal. Int. J. Commun. Netw. Syst. Sci.Commun. Netw. Syst. Sci. **10**(08), 108 (2017)
8. Ni, Y., Wang, Z., Huang, Q., Zhang, M.: High throughput rate-shift integrated system for joint radar-communications. IEEE Access **7**(1), 78228–78238 (2019)
9. Eedara, I.P., Amin, M.G., Hoorfar, A., Chalise, B.K.: Dual-function frequency-hopping MIMO radar system with CSK signaling. IEEE Trans. Aerosp. Electron. Syst.Aerosp. Electron. Syst. **58**(3), 1501–1513 (2022)

10. Gu, M.X., Lee, M.C., Liu, Y.S., Lee, T.S.: Design and analysis of frequency hopping-aided FMCW-based integrated radar and communication systems. IEEE Trans. Commun.Commun. **70**(12), 8416–8432 (2022)
11. Xu, J., Wang, X., Aboutanios, E., Cui, G.: Hybrid index modulation for dual-functional radar communications systems. IEEE Trans. Veh. Technol.Veh. Technol. **72**(3), 3186–3200 (2022)
12. Wang, X., Hassanien, A., Amin, M.G.: Dual-function MIMO radar communications system design via sparse array optimization. IEEE Trans. Aerosp. Electron. Syst.Aerosp. Electron. Syst. **55**(3), 1213–1226 (2018)
13. Hassanien, A., Himed, B., Rigling, B. D.: A dual-function MIMO radar-communications system using frequency-hop** waveforms. In: IEEE Radar Conference (RadarConf), pp. 1721–1725. IEEE, Seattle (2017)
14. Baxter, W., Aboutanios, E., Hassanien, A.: Dual-function MIMO radar-communications via frequency-hopping code selection. In: 52nd Asilomar Conference on Signals, Systems, and Computers, pp. 1126–1130. IEEE, Pacific Grove (2018)
15. Wang, X., Xu, J.: Co-design of joint radar and communications systems utilizing frequency hopping code diversity. In: IEEE Radar Conference (RadarConf), pp. 1–6. IEEE, Boston (2019)
16. Swerling, P.: Probability of detection for fluctuating targets. IRE Trans. Inf. Theory **6**(2), 269–308 (1960)
17. Shi, Q., Zhang, T., Yu, X., Liu, X., Lee, I.: Waveform designs for joint radar-communication systems with OQAM-OFDM. Signal Process. **195**(1), 108462 (2022)

A Construction of IoT Malicious Traffic Dataset and Its Applications

Yiping Zhang[1], Yiyang Zhou[2], Jie Yin[2(✉)], Yichen Yang[3], Wenwei Xie[4], Guangjun Liang[2], and Lanping Zhang[5]

[1] Department of Criminal Science and Technology, Jiangsu Police Institute, Nanjing, China
[2] Computer Information and Cyber Security, Jiangsu Police Institute, Nanjing, China
yinjie@jspi.cn
[3] Wuxi Public Security Bureau Intelligence Command Center, Wuxi, China
[4] Network Security, Trend Micro Incorporated, Nanjing, China
[5] College of Telecommunications and Information Engineering, NJUPT, Nanjing, China

Abstract. The current Internet of Things (IoT) malicious traffic dataset mainly relies on raw binary data at the traffic packet level and structured data at the session flow level for learning training and predictive classification, each of which has shortcomings and is not conducive to the construction of IoT malicious traffic classification models. In this paper, based on the packet and session stream datasets in the Iot-23 dataset, an importance-based feature extraction method is given, and a new malicious traffic dataset construction process is further proposed. The newly constructed malicious traffic dataset solves the shortcoming of mismatch between the number of features and the prediction performance in the traffic packet and session flow datasets. Experiments show that compared with a single dataset, the dataset constructed by the new method has richer and more important features, which improves the accuracy of malicious traffic classification and the generalization ability of the model, and is more conducive to the research of IoT malicious traffic classification.

Keywords: Machine learning · malicious traffic · feature engineering · datasets

1 Introduction

The malicious traffic of the Internet of Things (IoT) takes the growing IoT as the media and carries the attack load with the traffic, which has wide coverage and destructive harmful consequences. Applying machine learning classification to deal with potential network risks and classify malicious traffic in the Internet of Things is of great importance to effectively protect the security and reliability of network resources. At present, the number of IoT malicious traffic datasets is

H. Jin et al. (Eds.): IAIC 2023, CCIS 2058, pp. 304–318, 2024.
https://doi.org/10.1007/978-981-97-1277-9_23

small and the data structure is single. At the same time, most of the research experiments are based on only one aspect of traffic packages or session streams. In order to solve all kinds of problems that arise in the models trained using these datasets, this paper builds a IoT malicious traffic dataset with richer features, more accurate predictions and more generalization ability.

2   IoT Malicious Traffic Dataset and Research Status

2.1   Open Source IoT Malicious Traffic Dataset

Open source malicious IoT traffic datasets, IoT-23 [1] datasets and TON-IoT [2] dataset. Dataset generation activities are carried out in the laboratory, and the goal is to create a virtual security IoT environment. In a virtual environment, there are normal running IoT application nodes, malicious software attack nodes, and traffic capture nodes on the network boundary. The open source malicious IoT traffic dataset contains two formats of subdatasets, one is the traffic package dataset and the other is the session flow dataset. Traffic packet dataset is the original network traffic collected by the network traffic collector and stored in the binary format of pcap. Session stream datasets are structured data formed by Zeek tool parsing traffic package datasets and stored in csv format. Traffic packets and session flow datasets are essentially the same network traffic as they represent at different levels of a network protocol. The traffic packet dataset is based on the TCP/IP protocol's network layer, while the session stream dataset is based on the TCP/IP protocol's application layer.

2.2   IoT Research on the Classification of Malicious Traffic

The existing research on malicious traffic classification in the Internet of Things mainly focuses on two directions: model building and algorithm optimization, but there are few studies on dataset construction. Most malicious traffic classification studies only use session stream datasets, select specific fields in structured data, and use machine learning or neural network algorithms to build malicious traffic classification models [3–9]. Algorithms used include Nave Bayes [3], Distributed Random Forest (DRF) [4], Generalized Linear Model (GLM) [5], Gradient Boosting Machine (GBM) [6], Extreme Gradient Boost (XGBoost) [7], Artificial Neural Network (ANN) [8], and so on [9]. Some studies only use traffic packet datasets. In recent years, with the popularity of deep learning algorithms such as Convolutional Neural Network (CNN) [10] and Recurrent Neural Network (RNN) [11], many researchers have tried to bring them into the field of traffic monitoring and have achieved good prediction results. Researchers propose that network traffic characteristics [12] are not artificially designed and machine-selected traffic characteristics are achieved by using convolution and pooling of CNN networks. Initially, only network traffic was entered into the CNN network in a binary manner, and Vinayakumar first entered structured network traffic into the CNN network [13]. On this basis, some researchers have further implemented the CNN selection feature, replacing the subsequent full-connection layer

Table 1. Summary of dataset cybersecurity scenarios.

Network security scenarios	Malware	The number of traffic packets	The number of session flow
CTU-IoT-Malware- Capture-1-1	Hide and Seek	1,686,000	1,008,749
CTU-IoT-Malware- Capture-3-1	Muhstik	496,000	156,104
CTU-IoT-Malware- Capture-7-1	Linux. Mirai	11,000,000	11,454,723
CTU-IoT-Malware- Capture-8-1	Hakai	23,000	10,404
CTU-IoT-Malware- Capture-9-1	Linux. Hajime	6,437,000	6,378,294
CTU-IoT-Malware- Capture-20-1	Torii	50,000	3,210
CTU-IoT-Malware- Capture-21-1	Torii	50,000	3,287
CTU-IoT-Malware- Capture-34-1	Mirai	233,000	23,164
CTU-IoT-Malware- Capture-42-1	Trojan	24,000	4,427
...	...	...	...

change activation function with Long Short-Term Memory (LSTM) [14,15] and Gated Recurrent Unit (GRU) [16] implementations of RNN networks, as well as more variants such as bidirectional long-term and short-term memory and attention mechanism [17].

2.3 Introduction to the IoT-23 Dataset

IoT-23 is an Internet of Things dataset with malicious and benign traffic tagged. The dataset and its research were funded by Avast Software Company in Prague and created by Stratosphere Laboratory at CTU University in Czech Republic. The IoT-23 dataset provides different network security scenarios. Malicious traffic in each network security scenario comes from a specific malicious software sample that uses different network protocols and performs corresponding malicious actions. There are two types of IoT traffic in the scene: network traffic after malware infection and normal network traffic after benign infection. Malicious and benign scenarios are created in a controlled network environment, and experimental devices, like other real IoT devices, have unrestricted Internet connections. Details of the data in some network security scenarios in the loT-23 dataset are shown in Table 1.

Table 2. Session flow partial data.

ts	id.orig_p	conn_state	...	label	detailed_label
1533042913.90879	64923	OTH		Benign	–

In the corresponding network security scenarios generated by specific malware, there is data in the following formats: captured traffic packet data and tagged session flow data. Among them, traffic packet data is network traffic data stored in binary format, pcap format, and can be analyzed and viewed using tools such as Wireshark, Scapy. For the third traffic package in Scenario 8, the analysis of the Wireshark tool is shown in Fig. 1.

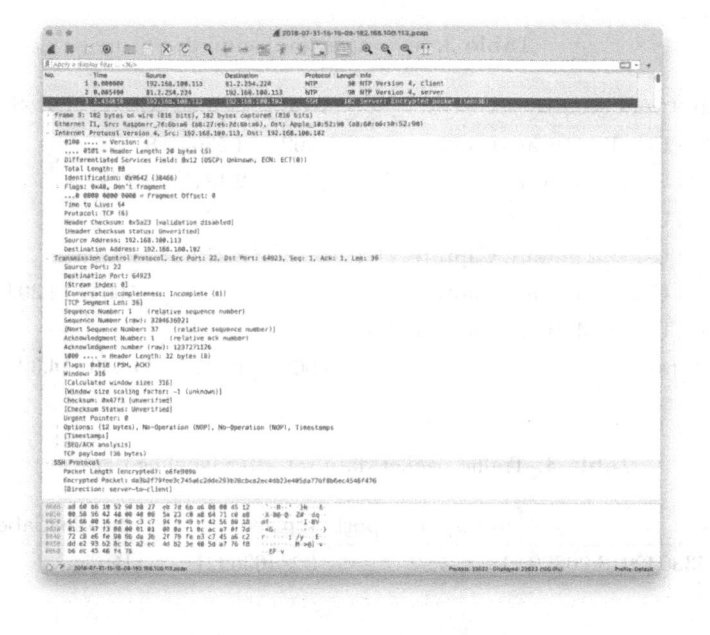

Fig. 1. Wireshark parses pcap traffic packets.

Session stream data is in csv format and is generated by Zeek tool's application layer protocol filtering and summary. For the third traffic packet in Scenario 8, the session flow portion of the data is shown in Table 2.

Each session flow data has a data classification label created by CTU University in addition to the Zeek tool's traffic resolution. As shown in Table 2, a two-class label named label indicates whether or not the traffic is malicious. If the data is malicious traffic, detailed-label records detailed malicious traffic classification label information.

3   Importance-Based Feature Extraction Methods

3.1   Preprocessing of IoT-23 Datasets

1) IoT-23 Dataset Sampling. Given the simplicity of a single network security scenario, where malicious traffic is generated by the same malicious program, normal traffic also has its fixed mode and characteristics. So it is necessary to select a variety of malicious traffic and normal traffic of different modes from different network security scenarios through data set sampling, to sample data sets with complete attributes, complete features, accurate values and true reality.

As shown in Table 3, the integrated dataset is sampled from the above six network security scenarios and contains one normal traffic type, Benign, and three malicious traffic types, Attack, C&C, and DDos.

Table 3. Types of flow sampled.

Network security scenarios	Attack	Benign	C&C	DDoS
CTU-IoT-Malware-Capture-3-1	5,962	4,536	8	–
CTU-IoT-Malware-Capture-8-1	–	2,181	8,222	–
CTU-IoT-Malware-Capture-20-1	–	3,193	–	–
CTU-IoT-Malware-Capture-21-1	–	3,272	–	–
CTU-IoT-Malware-Capture-34-1	–	1,923	6,706	14,394
CTU-IoT-Malware-Capture-42-1	–	4,420	–	–
Total	5,962	19,525	14,936	14,394

Table 4. Traffic packet data set after labeling data.

ts	ip_tos	tcp_sport	tcp_seq	...	detailed_label
1533042913.90879	18	22	3284636921	...	Benign

2) Processing of Data Labels. Traffic packages are binary flow data. The Scapy library extracts the structure of traffic packages in pcap format to generate a new csv format dataset. Unlike session flow, the pcap format is the most original traffic data. Label information for malicious traffic classification is missing and new datasets need to be labeled again using the existing label information in the session flow. Write a Python program that uses the memory-saving PcapReader method in the Scapy library to read the pcap file package by package, structurally parse the binary traffic package and store it in the corresponding dataframe of the field. Referring to the time information in the traffic packet, that is, the time stamp field in Unix timestamp format, and the ts field in the session flow data, the left join query is spliced into a detailed_label tag data table, and finally stored as a csv format file. For the third traffic package in Scenario 8, the data section after the label of the traffic package dataset is shown in Table 4.

3.2   Experiments on Iot-23 Dataset

On the experimental hardware, the CPU is Intel Core i9-9820X, the memory size is 128G, and the GPU is Nvidia GTX 1080Ti. On the experimental platform, the system is Ubuntu 20.04.3, the programming language is Python 3.8.10, and the machine learning framework is H_2O [18]. During model training, the implementation of related algorithms was introduced by the H_2O.estimators built into H_2O. Estimators were introduced. Because of the small amount of data, the training uses 4-Fold cross validation [19], dividing the validation set from the training set to avoid over-fitting caused by too small testing set [20]. A comparison of the predictive performance of machine learning and neural network algorithms on IoT-23 datasets is shown in Table 5 and Table 6.

Using the Accuracy and Weighed-F1 means as axes, respectively, plot scatter plots of the performance of different algorithms as shown in Fig. 2 and Fig. 3.

Based on the experimental results, the classification performance of models in the session flow test dataset is significantly better than that in the traffic package test dataset. DRF performs best in the traffic package test dataset, but GBM and XGBoost algorithms have the most stable prediction performance.

Table 5. Cassification performance on the traffic packet Iot-23 dataset.

Model	Accuracy	Marco-F1 mean	Weighed-F1 mean
Bayes	0.531745	0.382568	0.407899
DRF	**0.962987**	**0.949109**	**0.964253**
GLM	0.777279	0.629442	0.743140
GBM	0.958314	0.943811	0.960119
XGBoost	0.957581	0.941109	0.958964
ANN	0.929821	0.904852	0.931369

Table 6. Classification performance on the session flow Iot-23 dataset.

Model	Accuracy	Marco-F1 mean	Weighed-F1 mean
Bayes	0.741791	0.765111	0.736825
DRF	0.996990	0.997245	0.996991
GLM	0.960507	0.967669	0.960741
GBM	0.999270	0.999150	0.999270
XGBoost	**0.999726**	**0.999714**	**0.999726**
ANN	0.996808	0.997327	0.996809

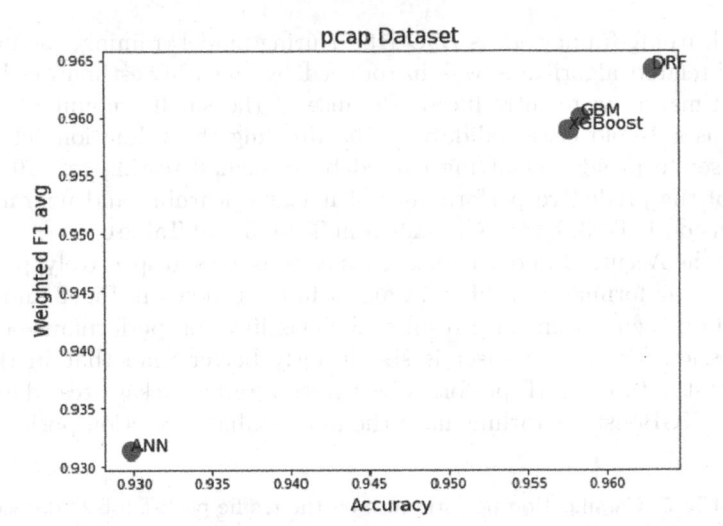

Fig. 2. Scatterplot of classification performance on the traffic packet test dataset.

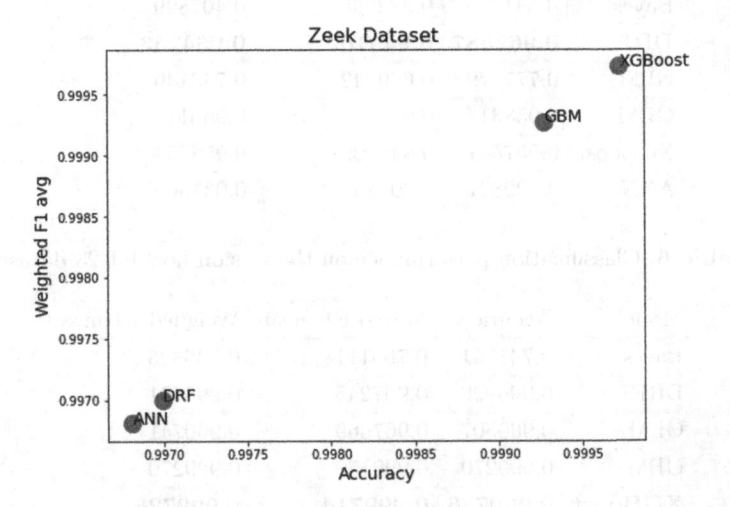

Fig. 3. Scatterplot of classification performance on the session flow test dataset.

3.3   Feature Correlation Calculation

The feature correlation calculation is based on Spearman correlation coefficient. Spearman correlation coefficient is a common correlation coefficient in statistics [21]. It has a good measurement effect in dealing with ranking problems and can be used to measure the closeness between two variables. Spearman method does not need to calculate the variance of the overall data, and can speculate the distribution of the overall data through the sample data. It is a non parametric test. In two-dimensional data, record the initial position of a group of variables as (X, Y), and the position of the sorted group of data as (X', Y'). For the i-th group of data, record the difference between the positions before and after sorting as d_i. Record the total data as n, as shown in formula

$$\rho = 1 - \frac{\sum d_i^2}{n(n^2 - 1)}, \tag{1}$$

This method is used to measure the feature correlation and draw the correlation heatmaps of the numerical features of the traffic packet data set and the session flow data set, as shown in Fig. 4 and Fig. 5.

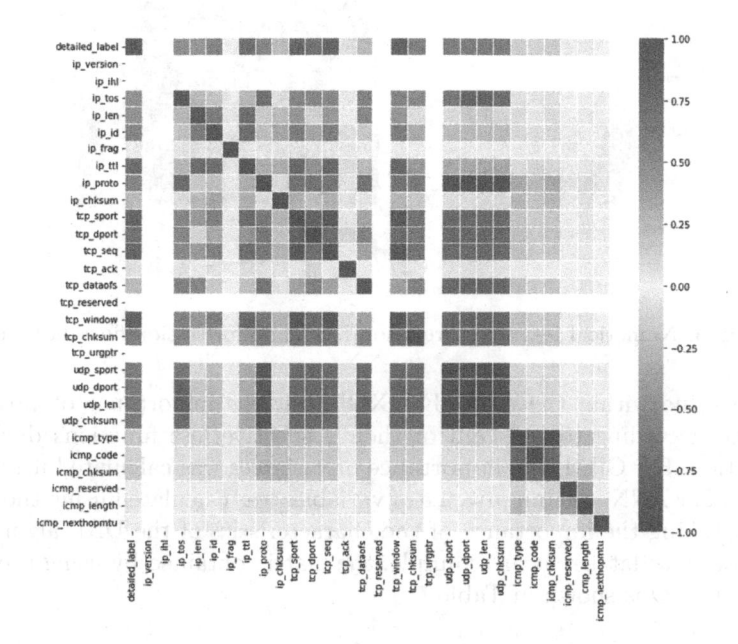

Fig. 4. Numerical feature correlation heat map for traffic packet dataset.

In Fig. 4 and Fig. 5, the color of each region represents the degree of correlation between the two features. The red area represents positive correlation, the blue area represents negative correlation, the gray area represents zero correlation, and the area without color represents no correlation.

3.4 Feature Importance Calculation

For the tree algorithm to which DRF and GBM belong, the importance of each variable is determined by calculating the relative influence of each variable: whether the variable is selected for splitting and the reduction of the square error of the result in the tree construction process. When a node is split based on a number or classification feature, the reduction of the square error of the feature is the difference between the square error of the node and its child nodes. The square error of each individual node is the reduction of the variance of the

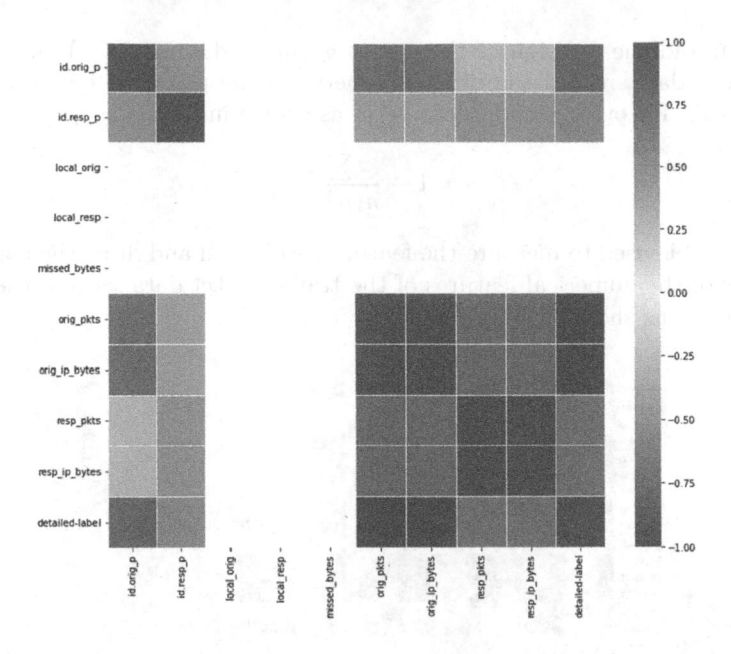

Fig. 5. Numerical feature correlation heat map for session flow dataset.

response value within the node. For XGBoost, the importance of variables is calculated according to the gain of their respective loss functions during tree construction. For GBM, the importance of variables was calculated using coefficient size. For ANN, the importance of variables was calculated using the Gedeon method. Taking the importance of the characteristics of the DRF algorithm on the session flow dataset as an example, the matrix obtained by using the varimp method of H_2O is shown in Table 7.

Table 7. Feature importance matrix of DRF on session flow dataset.

Features	Relative	Scaled	Percentage
history	420364.062500	1.000000	0.390739
id.orig_p	201310.250000	0.478895	0.187123
conn_state	154858.921875	0.368392	0.143945
id.resp_p	101601.226562	0.241698	0.094441
therefore	80541.390625	0.191599	0.074865
resp_ip_bytes	66421.546875	0.158010	0.061740
resp_bytes	24911.839844	0.059263	0.023156
resp_pkts	22012.664062	0.052366	0.020461
orig_pkts	1162.718018	0.002766	0.001081
orig_ip_bytes	1040.612671	0.002476	0.000967
orig_bytes	961.824951	0.002288	0.000894
duration	614.247192	0.001461	0.000571
missed_bytes	18.006281	0.000043	0.000017

The overall feature importance needs to be calculated by the respective feature importance matrices of different algorithms on the two data sets. A one-dimensional vector I representing the importance of features in N models is defined as shown

$$I = \frac{\sum I_i F_{1_i}}{N},$$ (2)

where I_n refers to a one-dimensional vector of feature importance (scaled) in the n-th model. Where, the value of F_1 is weighted-F_1 value. For the two types of data sets, the one-dimensional vector I of the feature importance of all models is calculated and combined, and the comprehensive ranking of the feature importance on the two types of data sets is shown in Table 8.

Table 8. Combined ranking of feature importance on both datasets.

Ranking	Features	Importance	Dataset
1	History	0.99878	Session flow dataset
2	tcp_flags	0.98461	Traffic package dataset
3	conn_state	0.67331	Session flow dataset
4	tcp_dport	0.60541	Traffic package dataset
5	tcp_window	0.56004	Traffic package dataset
6	tcp_sport	0.48426	Traffic package dataset
...	...	...	...

According to the ranking, the top five most important features are history, tcp_flags, conn_state, tcp_port, and tcp_window features. Where history and conn_state are generated by the session flow dataset, and other characteristics are generated by the traffic packet dataset.

3.5 Importance-Based Feature Extraction Methods

Importance-based features extraction method is divided into the following steps:

- Step 1: sort and sample data suitable for training according to network security scenarios.
- Step 2: through data preprocessing operations such as data integration, specification and cleaning, multiple scene data sets with the same malicious traffic classification are sorted into a unified template network traffic dataset.
- Step 3: divide the data set into traffic packet dataset and session flow dataset according to the file format and content, and select and process the characteristics of the two data sets respectively.
- Step 4: use the selected machine learning or neural network algorithm to train the model, verify the malicious traffic monitoring performance of the specific algorithm, and compare the performance of the algorithm from the accuracy index.

- Step 5: according to the classification results of multiple models, analyze the correlation and importance of the characteristics of the dataset to obtain the traffic characteristics with high importance.
- Step 6: optimize the flow characteristics of high importance to generate a new dataset, and verify the impact of the new input characteristics on the accuracy of model prediction.

The frame structure is shown in Fig. 6.

4   Construction and Analysis of Datasets

4.1   Dataset Construction

For the traffic packet test dataset, there are many features but the prediction performance of the model is not high enough. For the session flow dataset, the model has high prediction performance and few features. From the perspective of network level, traffic packets and session flow datasets represent the characteristics of malicious traffic at different network levels. In fact, since ZeeK tool can analyze based on session and history information, it can record history and conn_state. The two session flow information, state, provide great support for the

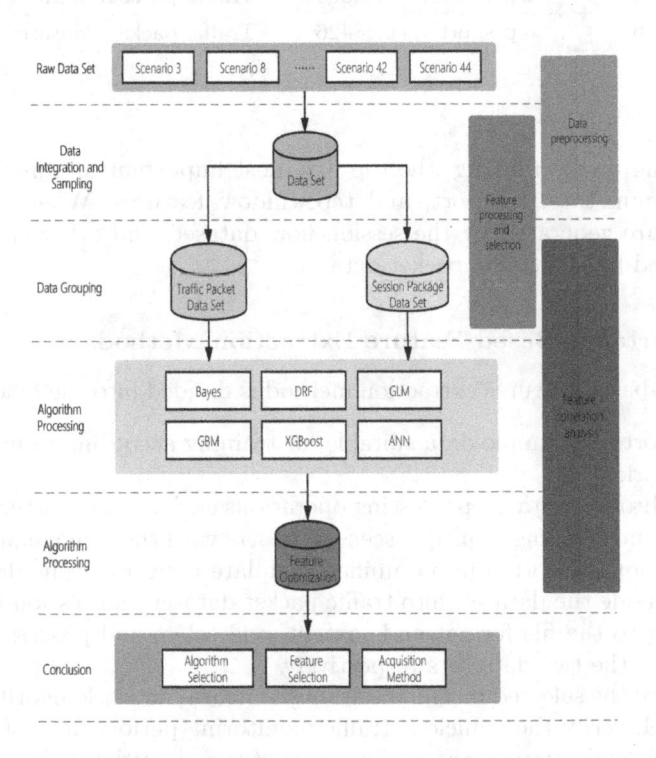

Fig. 6. Framework structure of feature importance-based extraction method.

classification of malicious traffic. However, in the original traffic packets, there are also tcp_flags, tcp_dport and other important features that can assist in malicious traffic classification. In order to improve the generalization of the model, so that the model is no longer disturbed by irrelevant features and does not rely on a few simple features, according to the feature correlation heat map, the features without correlation and the features with zero correlation are removed respectively, and the important features in the traffic packet dataset and the session flow dataset are integrated to construct the optimized dataset.

4.2   Experimental Results of the Dataset

Take the accuracy and weighted-F_1 mean as the coordinate axes respectively, and draw the scatter plots of the performance of different algorithms, as shown in Fig. 7. According to Fig. 7, it can be seen that the four algorithms with the best prediction effect have reached an accuracy of more than 99.99%, and the accuracy of XGBoost model with the best performance and the average value of Weighted-F_1 are both 99.9909%. Compared with the results before data feature optimization, the prediction effect on the comprehensive data set has been greatly improved. Taking the feature importance of DRF on the dataset as an example, the matrix obtained by using the varimp method of H_2O is shown in Table 9.

According to the table x of the original dataset and the optimized dataset and table x, the scatter diagram of DRF algorithm feature importance before and after the optimization of the data set is drawn with the scaled importance and percentage as coordinates, as shown in Fig. 8. It is found that the importance distribution of features on the comprehensive data set is more uniform, which

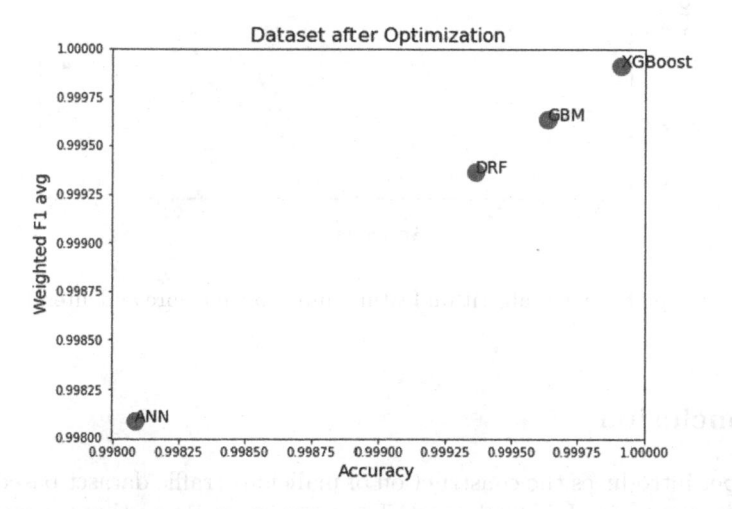

Fig. 7. Scatterplot of classification performance on the test dataset after feature optimization.

Table 9. Feature importance matrix of DRF on session flow dataset.

Features	Relative	Scaled	Percentage
conn_state	130419.367188	1.000000	0.111268
history	118598.656250	0.909364	0.101183
tcp_window	100236.328125	0.768569	0.085517
therefore	88913.875000	0.681754	0.075857
tcp_dport	87786.304688	0.673108	0.074895
tcp_sport	81293.140625	0.623321	0.069356
tcp_dataofs	78476.796875	0.601727	0.066953
tcp_flags	76353.218750	0.585444	0.065141
...	...	...	...

means that the prediction results of the model are determined by more features, rather than some specific features with larger weights.

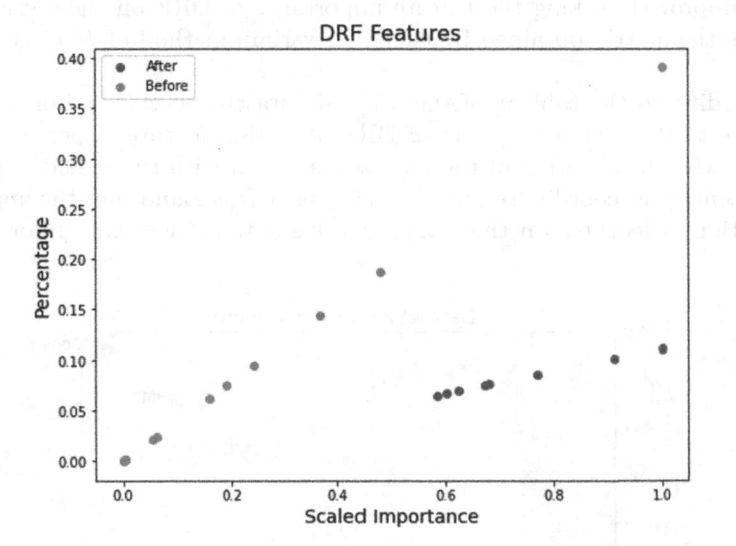

Fig. 8. Scatterplot of DRF algorithm feature importance before and after dataset optimization.

5   Conclusion

This paper introduces the construction of malicious traffic dataset based on IoT, and compares it with IoT-23 dataset. The experiment shows that the importance distribution of features on the comprehensive dataset is more uniform; XGBoost algorithm has the best prediction performance on the comprehensive data set.

However, there are some defects in IoT-23 dataset, and the data is duplicated in some areas and has obvious characteristics; The computational power consumed by data preprocessing is too high. The sampled data sets cover less network security scenarios and the amount of data is insufficient. These shortcomings cause the model training to be affected to a certain extent, showing the results of high prediction accuracy. However, the effect of improving the performance of the model is real and objective. With the decrease of data redundancy and the increase of sample size, it can be predicted that its generalization ability and robustness will be better than the model trained with the original dataset.

Acknowledge. This work is supported by Practical Innovation and Entrepreneurship Training Programme for College Students in Higher Educational Institutions in Jiangsu Province (202210329046Y, 202310329061Y).

References

1. Garcia, S., Parmisano, A., Erquiaga, M.J.: IoT-23: A labeled dataset with malicious and benign IoT network traffic (Version 1.0.0) [Data set], Zenodo (2020). https://doi.org/10.5281/zenodo.4743746
2. N. Moustafa, N.: A new distributed architecture for evaluating AI-based security systems at the edge: network TON-IoT datasets. Sustain. Cities Soc. **72**, 102994 (2021)
3. Moore, A.: Internet traffic classification using bayesian analysis techniques. In: ACM Sigmetrics Performance Evaluation Review (2005)
4. Gislason, p., Benediktsson, J., Sveinsson, J.: Random forests for land cover classification. Pattern Recognition Lett. **27**(4), 294–300 (2006)
5. Nelder, J., Wedderburn, R.: Generalized linear models. J. Roy. Stat. Soc. **135**(3), 370–384 (1972)
6. Friedman, J.: Greedy function approximation: a gradient boosting machine. Annal. Stat. **29**(5), 1189–1232 (2001)
7. Chen, T., Guestrin, C.: XGBoost: a scalable tree boosting system. In: 22nd ACM SIGKDD International Conference on Knowledge Discovery and Data Mining (2016)
8. Bishop, C., et al.: Neural Network for Pattern Recognition (1995)
9. Patcha, A., Park, J.: An overview of anomaly detection techniques: existing solutions and latest technological trends. Comput. Netw. **51**(12), 3448–3470 (2007)
10. Lawrence, S., Giles, C., Tsoi, A., et al.: Face Recognition: a convolutional neural network approach. IEEE Trans. Neural Netw. **8**(1), 98–113 (1997)
11. Williams, R., Zipser, D.: A learning algorithm for continually running fully recurrent neural networks. Neural Comput. **1**(2), 270–280 (1998)
12. Wang, W.: Research on network traffic classification and anomaly detection based on deep learning. University of Science and Technology of China (2018)
13. Vinayakumar, R., Soman, K., Poornachandran, P.: Applying convolutional neural network for network intrusion detection, In: International Conference on Advances in Computing, Communications and Informatics (ICACCI), pp. 1222–1228 (2017)
14. Thapa, K.K., Duraipandian, N.: Malicious traffic classification using long short-term memory (LSTM) model. Wirel. Pers. Commun. **119**(3), 2707–2724 (2021)

15. Lu, X., Liu, p., and Lin, J.: Network traffic anomaly detection based on information gain and deep learning. In: Proceedings of the 2019 3rd International Conference on Information System and Data Mining, pp. 11–15 (2019)

16. Liu, X., Liu, J.: Malicious traffic detection combined deep neural network with hierarchical attention mechanism. Sci. Rep. **11**(1), 1–15 (2021)

17. Su, T., Sun, H., Zhu, J., et al.: BAT: deep learning methods on network intrusion detection using NSL-KDD dataset. IEEE Access **8**, 29575–29585 (2020)

18. Cook, D.: Practical machine learning with H2O (2017)

19. Fan, Y.: Overview of Cross Validation Methods in Model Selection. Shanxi University (2013)

20. Blum, A., Kalai, A., Langford, J.: Beating the hold-out: bounds for K-fold and progressive cross-validation. In: Conference on Learning Theory (1999)

21. Borkowf, C.: Computing the nonnull asymptotic variance and the asymptotic relative efficiency of Spearman's rank correlation. Comput. Stat. Data Anal. **39**(3), 271–286 (2002)

Meta-learning Based on Multi-objective Optimization

Xin Zhong[1], Jiahao Wang[1], Zutong Sun[1], YuHeng Ren[3],
and Kuihua Li[2](✉)

[1] University of Electronic Science and Technology of China, Sichuan 610056, China
[2] Chongqing Three Gorges Medical College, Chongqing 404020, China
y3k1980@163.com
[3] Belarusian State University, 220071 Minsk, Republic of Belarus

Abstract. Meta-learning aims to enable machine learning systems to learn how to learn, which makes it widely applicable in the field of few-shot learning. In common situations, a meta-learner learns meta-knowledge from the training experiences of a set of related tasks and utilizes this meta-knowledge to enable the model to quickly adapt to new tasks. However, the varying impact of different tasks on meta-knowledge, and the potential conflicts that arise from certain tasks, make it difficult to attain an optimal meta-knowledge acquired from multiple tasks. To overcome these limitations, we introduce the concept of multi-objective optimization to the classical meta-learning algorithm, Model-Agnostic Meta-Learning (MAML), aiming to train and obtain the optimal meta-knowledge. Specifically, we employ a gradient-based multi-objective optimization algorithm to efficiently determine a Pareto-optimal set of tasks. Subsequently, the meta-learner effectively utilizes this set of optimal solutions to establish an optimal initial model. Experimental results on the benchmark dataset demonstrate significant performance improvements in the proposed algorithm. Moreover, even with a limited number of tasks, the acquired meta-knowledge approaches the optimal solution and exhibits good generalization capabilities.

Keywords: Few-shot learning · Meta-Learning · Multi-objective optimization · Pareto-optimal

1 Introduction

Meta-learning is a machine-learning algorithm that is inspired by the learning patterns observed in humans. Its objective is to enable machines to automatically learn how to learn, in other words, to acquire the ability to learn [23]. In the current context of machine learning models heavily reliant on vast amounts of high-quality labeled data, this unique capability becomes especially crucial. More precisely, meta-learning generalizes training experiences from a set of tasks also referred to as prior knowledge, and applies them when the model is faced with new tasks [1]. This enables the model to rapidly learn specific task knowledge

H. Jin et al. (Eds.): IAIC 2023, CCIS 2058, pp. 319–332, 2024.
https://doi.org/10.1007/978-981-97-1277-9_24

with minimal data samples. Indeed, due to the aforementioned characteristics, meta-learning has emerged as one of the crucial solutions in the context of small-sample scenarios.

Model-agnostic meta-learning (MAML) holds a prominent position among the algorithms in the field of meta-learning, owing to its exceptional significance [4]. MAML algorithm comprises an inner and outer optimization loop. The outer loop aims to find the meta-initialization parameters, while the inner loop applies these initialization parameters and facilitates rapid adaptation. The effectiveness of the meta-initialization learned in the outer loop can be attributed to the phenomenon of feature reuse [15, 16, 26]. Optimal feature reuse can significantly elevate model performance, whereas inadequate feature reuse can lead to adverse effects. However, in the case of MAML, the feature reuse originates from the meta-tasks within the inner loop, which introduces inherent challenges related to balancing data and task difficulty [7, 8, 10].

This challenge manifests specifically in the meta-training process, where due to variations in training difficulties between tasks and differences in data samples for each task, the training of initial model may become biased towards certain tasks. The presence of such biased initial models, which are influenced by this inherent imbalance, can result in poor generalization when facing new tasks. To mitigate the bias in initial model, one straightforward statistical approach is to increase the number of tasks trained simultaneously. However, this approach leads to a linear increase in training time, which is impractical when dealing with complex models.

To tackle this challenge, we propose a meta-learning algorithm based on multi-objective optimization. Specifically, we consider each meta-task as a separate objective function, and the overall performance of multiple meta-tasks after inner adaptation is treated as the optimization objective. By employing multi-gradient descent, a set of Pareto optimal solutions for conflicting objectives can be obtained. These solutions strike a balance and trade-off among multiple objectives, achieving optimality or near-optimality under the current constraints. Taking a global perspective, this method effectively mitigates the adverse effects of task imbalance and produces an initial model with stronger generalization capabilities.

In this paper, we conducted an in-depth exploration and analysis of this problem. Our main contributions are as follows:

- We approached the optimization problem in the meta-learning process as a multi-objective optimization problem, where each meta-task corresponds to an objective. This provides better guidance and measures for addressing subsequent task balancing issues.
- We proposed a meta-learning algorithm based on multi-objective optimization, aiming to obtain Pareto optimal solutions between meta-tasks. By training with this approach, we achieved a more generalized and higher-performing meta-initialization model.
- We validated our model on baseline datasets and experimental results demonstrated its significant superiority over existing meta-learning models.

2 Related Work

In this section, we reviewed some milestone literature in the fields of meta-learning and multi-objective optimization.

2.1 Meta-learning

The concept of meta-learning was introduced several decades ago [13,23]. After years of development, meta-learning has branched out into various subfields, including metric-based, memory-based, and optimization-based approaches. However, they all share a common goal, which is to acquire the ability to learn from a set of meta-learning tasks by capturing shared features, knowledge, or strategies. This enables the model to adapt more effectively and efficiently to new tasks.

Model-agnostic meta-learning (MAML) is one of the groundbreaking algorithms in the field of meta-learning, which is applicable to any model that uses gradient descent optimization [4]. It trains a model on a set of meta-tasks, inducing and summarizing meta-knowledge to obtain well-initialized parameters. This enables the model to quickly adapt to new tasks with only a few samples. MAML++ is an enhanced version of MAML that addresses challenges such as training instability, high computational cost of second-order derivatives, and fixed learning rates in inner and outer loops. These improvements make MAML++ a more comprehensive and refined algorithm [1]. ANIL (Almost No Inner Loop) argues that the effectiveness of MAML primarily lies in feature reuse. As a result, ANIL simplifies the algorithm by removing the inner loop component [16]. TAML addresses the problem of task bias during the meta-training process by utilizing entropy-based and inequality-minimization methods. It aims to train an unbiased initial model that considers task-specific information and minimizes bias [7]. DAML proposes a dynamic task modulation function that adjusts the weights of learning tasks. Specifically, this function focuses more on challenging tasks by assigning them higher weights, allowing the model to pay greater attention to these tasks during the meta-learning process [10]. The Bayesian TAML model utilizes a Bayesian framework to adaptively adjust parameters during the meta-learning process. This approach effectively mitigates the challenges of class imbalance and task imbalance in real-world scenarios [8].

Recent studies have shown that the task balancing problem in meta-learning has a significant impact on the model's generalization and effectiveness. The achievement of task adaptation and training an unbiased model has become an important research direction. In the upcoming section, we will approach this problem from another perspective, specifically through the lens of multi-objective optimization.

2.2 Multi-objective Optimization

The objective of multi-objective optimization problems is to find a set of solutions that can achieve good performance across multiple potentially conflicting objective functions [9,12,19]. This set of solutions, also known as Pareto optimal solutions, has been widely applied in various fields such as multi-task learning [12,19], reinforcement learning [24], Bayesian optimization [6,20], etc.

The solution methods for multi-objective optimization problems can be broadly categorized into two main classes: analytical method and numerical method [27]. In previous research, it has been observed that analytical methods are not suitable for most scenarios due to rigorous mathematical proofs and formula derivations. While, numerical method, primarily composed of various heuristic algorithms inspired by nature, have been extensively employed in practice [2].

In machine learning, when it comes to solving a large amount of gradient information, traditional methods are no longer applicable. In such cases, utilizing the method of multi-gradient descent proves to be highly effective [5]. Ozan Sener and Vladlen Koltun utilized the MGDA (Multi-Gradient Descent Algorithm) based on the Karush-Kuhn-Tucker (KKT) conditions to solve for the Pareto optimal solutions [19]. Xi Lin et al. further extended the approach by decomposing the multi-objective optimization problem into a series of subproblems. By leveraging parallel computation to solve these subproblems, they obtained a set of Pareto optimal solutions and devised different solution strategies for various scenarios [12]. Our work focuses on utilizing a gradient-based multi-objective optimization algorithm to effectively obtain a set of Pareto optimal solutions for a given set of meta-tasks.

3   Proposed Method

In this section, we provided a detailed explanation of the proposed method, addressing the task imbalance issue in meta-learning from the perspective of multi-objective optimization. The proposed method considers a batch of meta-tasks in MAML as multiple objectives. It leverages the gradient information obtained after adapting to each task to solve for corresponding task weights. This approach enables the meta-learner to optimize towards the global optimum direction, mitigating the issue of bias towards specific tasks.

3.1   Problem Definition

In this paper, we represent the model as a function $f_\theta(.)$, where θ denotes the model parameters. It can be any classification or regression model, implying that the model is capable of producing corresponding outputs when given input samples. Taking the example of a classification task, the model's input consists of tasks sampled randomly from the task distribution $p(\mathcal{T})$. Each task is formulated as an N-way K-shot setup, where N represents the number of classes, and K represents the number of training samples. MAML involves sampling individual tasks $p(\mathcal{T})$ from the task distribution $\mathcal{T}_i$ for meta-training. The model is then evaluated and optimized on test set to measure its performance.

During the meta-training, the model's initial parameters, denoted as θ, are utilized for training on task $\mathcal{T}_i$ using gradient descent. For a classification task, the specific gradient update formula is as follows:

$$\theta_i' = \theta - \alpha\nabla_\theta\mathcal{L}_{\mathcal{T}_i}\left(f_\theta\right) \tag{1}$$

The variable α represents a fixed hyperparameter that serves as the learning rate for the inner loop.

To facilitate meta-optimization across tasks, the authors utilize stochastic gradient descent (SGD) to update the model parameters θ. The update process is formulated as follows:

$$\theta \leftarrow \theta - \beta\nabla_\theta \sum_{\mathcal{T}_i\sim p(\mathcal{T})} \mathcal{L}_{\mathcal{T}_i}\left(f_{\theta_i'}\right) \tag{2}$$

The variable β represents a fixed hyperparameter that serves as the learning rate for the outer loop.

As observed, the meta-optimization in MAML is performed based on the updated parameter θ_i' computation on task $\mathcal{T}_i$. It can be stated that MAML discovers the initialization parameters in $p(\mathcal{T})$ that are adaptable to all tasks. So the optimization objective is defined as follows:

$$\min_\theta \sum_{\mathcal{T}_i\sim p(\mathcal{T})} \mathcal{L}_{\mathcal{T}_i}\left(f_{\theta_i'}\right) = \sum_{\mathcal{T}_i\sim p(\mathcal{T})} \mathcal{L}_{\mathcal{T}_i}\left(f_{\theta-\alpha\nabla_\theta\mathcal{L}_{\mathcal{T}_i}(f_\theta)}\right) \tag{3}$$

Indeed, the optimization objective may not fully capture the impact of conflicting tasks on the θ. Consequently, the updated θ may exhibit a bias towards specific tasks, deviating from the global optimum under $p(\mathcal{T})$. This limitation can lead θ deviating from the global optimum.

Our goal is to add optimal weight coefficients into a diverse set of meta-tasks, facilitating the update of meta-initialization parameters θ towards the global optimum direction. By doing so, we can effectively alleviate the negative consequences resulting from task imbalance and conflicts. In particular, we treat the conflict issues in meta-tasks as a multi-objective optimization problem. We define the global optimum solution for the multiple objectives using the Karush-Kuhn-Tucker (KKT) conditions. Subsequently, we employ MGDA algorithm to solve for a set of weight coefficients for the meta-tasks. Finally, we update the initialization parameters in the direction of global optimality by applying task weighting based on these coefficients, as shown in Fig. 1. The gradient descent formula with weighting is as follows:

$$\theta \leftarrow \theta - \beta\nabla_\theta \sum_{\mathcal{T}_i\sim p(\mathcal{T})} \alpha^i\mathcal{L}_{\mathcal{T}_i}\left(f_{\theta_i'}\right) \tag{4}$$

In contrast to other task-agnostic meta-learning algorithms, our approach stands out in two key aspects. Firstly, we utilize a multi-objective optimization algorithm to define and optimize our objectives. By employing the Method of Multiple Gradient Descent (MGDA) for iterative solving, we effectively reduce the computational cost during the optimization process. This not only enhances the efficiency of our approach but also allows for better scalability. Secondly, our adoption of multi-objective optimization enables us to discover the global optimum solution for updating the meta-learner, thereby avoiding the bias that can arise from being trapped in local optima. This ensures that our meta-learner is not hindered by suboptimal solutions and maintains a broader perspective when adapting to new tasks. Furthermore, the multi-objective optimization approach reduces the requirement for a large number of parallel tasks during model training, resulting in significant reductions in training overhead. This aspect is particularly beneficial for the further advancement of meta-learning techniques, as it makes the training process more accessible and efficient.

3.2   Multi-objective Optimization Algorithm

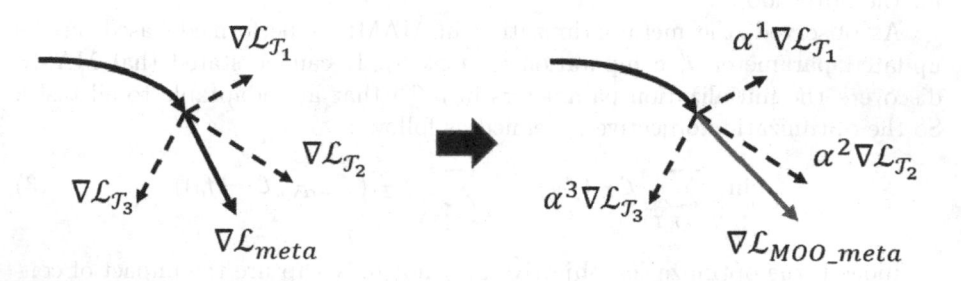

Fig. 1. The left side illustrates the principle of the MAML algorithm, which aims to optimize a parameter θ capable of quickly adapting to various tasks by optimizing across multiple tasks. On the right side, the algorithm based on multi-objective optimization assigns weights to the gradients obtained from multiple meta-tasks, enabling the updating of θ towards the direction of the global optimum. This process results in the attainment of a Pareto optimal θ that adapts effectively to all tasks.

The multi-objective optimization algorithm is capable of solving optimization problems involving conflicts and trade-offs among multiple objective functions. In our study, the multi-objective optimization algorithm is specifically employed to find a set of solutions that address the conflicts and trade-offs among the meta-tasks. The objective is to ensure that the updated meta-learner parameters achieve favorable performance levels across all individual meta-tasks. Thus, we transform the optimization objectives 3 into a vector representation, as outlined below:

$$\min_{\theta} \sum_{\mathcal{T}_i \sim p(\mathcal{T})} \mathcal{L}_{\mathcal{T}_i}\left(f_{\theta_i'}\right) \sim min_{\mathcal{T}_i \sim p(\mathcal{T})} \left(\mathcal{L}_{\mathcal{T}_1}\left(f_{\theta_i'}\right), ..., \mathcal{L}_{\mathcal{T}_T}\left(f_{\theta_T'}\right)\right)^{\top} \quad (5)$$

The set of optimal solutions in multi-objective optimization is referred to as the Pareto optimal solutions, which are not dominated by any other solutions (In other words, they are not inferior to other solutions in all objective functions). This set of solutions forms the Pareto frontier, as shown in Figure.

To determine the optimal solution for multi-objective optimization based on gradient information, we employ the MGDA (Multi-Gradient Descent Algorithm) method [3]. MGDA utilizes the KKT (Karush-Kuhn-Tucker) conditions. In the context of the current problem, the definition of these conditions is as follows:

- There exist $\alpha^1, \ldots, \alpha^T \geq 0$ and $\sum_{t=1}^{T} \alpha^t = 1$ and $\sum_{i=1}^{T} \alpha^i \nabla_\theta \mathcal{L}_{T_i} \left(f_{\theta'_i} \right)$
- For all tasks i, $\nabla_{\theta'} \mathcal{L}_{T_i} \left(f_{\theta'_i} \right) = 0$

A solution that meets either of the aforementioned conditions is known as a Pareto optimal solution. The objective of optimizing towards this goal can be represented by the following equation:

$$
\min_{\alpha^1, \ldots, \alpha^T} \left\{ \left\| \sum_{i=1}^{T} \alpha^i \nabla_\theta \mathcal{L}_{T_i} \left(f_{\theta'_i} \right) \right\|_2^2 \mid \sum_{t=1}^{T} \alpha^t = 1, \alpha^t \geq 0 \right\} \tag{6}
$$

Désidéri's research indicates that when solving this optimization problem, the solution set either satisfies the KKT conditions or provides a descent direction that improves all tasks. As a result, upon obtaining the Pareto solution for Eq. 6, we can proceed to update the initialized parameters θ, using Eq. 4 through the corresponding gradient descent. The gradient direction for updating the initialized parameters θ is adjusted through a multi-objective optimization algorithm, which leads to a Pareto optimal solution that simultaneously satisfies all tasks.

The essence of solving equation lies in searching for the point of minimum norm within a convex space. In this context, we utilize the optimized MGDA proposed by Ozan Sener to solve the optimization objectives. Specifically, the algorithm utilizes the Frank-Wolfe method to address the constrained optimization problem, with a sub-routine of 2-target optimization serving as the linear search procedure. The optimization problem with two objectives, denoted as $\min_{\alpha \in [0,1]} \left\| \alpha \nabla_\theta \mathcal{L}_{T_1} \left(f_{\theta'_1} \right) + (1 - \alpha) \nabla_\theta \mathcal{L}_{T_2} \left(f_{\theta'_2} \right) \right\|_2^2$, can be solved by the following equation:

$$
\hat{\alpha} = \left[\frac{\left(\nabla_\theta \mathcal{L}_{T_2} \left(f_{\theta'_2} \right) - \nabla_\theta \mathcal{L}_{T_1} \left(f_{\theta'_1} \right) \right)^\top \nabla_\theta \mathcal{L}_{T_2} \left(f_{\theta'_2} \right)}{\left\| \nabla_\theta \mathcal{L}_{T_1} \left(f_{\theta'_1} \right) - \nabla_\theta \mathcal{L}_{T_2} \left(f_{\theta'_2} \right) \right\|_2^2} \right]_{+, \frac{1}{T}} \tag{7}
$$

where $[.]_{+, \frac{1}{T}}$ represents clipping to $[0, 1]$ as $[a]_{+, \frac{1}{T}} = max \left(min \left(a, 1 \right), 0 \right)$. Figure 2 is a visual description of how to minimize the norm coefficient when two tasks conflict, and the solution process. The algorithm flow based on multi-objective optimization meta learning is shown in Algorithm 1.

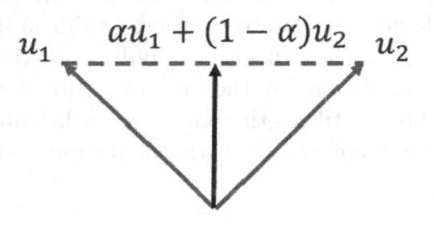

Fig. 2. The figure represents a visual depiction of Formula 7, where the coefficients are determined by solving for the minimum norm when two tasks conflict ($u_1^\top u_2 \leq u_1^\top u_1$ and $u_1^\top u_2 \leq u_2^\top u_2$).

Algorithm 1. Meta-learning based on multi-objective optimization

---

Require: $p(\mathcal{T})$:distribution over tasks
Require: α, β:step size hyperparameters
1: randomly initialize θ
2: **while** not done **do**
3:     Sample batch of tasks $\mathcal{T}_i \sim p(\mathcal{T})$
4:     **for all** $\mathcal{T}_i$ **do**
5:         Sample K samples from $\mathcal{T}_i$
6:         Evaluate $\nabla_\theta \mathcal{L}_{\mathcal{T}_i}(\mathbf{f}_\theta)$ and $\mathcal{L}_{\mathcal{T}_i}(\mathbf{f}_\theta)$ using
            K samples.
7:         Compute adapted parameters with gradient
            descent:$\theta_i' = \theta - \alpha \nabla_\theta \mathcal{L}_{\mathcal{T}_i}(f_\theta)$
8:         Compute $\mathcal{L}_{\mathcal{T}_i}(f_{\theta_i}')$ and $\nabla_\theta \mathcal{L}_{\mathcal{T}_i}(f_{\theta_i}')$ using D_{val}
            from $\mathcal{T}_i$
9:     **end for**
10:    Define $\nabla = \left(\nabla_\theta \mathcal{L}_{\mathcal{T}_1}(f_{\theta_1}'), \ldots, \nabla_\theta \mathcal{L}_{\mathcal{T}_T}(f_{\theta_T}')\right)$
11:    Compute a batch of tasks weight using multi-
       objective optimization algorithm:
       $\alpha^1, \ldots, \alpha^T = FrankWolfe(\nabla)$
12:    update $\theta \leftarrow \theta - \beta \nabla_\theta \sum_{\mathcal{T}_i \sim p(T)} \alpha^i \mathcal{L}_{\mathcal{T}_i}\left(f_{\theta_i'}\right)$
13: **end while**

14: **procedure** FRANKWOLFE(∇)
15:    Initialize $\alpha = \left(\alpha^1, \ldots, \alpha^T\right) = \left(\frac{1}{T}, \ldots, \frac{1}{T}\right)$
16:    Precompute M st.$M_{i,j} = (\nabla_i)' \nabla_j$
17:    **repeat**
18:        $\hat{t} = \arg\min_r \sum_t \alpha^t \mathbf{M}_{rt}$
19:        $\hat{\gamma} = \arg\min_\gamma ((1-\gamma)\alpha + \gamma \mathbf{e}_{\hat{t}})^\top \mathbf{M} ((1-\gamma)\alpha + \gamma \mathbf{e}_{\hat{t}})$
20:        $\alpha = (1-\hat{\gamma})\alpha + \hat{\gamma}\mathbf{e}_{\hat{t}}$
21:    **until** $\hat{\gamma} \sim 0$ **or** Number of Iterations Limit
22:    **return** $\alpha^1, \ldots, \alpha^T$
23: **end procedure**

---

4 Experiments

Our experiment aims to validate the following hypotheses: 1) Can meta-learning algorithms based on multi-objective optimization effectively attain globally optimal meta-knowledge? 2) Can multi-objective optimization algorithms demonstrate enhanced generalization performance in scenarios involving a limited number of meta-tasks?

In this section, we present a detailed description of the experimental setup, elaborate on the experimental results, and provide corresponding analysis. Furthermore, through additional evaluation, we demonstrate the efficacy and advancement of our algorithm.

4.1 Experimental Design

Datasets. We conducted our experiments using the benchmark dataset MiniIma-genet [25] and Omniglot. The Omniglot dataset consists of 1,623 unique handwritten characters representing 50 different alphabets. Each character within the dataset is drawn by 20 distinct individuals. MiniImagenet is a subset of ImageNet, consists of 60,000 color images belonging to 100 classes. Each class contains 600 samples, and the images have a resolution of 84×84 pixels. In this experiment, we divide the data into training, validation and test set with the ratio of $64/16/20$. Subsequently, we randomly sampled meta-tasks from the dataset to create the corresponding experimental setups for 5-way-1-shot, 5-way-5-shot, 20-way-1-shot, and 20-way-5-shot scenarios.

Experimental Setup. We use conventional 4-block convolutional neural networks with 32 channels [4] (It consists of four repeated convolutional blocks, each consisting of a convolutional layer, an activation function, a BatchNorm layer, and a max pooling layer. The convolution kernel size is 32*3*3, and the window size of the max pooling layer is 2*2. The final output of the classification results is obtained through a linear layer.), and set the number of gradient update steps for both meta-training and meta-testing to 5. The learning rate $\alpha = 0.001$ and $\beta = 0.01$ are set in all experiments. We conducted a dedicated comparative analysis specifically focusing on the batch size, also known as the number of meta-tasks. This analysis aimed to explore the relationship between batch size, time consumption, and accuracy. We compared the time overhead and accuracy for batch sizes ranging from 2 to 8.

4.2 Results

We have conducted relevant experiments on few shot algorithms using the Omniglot dataset. The experimental results demonstrating the favorable performance of MLMOO in the 5-way and 20-way settings on the Omniglot dataset are presented in Table 1. However, due to the relatively simplistic nature of the dataset, the observed performance enhancements may not be substantial.

Table 1. The performance of few shot classification tasks on the Omniglot dataset is assessed using specific task configurations, namely 5-way-1-shot, 5-way-5-shot, 20-way-1-shot, and 20-way-5-shot. We report mean accuracy and 95% confidence interval.

Methods	5-way		20-way	
	1-shot	5-shot	1-shot	5-shot
MANN, no conv [18]	82.8%	94.9%	–	–
MAML, no conv [4]	89.7 ± 1.1%	97.5 ± 0.6%	–	–
Matching Nets [25]	98.1%	98.9%	93.8%	98.5%
Prototypical Nets [21]	98.8%	99.7%	–	–
MAML [4]	98.7 ± 0.4%	99.9 ± 0.1%	95.8 ± 0.3%	98.9 ± 0.2%
MLMOO(ours)	**99.5±0.16%**	99.82±0.10%	**96.12±0.22%**	98.68±0.13%

Table 2. The performance of few shot classification tasks on the Mini-Imagenet dataset is assessed using specific task configurations, namely 5-way-1-shot, 5-way-5-shot, 20-way-1-shot, and 20-way-5-shot. We report mean accuracy and 95% confidence interval.

Methods	5-way		20 way	
	1-shot	5-shot	1-shot	5-shot
Fine-tune	28.86 ± 0.54%	49.79 ± 0.79%	–	–
Nearest Neighbors	43.56 ± 0.84%	55.31 ± 0.73%	17.31 ± 0.22%	22.69 ± 0.20%
Matching Nets [25]	43.56 ± 0.84%	55.31 ± 0.73%	17.31 ± 0.22%	22.69 ± 0.20%
Meta-Learn LSTM [17]	43.44 ± 0.77%	60.60 ± 0.71%	16.70 ± 0.23%	26.06 ± 0.25%
MAML [4]	48.70 ± 1.84%	63.11 ± 0.92%	16.49 ± 0.58%	19.29 ± 0.29%
Meta-SGD [11]	50.47 ± 1.87%	64.03 ± 0.94%	17.56 ± 0.64%	28.92 ± 0.35%
Prototypical network [21]	46.61 ± 0.78%	65.77 ± 0.70%	–	–
Reptile [14]	49.97 ± 0.32%	65.99 ± 0.58%	–	–
Relation Network [22]	50.44 ± 0.82%	65.32 + 0.70%	–	–
MLMOO(ours)	**52.77±1.13%**	**66.00±1.07%t**	**21.73±0.47%**	**30.81±0.42%**

When faced with the more complex Mini-Imagenet dataset, the algorithm presented in this paper outperforms other methods across all four experimental settings. Table 2 presents the experimental results, showcasing the superior performance of our proposed algorithm compared to MAML and other small-sample algorithms in both the 5-way and 20-way settings. Importantly, in the more demanding 20-way experimental setting, our proposed algorithm exhibits even more significant performance enhancements. Specifically, it achieves an impressive improvement of approximately 5% and 11% in the 1-shot and 5-shot scenarios, respectively.

We also conducted experiments to investigate the impact of batch size (task number) on the performance of meta-learning. The detailed experimental results can be found in Tables 3 and 4. The experimental results indicate that as the task number increases, the meta-learner is capable of acquiring more generalized prior knowledge. However, this improvement comes at the cost of increased training time expenses, which exhibit a linear growth pattern. In contrast, MLMOO leverages the search for a set of Pareto optimal solutions across tasks to obtain a more cohesive and generalized representation that accommodates the differences between tasks. This enables MLMOO to showcase good performance even under conditions with a minimal number of tasks.

Table 3. The training time costs of MAML and MLMOO across different task quantities were recorded in this table. The experiments were conducted on a computing cluster, where each experiment utilized one Nvidia GeForce 3090 GPU and 8 CPU cores (Intel Xeon Silver 4210R 2.4G).

	Training time(hour)						
	2 tasks	3 tasks	4 tasks	5 tasks	6 tasks	7 tasks	8 tasks
MAML	23.20	31.02	38.66	45.69	53.36	62.38	72.06
MLMOO	23.98	31.63	39.77	48.27	56.30	64.30	71.61

Table 4. MAML and MLMOO's performance in terms of accuracy across different task quantities was systematically investigated in this study. The experimental equipment is detailed in Table 3.

	5-way 1-shot Accuracy(%)						
	2 tasks	3 tasks	4 tasks	5 tasks	6 tasks	7 tasks	8 tasks
MAML	47.90±1.13%	48.22±1.13%	48.32±1.13%	47.98±1.13%	48.47±1.13%	48.53±1.13%	48.66±1.13%
MLMOO	51.25±1.13%	51.46±1.13%	51.63±1.13%	51.97±1.13%	51.64±1.13%	52.48±1.13%	52.77±1.13%

4.3   Analysis

In the aforementioned experiments, we can observe that for meta-learning to effectively learn a generalized representation for a set of related tasks, it is crucial to induce and summarize knowledge from a larger number of tasks. This approach ensures that the acquired prior knowledge is more representative and avoids biases towards specific tasks.

According to Tables 3 and 4, it can be observed that increasing the number of meta-tasks effectively enhances the generalization of meta-knowledge. However, the linear increase in time costs makes it more challenging to elevate the complexity of the backbone network, thereby restricting the application of meta-learning in various domains. However, Table 4 demonstrates that even under fewer meta-task conditions, MLMOO exhibits stronger capabilities and is less affected by the number of meta-tasks. This indicates that MLMOO effectively addresses the potential conflicts and imbalances among meta-tasks through multi-objective optimization algorithms, enabling the meta-learner to acquire better meta-knowledge.

In summary, MLMOO aims to find the optimal solution for a set of tasks, which closely approximates the essence of prior knowledge obtained from learning across a large number of tasks. This approach allows MLMOO to capture a more comprehensive and representative generalized representation. Extensive ablation experiments presented in Tables 1 and 2 further validate this viewpoint, indicating that we can indeed learn superior prior knowledge through multi-objective optimization. Importantly, this approach does not necessarily require an increase in the number of tasks, thereby enhancing the potential applications of meta-learning across various domains.

5 Conclusion

We present a meta-learning approach based on multi-objective optimization(MLMOO), which achieves the Pareto optimal meta-learner by finding the optimal gradient update direction for multiple meta-tasks, effectively mitigating the adverse effect of task bias on the meta-learner. We note that even with small batch size, the meta-learner based on multi-objective optimization algorithm can improve in a favorable direction and achieve a solution with high generalization. This allows the complexity of the basic modules of meta-learning algorithms to be higher. Experiments also indicate that MLMOO exhibits superior ability in generalizing and summarizing meta-knowledge compared to previous meta-learning approaches. It is capable of acquiring more generalized prior representations across tasks of varying difficulty levels. We firmly believe that our work has contributed meaningful insights to the advancement of meta-learning.

Acknowledgements. This work was supported by the UESTC-ZHIXIAOJING Joint Research Center of Smart Home (H04W210180), Neijiang Technology Incubation and Transformation Funds (2021KJFH004), the Science and Technology Program of Sichuan Province, China (2022YFG0212, 2021YFG0024).

References

1. Antoniou, A., Edwards, H., Storkey, A.: How to train your MAML. arXiv preprint arXiv:1810.09502 (2018)
2. Cui, Y., Geng, Z., Zhu, Q., Han, Y.: Multi-objective optimization methods and application in energy saving. Energy **125**, 681–704 (2017)
3. Désidéri, J.A.: Multiple-gradient descent algorithm (MGDA) for multiobjective optimization. C.R. Math. **350**(5–6), 313–318 (2012)
4. Finn, C., Abbeel, P., Levine, S.: Model-agnostic meta-learning for fast adaptation of deep networks. In: International Conference on Machine Learning, pp. 1126–1135. PMLR (2017)
5. Fliege, J., Vaz, A.I.F.: A method for constrained multiobjective optimization based on SQP techniques. SIAM J. Optim. **26**(4), 2091–2119 (2016)
6. Hernández-Lobato, D., Hernandez-Lobato, J., Shah, A., Adams, R.: Predictive entropy search for multi-objective Bayesian optimization. In: International Conference on Machine Learning, pp. 1492–1501. PMLR (2016)

7. Jamal, M.A., Qi, G.J.: Task agnostic meta-learning for few-shot learning. In: Proceedings of the IEEE/CVF Conference on Computer Vision and Pattern Recognition (CVPR) (2019)

8. Lee, H.B., et al.: Learning to balance: Bayesian meta-learning for imbalanced and out-of-distribution tasks. In: International Conference on Learning Representations (2020). https://openreview.net/forum?id=rkeZIJBYvr

9. Lee, S., Son, Y.: Multitask learning with single gradient step update for task balancing. Neurocomputing **467**, 442–453 (2022)

10. Li, X., Yu, L., Jin, Y., Fu, C.-W., Xing, L., Heng, P.-A.: Difficulty-aware meta-learning for rare disease diagnosis. In: Martel, A.L., et al. (eds.) MICCAI 2020. LNCS, vol. 12261, pp. 357–366. Springer, Cham (2020). https://doi.org/10.1007/978-3-030-59710-8_35

11. Li, Z., Zhou, F., Chen, F., Li, H.: Meta-SGD: learning to learn quickly for few-shot learning. arXiv preprint arXiv:1707.09835 (2017)

12. Lin, X., Zhen, H.L., Li, Z., Zhang, Q., Kwong, S.: Pareto multi-task learning. In: Proceedings of the 33rd International Conference on Neural Information Processing Systems, pp. 12060–12070 (2019)

13. Naik, D.K., Mammone, R.J.: Meta-neural networks that learn by learning. In: [Proceedings 1992] IJCNN International Joint Conference on Neural Networks, vol. 1, pp. 437–442. IEEE (1992)

14. Nichol, A., Achiam, J., Schulman, J.: On first-order meta-learning algorithms. arXiv preprint arXiv:1803.02999 (2018)

15. Oh, J., Yoo, H., Kim, C., Yun, S.: BOIL: towards representation change for few-shot learning. In: 9th International Conference on Learning Representations, ICLR 2021, Virtual Event, Austria, May 3–7, 2021. OpenReview.net (2021). https://openreview.net/forum?id=umIdUL8rMH

16. Raghu, A., Raghu, M., Bengio, S., Vinyals, O.: Rapid learning or feature reuse? Towards understanding the effectiveness of MAML. In: International Conference on Learning Representations (2020). https://openreview.net/forum?id=rkgMkCEtPB

17. Ravi, S., Larochelle, H.: Optimization as a model for few-shot learning. In: International Conference on Learning Representations (2017). https://openreview.net/forum?id=rJY0-Kcll

18. Santoro, A., Bartunov, S., Botvinick, M., Wierstra, D., Lillicrap, T.: Meta-learning with memory-augmented neural networks. In: International conference on machine learning, pp. 1842–1850. PMLR (2016)

19. Sener, O., Koltun, V.: Multi-task learning as multi-objective optimization. In: Proceedings of the 32nd International Conference on Neural Information Processing Systems, pp. 525–536 (2018)

20. Shah, A., Ghahramani, Z.: Pareto frontier learning with expensive correlated objectives. In: International conference on machine learning, pp. 1919–1927. PMLR (2016)

21. Snell, J., Swersky, K., Zemel, R.: Prototypical networks for few-shot learning. In: Advances in Neural Information Processing Systems, vol. 30 (2017)

22. Sung, F., Yang, Y., Zhang, L., Xiang, T., Torr, P.H., Hospedales, T.M.: Learning to compare: relation network for few-shot learning. In: Proceedings of the IEEE Conference on Computer Vision and Pattern Recognition, pp. 1199–1208 (2018)

23. Thrun, S., Pratt, L.: Learning to learn: introduction and overview. In: Thrun, S., Pratt, L. (eds.) Learning to Learn, pp. 3–17. Springer, Boston (1998). https://doi.org/10.1007/978-1-4615-5529-2_1

24. Van Moffaert, K., Nowé, A.: Multi-objective reinforcement learning using sets of pareto dominating policies. J. Mach. Learn. Res. **15**(1), 3483–3512 (2014)
25. Vinyals, O., Blundell, C., Lillicrap, T., Wierstra, D., et al.: Matching networks for one shot learning. In: Advances in Neural Information Processing Systems, vol. 29 (2016)
26. Von Oswald, J., et al.: Learning where to learn: gradient sparsity in meta and continual learning. Adv. Neural. Inf. Process. Syst. **34**, 5250–5263 (2021)
27. Xu, H.: Research on Multiobjective Particle Swarm Optimization Algorithms. Shanghai Jiao Tong University, Shanghai (2013)

Research and Implementation of Cooperative Utilization of Network Security Devices for Typical Security Events

Jia Du[✉], Junqiang Yang, Wanli Kou, and Huaizhe Zhou

Test Center, The University of Defense Technology, Xi'an 710106, Shanxi, China
dujia0227@126.com

Abstract. Aiming to improve the efficiency of network security protection, the collaborative utilization of network security devices is studied. Firstly, the utilization status quo of network security devices was analyzed. Then, the system architecture is designed, and the main modules and capabilities of each layer which shows the work flow of the system are described. Finally, the collaborative utilization process of the functional verification is designed and carried out for security vulnerability events. The verification results show that the automatic collaborative utilization system designed can effectively realize the collaborative detection and collaborative disposal of a variety of network security devices, and provide ideas and means for systematic network security protection.

Keywords: Network Security Protection · Cooperative Detection · Cooperative Disposal · Automatic Collaboration · Human-machine Collaboration

1 Introduction

Aiming to address the challenges of network threat detection and security incident disposal, this study focuses on the automatic collaborative utilization of security devices [1]. Firstly, an analysis is conducted on utilizations status of security devices. Then, a collaborative utilization system of network security devices is designed based on system architecture design, technical architecture design, and workflow design. Finally, the capability verification of this collaborative system is performed through vulnerability events. The verification results demonstrate that the collaborative utilization system designed enables the collaboration in detection and disposal among different network security devices effectively, and enhances the efficiency of network security protection.

2 Status Quo of Network Security Devices

The network security protection system mainly is composed of network security, computing security, utilization security, data security and security management. Network security protection capability depends on a variety of network security devices, but with which there are the following problems.

2.1 Low Accuracy of Network Threat Detection

The situation awareness platform collects alarms from network security devices, and the types of security event are mainly analyzed by technical personnel. It is urgent to conduct the association analysis for various alarms and network threat [2], which can detect network threat automatically or by human-in-the-loop, so as to improve the detection efficiency of security events.

2.2 Low Disposal Efficiency of Network Security Devices

There are many security devices deployed in the network security protection system. When dealing with security incidents, it is necessary to configure each network security device. It is urgent to study the script of security incident disposal [3, 4], with which can handle security incidents automatic or semi-automatic by the execution of the script.

2.3 The Difficulty in Collaborative Utilization of Network Security Devices

There are many kinds of network security devices deployed currently, and the problems of segmented protection and single point defense are prominent. The collaborative effect between security devices is not obvious and the overall defense capability is weak. It is urgent to carry out multi-device linkage through correlation analysis and collaborative disposal [5–8], which can improve the efficiency of network security protection.

3 The Design of Automatic Collaborative Utilization System of Network Security Devices

Based on the analysis of the status quo of network security devices, an automatic collaborative utilization system of network security devices is designed, which is composed of device centralized management level, event collaborative detection level, disposal policy construction level and event collaborative response level, as shown in Fig. 1.

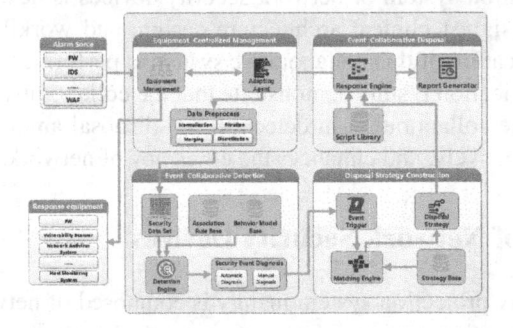

Fig. 1. The architecture of automatic collaborative utilization system of network security devices.

3.1 Device Centralized Management Layer

The model provides device performance, configuration, connection and other information, and generates a unique device resource vector for each device to achieve unified management for registered devices.

The security data is received from the registered devices by calling different protocol agents and format parsers, and sent to the data preprocess module. When the cooperative response flow is received from event cooperative disposal layer, the adapting agent module controls the security devices registered.

The security data from the adapting agent module is processed to generate a standardized reduced security data set in data preprocess module, which uploads the security data set to event correlative detection layer.

3.2 Event Collaborative Detection Layer

The association rule base of known security events is constructed manually, and the abnormal behavior models of unknown security events are established by using machine learning technology.

Based on the association rule base and the abnormal behavior model base, the data in the security data set is detected to determine whether there is a security event.

3.3 Disposal Policy Construction Layer

The security events are received from the event collaborative detection layer, and activate the matching engine module through event trigger module.

The collaborative disposal strategy is quickly matched from the strategy base module and uploaded to collaborative disposal layer.

Through the strategy base module, the model, the incremental storage, and the rapid query of cooperative disposal policy for 10 kinds of typical network security events (such as computer virus events, worm events, and Trojan horse events) are realized.

3.4 Event Collaborative Disposal Layer

For 10 kinds of typical network security events, response scripts containing a series of disposition actions are generated based on cooperative disposal strategies. The response script is transformed into the operation command set, and sent to the adapting agent module of device centralized management layer. The incident handling processes and results are summarized and archived.

4 The Collaborative Utilization Verification of Network Security Devices

Based on technical research of device centralized management model, event collaborative detection model, disposal policy construction model and event collaborative response model above, a prototype system is built and the collaborative utilization function is verified.

4.1 Event Description

The test network is mainly divided into three areas: technical support area (1.1.1.0/24), utilization service area (4.4.4.0/24) and operation area (3.3.3.0/24). The collaborative utilization system of network security devices, host monitoring server, anti-virus server and traffic analysis server are deployed in technical support area. Web server, ftp server and data server are deployed in utilization service area. A number of PCs are deployed in operation area. Firewalls and WAF implement access control between different areas. Intrusion Detection System (IDS) and vulnerability scanning system access core switches to detect data flowing through the entire network. When the warnings of high risk vulnerability are released and the remote overflow attack is happened, serval PCs and servers were affected by this attack, which results in abnormal use of the information system. In order to find out the causes of the accident and restore the information network as soon as possible, the network security devices are used to cooperate with the collaborative utilization system of network security devices to detect the type of security event and carry out event disposal. The topology of the whole network is shown in Fig. 2.

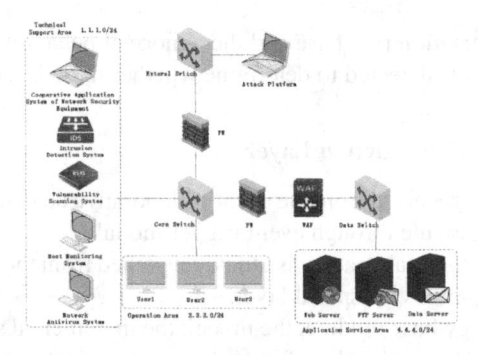

Fig. 2. The network topology.

4.2 The Process of Collaborative Utilization

The process of collaborative utilization is consisted of alarm collection, collaborative detection and collaborative disposal, as shown in Fig. 3.

1. Alarm Collection

Based on the underlying plug-in by using API interfaces, real-time security data can be automatically collected from IDS, anti-virus system, EDR and other security devices, and the original multi-source security data can be converted to standardized alarm information according to the pre-processed rules and alarm receiving rules of the collaborative utilization system.

2. Collaborative Detection

The system automatically performs association analysis on pre-processed standardized alarm information, matches the alarm types, and obtains the simplest safety data set by merging the associated alarm information.

Every once in a while, the alarm information in the minimal security data set is matched the workflow in detection engine automatically, and the vulnerability event is detected through the automatic execution of the event detection workflow.

In order to improve the accuracy of event diagnosis, human-machine interface is used to implement manual decision-making after automatic detection.

3. Collaborative Disposal

(1) Policy matching

According to the event type detected and related information obtained, the policy matching mechanism is activated. The best matching cooperative disposal policy is retrieved from the cooperative disposal strategy base automatically to determine the cooperative disposal action for the event.

(2) Report the situation

The event type and related alarm information detected are extracted automatically, and the disposal report is sent to the relevant units manually.

(3) Emergency response

By invoking the executable script in disposal strategy base, the API interface of the firewall is invoked automatically to block the IP and the network attack path in the first time.

(4) Tracing the source

According to analysis and judgment of vulnerability information and the related alarms collected by the system, the geographical location of attacks is filled in and the threat intelligence of the source IP is searched manually. Then the hacker portrait of the source IP is completed to realize event traceability.

(5) Attack clearance

Security vulnerabilities are removed by the way of man-machine combination. Firstly, the vulnerability scanning task is set manually. Then, the vulnerability scanning system is called to automatically obtain the vulnerability scanning task and view the vulnerability scanning results through the executable script in script library. Finally, the host IP of the vulnerability repaired is confirmed manually.

(6) System repair

The database of the vulnerability scanning system is manually upgraded after confirming the removal of the attack risks. Through the executable script in the script library, the API interface of the firewall is called automatically to unlock the IP and repair the network connection of the damaged terminal.

(7) Disposal assessment

After the collaborative disposal to the vulnerability event, the disposal report is generated and reported to the superior department automatically. The use of personnel, the utilization of protection means, the implementation of security strategy, the scanning of vulnerability and the repair of vulnerability are evaluated. The collaborative disposal report of harmful vulnerability events can also be sent to domain experts on demand and the comprehensive evaluation and feedback of incident disposal schemes can be realized.

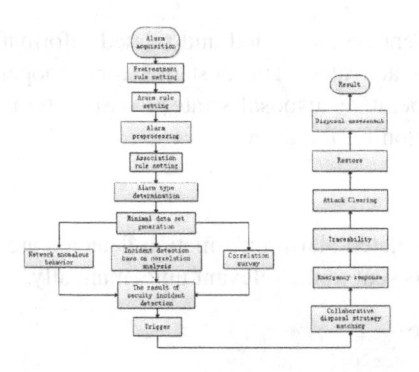

Fig. 3. The verification process of collaborative utilization system.

4.3   The Comparison of Collaborative Utilization Efficiency

Compared with the traditional network security system, the automatic collaborative utilization of network security devices has significant advantages in centralized management of security devices, collaborative detection and collaborative treatment of security events, which can effectively improve the efficiency of network security protection, as shown in Fig. 4.

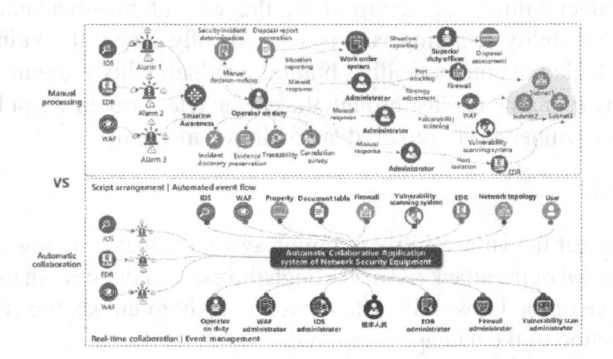

Fig. 4. The efficiency comparison of two network security system.

5 Conclusion

Aiming at the practical problem of insufficient collaborative utilization of network security devices, which cause low efficiency of network threat detection and security event disposal, a collaborative utilization system of network security devices is designed and the verification was carried out for typical network security events. The verification results show that the collaborative utilization system of network security devices proposed can realize collaborative detection and collaborative disposal among different network security devices effectively, and promote the improvement of systematic protection efficiency [5].

References

1. Du, J., Yang, J., Wang, Z., Song, X.: Research on automatic cooperative utilization of network security devices. In: International Conference on Robots & Intelligent System, pp. 449–452. IEEE-CPS, China (2020)
2. Yang, Y., Yan, Y., Shen, F., Yuheng, G.: Review of intrusion detection based on machine and deep learning. Sci. Technol. Eng. 23(18), 7607–7621 (2023)
3. Yan, Z., Yuanbo, G.: SOAR cybersecurity emergency response case revision based on bilinear rule learning. J. Inf. Eng. Univ. 24(4), 484–491 (2023)
4. Xiaoxiao, C., Peng, Z., Jingwei, Q., Huikai, Z.: Research on key technology construction and future evolution direction of security operation automation based on SOAR. Ind. Technol. Innov. 5(3), 82–84 (2023)
5. Feng, Z.: Research and utilization on network security device linkage policy. Hebei: School of Control and Computer Engineering (2014)
6. Liu, H.: Research and design on the log pretreatment of network security devices linkage system. Hebei: School of Control and Computer Engineering (2014)
7. Cao, L.: Research and utilization of event correlation model in network security devices linkage system. Hebei: School of Control and Computer Engineering (2014)
8. Wang, S.: Research and utilization of access control based on policy in network security devices linkage system. Hebei: School of Control and Computer Engineering (2015)

Application of T-S Fuzzy Neural Network
in Water Quality Classification and Evaluation

Lingyu Song[1], Yifu Sheng[1], Hao Li[1], Jianjun Zhang[1(✉)], Ziwen Yu[1], Enling Zhou[1],
Haijun Lin[1], and Ziqiu Zhang[2]

[1] College of Engineering and Design, Hunan Normal University, Changsha, China
87890878@qq.com
[2] LIHERO Technology (Hunan) Co., Ltd., Changsha 410205, China

Abstract. Water quality classification and evaluation is an important part of water resources protection and development. In order to further improve the accuracy and convenience of water quality evaluation classification, combined with the strong fuzzy information learning ability of T-S fuzzy neural network, the water quality monitoring data of a monitoring station in 2019 was used as the training set and validation set, and the 2020 data was used as the testing set, The T-S fuzzy neural network was used to construct a classifier. In view of the problem of data imbalance, The SMOTE oversampling method was used to generate data, and the model effects obtained by training before and after data balance were compared and analyzed. Finally, the trained model was used to classify and evaluate all the stations in the monitoring station and the results were analyzed. Experimental results showed that T-S fuzzy neural network had good results in water quality classification and evaluation.

Keywords: T-S fuzzy neural network · water quality monitoring data · water quality classification · water quality evaluation

1 Introduction

The earliest water quality evaluation research in the world can be traced back to the 1910s. German scientist Kirkewitz and others proposed the water quality evaluation method [1]. However, this method has major limitations and only conducts simple water quality evaluation classification, definition of water quality evaluation is not made.

In recent years, with the rapid development of artificial intelligence and neural networks, an increasing number of scholars and researchers have endeavored to apply algorithms such as neural networks to the field of aquatic ecological environments. Ni Shenhai measured six indexes for the underground self-supply water wells in Feicheng City and used the measured data to establish a BP neural network model, which improved the evaluation accuracy compared to traditional methods [2]. However, it was prone to local convergence and had a longer training time. In view of the limitations of the traditional BP algorithm, many scholars have proposed improvement ideas. Huang Shengwei, for example, applied an Adaptive Variable Learning Rate algorithm (ABPM) to enhance

H. Jin et al. (Eds.): IAIC 2023, CCIS 2058, pp. 340–352, 2024.
https://doi.org/10.1007/978-981-97-1277-9_26

the traditional BP neural network using monitoring data from the Dawen River basin. They compared the evaluation results of the ABPM neural network with those of the traditional BP neural network and single-factor evaluation results. The analysis showed that the ABPM neural network's evaluation results are more scientific, with improved convergence speed and prediction accuracy [3]; Feng Dongqing combined the genetic algorithm and the neural network to propose the GA-BP algorithm to evaluate the impact plain of the middle reaches of the Yellow River [4], the model to some extent resolved the issue of local convergence. However, the limited sample size and uneven class distribution in water quality data hindered the genetic algorithm-optimized BP neural network from achieving the expected performance.

The T-S fuzzy neural network is a type of network that combines neural networks with fuzzy mathematics. It is a network with strong learning capabilities for handling fuzzy information, possessing robust adaptability and approximation abilities [5]. It retains the principles of fuzzy logic while also incorporating the learning capabilities of neural networks, making it suitable for handling various unclear concepts in real-world scenarios. Currently, researchers primarily apply T-S fuzzy neural networks in areas such as electricity forecasting [6], fuzzy control [7], economic research [8], among others, and have achieved promising results. Building upon this, this paper proposed a water quality classification evaluation model based on the T-S fuzzy neural network, and applied it to the classification and evaluation of water quality data from various sites in the monitoring station. It greatly simplified the workflow of water quality assessment and had achieved a good classification accuracy.

2 Data Set Construction and Data Preprocessing

This paper focused on a certain river basin and downloaded data from the National Water Quality Monitoring Platform for the years 2019–2020. The data includes information from 12 monitoring stations in this river basin, namely Qinglan ditch, Banister weir next to Jiannan Avenue, Jiangan River Shuangnan Road Bridge, Jinjiang Shuangba Bridge, Jinjiang Xia Jiatuo Exit, Entrance of Huanglong River in Luxi River, Guihuayan Community Sanshiyan Entry, Jinjiang Yong'an Bridge, Huangyan River Entrance, Jiang'an River Jiannan Avenue Bridge, Jinmahe Seven Branches and Seven Dous Entry, and Yellow Yan River Exit. The original data collected by the water quality monitoring platform includes 8 indexes. The collection was carried out at a frequency of sampling every 2 h to ensure the comprehensiveness of the data to the greatest extent. Some water quality data are shown in Fig. 1.

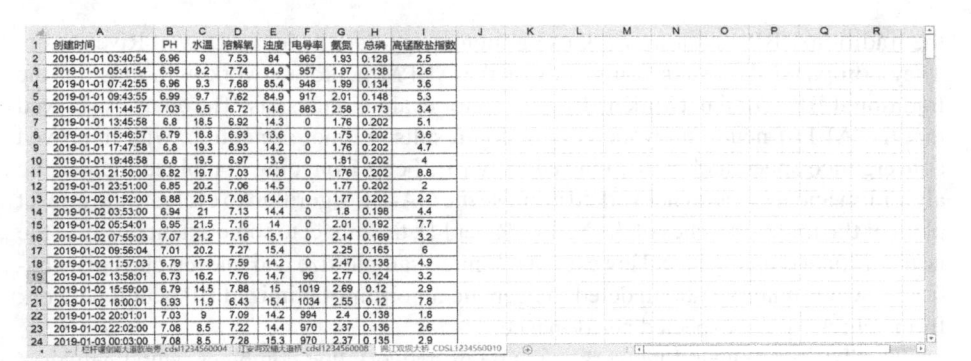

Fig. 1. Partial data from the monitoring site

2.1 Data Indicator Selection

Comparing the collected data set with the "Surface Water Environmental Quality Standard (GB3838–2002)", in the 8 water quality indexes, turbidity and conductivity have no corresponding items in the given standards. Although pH and water temperature have corresponding items in the standards, they are range indexes without specific grade dimensions and cannot be used as classification indexes. Therefore, the data used in the subsequent evaluation and classification experiments only retained the four indexes of dissolved oxygen (DO), permanganate index (COD_{MN}), ammonia nitrogen (NH_3-N), and total phosphorus (P).

2.2 Data Preprocessing

Data preprocessing has two main purposes, one is to improve data quality, and the other is to make the data better adapt to specific mining needs. The data is processed as follows according to actual needs: data cleaning, sample category labeling, and data set division.

The data source of the online monitoring platform relies on water quality sensing equipment deployed on-site at water quality monitoring stations in various river basins. Due to some uncontrollable reasons (such as power outages, equipment failures, etc.), there may be missing and abnormal data, necessitating a data cleaning process. In the data cleaning process, the two major Python data analysis libraries, pandas and numpy, were used for processing.

T-S fuzzy neural network is a supervised learning network. It needs to know the clear category of each sample data in advance to learn the relationship between pure numerical samples and water quality grades. However, real-time data collected by water quality monitoring platforms only provide numerical values for various indexes and do not include clear category labels. Therefore, it is necessary to annotate the dataset with category labels. The fuzzy comprehensive evaluation method was selected to label the data set.

In order to better train and test the model, the data of the 12 sites was divided into two parts. The first part was used to train and test the T-S fuzzy neural network, and the second part was used as the T-S neural network model classification data set. The first part of the data was based on two years of data from Jinjiang Shuangba Bridge Station. The data in 2019 was used as the training and validation set of the model, and the data in 2020 was used as the test set. The second part taked the data of the remaining 11 sites for subsequent classification by the model.

3   Construction and Evaluation of T-S Fuzzy Neural Network Model

In 1985, Takagi and Sugeno jointly proposed the fuzzy system T-S model, which can utilize the membership function to linearize the system, effectively approximating any nonlinear function [9], thereby describing complex processes or systems. The greatest advantage of this model lies in its ability to determine system parameters and order through fuzzy system estimation, resulting in higher overall system accuracy.

3.1   Network Model Construction

The model construction of a T-S Fuzzy Neural Network begins with establishing the structure based on the dimensions of the input samples and the output results. In this research, the primary focus is on four parameters: permanganate index, total phosphorus, ammonia nitrogen, and dissolved oxygen. Therefore, the network structure was configured with 4 input nodes. The output was the single quantity of the grade of the water quality, so the output node was set to 1. The layer of membership functions corresponded to two language variables for each input. Thus, the number of nodes in the rule layer was set to twice the number of input nodes, resulting in 8 nodes. The model network, developed based on the actual sample features, is depicted in Fig. 2. Among them, "s" represents accumulation, "p" represents cumulative multiplication, "•" represents logical operators, "NL" and "PL" represent two fuzzy language variables, and "p_j^i" represents fuzzy system parameters.

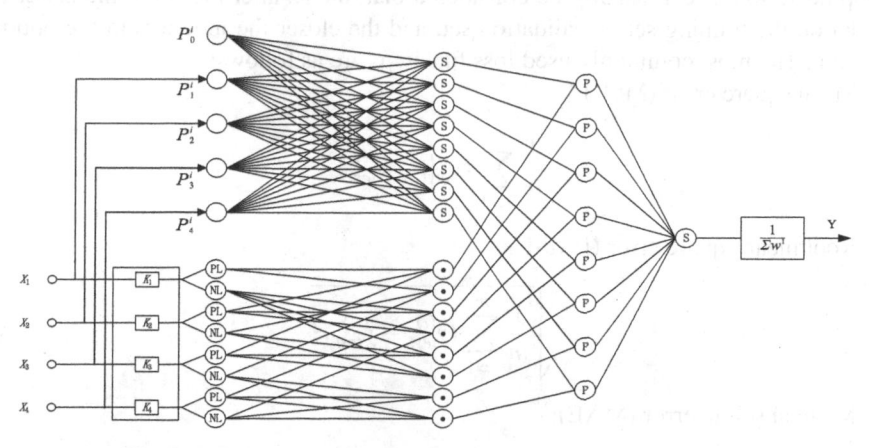

Fig. 2. T-S fuzzy neural network structure with four input nodes

The parameter learning algorithm used backpropagation to specifically describe a single neuron. The q-th layer and the j-th neuron of the fuzzy neural network are as shown. The activation function used the sigmoid function, and the single-point structure is shown in Fig. 3.

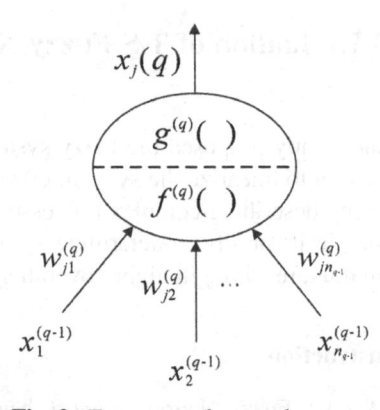

Fig. 3. Fuzzy neural network neurons

From the above figure, it can be seen that the input of a single neuron node is $f^{(q)} = \sum_{i=1}^{n_q-1} \omega_{ji}^{(q)} x_i^{q-1}$, the output is $x_j^{(q)} = g^{(q)}(f^{(q)})$. According to the sigmoid activation function, it can be obtained $x_j^{(q)} = \dfrac{1}{1+e^{-\sum_{i=1}^{n_q-1} \omega_{ji}^{(q)} x_i^{q-1}}}$.

3.2 Model Effect Evaluation

There are many indicators to judge the effectiveness of the model [10]. Error is the goal of optimization. It can usually be considered that, the smaller the error, the better the model on the training set or validation set, and the closer the model is to the optimal solution. The most commonly used loss functions are as follows:

Mean square error (MSE):

$$\frac{1}{m} \sum_{i=1}^{m} (y_{test}^{(i)} - \hat{y}_{test}^{(i)})^2 \tag{1}$$

Root mean square error (RMSE):

$$\sqrt{\frac{1}{m} \sum_{i=1}^{m} (y_{test}^{(i)} - \hat{y}_{test}^{(i)})^2} \tag{2}$$

Mean absolute error (MAE):

$$\frac{1}{m} \sum_{i=1}^{m} |y_{test}^{(i)} - \hat{y}_{test}^{(i)}| \tag{3}$$

In the above formula, "m" represents the total number of samples, "$y_{test}^{(i)}$" represents the true value of the sample, and "$\hat{y}_{test}^{(i)}$" represents the predicted value.

In the water quality classification evaluation, the most commonly used evaluation method of T-S fuzzy neural network are the classification accuracy and training error fitting curve to judge the quality of the model [11, 12]. Therefore, this will be used as the standard for testing the model in subsequent experiments.

4 Water Quality Classification Evaluation Based on T-S Fuzzy Neural Network

4.1 Neural Network Training

The T-S fuzzy comprehensive evaluation method was used to evaluate and label the data of Jinjiang Shuangba Bridge Station in 2019 and 2020, and the water quality category of the data sample was obtained. Then the 2019 data was used as the experimental sample. First, the 2019 data set was randomly shuffled, and then divided into a training set and a validation set in a ratio of 8:2.

MATLAB R2018a software was used for programming and training. The maximum number of iterations (maxgen) was set to 200, and the learning rate (rate) was set to 0.01. The rounding-off method was applied to ensure that the output corresponds to specific water quality levels, represented as I to V.

The T-S fuzzy neural network was model trained using the divided training data, and the loss curve is depicted in Fig. 4. From the curve, it can be observed that the T-S fuzzy neural network converged to its optimal state after approximately 40 iterations, with a loss of around 0.14. The loss curve did not converge to a very satisfactory level.

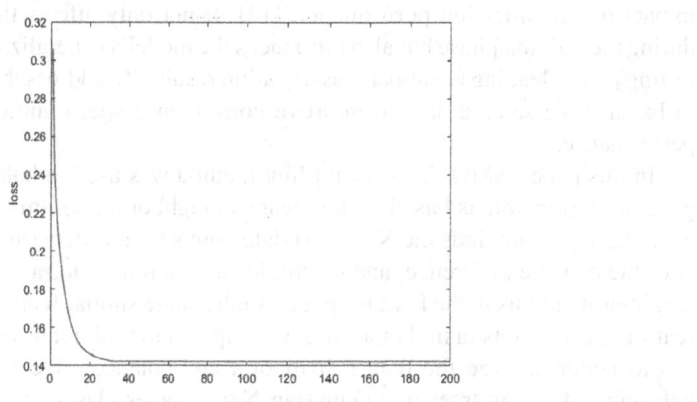

Fig. 4. Training loss curve chart

The fitting effect of the training set and the verification effect of the validation set are shown in Table 1.

Table 1. Accuracy table of measured data training set and validation set

Sample type	Correct classification number	Total sample size	Classification accuracy
Training set	247	292	84.58%
Validation set	53	73	72.60%

As can be seen from Table 2, the model trained by the 2019 Jinjiang Shuangba Bridge Station data set did not achieve ideal results. To this end, we analyzed the composition of the 356 sample data used in training the model in 2019, as shown in Tables 2, 3 and 4. Among the sample data, the number of Class III water samples reached 244, accounting for a huge proportion. The number of Class I, II, IV, and V water quality samples were 16, 52, 21, and 30 respectively, accounting for only a small part of the entire sample data. This may be the reason for the unsatisfactory model effect.

Table 2. Number of water quality samples of various types throughout the year

Water quality category	I	II	III	IV	V
Sample size	16	54	244	21	30

4.2 Data Balancing Processing

In neural networks, the imbalance of data sample categories can have a significant impact on classification performance [13]. It not only affects the convergence speed during the training phase but also influences the model's generalization ability during the testing phase, leading to subpar classification results. To address this issue, it's necessary to balance the sample data to improve convergence speed and enhance classification performance.

In this paper, SMOTE oversampling method was used to balance the data. Its data generation principle is based on the nearest neighbor algorithm [14]. When generating data, the algorithm finds the K nearest data points for any data sample, randomly selects N of them as the difference, and multiplies it by a threshold ranging from 0 to 1. Since neighboring points in the feature space usually share similar features, sampling from the feature space results in higher accuracy compared to traditional sampling methods [15].

To better analyze the data distribution and balance effects before and after data balancing, decision trees and Gaussian Naive Bayes classifiers were used as tools to compare the performance of the original class I data and the data sampled by the SMOTE algorithm using ROC curves and AUC values as criteria. The results are shown in Figs. 5 and 6. From the graphs, we can observe that the AUC values for the original data in both classifiers are 0.77 and 0.87, with an average AUC of 0.82. When using SMOTE algorithm for oversampling, the AUC values for the sampled data in both classifiers are 0.96 and 0.95, with an average AUC of 0.955. Additionally, it is evident from the graphs that the AUC area of the SMOTE-sampled data is greater than that of the original data, indicating better overall performance and robustness. This ensured that each class had a relatively equal representation, resulting in sparser and more evenly distributed data.

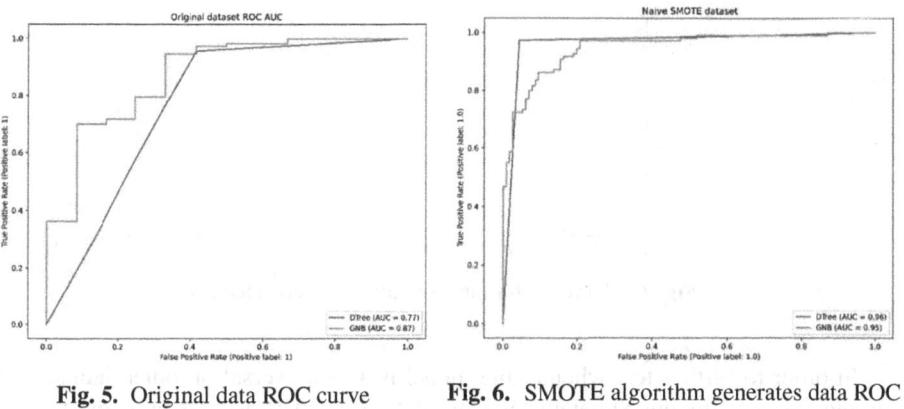

Fig. 5. Original data ROC curve   **Fig. 6.** SMOTE algorithm generates data ROC curve

By analyzed the effect of SMOTE algorithm on data generation, we can see that the SMOTE algorithm can achieve the purpose of balance data. To this end, the data of class II, IV and V are generated by SMOTE algorithm one by one, so as to balance the data.

4.3   Water Quality Classification Evaluation After Data Enhancement

Firstly, the randperm function was used to upset all the indexes of the generated data set, so as not to let some accidental rules of the data affect the objectivity of the model. The balanced data was divided into training set and validation set according to the ratio of 8:2, with 975 samples for training set and 245 samples for validation set. Secondly, the number of iterations was set to 200 and the learning rate was set to 0.01. The loss iteration curve obtained by model training is shown in Fig. 7. It can be seen from the figure that the loss curve trained with the balanced data converges faster. The T-S fuzzy neural network tends to converge to the best state after about 20 iterations, and the loss value is also significantly improved, approaching to 0 indefinitely.

The 975 samples of the training set were substituted into the trained model for fitting. The model fitting had achieved good results, with the classification accuracy reached 96.20%. The 245 validation set samples were reserved in advance for model testing, and the classification accuracy reached 90.61%.

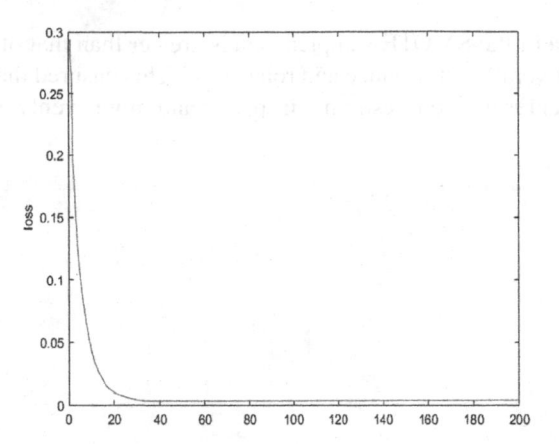

Fig. 7. Curves of training loss after data enhancement

In order to further test whether the model is also universal on other data sets, the sample data of Jinjiang Shuangba Bridge Station in 2020 was taken as the test data set, and 15 samples of class I to V were randomly selected to test the classification effect of the model trained by T-S fuzzy neural network on each class of samples. The accuracy of classification results is shown in Table 3. The T-S fuzzy neural network model correctly classified 68 samples, with a classification accuracy of 90.66%, and most of the sample points could be accurately classified successfully, proving that after sample data enhancement and class balancing, the trained model has stronger generalization ability than the previous training model with unbalanced data, and its classification is more accurate. Therefore, T-S fuzzy neural network can learn the characteristics of all kinds of water quality sample data well, so that it can make more accurate prediction to the new data.

Table 3. Precision table of balanced data training set, validation set and test set

Sample type	Correct classification number	Total sample size	Classification accuracy
Training set	938	975	96.20%
Validation set	222	245	90.61%
Testing set	68	75	90.66%

The experimental results showed that the classification accuracy of training set, validation set and test set was greatly improved, which showed that balancing data categories was an effective method.

4.4   Classification Evaluation of 12 Sites in the Basin

After verifying the performance of the T-S fuzzy neural network trained on the balanced data, the model was applied to the data of all stations to uniformly evaluate the classification of 12 stations in the monitoring station.

The data of 12 stations were sampled according to the monthly average, and then the sampled data was used to read the model by load based on the T-S fuzzy neural network model obtained in the previous training for classification and evaluation. The specific monthly water quality of 12 stations in the monitoring station in 2019 was obtained, as shown in Table 4.

Table 4. Types of water quality at various stations in the monitoring station in 2019

Monitoring station	Month											
	1	2	3	4	5	6	7	8	9	10	11	12
Qinglan ditch	V	V	V	V	IV	IV	III	II	II	II	II	I
Banister weir next to Jiannan Avenue	V	V	V	IV	II	III	IV	III	III	I	III	II
Jiangan River Shuangnan Road Bridge	II	II	II	II	II	I	I	I	I	I	I	I
Jinjiang Shuangba Bridge	II	II	III	V	V	III	III	III	III	III	III	II
Jinjiang Xia Jiatuo Exit	V	V	V	V	V	V	V	II	III	III	II	II
Entrance of Huanglong River in Luxi River	II	III	V	III	II	II	III	II	II	II	II	III
Guihuayan Community Sanshiyan Entry	I	II	V	V	V	V	II	II	IV	I	I	V
Jinjiang Yong'an Bridge	II	II	II	II	II	II	III	III	II	II	I	II
Huangyan River Entrance	V	IV	V	II	III	II	II	IV	IV	III	III	III
Jiang'an River Jiannan Avenue Bridge	I	III	I	I	II	I	II	II	II	I	II	II
Jinmahe Seven Branches and Seven Dous Entry	V	V	V	V	V	V	II	II	I	I	V	V
Yellow Yan River Exit	V	III	V	V	II	III	V	IV	IV	I	I	III

In 2019, the total number of unqualified water quality (water quality categories are IV and V) reached 46, accounting for 31.94%. Among the substandard water quality, 36 were mainly in category V, accounting for 80% of the substandard water quality, indicating that the overall level of water quality in 2019 was not very good. Throughout the year, a high proportion of monitoring sites had experienced more serious pollution. The reason was found that there are a number of large-scale aquaculture farms in the Jinma River Qizhi Qidou entry station, and it is located in residential areas where two nearby schools have daily sewage discharge.

Similarly, the trained T-S fuzzy neural network model was applied to the same 12 stations in the monitoring station in 2020, and the monthly water quality evaluation resulted in 2020 were obtained, as shown in Table 5.

Table 5. Categories of water quality by station in the monitoring station in 2020

Monitoring station	Month											
	1	2	3	4	5	6	7	8	9	10	11	12
Qinglan ditch	IV	I	II	II	I	II	II	II	II	I	I	I
Banister weir next to Jiannan Avenue	II	I	I	V	II	II	II	II	III	V	I	V
Jiangan River Shuangnan Road Bridge	I	I	I	I	I	I	I	II	I	I	I	I
Jinjiang Shuangba Bridge	II	II	II	III	III	III	III	II	II	IV	II	II
Jinjiang Xia Jiatuo Exit	II	II	III	V	V	V	V	II	II	II	I	II
Entrance of Huanglong River in Luxi River	III	IV	II	II	II	II	II	II	III	II	I	II
Guihuayan Community Sanshiyan Entry	I	II	I	III	V	II	II	II	I	I	II	I
Jinjiang Yong'an Bridge	I	I	II	II	II	II	II	II	II	I	I	I
Huangyan River Entrance	IV	I	II	V	V	II	II	III	IV	II	II	II
Jiang'an River Jiannan Avenue Bridge	II	I	II	I	I	I	II	II	III	I	I	III
Jinmahe Seven Branches and Seven Dous Entry	V	II	III	V	V	II	II	II	II	II	V	V
Yellow Yan River Exit	III	I	III	III	V	IV	II	II	II	IV	IV	IV

In 2020, the monthly water quality of almost all stations had been improved to some extent. According to statistics, there were only 23 water months of class IV and V in 2020, accounting for 15.97% of the total, which was a full double of that in 2019. The proportion of class I and II water rose from 15.27% in 2019 to 28.47% and 33.33% to 45.14% respectively, and even the two sites with the most serious pollution in 2019 also improved significantly in 2020. This showd that as the state attaches great importance to the water ecological environment, it had successively introduced water environmental protection policies such as the river chief system and the lake chief system, which had made the water area management more perfect. The local water bureau and the Water resources Bureau had taken the lead in cleaning up the rivers and canals of each water area to reduce the existing pollutants. And all sewage outlets in the region were included in the daily inspection and supervision category to further prevent the input of foreign pollutants, so that the water quality in 2020 had been significantly improved.

5 Summary

Water quality classification and evaluation is an important link in the development and utilization of water resources. In order to further improve the convenience of water quality classification and evaluation, combined with the strong fuzzy information processing capability of T-S fuzzy neural network, the water quality monitoring data of a monitoring station in 2019 was taken as the training set, and T-S fuzzy neural network was used for training. At the same time, in view of the imbalance of data categories leading to the unsatisfactory experimental effect, SMOTE oversampling method was used to enhance the data, generating class I, II, IV, V sample data, so that the samples of each class of water quality were basically balanced, and T-S fuzzy neural network was used to train the balanced data. The experimental results showed that the accuracy of the enhanced data on the training set is 96.20%, the validation set is 90.61%, and the test set is 90.66%. The model trained by T-S fuzzy neural network was used to classify and evaluate the data of all monitoring stations in the region from 2019 to 2020, and it was concluded that the overall situation of water quality in 2020 was better than that in 2019, which was consistent with the actual situation under the current environmental policy governance, and further explained the rationality of applying T-S fuzzy neural network to water quality classification and evaluation.

Acknowledgement. This research was funded by the National Natural Science Foundation of China (No.51775185), Natural Science Foundation of Hunan Province (2022JJ90013, 2023JJ60157), Hunan Province Intelligent Environmental Monitoring Technology Postgraduate Joint Training Base Project, and Hunan Normal University University-Industry Cooperation.

References

1. Qing, X.Y.F.: Evaluation and Characteristics of Water Environment Pollution of Secondary River in Chongqing, pp. 1–66. Chongqing Technology and Business University, Chongqing (2015)
2. Ning, S.H.F., Bai, Y.H.S.: Application of BP neural network model in groundwater quality evaluation. Syst. Eng. Theory Pract. (08), 124–127 (2000)
3. Huang, S.W.F., Dong, M.L.S.: Application of adaptive variable step length BP neural network in water quality evaluation. J. Hydraul. Eng. (10), 119–123 (2002)
4. Feng, D.Q.F., Guo, Y.S.: Application of BP neural network improved by genetic algorithm in groundwater quality evaluation. J. Zhengzhou Univ. **30**(03), 126–129 (2009)
5. Yang, B., et al.: Synergistic effect of ball-milled Al micro-scale particles with vitamin B12 on the degradation of $2,2',4,4'$-tetrabromodiphenyl ether in liquid system. Chem. Eng. J. **333**, 613–620 (2018). https://doi.org/10.1016/j.cej.2017.09.183
6. Wen, Z., Xie, L., Fan, Q., Feng, H.: Long term electric load forecasting based on TS-type recurrent fuzzy neural network model. Electr. Power Syst. Res. **179**, 106106 (2020). https://doi.org/10.1016/j.epsr.2019.106106
7. Pang, H., Liu, F., Zeren, X.: Variable universe fuzzy control for vehicle semi-active suspension system with MR damper combining fuzzy neural network and particle swarm optimization. Neurocomputing **306**, 130–140 (2018). https://doi.org/10.1016/j.neucom.2018.04.055
8. de Campos Souza, P.V.: Fuzzy neural networks and neuro-fuzzy networks: a review the main techniques and applications used in the literature. Appl. Soft Comput. **92**, 106275 (2020). https://doi.org/10.1016/j.asoc.2020.106275
9. Tang, J., Liu, F., Zou, Y., Zhang, W., Wang, Y.: An improved fuzzy neural network for traffic speed prediction considering periodic characteristic. IEEE Trans. Intell. Transp. Syst. **18**(9), 2340–2350 (2017). https://doi.org/10.1109/TITS.2016.2643005
10. Chinchor, N.F., Sundheim, B.M.S.: MUC-5 evaluation metrics. In: 5th Message Understanding Conference (MUC-5), pp. 25–27. Baltimore, Maryland (1993)
11. Hu, H.F., Wang, D.Q.S.: Application of T-S fuzzy neural network in water quality evaluation. J. Inner Mongolia Agric. Univ. **36**(04), 128–132 (2015)

12. Liu, M.F., Nie, L.S., Zhou, Z.Q.T.: Application of T-S fuzzy neural network in intelligent diagnosis of coronary heart disease. Sci. Guide **36**(17), 91–96 (2018)
13. Chawla, N.V., Bowyer, K.W., Hall, L.O., Kegelmeyer, W.P.: SMOTE: synthetic minority over-sampling technique. J. Artif. Intell. Res. **16**, 321–357 (2002). https://doi.org/10.1613/jair.953
14. Wang, H.Y.F., Fan, H.K.S., Yao, Z.A.T.: Imbalance data set classification using SMOTE and Biased-SVM. Comput. Sci. (05), 174–176 (2008)
15. Feng, H., Li, M., Hou, X., Xu, Z.: Study of network intrusion detection method based on SMOTE and GBDT. Appl. Res. Comput. **34**(12), 3745–3748 (2017)

Grid-Level Data Oriented Real-Time Automatic Calculation and Storage Method

Yuanhe Tang[✉], Shiming Sun, Xin Shan, and Tong Tai

NARI Group Corporation (State Grid Electric Power Research Institute), Nanjing 211106, China
tangyuanhe@sgepri.sgcc.com.cn

Abstract. The intelligent grid dispatch automation system has a high demand for system computing capabilities. The current system lacks automatic calculation and storage capabilities for real-time monitoring objects at the grid level. The model construction and automatic calculation of the calculation and analysis of grid monitoring objects require extensive manual maintenance, which fails to meet the demands of an expanding grid scale. We propose a grid object-oriented automatic calculation method based on abstract rule grammar. First, according to the object-oriented modeling principle, the calculation logic and calculation result storage model are designed, providing a modeling foundation for calculation data; second, an abstract rule grammar is designed to support the electronic description of calculation logic; finally, the parsing algorithm of abstract description of object calculation rules is discussed, and the common statistical algorithms for grid objects are presented to demonstrate the completeness and feasibility of the abstract rule grammar.

Keywords: Power Grid Dispatch Automation · Abstract Rule Grammar · Attribute Calculation · Object-Oriented Modeling

1 Introduction

With the continuous expansion of the number of UHV AC and DC in China, the integrated operation characteristics of UHV AC and DC hybrid power grid are obvious, the receiving end of the power grid is closely coupled, the security and stability characteristics of the system are complex [1][2], and the power grid dispatching and control is facing new challenges. UHV grid transmission and receiving end, AC and DC are strongly coupled, resulting in overall security risks affecting local faults. At national and provincial levels, it is urgently necessary to grasp the operation situation of the power grid simultaneously, and it needs to carry out inter-regional integrated security warning and risk prevention and control [3] at the whole network level. Multi-level dispatching is also the system advantage to ensure the safe operation of power grid over the years. However, in order to solve the contradiction between the integrated decision-making needs of the whole network and the decentralized control of dispatching at all levels, State Grid Corporation of China designed a new dispatching control system architecture [4] supporting the integrated large power grid in 2017, and built the [5] of national classification and provincial and prefectural integrated analysis center. The classification analysis center collection cut in 220kV and above voltage level of real-time data, provincial analysis decision center province 10kV and above voltage level real-time data [6], data scale is previous system ten, ten times, which involves a large number of power grid, such as real-time statistical results data need to provide through the power grid dispatching control system calculation function.

In the traditional smart grid dispatching control system [7], the calculation function usually uses the calculation object and calculation manually, and it is carried out by the formula, which is not suitable for the statistical calculation demand under the large-scale power grid. There are three reasons: first, the calculation model needs to manually define the storage location, a new class of power grid level meter, calculation requirements usually need to define a large number of calculation points, definition and tedious work; the second, the power grid topology structure and calculation formula need manual maintenance, new and old plant changes, the calculation formula must be manually modified, maintenance workload is huge and easy; error is the calculation logic easily understand and difficult to reuse, new calculation requirements, operation and maintenance personnel to understand whether the system has the corresponding calculation results, usually repeatedly define the new formula, resulting in a waste of computing resources. In view of the above three problems, it is urgent to improve the existing calculation methods, and study the automatic modeling and method of calculation data, as well as the expression of statistical logic and automatic analytical calculation methods of power grid level calculation results.

Under the regulation cloud platform technology architecture system, [8] classification and provincial analysis center has realized the basic model of data concentration, according to the power dispatching general data object structured design [9] principle (hereinafter referred to as "structured design"), the power dispatching general data object for unified modeling design, including power grid, plant, abstract container object, and generator, transformer equipment such as object, provides the theoretical and practical basis for calculation data automatic modeling.

Domestic researchers to flexible custom calculation requirements of automatic calculation technology has been some research, literature [10] by analyzing the existing grid regulation data acquisition and monitoring processing system model function limitations, for area, power supply and plant example, designed the power grid network model, and use the template formula to realize the calculation formula automatically generated. This method can automatically sense the changes of power grid structure and dynamically adjust the calculation model in real time, laying the foundation for the device object monitoring to power grid network object monitoring. However, the template formula only reduces the maintenance of formula definition, and its core mechanism is still the formula calculation for the measurement point level. When the calculation component increases, the actual calculation formula generated by the template is lengthy, and can only support ordinary summation formula, so it is difficult to support the calculation requirements of complex screening conditions. It is necessary to improve the calculation rule expression form, combined with the expression evaluation algorithm [11], to realize the calculation formula of object attribute level.

The compiled formula needs to have the analytical computing ability. Usually, the logical expression evaluation process is solidified in the software for [12, 13], which is not conducive to transplantation and secondary development. Literature [14] designs a scalable logical expression evaluation system, but it only stays at the basic level of theoretical research and lacks practical application.

Based on the research status, this paper proposes an automatic computing technique based on abstract rule grammar. First of all, establish a multi-level monitoring data storage model, according to, equipment, plant, area and the hierarchical relationship structure between power grid, the discrete calculation point into topological relationship, grid object, to meet the statistical analysis and object calculation, the results access requirements, solve the problem of manual definition calculation data model. Secondly, the object calculation gauge supporting the functions of four operation, function operation and aggregate calculation is designed to abstract the description grammar to meet the requirements of arbitrary custom calculation in the scheduling automation business. Finally, we discuss the analysis and calculation method of the abstract rule description grammar in detail and verify the feasibility in power dispatching automation system.

2 Overall Objectives and Architecture

2.1 Overall Goals

To realize the power grid-oriented object attribute level automatic calculation, the following problems should be solved:

1) It is necessary to have the automatic generation and storage capacity of multi-level monitoring data storage model, dynamically generate results according to objects and algorithms, and meet the requirements of "one setting, meter and multiple construction". There is no need to modify the storage structure when new statistical calculation requirements are added.

2) It is necessary to design object calculation rules that support the functions of four operations, function operations, aggregate computing and other functions, like description grammar, which can meet the requirements of any customized computing in scheduling automation business.

3) It is necessary to be able to realize the parsing and calculation method of abstract rule description grammar, to automatically execute the corresponding meter and calculation logic according to the module, type dependency and attribute values of power grid objects, and to store the calculation results in the multi-level monitoring data storage model.

4) It is necessary to be able to verify the feasibility of the technology, and to compile the corresponding abstract rule description syntax according to the common computing requirements of power grid objects in the power dispatching automation system, and to test and evaluate it.

2.2 Architectural Design

Object modeling and automatic calculation are self-generated and dynamic calculation technologies oriented to the statistical analysis results of equipment, plant and power grid according to the hierarchical network relationship of equipment, plant and power grid, so as to replace the complex calculation formula in the traditional dispatching automatic system.

The object attribute-level automatic computing technology includes three aspects: computing data standard modeling, abstract description design of object calculation rules and automatic computing technology. The overall architecture is shown in Figure Fig. 1. Computational class data modeling is based on the Structural Design principle, Prepared the calculation type coding, Designed a identification of traceable calculation object, Enable the automatic generation of the computational models, Solve the storage problem of the objectified calculation results; Abstract description of the object calculation rules (Object Calculation Rule Abstract Description, OCRAD) by providing a metadata illustration of the underlying components of the grammar, And designed the syntax of the interfix expression form, Support the electronic description of the power grid-level monitoring and computing logic; Automatic computational techniques describe the analytical algorithm for OCRAD, The feasibility of OCRAD is verified by the statistical demand of common power grid.

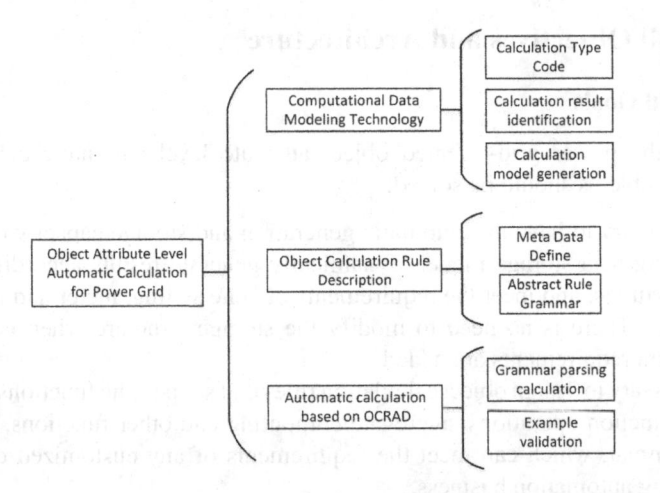

Fig. 1. Overall architecture of object property-level automatic computation technology

3   Key Technology

3.1   Standard Modeling Techniques for Computational Class Data

In the smart grid dispatching technical support system, the corresponding storage mode and type should be established for the calculation data, so that the calculation results can be organically related with the calculation object. Calculation data includes three parts: calculation object, calculation logic and calculation results. The calculation object has been modeled in Structures, meter, usually abstract container objects and primary equipment objects, such as power grid, power plant, generators, transformers, etc. In this paper, it is necessary to solve the modeling mode of calculation logic and calculation result. Firstly, we model the calculation logic and use the calculation type code to express the first and the calculation logic, then design the calculation result with the mutual conversion relationship with the calculation object identification. Finally, we build the storage model of calculation results to support the growing demand of dynamic expansion of calculation type.

1) Calculation type encoding

Computational class data modeling first requires the modeling and storage of computational logic. The calculation logic is the calculation requirement of a user, such as the total installed capacity of all power plants, the connection of the statistical system, the sum of line exchange functions, and the total output of distributed photovoltaic in the statistical system, etc. Each calculation requirements, can be defined as a calculation index code, in the power dispatching automation system software can be used in four decimal positive integer, expression, such as using digital 1001 said alignment caliber power plant total installed capacity, digital 2301 said system distributed photovoltaic total output, etc., can store a total of 10000 kinds of calculation requirements, can meet the single scheduling of real-time monitoring and calculation requirements of automation system. According to this idea, for the calculation logic, the object calculation type table can be established for storage. The table structure consists of encoding, name, calculation rules and other fields. For the abstract grammar expression which should be calculated by the calculation logic, the calculation factor can be obtained through the syntax analysis.

2) Identification and design of the calculation results

Another challenge to be solved in computational-like data modeling is the storage of computational results. For device-like objects, the calculation indicators are usually limited, and they can be stored directly in the device object data table by expanding the fields. However, for abstract container objects, the statistical requirements are variable, the use of expanded fields is not flexible enough, and it is difficult to upgrade and deploy the online system. Therefore, the storage mode of computing results needs to be changed to cope with the unexpected growth requirements of computing logic. For any kind of calculation logic, the corresponding calculation results need to include the identification of the calculation object, calculation type coding, calculation result value, calculation, status, calculation time scale, etc. Moreover, the calculation results themselves also need a unique identification, so as to facilitate the index positioning of the calculation results.

The calculation result identification can be generated as a natural main key by sequence, but it can not play an effective role and can only ensure its uniqueness. This paper designs a generation method of calculation result identification, which contains the key information such as object identification and calculation type, which has certain logical meaning. The identification design of the calculation results is shown in Figure Fig. 2.

3) Computational model generation and storage

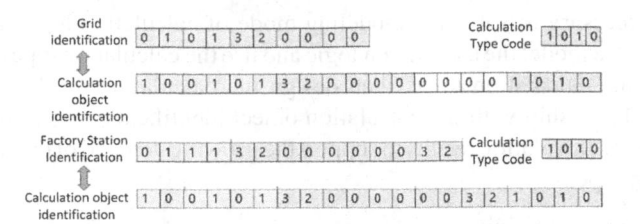

Fig. 2. Design of calculation result markers

As mentioned above, the results of materialized calculation should include the identification of calculation results, identification of calculation object, calculation type, coding, calculation result value, calculation status, calculation time scale, etc. The automatic calculation software can cross the object table of the power grid and the plant and the calculation coding table. According to the corresponding algorithm to complete all the calculation for each object, automatically generate the result identification according to the design principle of the calculation result identification, and deposit the calculation result and calculation time and other information into the calculation result table of the object calculation together. For different abstract containers, the calculation results can be stored in tables, easy to use and manage.

3.2   Abstract Description of Object Calculation Rules (OCRAD)

1) Metadata definition

OCRAD Expression syntax includes three parts: constant, variable and operator. It describes the expression at the attribute level through the combination of four operations, Boolean operations and function operations. The sections are designed as follows:

a) constant. It is used to express the unchangeable data in the operation process, which is mainly divided into menu constant, equipment identification constant, quantity, constant floating point constant, remote communication value constant, string constant, etc.

b) variable.used to express any attribute of the device object, you need to trathrough each record of the device and automatically replace it with the corresponding attribute of one record. The variable expression is point split, which can be continuously nested. The first layer is the device, the type name (including abstract container), followed by the corresponding attribute name of the device. The expression range of variables includes equipment measurement, equipment parameters, model affiliation, etc.

c) operator. It is used to connect constants and variables to form expressions, mainly including assignment operation, arithmetic operation, set operation, Boolean operation, bit operation, relation operation, function operation, string matching operation, etc.

d) key word. For expressing fixed meanings in syntax, usually for expressing calculation objects.

2) Abstract rule grammar

OCRAD The body is an assignment expression, and an assignment statement describes the object to be calculated and the location where the result is stored. The left value is the abstract container attribute variable, representing the calculated target object, which can be any device object (including the power container object); the right value represents the specific calculation rules, which can be nested with four operation, Boolean operation, function operation and other arbitrary operators, and written with medium fix expression to realize flexible automatic calculation requirements. For instance:

$$GRID.load = REF.gen + REF.recv \tag{1}$$

Two aggregate computation functions of sumif and countif are added to OCRAD to calculate the numerical sum and number sum, respectively. The function usage is sumif (device, attribute, conditional expression), indicating that the device attributes that meet the conditional expression are added up. For example, to calculate the total power generation of the plant, you can write the following expressions:

$$PLANT.sum\_p = \\ sumif\left(GENERATOR, p,'' st\_id == REF''\right) \tag{2}$$

Through the combination of functional operation and four operations, for power grid scheduling monitoring scenario of all kinds of calculation requirements, has basically can realize all calculation logic can edit, no longer like traditional scheduling automation system generally cost manpower, maintain huge power balance calculation formula library, use, only need to focus on the correctness of the calculation logic itself, be able to adapt to the power production changes, new plant operation, the old factory retired without the need to modify the calculation formula.

3.3 Automatic Computing Technique Based on OCRAD

OCRAD The left value is the expression of the calculation result, without the complicated operation process. The right value expression is more complex, we need to use some algorithm to solve the results and complete the calculation process. Midfix expressions are consistent with comprehension habits, but they are not suitable for program resolution. Usually, they need to be converted to suffix expression (inverse, Polish), combined with stack data structure to realize expression resolution and evaluation process [11].

The parsing method of OCRAD is different from the traditional parsing algorithm. On the one hand, all variables need to be swept to trace all records of the object, and on the other hand, for the aggregation functions of sumif and countif, the parameters include a nested conditional expression, described in the form of string as the whole parameter, and evaluate the sumif operation. Figure 3 shows a complete process for the OCRAD parsing process with sumif syntax. The core idea is to implement the parsing process in the two-stack mode, stack S1 is used to process the complete OCRAD, and stack S2 is used to process the embedded expressions in sumif.

The general processing process for the OCRAD is described as follows:

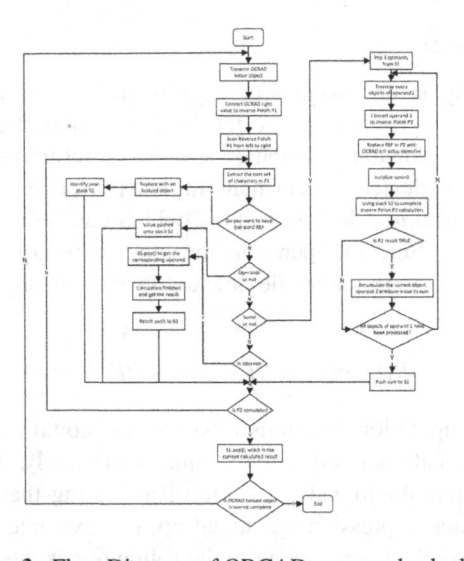

Fig. 3. Flow Diagram of ORCAD parse and calculate

1) Analyse the left value, go through the object described by the left value expression, and execute the calculation process of the right value expression for each object;

2) the right value expression into inverse Poland P1, in the conversion process can replace all REF reserved words for the current through the left value object identification, and use the stack S1 for inverse Poland P1 start evaluation process, namely: scan to the operator and the stack, scan to the operator according to the operator, the operator, the operator number of the stack, after the expression operation result.

3) When sumif is found in scanning P1, the corresponding operand is displayed, and the sum temporary variable is initialized, using stack S2 to evaluate the conditional expression P2; when the evaluation result is true, the corresponding attribute value of the record is added to the sum variable. After all records are processed, the value is the evaluation result of sumif function.

4 Conclusion

First, this paper presents a power grid-oriented object-attribute-level automatic computing technique, The technique follows the principles of the Structured Design, Realize the modeling of the power grid-oriented computational data, And the calculation results have compiled the global unique identification and realize the elastic storage; next, Based on the modeling features, In this paper, we have designed the OCRAD syntax, Upgrade the measurement point-level statistical calculation of the traditional scheduling automation system to the statistical calculation of the attribute level, Waiving a lot of formula maintenance work, Make the automated operation and maintenance can focus on the preparation of calculation rules, Do not care about the grid model changes; last, In this paper, we propose the parsing and calculation method of OCRAD grammar, The proposed method can be applied to the power grid dispatching automation system at all levels, Realize the power grid-oriented object computing. There is still some research space for this work, and in-depth research will be carried out in the following two aspects:

1) Servitization of calculation results, The object attribute level automatic computing technology can meet the computing requirements of scheduling automation system, However, the current monitoring centers at all levels are limited by the scope of collection, The monitoring system below the provincial level does not realize the data access of the full voltage level, Generally, the regional level monitoring system, Therefore, the use of the technology studied in this paper in the monitoring system below the provincial level is not enough to fully solve the whole network computing problem, Service sharing of automatic calculation data among various monitoring centers, You can subscribe the results on demand, Realize the fusion calculation of all levels of scheduling system.

2) computational efficiency, the OCRAD analytical algorithm verified by practice, but need to each, a calculation object are a analytical calculation process, involved, and total aggregate class function calculation, each analytical calculation when need to a full scan equipment model, when the system model scale is larger, complete a calculation time long, need to study more efficient analytical calculation method, improve the efficiency of calculation.

References

1. Shu, Y., Chen, G., He, J., et al.: Research on the framework of building a new power system with new energy as the main source. Strateg. Study CAE **23**(6), 3–11 (2021)
2. Kang, C.: New Power System Technology Study Report. Department of Electrical Engineering and Applied Electronics Technology, Tsinghua University, Beijing (2021)
3. Xu, H., Yao, J., Yu, Y., et al.: Dispatch control system architecture and key technologies to support integrated large power grids. Autom. Electr. Power Syst. **42**(6), 8 (2018)
4. Xu, H., Yao, J., Nan, G., et al.: New features of future grid dispatch control system application functions. Autom. Electr. Power Syst. **42**(1), 7 (2018)
5. Feng, S., Yao, J., Yang, S., et al.: Overall design of integrated analysis center for dispatching system under "physically distributed, logically centralized" architecture. Electr. Power Autom. Equip. **35**(12), 7 (2015)
6. Xu, H., Sun, S., Ge, C., et al.: Research and application of real-time data platform architecture and key technologies for power grid regulation and control. Autom. Electr. Power Syst. **43**(22), 8 (2019)
7. Xin, Y., Nan, G., Liu, J., et al.: Smart grid dispatch control system status and technology outlook. Autom. Electr. Power Syst. **39**(1), 7 (2015)
8. Zhang, Y., Guo, J., Liu, J., et al.: Regulatory cloud platform IaaS layer technology architecture design and key technologies. Autom. Electr. Power Syst. **045**(002), 114–121 (2021)
9. Xu, H.: Structured design and application of common data objects of power dispatch for regulation and control cloud. Power Syst. Technol. **42**(7), 7 (2018)
10. Jin, F., Zhou, T., Yao, M., et al.: Object-oriented modeling and application research for grid networks. Electr. Autom. **43**(2), 6–8 (2021)
11. Tang, N., You, H., Zhu, H.: Data structures and Algorithms (C++ version). Tsinghua University Press (2009)
12. LI, W., Mu, D., Ma, D., et al.: Application of algebraic expression evaluation in data simulator design. Comput. Eng. Des. **32**(6), 2187–2190 (2011)

362   Y. Tang et al.

13. Cao, S., Liu, J.: Expression finder in workflow. Comput. Knowl. Technol. Acad. Edn. **3**, 3 (2006)
14. Xiong, F., Kuang, L., Han, Y.: Design and implementation of a scalable logic expression valuation system. Comput. Eng. Des. **33**(10), 3858–3861+3958 (2012)

FedMLP4SR: Federated MLP-Based Sequential Recommendation System

Zhi Yuan ⓘ and Yongli Wang (✉) ⓘ

Nanjing University of Science and Technology, Nanjing 210000, China
{Beetroot,yongliwang}@njust.edu.cn

Abstract. Sequential Recommendation predicts users' next possible item by modeling their historical interaction sequences. Transformer-based models can efficiently process chronological interaction sequences, but require quadratic model complexity. Traditional sequential recommendation system often adopts a centralized architecture, which easily leads to the leakage of user privacy. In our work, we propose a new sequential recommendation model based on Federated Learning called FedMLP4SR. On the one hand, MLP-based (Multi Layer Perceptrons) models have linear model complexity and can process sequential data without adding additional position embeddings. On the other hand, Federated Learning can meet the requirement of training the global model without collecting the private data from users. The MLP-based models have fewer parameters and less computational cost, making it highly adaptive to Federated Learning scenarios. Specifically, in our work clients utilize cross-sequence and cross-channel MLP blocks to capture correlational information in interaction sequences and complete the training of the local model; The server collects model parameters from clients to complete the updating of the global model. Besides, we use clients' local graphs to extract high-order interactions between local items and fully consider item ratings. Finally, experiments validated the effectiveness and the feasibility of the proposed model over five representative baselines on two public datasets.

Keywords: Federate Learning · Sequential Recommendation · Multi Layer Perceptrons

1 Introduction

In the era of big data, people need to face thousands of information like texts, pictures and short videos every day, which generates a large amount of potentially valuable user-item interactions. The recommendation system aims to model user-item interactions and extract users' interests and preferences, then predicts and recommends content that users may be interested in. On the one hand, the recommendation system can alleviate the information-overload problem. On the other hand, accurate recommendations can increase the retention rate of users. In early recommendation system, methods such as Collaborative Filtering [1, 2] were often used to complete simple recommendation tasks, which are usually based on matrix factorization. The DNN (Deep Neural Network) [3]

H. Jin et al. (Eds.): IAIC 2023, CCIS 2058, pp. 363–375, 2024.
https://doi.org/10.1007/978-981-97-1277-9_28

allows recommendation systems extract more important hidden information from user-item interactions, greatly improving the accuracy of recommendation systems. With the development of recommendation system, sequential recommendation system [4], as a branch of recommendation system, aims to better understand users' interests and preferences by analyzing users' chronological interaction sequences. Compared with traditional recommenders, sequential recommendation can capture dynamic evolution in users' interests and preferences. RNN-based (Recurrent Neural Networks) [5] methods can easily handle sequential data, but have gradient-vanishing problem. The self-attention methods can focus on important information in the sequences, Transformer-based models which are based on self-attention have achieved state-of-the-art performance in sequential recommendation tasks. For example, [6] uses a bidirectional model based on Transformers to complete sequential recommendation.

Existing models have achieved competitive performance in sequential recommendation, but they still suffer from the following three issues: Firstly, the high model complexity limits models to process long sequences. The complexity of Transformer-based models often increase quadratically as the length of sequences increases. Secondly, the architectures applied in traditional recommendation systems are mostly centralized, which require users to upload their private data to the server for model training. On the one hand, centralized models easily lead to the leakage of user privacy [7]. On the other hand, the server suffers from high computational cost during the model training. Thirdly, sequential recommenders often only model sequential information without considering extra information like item ratings which may contain users' preferences. For example, if user A and user B have the same interaction sequence, traditional sequential recommenders may tend to offer the same result to A and B. However, even for the same items in the sequence, A and B probably have different ratings, which means two identical interaction sequences have different importance to two different users.

Different solutions have been proposed to address the above issues. MLP-based models have received widespread attention due to their linear computational cost, and [8, 9] has showed competitive performance in computer vision. MLP-based models have already showed its effectiveness in sequential recommendation tasks. Based on MLP-MIXER, [10] uses across-sequence and cross-channel fusion method to capture sequential information, and proposed MOI (Multi Order Interaction) to capture high-order interactions between sequences and channels. The MLP-based models only require linear time and memory complexity, and MLPs are naturally sensitive to sequential order of the input, while other sequential recommenders need additional positional embeddings. In 2016, the Google team [11] proposed a distributed model training architecture called Federated Learning and the training algorithm called FedAvg. Different from traditional distributed architectures, Federated Learning does not require user information to be centrally collected, but rather updates global model by collecting model parameters from local clients. Regarding the different importance of items in sequences, some works model different behaviors separately or adopt different weights to different items according to their importance.

To overcome the three issues above, we propose a new federated sequential recommendation model called FedMLP4SR, which contains three main parts, Local Graph Layers, MLP Layers and a Predication Layer. Firstly we propose Local Graphs for every client to extract high-order interactions among local interacted items, and we take item ratings into consideration while updating local graphs. Our Local Graph Layers are similar to [12, 13], but our local graphs focus on the local item high-order interactions and side information like item ratings, while [12, 13] focus on Partial Federated Learning. Secondly, we use MLP blocks in MLP Layers to capture cross-sequence and cross-channel correlations in user-item interaction sequences, MLP blocks are sensitive to the sequential order of the input data which means these is no need for adding extra position embeddings, and MLP blocks only require linear complexity. Due to their simplicity, MLP blocks can alleviate the computational cost of the clients and the communicational cost between the server and clients. We summarize our contributions as follows:

1. We propose a new model based on MLPs called FedMLP4SR for federated privacy-preserving sequential recommendation.
2. We propose Local Graphs to capture high-order item feature interactions and take item ratings as its importance in the sequences for graph updates.
3. We validate the effectiveness and the feasibility of the proposed model over five representative baselines on two public datasets.

2 Related Works

2.1 Sequential Recommendation Systems

Recommendation systems aim to model user-item interactions, overcome information-overload problem, and recommend items that users may be interested in. Early recommendation systems can be divided into content-based filtering and collaborative filtering [14]. Collaborative filtering mainly decomposes user-item rating matrix into user vector and item vector. The dot product of user and item vectors can be used to predict users' ratings for non-interacted items. The emergence of neural networks has further enhance data-processing ability of recommendation systems. NCF [15] has achieved good results by using multi-layer neural networks to simulate traditional matrix factorization. NGCF [16] completes high-order information extraction by constructing a bipartite graph of users and items. As a branch of recommendation systems, sequential recommendation systems model users' chronological interactions and capture the sequential information in these interactions. Due to the structure of RNN, RNN-based models can effectively capture sequential information in sequences, GRU4REC [5] uses RNN and optimizes RNN structure and shows competitive performance among RNN-based models. While self-attention models focus on important information in sequences and capture the dynamic changes in sequences, Transformer-based model [6, 17] has already achieved state-of-the-art performance in sequential recommendation tasks. Considering different behaviors and side information which may influence recommendation results, [18] proposed global graphs and local graphs to learning the behavior transitions and user intentions. ARD [19] proposed an item-embedding re-distribution method for learning better item embeddings. However, sequential recommendation still has many issues like cold start, high computational cost.

2.2   Federate Recommendation Systems

Federated Learning, a distributed learning framework, was first proposed by Google team in 2016, which can train a global model without centrally collecting users' private and original data. Federated Recommendation Systems aim to train recommendation systems in a safe and privacy-preserving way. FCF [20] first applied Federated Learning to collaborative filtering. Clients keep and update user embeddings locally, and upload gradients to the server which then merges the gradients and updates item embeddings. FNCF [21] is a federated version of NCF. FedRec [22] proposes a federated recommendation model based on explicit feedbacks. In order to protect the private information of which items each user has rated, FedRec proposed user averaging and hybrid filling strategies to randomly sample some unrated items and assign the items with some virtual ratings accordingly. FedFast [23] extends the FedAvg framework in order to complete fast training while achieving good accuracy for every client. FedFast proposed ActvSAMP to sample from a diverse set of clients in each training round, and ActvAGG propagates the updated model to the clients. FedGNN [24] applies GNN (Graph Neural Networks) [25, 26] to the federated recommendation system. FedGNN captures user-item high-order interactions by graph expansion and a trusted third party, and uses pseudo interacted items to protect users' privacy. There are few works done in the federated sequential recommendation scenarios. On one hand, sequential recommendation basically requires complex models to achieve competitive performance. On the other hand, the amount of data in some clients may be not enough to train a local model. FedSeqRec [27] proposes a federated sequential recommendation framework for organizations rather than individual clients. On the one hand, it proposed a new parameter-merging approach to alleviate the problems caused by NON-IID data, and on the other hand, it uses the mathematical method of low-rank tensor mapping to complete the modeling of users' long-term sequence behavior. However, Federated recommendation systems still has many issues like user privacy protection, high communicational cost, model robustness.

3   Framework

In this section, we introduce the detailed framework of the proposed FedMLP4SR. FedMLP4SR consists of three main part: (a) Local Graph Layers; (b) MLP Layers; (c) a Predication Layer. We first introduce the basic structure of FedMLP4SR, which each client needs to train locally. Then we introduce how clients updates their parameters of MLP Layers and Predication Layer according to the FedAvg algorithm. The basic structure of FedMLP4SR is showed in Fig. 1, and the federated model training steps are illustrated in Fig. 2.

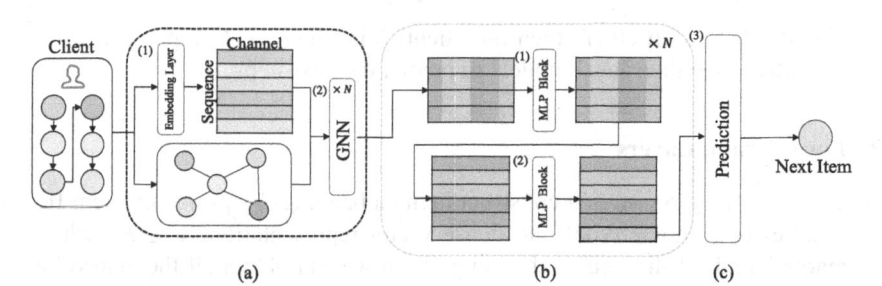

Fig. 1. The client-side architecture of FedMLP4SR which consists of three parts: (a) Local Graph Layers; (b) MLP layers; (c) Predication Layer. The model takes client's local sequential interactions as input and outputs the predicted next-item.

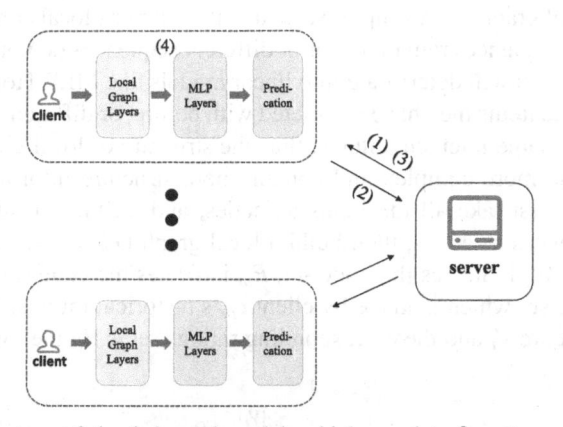

Fig. 2. The architecture of the federated model, which consists five steps. (1) Initializing the model; (2) Training and Updating the mode; (3) Merging and distributing the model (note that step 2 and 3 will be repeated for several times); (4) Applying the model.

3.1 Problem Formulation

According to the definition of federated sequential recommendation, We denote clients in Federated Learning as $C = \{c_1, c_2, ..., c_M\}$, where M indicates the number of the clients. Items as $I = \{i_1, i_2, ..., i_N\}$, where N indicates the number of the items. For clients c_m, it has its own private interaction sequence $S_m = \{s_{m,1}, s_{m,2}, ...s_{m,K}\}$, where $s_{m,k} \in I, m \in [0, M], k \in [1, K]$ indicates the k-th interacted item in sequence S_m. With the definitions above, the problem of next-item recommendation can be defined as follows: Given client c_m's interaction sequence S_m, client c_m need to train a local model F_m, which take S_m as input and output the most probable next-item i_{next} that client c_m may interact with. F_m can be denoted as $F_m : S_m \rightarrow i_{next}$.

In FedMLP4SR, Clients only upload the parameters of MLP Layers and the Predication Layer. So we define clients' local parameters of there two parts in the t-th training round as w_m^t. The server collect the clients' parameters in every round and follow the FedAvg algorithm to merge parameters and then update the global model F. After the global model F converges or updates for prescribed rounds, the server distribute the global model F to each client, then the clients joint their local graphs and the global model F rather than their local models to predict the next-item.

3.2 Local Graph Layers

In Local Graph Layers, we first construct item embeddings by projecting item IDs and item features (e.g. movie type) into dense vector representations $x \in R^h$, where h is the tunable length of the item embedding. Then we can obtain all the embeddings of the items in sequential interactions and overlap all of the item embeddings into a two-dimension table $X \in R^{K \times h}$ according to their order in the sequence, as shown in Fig. 1 (a)(1), where K indicates the length of the sequence.

Local graphs are constructed by clients' local interaction sequences. As a part of the global graph, local graph has a simpler structure and contains local context information. If the interaction sequences remain linear or different sequences do not contain the same items, the local graph will deteriorate into linear models like MLP. However users often re-interact with the items they have interacted with before, or different sequences in one client contain the same interacted items, thus the structure of local graph will not keep linear but become more complex and contain more structure information. According to client c_m, We first take all the items as nodes, and all the transition behaviors in interaction sequences as edges, then build a local graph $G_m = (E_m, V_m)$, as shown in Fig. 1 (a), where V_m indicates the node set, E_m indicates the edge set. Then we set R_m as the item rating set which is made by client c_m's historical ratings. For node v_i, if its adjacent node set are V_i and the corresponding ratings set is R_i, then we update node v_i as:

$$h_i = v_i + w_i \sum_{j=1}^{|V_i|} \frac{R_j}{\sum R_i} v_j \tag{1}$$

where w_i is learning weight, h_i is the updated item embedding of v_i. If we put the two-dimension table X into the graph layer, as shown in Fig. 1 (a)(2), we can obtain a new updated two-dimension table $H_{gnn} \in R^{K \times h}$, which is the input of the following MLP Layers.

3.3 MLP Layers

The MLP Layers consist of cross-sequence and cross-channel feature fusion blocks, as shown in Fig. 1 (b) (1), (2). In [28], three-dimension fusion was proposed, which includes cross-sequence, cross-channel and cross-feature fusion, but the three-dimension fusion make the model more complex and it disregards the user privacy. In [10], cross-sequence and cross-channel fusion was proposed, and it can extract high-order interactions through the proposed MOI blocks. However it fails to capture item feature interactions and disregards user privacy as well. In our work, we still fuse cross-sequence and cross-channel information. On the one hand, it can reduce model complexity. On the other hand, we propose local graphs to complete the fusion of item features and utilize item ratings as weights to update item features, Meanwhile these local graphs are kept locally, which will not increase communicational overhead. Cross-sequence and cross-channel fusion are similar, only the size of input date is different. And they are both based on MLP blocks. We take columns of H_{gnn} as input, then the output of cross-sequence fusion block can be denoted as:

$$x_s^j = W_2^T LayerNorm(\sigma(W_1^T x_g^j)) \tag{2}$$

where W_1, W_2 are learnable weight matrices, $\sigma(\cdot)$ is nonlinear active function, $LayerNorm(\cdot)$ [29] is the layer normalization function, x_g^j is the j-th column of H_{gnn}, x_s^j is the output of cross-sequence fusion of x_g^j. We still restore the output vectors into matrix like H_{gnn} which is denoted as $H_{sequence} \in R^{K \times h}$ then we take the rows of $H_{sequence}$ as input, then the output of cross-channel fusion block can be denoted as:

$$x_c^k = W_4^T LayerNorm(\sigma(W_3^T x_s^k)) \tag{3}$$

where W_3, W_4 are learning weight matrices, $\sigma(\cdot)$ is nonlinear active function, $LayerNorm(\cdot)$ is the layer normalization function, is the k-th row of $H_{sequence}$, x_c^k is the output of cross-sequence fusion of x_g^j. We still restore the output vectors into matrix like $H_{sequence}$ which is denoted as $H_{channel}$. If MLP Layers are more than two layers, then the output of cross-channel fusion block will be put into cross-sequence fusion block again. After that, we take the last row of $H_{channel}$ as output which is the input of predication layer.

3.4 Prediction Layer

To recommend the next-item, we apply a predication layer and use softmax to calculate the predicted probabilities of all the candidate items as below:

$$y = softmax(W_o x + b_o) \tag{4}$$

where W_o is the weight matrix, b_o is the bias, x is the output of MLP layers, and the output $y = \{y_1, y_2, ..., y_N\}$ represents the predicted scores of all the candidate items. Then we take cross-entropy as loss function for model training:

$$L = -\sum_{i=1}^{N} \lambda_i \log(y_i) + (1 - \lambda_i) \log(1 - y_i) \tag{5}$$

where $y_i \in y$, $\lambda_i = 1$ if the i-th item is the true next-item, $\lambda_i = 0$ if the i-th candidate is not the true next-item.

3.5 Federated Model

We next discussed the federated parts of FedMLP4SR, we will introduce how clients train and upload their local model, and how the server update the global model. The federated model is represented as Fig. 2. As shown in Fig. 2, each client has its own local FedMLP4SR model which consists of three aforementioned parts. The Local Graph

Layers are personalized layers which are unique to each client. That means by inputting even the same interaction sequences, clients can extract different and personalized representations of the item embeddings. And as described above, item embeddings will be updated according to the clients' local interactions and the item ratings and that is how Local Graphs work. Then for the MLP Layers and Predication Layer, clients need to upload the model parameters each round they update these parts. MLP Layers and Predication Layser functioning as global layers can extract global sequential information. Then we set the global training round as t, the local epoch as e, the local batch size as b. According to the FedAvg algorithm, the federation of the model can be divided into four steps as follows: (1) The server initializes the global model F, and distributes the initialized parameter w^0 to all clients involved. (2) Each client receives the initialized w^0 to as their initial local model. And then clients train their local model by gradient descent according to the local epoch e and local batch size b. After one round of local training, clients keep the Local Graphs themselves, and upload the parameter of MLP Layers and the Predication Layer which is denoted as w_m^t, where t indicates the t-th round and m indicates the m-th client c_m. (3)The server merges all the uploaded model parameters according to the FedAvg algorithm which can be denoted as:

$$w^{t+1} = \sum_{i=1}^{M} \frac{n_i}{\sum n} w_i^t \tag{6}$$

where n_i indicates the number of samples client c_i. Then the server send the new model parameters w^{t+1} to all the clients involved again and repeat the step (2) and (3) t times. (4)Finally, After t rounds of training, the server get the final well-trained global model w_g, and distribute w_g to the clients. The client replace their MLP Layers and Predication Layer with w_g. And each client can predict the next-item via their Local Graphs and the global MLP Layers and the global Predication Layer.

4 Experiments

4.1 Experimental Setup

Datasets: In our work, we choose two widely used datasets to evaluate FedMLP4SR's performance. The detailed statistics are represented as Table 1. (1) Movielens Dataset: Movielens is one of the most widely used dataset in recommendation systems. It consists of users, movies and users' ratings. And we use Movielens-1m in our experiments. 1m means there are one million interactions in the dataset. (2) Amazon Beauty: Amazon collect user-item interactions on their online shop, and release Amazon datasets which contains a number of kinds of datasets according to the categories. We use Amazon Beauty in our experiment.

For these two dataset, we set the maximum interaction sequence length as 50, and we remove those interactions whose length is less than 5. For those whose length is between 5 and 50, we use zero to pad these sequences.

Table 1. Statistics of the datasets.

Datasets	Users	Items	Interactions	Ratings
Movielens-1m	6,040	3,900	1,000,209	1,...,5
Beauty	1,210,271	249,274	2,023,070	1,...,5

Metrics: In our work, we use three commonly used metrics to evaluate our proposed model: HR (Hit Rate), NDGC (Normalized Discounted Cumulative Gain) and MRR (Mean Reciprocal Rank). HR is a metric for recommendation accuracy, which measures the probability of the ground-truth item that appears in model's top-k recommendation. NDGC and MRR are metrics to evaluate the order of the top-k recommendations. In next-item recommendation tasks, the higher order the ground-truth items get in the top-k list, the bigger NDGC and MRR are. We set k as 10 in our work.

Baselines: We will compare our proposed model with five baselines as follows: First we pick three unfederated models: NGCF [16] are Collaborative Filtering-based model which simulate matrix factorization via Neural Networks and capture high-order interaction information through GNNs. GRU4Rec [5] utilize RNNs to capture sequential interactions. BERT4Rec [6] is a Transformers-based model which utilize bidirectional self-attention layer to complete sequential recommendation tasks and achieves state-of-the-art performance. These three baselines are all centralized architectures, then We pick two most famous decentralized recommendation model FCF [20] and FNCF [21] which are all based on Federated Learning.

For NGCF, GRU4Rec and BERT4Rec, we user the official code and make a slight modification. And for FCF and FNCF we reproduced the experiments according to their original settings.

Implementation: For federated setting, we trained our model by changing the client number from {10, 20}, the local epoch as {10, 20}, and we set the global training round as 200, and the local batch size as 256. For the model training and the model layers, we set the local graph layers as 2, which means it can capture 2-order interaction information by the local graphs, MLP layers as 3, the hidden dimension of item embeddings as 256 and dropout rate as 0.2. And for the model optimizer, we user Adam optimizer with cosine learning rate decay [30].

4.2 Performance Comparison

We compare the performance of all the models accordingly and discuss the potential reasons leading to the results.

As shown in Tables 2 and 3, we can make several observations as follows: (1) Compared to traditional CF-based methods like NGCF, GRU4Rec, BERT4Rec and our proposed model are superior to NGCF. The reason may be that these three models capture nonlinear hidden feature interactions as well as sequential information, whereas NGCF is based on matrix factorization and fails to capture sequential interactions. However, NGCF outperforms FCF and FNCF, because NGCF is centralized and uses Graph

Table 2. Overall performance comparison on Movielens-1m.

Methods	HR@10	MRR@10	NDCG@10
NCF	0.4545	0.1846	0.2465
GRU4Rec	0.5712	0.3304	0.3970
BERT4Rec	**0.6629**	**0.4130**	**0.4543**
FCF	0.3531	0.2027	0.2327
FNCF	0.4428	0.1698	0.2230
FedMLP4SR	0.6012	0.3398	0.4076

Table 3. Overall performance comparison on Amazon Beauty.

Methods	HR@10	MRR@10	NDCG@10
NCF	0.2637	0.1281	0.1813
GRU4Rec	0.3362	0.1737	0.2124
BERT4Rec	0.3214	0.1614	**0.2160**
FCF	0.1937	0.1323	0.1446
FNCF	0.2443	0.1242	0.1721
FedMLP4SR	**0.3412**	**0.1825**	0.2152

Neural Network to extract high-order user-item interactions which enhance the expressive ability of the model. (2) compared to GRU4Rec, FedMLP4SR perform better in both datasets. On the one hand, GRU4Rec uses RNNs to capture sequential information, which fail to process long sequences. And on the other hand, FedMLP4SR takes item feature interactions and item ratings into consideration while GRU4Rec fails to, And cross-sequence and cross-channel fusion better capture interactions among sequences and channels. (3) BERT4Rec is a Transformers-based model which utilize item context and sequential information and achieves SOTA performance. Self-attention allow BERT4Rec focus on important information in the unknown sequences. BERT4Rec outperforms our proposed model in the Movielens dataset, where the reason may be that FedMLP4SR is federated, and there is naturally a accuracy gap between federated and unfederated model. Besides, it is possible that BERT4Rec can be better able to take advantage of Transformers on datasets of non-sparsity and sufficiency. However, our proposed model outperforms BERT4Rec on dataset Beauty, it shows FedMLP4SR is suitable for sparse datasets like Beauty, and FedMLP4SR takes high-order item feature interactions and item ratings into consideration, whereas BERT4Rec only process sequential interactions without item features. (4) Compared to the federated models FCF and FNCF, our proposed model performs better. On the one hand, though these models are Federated Learning-Based, FCF and FNCF are based on simple and traditional recommendation systems, which fail to capture hidden interactions. On the other hand,

FedMLP4SR combines cross-sequence and cross-channel MLP-blocks for sequential information and utilizes local graphs to fuse item ratings and item features. So our FedMLP4SR is more complex and more accurate.

4.3 Complexity Comparison

Table 4. The complexity comparison over BERT4Rec and FedMLP4SR on two datasets. (M indicates Movielens, B indicates Beauty).

Methods	Params (M)	NDCG@10(M)	Params (B)	NDCG@10 (B)
BERT4Rec	2.3 m	0.4543	2.3 m	0.2160
FedMLP4SR	1.4 m	0.4076	1.4 m	0.2152

As shown in Table 4, our proposed model has fewer model parameters than BERT4Rec on two datasets, meanwhile FedMLP4SR achieves competitive performance compared to BERT4Rec. This shows the suitability of MLP-based model for Federated Sequential Recommendation since the fewer the model parameters, the lower the client-server communicational overhead. It is known that one of the major bottlenecks of Federated Sequential Recommendation is the high communicational overhead. If MLP-based model can be in conjunction with other methods such as Knowledge Distillation, then then communication overhead can be greatly reduced and make Federated Recommendation Systems more generalizable.

5 Conclusions

In this paper, we proposed a new federated sequential recommendation model called FedMLP4SR, which can train a MLP-base sequential recommender in a privacy-preserving way. We leverage MLP Layers to capture cross-sequence and cross-channel correlations. With regard to the item feature interactions, we utilized Local Graphs to capture high-order item feature interactions. And our proposed FedMLP4SR achieves competitive performance against Transform-based BERT4Rec, and traditional federated recommendation systems. Our proposed FedMLP4SR efficiently alleviate three afore-mentioned issues: (1)FedMLP4SR utilizes Federated Learning to protect users' privacy; (2) FedMLP4SR utilizes simple yet efficient MLPs to complete the sequential recommendation tasks; (3) Local Graphs can take side information such as item ratings into consideration, thus improving the accuracy of the model.

References

1. Su, X., Khoshgoftaar, T.M.: A survey of collaborative filtering techniques. In: Advances in Artificial Intelligence (2009)

2. Hu, Y., Koren, Y., Volinsky, C.: Collaborative filtering for implicit feedback datasets. In: Proceedings of the IEEE International Conference on Data Mining, pp.263–272. IEEE (2008)
3. Jiaxi, T., Ke, W.: Personalized top-N sequential recommendation via convolutional sequence embedding. In: WSDM 2018, pp. 565–573 (2018)
4. Wang-Cheng, K., McAuley, J.: Self-attentive sequential recommendation. In: ICDM 2018, pp. 197–206 (2018)
5. Wang-Cheng, H., McAuley, J.: Session-based recommendations with recurrent neural networks. In: Proceedings of the IEEE International Conference on Data Mining (2018)
6. Fei, S., et al.: BERT4Rec: sequential recommendation with bidirectional encoder representations from transformer. In: Proceedings of the ACM Conference on Information and Knowledge Management (2019)
7. Zhang, B., Wang, N., Jin, H.: Privacy concerns in online recommender systems: influences of control and user data input. In: Proceedings of SOUPS, pp. 159–173 (2018)
8. Hanxiao, L., Zihang, D., David, R.S., Quoc, V.L..: Pay attention to MLPs. arXiv preprint arXiv:2105.08050 (2021)
9. Tolstikhin, I., et al.: MLP-mixer: an all-MLP architecture for vision. arXiv preprint arXiv: 2105.01601 (2021)
10. Hojoon, L., Dongyoon, H., Sunghwan, H., Changyeon, K., Seungryong, K., Jaegul, C.: MOI-mixer: Improving MLP-mixer with multi order interactions in sequential recommendation. arXiv preprint arXiv:2108.07505 (2021)
11. Brendan, M., Eider, M., Daniel, R., Seth, H., Aguera, A.B.: Communication-efficient learning of deep networks from decentralized data. In: Artificial Intelligence and Statistics, 2017, pp. 1273–1282 (2017)
12. Arivazhagan, M.G., et al.: Federated Learning with Personalization Layers (2019)
13. Le, Q., et al.: Personalized federated recommender systems with private and partially federated autoencoders. In: 2022 56th Asilomar Conference on Signals, Systems, and Computers, pp. 1157–1163 (2022)
14. Rocca, B.: Introduction to recommender systems: overview of some major recommendation algorithms. Towards Data Science (2019)
15. He, X., et al.: Neural Collaborative Filtering. In: International World Wide Web Conferences Steering Committee (2017)
16. Mizem-Pietraszko, J., Wroclawia, L.K.: Neural graph collaborative filtering. Comput. Rev. 2(2021)
17. Vaswani, A., et al.: Attention is all you need. In: Advances in Neural Information Processing Systems, pp. 5998–6008 (2017)
18. Weixin, C., Mingkai, H., Yongxin, N., Weike, P., Li, C., Zhong, M.: Global and personalized graphs for heterogeneous sequential recommendation by learning behavior transitions and user intentions. In: Proceedings of the 16th ACM Conference on Recommender Systems (RecSys 2022), pp. 268–277. Association for Computing Machinery, New York, NY, USA (2022)
19. Wei, C., Weike, P., Jingwen, M., Zhechao, Y., Congfu, X.: Aspect re-distribution for learning better item embeddings in sequential recommendation. In: Proceedings of the 16th ACM Conference on Recommender Systems (RecSys 2022), pp. 49–58. Association for Computing Machinery, New York, NY, USA (2022)
20. Muhammad, A.-U.-D., et al.: Federated Collaborative Filtering for Privacy-Preserving Personalized Recommendation System (2019)
21. Vasileios, P., Efraimidis, P.S.: Federated neural collaborative filtering. arXiv preprint arXiv: 2106.04405 (2021)
22. Lin, G.Y., Liang, F., Pan, W.K., et al.: FedRec: federated recommendation with explicit feedback. IEEE Intell. Syst. 36, 21–30 (2020)

23. Muhammad, K., et al.: FedFast: going beyond average for faster training of federated recommender systems. In: Proceedings of the 26th ACM SIGKDD International Conference on Knowledge Discovery & Data Mining (KDD 2020), pp. 1234–1242. Association for Computing Machinery, New York (2020)
24. Wu, C., et al.: FedGNN: federated graph neural network for privacy-preserving recommendation. arXiv preprint arXiv:2102.04925 (2021)
25. Zhou, J., et al.: Graph neural networks: a review of methods and applications. arXiv preprint arXiv:1812.08434 (2018)
26. Wang, X., He, X., Cao, Y., Liu, M., Chua, T.S.: KGAT: knowledge graph attention network for recommendation. In: KDD, pp. 950–958 (2019)
27. Li, L., Lin, F., Xiahou, J., Lin, Y., Pengcheng, W., Liu, Y.: Federated low-rank tensor projections for sequential recommendation. Knowl.-Based Syst. 255, 109483 (2022)
28. Li, M., et al.: MLP4Rec: a pure MLP architecture for sequential recommendations. arXiv preprint arXiv:2204.11510 (2022)
29. Ba, J.L., Kiros, J.R., Hinton, G.E.: Layer Normalization (2016)
30. Loshchilov, I., Hutter, F.: SGDR: stochastic gradient descent with warm restarts. arXiv preprint arXiv:1608.03983 (2016)

Attribute Encryption Information Sharing Scheme Based on Blockchain Technology

Ke Zhang(✉) 🆔 and Yongli Wang 🆔

Nanjing University of Science and Technology, Nanjing 210000, China
121127223752@njust.edu.cn

Abstract. This article proposes a new attribute encryption (ABE) database sharing scheme based on blockchain technology. The data in the database is controlled more precisely by encrypting each column of data, enabling users to request data under their own permissions on demand. Implementing access control between data owners and users through blockchain and smart contracts. Data protection is achieved with the help of cloud service storage as well as blockchain. In order for transactions to proceed, the data owner assigns the necessary permissions to the user, which allows the user to access data on cloud storage and then access the data in the corresponding columns based on their permission level.

Keywords: Privacy Protection · Attribute Encryption(ABE) · Blockchain and Smart Contracts

1 Introduction

Data sharing refers to making data available to other people or organizations so that they can use, analyse or process that data. Data sharing can occur between individuals, organizations or across organizations and can involve different purposes and modalities. However, the following issues need to be kept in mind when data sharing takes place:

- Privacy protection: Ensure that the data shared does not reveal sensitive information, take measures to protect the identity of individuals, and comply with relevant laws and regulations.
- Data security: Adopting different security measures to prevent illegal access, tampering, or data leakage.
- Data quality: Ensure that the data shared are accurate, complete and reliable, providing the necessary documentation and instructions.

The use of encryption algorithms to prevent unauthorized access to confidential (private) information is the main solution in the field of data sharing. However, in the process of data encryption, there are also a series of challenges, and each algorithm that attempts to solve or mitigate the impact of these issues will increase the overall system complexity. In the previous work, the above issues are also discussed, and the related work in this section is explained in detail in the next section. In this paper, the data in the database is introduced into the field of secure sharing and the sharing scheme of database

H. Jin et al. (Eds.): IAIC 2023, CCIS 2058, pp. 376–388, 2024.
https://doi.org/10.1007/978-981-97-1277-9_29

data is given by analyzing the above problems. The main work of attribute encrypted information sharing scheme based on blockchain technology is as follows:

1. Each column of data in the database is encrypted using an attribute-based encryption scheme to provide more precise permission control over each column of data. Attribute based encryption is also used to respond to transactions to enforce access policies on data and maintain user anonymity.
2. Access polynomials are used to request and change permissions.
3. Utilize blockchain and smart contract to realize the access control between data owner and user.

2 Related Work

In blockchain system, transaction consists of two parts: transaction identity and transaction content [1]. The transaction identity is used to identify the owner of the digital asset, and the transaction content is used to represent the attributes of the digital asset itself. Blockchain transaction content privacy refers to the fact that for a transaction in blockchain system, and observers cannot confirm the specific content of the transaction, such as the transaction amount, transaction type and so on. That is mean that for the observer cannot access the content of the transaction, or cannot obtain the explicit information of the transaction content. The literature related to this is discussed below:

Reference [2] constructs an identity-based linkable ring signature scheme on the lattice based on original image sampling and rejection sampling techniques. However, this scheme focuses more on protecting the anonymity of user identities and identities in the process of privacy protection, while lacking in content. At the same time, storage performance can also be optimized. Reference [3] designed and implemented a blockchain based public key searchable encryption system. However, the method encrypts the content is stored in the cloud server, although the content privacy is protected, the privacy leakage problem on the blockchain may still occur when stored on the cloud. Reference [4] proposes a random number reusable multi-receiver public key encryption scheme based on the national encryption algorithm SM2. But this scheme generates the ciphertext generation process for each recipient, and in the process of data sharing, it is necessary to ensure the efficiency of sharing, only encrypting, storing, and sharing the relevant content. Reference [5] proposes ring signature and linkable ring signature based on SM2 digital signature algorithm. However, this scheme still involves encrypting the user's identity and hiding it in a ring, with less involvement in content protection. Reference [6] proposes a group signature based anonymous endorsement scheme for permission blockchain. This scheme mainly solves the problems of revealing the identity of the endorsing node, low fault tolerance of the endorsement consensus, and leakage of the endorsement strategy in the "execute-sort-verify" endorsement model of the permission blockchain. Reference [7] designed a blockchain based medical service system solution under group intelligence networks. This scheme mainly protects the user's privacy by share partitioning and share recovery, but what is needed for the data is to protect the related sensitive data. Reference [8] proposes a digital image batch authentication and fair-trading scheme based on digital watermarking, linear homomorphic signature, and blockchain technology. However, this scheme mainly targets digital images and does not have universality. However, the scheme is mainly for digital images and is not generalizable.

Reference [9] designed an inter-hospital electronic medical record sharing system based on blockchain technology and symmetric key hidden vector encryption (SHVE) algorithm. However, this method is still limited to a single complete piece of data. Reference [10] proposes an attribute-based searchable encryption with verifiable database (VDB-ABSE) scheme with both attribute management and verifiability. The scheme constructs a verifiable database VDB that can provide users with proofs of search results. Although the scheme provides reasonable control over permissions, it uses the concept of a verifiable database, which mainly validates the search of documents, and lacks finer-grained control over content attributes such as name, occupation, tenure, and other personal information in the personal information. Reference [11] proposes an attribute-based encryption and blockchain based personal privacy data protection scheme. This scheme has flexible control permissions that can effectively protect data, but its permissions are related to third-party applications, which relatively limits the efficiency of data sharing.

Reference [12] proposes a safe and efficient UAV-assisted disaster rescue information sharing scheme, Rescue Chain. The scheme realizes rescue information sharing, while it is relatively lacking in this aspect of smart contracts. Reference [13] proposed a new access control framework for Attribute based signature encryption. This solution only encrypts files and introduces the concept of hypergraphs to achieve permission control. However this scheme does not generate auditable logs. Reference [14] used CP-ABE to encrypt files and stored the encrypted files on a cloud server. To obtain these encrypted data, users need to send the request to the blockchain. Reference [15] proposes a collusion-resistant multi-authority ABE model for protecting private data, which is mainly targeted at the privacy data of a specific user. Reference [16] proposes a traceable attribute-based encryption with dynamic access control (TABE-DAC) scheme for achieving fine-grained sharing of private data. Due to its traceability, this scheme can monitor the behavior of abnormal users.

3 Technical Proposal

3.1 Basic Knowledge

3.1.1 Accessing Polynomial

Access polynomial [17] is used for permission control between data owners and users, which is implemented using a key. The entire process is as follows:

Firstly, the group controller sends the key K_i to the first, second,..., and nth group members through secure channels.

The controller then generates the following access polynomial using the key K_i and the system master key K_C. Then, the coefficients of the access polynomial $(\alpha_0, \alpha_1, \cdots, \alpha_{n-1})$ can be obtained and shared with all group members:

$$f(x) = K_C + \prod_{i=1}^{n}(x - K_i) = \alpha_0 + \alpha_1 x + \cdots + \alpha_{n-1}x^{n-1} + x^n \qquad (1)$$

The i-th member can retrieve the master key by $K_C = f(K_i)$ after obtaining the coefficients.

3.1.2 Attribute-Based Information Encryption

Attribute-based encryption (ABE) is an extension of id-based encryption algorithms in which the ciphertext as well as the user key depends on a set of attributes. This scheme implies that the user can decrypt the ciphertext if and only if the attributes of the ciphertext satisfy the access policy. There are two types of ABE schemes: key-policy ABE (KP-ABE) [18] and ciphertext-policy ABE (CP-ABE) [19]. This scheme uses the CP-ABE scheme, which consists of the following algorithms:

- Setup: In addition to the implicit security parameters, it describes the assignment of weights to the relevant weights for each column of the database. It generates the public parameter P, the master key K, and the corresponding data D according to each row of data in the database.
- Key generation (K, A): It uses K and an attribute group A that describes the key, which generates the private key PK.
- Encrypt(K, D, A): It takes K, data D and an access strategy on the set of attributes AS. It encrypts D and produces the ciphertext CT.
- Decrypt(K, CT, PK): It takes K, the ciphertext CT containing the access policy AS, and the private key PK for set A. If A satisfies AS, the algorithm decrypts the ciphertext and returns information about the corresponding columns in the database under the corresponding privileges.

3.2 Scheme Design

This scheme utilizes blockchain technology and attribute encryption technology to achieve data storage and sharing security, and the scheme model is shown in Fig. 1, including five participants: data service provider, data demander, cloud server, blockchain, and attribute agency.

- Data service providers who want to share data.
- Users who want to subscribe to use data from the data service provider.
- A cloud space for storing data service provider data.
- A smart contract created by each data service provider on the blockchain to achieve permission control.
- The Attribute Authority (AA) mainly creates key materials for each attribute and distributes them to users.

Both the data service provider and the demand side interact with the cloud server. The data service provider uploads data to the cloud server, while the demand side obtains data from the cloud server. In addition, both the data service provider and the demand side communicate [20] with the blockchain by invoking the function of smart contracts. It should be noted that each data service provider has its own smart contract, which is created on the blockchain during data upload. Then, the demand side contacts AA to obtain the private key corresponding to its own attributes. AA is responsible for providing attributes to users.

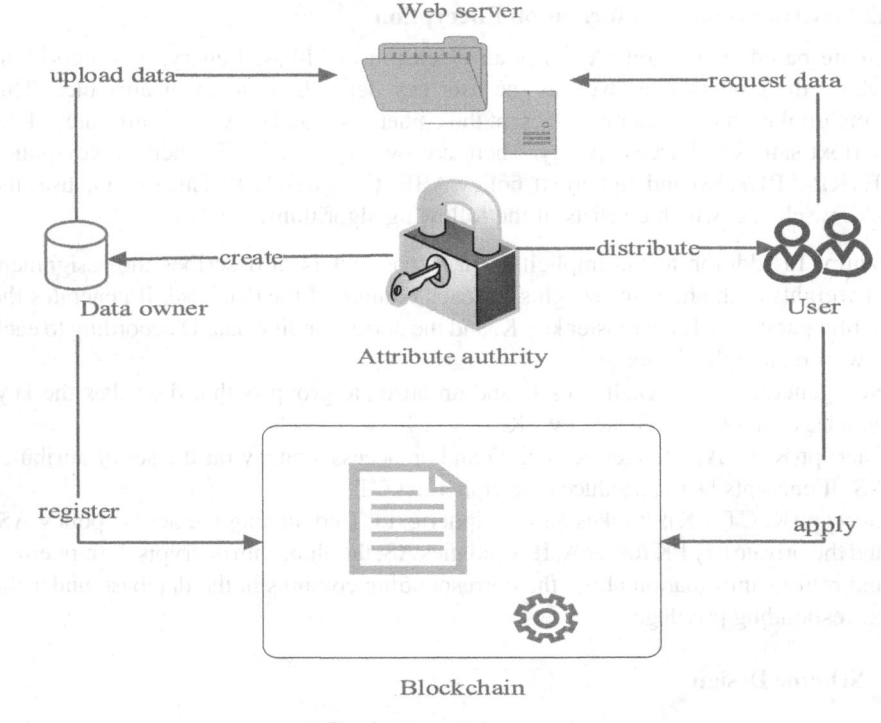

Fig. 1. Overall framework.

3.3 Scheme Description

This program includes five stages: user data preprocessing, startup, data upload, data request, and data acquisition.

User data preprocessing: Give the permission level of the data columns in the database table, encrypt each column of data in each row using symmetric encryption key, generate encrypted data and key, organize these data to generate the data to be stored.

Startup: Generate global parameters and system master key using security parameters, each node calculates partial private key according to user attributes and sends it to the user, who calculates the complete private key using the partial private key. Meanwhile, corresponding smart contracts are generated for each data owner.

Data upload: Data users utilize symmetric key to symmetrically encrypt the preprocessed data, and the encrypted data is uploaded to the cloud server.

Data request: The data demander sends its own request according to its own needs (the authority of the database columns it needs to view) in order to obtain the authority of the relevant data.

Data acquisition: The data demand side uses its own private key, attribute set and the corresponding permission level, if it meets the access structure, it can be decrypted into encrypted data in a line of the database, and at the same time give the key corresponding to the data columns that can be accessed, and it can be returned to the corresponding plaintext information, to obtain the data that can be viewed under its authority.

3.4   Concrete Step

3.4.1   User Data Preprocessing

This step is to give the process to each row of data in the database. This encryption process is relatively simple as this solution needs to refine the data to the level of each column in the database, so it is necessary to encrypt each column of data in each row, while at the same time each of the columns is a relatively small amount of data.

For each line of data in the database, each of its columns can be set to a different permission level, the level from 0 to 1 in ascending order, the default is 0, for the lowest authority, as long as the demand side to meet the corresponding attributes can be viewed. Other data is set according to the degree of confidentiality, in this process use the tuple form to represent the permission level,$(column_1, column_2, ..., column_n)$,where $column_i$ denotes the permission level of a column with a value of one of 0 to 1. Each row of data is then encrypted using a symmetric encryption algorithm to produce the ciphertext and the corresponding key, $(enM_1, enM_2, ..., enM_n),(enMK_1, enMK_2,..., enMK_n)$, where enM_i denotes the ciphertext corresponding to the encryption of a column in the database, $enMK_i$ is the key of a column in the database.

Subsequently, the encrypted ciphertext, key, and permission level are used to form a piece of data D for sharing. The algorithm operation process is shown in Alg1, $origin\_c_i$ represents the original information of a column in the database:

Alg1. Data preprocessing

Input： $(origin\_c_1, origin\_c_2 ,..., origin\_c_n), (column_1, column_2 ,..., column_n)$
Output： D

1.$(enMK_1, enMK_2 ,..., enMK_n) \leftarrow (column_1, column_2 ,..., column_n)$
2.$(enM_1, enM_2 ,..., enM_n) \leftarrow (enMK_1, enMK_2 ,..., enMK_n)|\ (origin\_c_1,$
$origin\_c_2 ,..., origin\_c_n)$
3.$M \leftarrow (enM_1, enM_2 ,..., enM_n)|(enMK_1, enMK_2 ,..., enMK_n)$

3.4.2   Startup

This step mainly involves the following process:

Notify AA to generate the corresponding key based on the attributes of the requesting party and distribute it to the corresponding requesting party. Each data service provider will create smart contracts on the blockchain. It should be noted that each data service provider will have a corresponding smart contract and notify the data demander after creation is completed. The smart contract is invoked by both the data service provider and the demand side.

3.4.3   Data Upload

The data service provider uses a symmetric algorithm that encrypts the data D using a random key K_C. In addition, the data service provider signs the encrypted data using its private key utilizing a public key signing algorithm that provides end users with data source authentication. The signature field can be inserted at the end of the encrypted data. Then, save the encrypted data to the cloud server.

3.4.4 Data Request

This algorithm is run when each demand side wants to access fields in a specific database. The operation process is shown in Alg2.

$RP_{id}, D_{id}, T_{expiry}$ and RC_{id} are the input conditions for the algorithm, which represent the requesting party id, the Data address, permission expiration time, and the database column permissions requested by requesting party. K_{temp} is a temporary key of the demand side, which is randomly generated by the demand side. Use the public key of the data service provider to encrypt the above parameters and K_{temp} to generate the request field, where $K_{pu}(ds)$ is the public key of the data service provider.

Alg2. Request access to specific data

Input: $RP_{id}, D_{id}, T_{expiry}, RC_{id}$
Output: NULL

1. $K_{temp} \leftarrow$ A randomly generated number
2. request_info $\leftarrow RP_{id}|D_{id}|T_{expiry}|RC_{id}|K_{temp}$
3. request $\leftarrow Enc_{PKI}(K_{pu}(ds),$ request_info$)$
4. request_id $\leftarrow h(RP_{id}|D_{id}|T_{expiry}|RC_{id}|K_{temp})$
5. Register (request_id, request) to the smart contract

At the same time, these parameters will also obtain the unique identifier of the request through the hash algorithm, which is the request_ id. Then, the requesting party registers the request information into the blockchain.

3.4.5 Granting Access Rights by the Data Service Provider

After the requesting party sends the request, the data service party will respond to the requesting party's request, and this process is referred to as Alg3 below. And after receiving the request from the requesting party, the data service provider decrypts the request to obtain $RP_{id}, D_{id}, T_{expiry}, RC_{id}$ and K_{temp}. Then, the data service provider generates two random numbers based on the random function, namely r_{new} and K_{id}. The data server generates a piece of data (RP_{id}、D_{id}、K_{id}、T_{expiry}、RC_{id}) and inserts it into the authorization table. The data service provider updates the access polynomial based on r_{new}, cryptographic key K_C, K_{id}, and the existing subscription key. Enable the demand side to access the data under their authority, the metadata of D_{id} is replaced by r_{new} and a new polynomial coefficient. At the same time, in order to enable the demand side to obtain data, the randomly generated K_{id} is encrypted using the demand side's temporary key. Subsequently, the encrypted results and access policy attributes are encrypted again using the system master key. Register the result res generated by the encryption to the blockchain along with the resuest_id sent by the demand side.

Alg3. Granting access rights by the data service provider

Input: request, request_id

Output: NULL

1. request_info $\leftarrow Dec_{PKI}(K_{pu}(ds)$, request)
2. Decode request_info to obtain $(RP_{id}、 D_{id}、 T_{expiry}、 RC_{id}、 K_{temp})$
3. $r_{new} \leftarrow$ a random number
4. $K_{id} \leftarrow$ a random number
5. Insert a piece of data $(RP_{id}, D_{id}, K_{id}, T_{expiry}, RC_{id})$ to authorization table
6. Select all the keys for D_{id} as $(K_1, K_2,..., K_n)$
7. Let K_C denote the cipher-key of D_{id}
8. Get $(\alpha_0, \alpha_1, \cdots, \alpha_{n-1})$ by alg. 6 with K_C, r_{new} and $(K_1, K_2,..., K_n)$
9. Update the metadata of D_{id} by $(\alpha_0, \alpha_1, \cdots, \alpha_{n-1}, r_{new})$
10. Let Γ be the access policy for getting D_{id}
11. res $\leftarrow Enc_{PKI}(PK, Enc_{SYM}(K_{id}|request\_id, K_{temp}), \Gamma)$
12. Register (res, request_id) to the smart contract

3.4.6 Data Acquisition

After obtaining permission, the demand side can obtain data, and the process of obtaining data is shown in Alg4. Firstly, the demand side decrypts the data obtained from the cloud server based on its own attributes. If the attributes of the demand side meet the access policy of the data to be accessed, the demand side can decrypt the res in the smart contract to obtain TK. Then, the requesting party decrypts the obtained TK using his/her temporary key K_{temp} to obtain req_{id} and subscription key K_{id}. Then combine the obtained req_{id} with request_id comparison, request_id is the request identifier obtained by the requesting party through the hash algorithm. If matched, the demand side will obtain encrypted data and perform signature verification operations to ensure the correctness of the data. If the signature verification is successful, the demand side can obtain the coefficients of the access polynomial to obtain the key for encrypting the data. The constructed polynomial is as follows:

$$F(x) = ((\alpha_0 + \alpha_1 x + \cdots + \alpha_{n-1}x^{n-1} + x^n)(modp)) \tag{2}$$

The process of obtaining the key can be completed through Alg5. Next, the demand side can decrypt the data based on the key to obtain the pre-processed data D. Subsequently, the corresponding columns of $enMK_i$ as well as the corresponding encrypted column data enM_i are obtained according to the access level of the user, and the data is decrypted by utilizing $enMK_i$ to obtain the data of the corresponding columns in the database to which the user has access.

Alg4. data acquisition
Input: res, request_id
Output: NULL

1. $TK \leftarrow Dec_{ABE}(K, res, PK)$, return if failed
2. $(K_{id}, req_{id}) \leftarrow Dec_{SYM}(K_{temp}, TK)$
3. if $(req_{id} \neq \text{request\_id})$ then
4. Mismatching return end if
5. Download the encrypted data D_{id}
6. Obtain sign and verify it using $K_{pu}(ds)$, return if not OK
7. Obtain the metadata of D_{id} as $(\alpha_0, \alpha_1, \cdots, \alpha_{n-1}, r)$
8. Compute K_C using Alg5 and the values $(K_{id}, \alpha_0, \alpha_1, \cdots, \alpha_{n-1}, r)$
9. Decrypt data D using K_C
10. $origin\_c_i \leftarrow Dec(enMK_i, enM_i)$

Alg5. Evaluation of cipher-key
Input: $K_{id}, r, \alpha_0, \alpha_1, \cdots, \alpha_n$
Output: K_C

1. $a \leftarrow h(K_{id}, r)$
2. $b \leftarrow 1$
3. $T \leftarrow \alpha_0$
4. for $i \leftarrow 1$ to n
5. $y \leftarrow b*a \pmod p$
6. $T \leftarrow (T + \alpha_i*b) \pmod p$
7. end for return T

3.5 Computations Related to Access Polynomial

This part is mainly determined and modified by the data service provider for accessing polynomial coefficients. And this is the main part of the access authorization operation, so it is quite important. The calculation process of this part is shown in Alg6:

In the outer loop, multiplying the current polynomial by $(x - h(K_i, r))$ yields a new polynomial coefficient:

$$F(x) = \left(\alpha_0 + \alpha_1 x + \cdots + \alpha_{i-1} x^{i-1}(x - h(K_i, r))\right)$$

$$= -\alpha_0 h(K_i, r) + (\alpha_0 - \alpha_1 h(K_i, r))x + \cdots + (\alpha_{i-2} - \alpha_{i-1} h(K_i, r)x^{i-1} + \alpha_{i-1} x^i \quad (3)$$

It should be noted that all calculation results must be modulo for p. Finally, the modulo will be added to and the final polynomial coefficients $(\alpha_0, \alpha_1, \cdots, \alpha_{n-1})$ will be returned.

Alg6. Computing the coefficients of access polynomial

Input: $K_1, K_2, ..., K_n, r, K_C$

Output: $\alpha_0, \alpha_1, \cdots, \alpha_{n-1}$

1. $\alpha_0 \leftarrow -h(K_1, r) \pmod{p}$
2. $\alpha_1 \leftarrow 1$
3. for $i \leftarrow 2$ to n
4. $\beta \leftarrow h(K_i, r)$
5. $\alpha_i \leftarrow 1$
6. for $j \leftarrow i-1$ to 1
7. $\alpha_j \leftarrow (\alpha_{j-1} - \beta * \alpha_j) \pmod{p}$
8. end for
9. $\alpha_0 \leftarrow (-\beta * \alpha_0) \pmod{p}$
10. end for
11. $\alpha_0 \leftarrow (\alpha_0 + K_C) \pmod{p}$
12. return $(\alpha_0, \alpha_1, \cdots, \alpha_{n-1})$

---

4 Security Proof

4.1 Key Security

The security of the key generation and update process is an important evaluation index of the security of the data sharing scheme, node failure or being attacked may leak part of the key information, there is a certain probability of indirectly leading to the node key information leakage, in order to better analyze the key security of this scheme to define anti-attacking index parameter anti_attacking.

Definition If z nodes fail or are attacked, when z partial keys are leaked, the following equation is the probability formula for the complete key leakage of a node.

$$p(z) = \begin{cases} 0, 0 \leq z < N \\ 1 - \sum_{i=1}^{N-1} C_N^i / C_S^z, z \geq N \end{cases} \tag{4}$$

Proof: Assume that z nodes in a distributed network of S nodes are out of trust or attacked, for any normal node P there may exist z partial keys that are all compromised, and the number of all partial keys of node P is N. When all partial keys of node P are compromised, the complete key of node P is insecure.

When $z < N$, the number of partial key leaks is less than the number of all partial keys N of node P. At that time, the complete key of node P is secure, and the probability of complete key leak is 0.

When $z \geq N$, calculate the probability that the key is secure. The key consists of N partial keys, the compromised partial key P has a total of C_S^z possibilities, to ensure the security of the complete key of node P it is necessary to ensure that z partial keys do not include all the partial keys of node P. There are a total of $\sum_{i=1}^{N-1} C_N^i C_{S-N}^{z-i}$ cases in which the compromised partial key z contains the partial key of node P.

The probability of key security of node P is $\sum_{i=1}^{N-1} C_N^i C_{S-N}^{z-i}/C_S^z$ when z nodes are out of trust or attacked, so the probability of complete key leakage of node P is p(z) $= 1-\sum_{i=1}^{N-1} C_N^i C_{S-N}^{z-i}/C_S^z$. . The node's key anti-attacking anti_attacking(P) = p(z) can be adjusted to satisfy different security requirements by adjusting the number of partial keys that generate the complete key.

4.2 Data Storage Security.

In this paper, the security of the user's encrypted data integrity depends on the blockchain structure, the blockchain is a chain storage structure, the current block has the hash value of the previous block and the next block, to ensure that the data can be traced and cannot be tampered with. To achieve the purpose of tampering with block data, the attacker needs to attack all historical blocks. Assuming that the attacker needs to tamper with the block height h, the probability that the attacker generates a block per second is p, the probability that the honest party generates a block per second is q. If the attacker generates a block a times, the honest node generates a block b times, and fails to generate a block m-a-b times in m seconds, the success of the attacker in tampering with the block implies that $a \geq b + h$, then the following equation represents the probability of the attacker in successfully tampering with the block. Where b = q(1-p),c = p(1-q). It can be found that the probability of the blockchain being modified is lower when m is larger, other conditions being certain. The height of the block to be attacked by the actual attacker to achieve the purpose of tampering is very large, so the blockchain can guarantee the security of the data on the chain.

$$p = \sum_{a=0}^{\frac{m-1-h}{2}} \sum_{i=1}^{m-h-2a} b^a c^{a+i+h} (1-b-c)^{m-i-h-2a} \frac{m!}{a!(a+i+h)!(m-i-h-2a)!} \tag{5}$$

5 Comparison

In this section, a brief comparison is made between the proposed scheme and previous data sharing schemes. The first indicator for comparison is whether smart contracts have been used, which can simplify permission control. Only the permission control framework introduced in reference [20] did not use smart contracts, while all other methods used smart contracts. The second comparative indicator is the level of data sharing, mainly considering the level of data control. Only our solution has introduced the level of columns in the database. The third indicator for comparison is the storage space used, which mainly comes in two ways: cloud and blockchain, but the cost of blockchain is higher. The fourth comparative indicator is whether the scheme has anonymity, mainly examining whether it can protect user information. The fifth indicator is auditability, which is used for review, tracking, and can also record the usage of malicious users (Table 1).

Table 1. Comparison.

Compare items	[13]	[14]	[15]	[16]	This scheme
Smart Contract	No	Yes	Yes	Yes	Yes
Data Sharing Level	File	File	File	File	Database
Storage Space	Cloud	Blockchain	Cloud	Cloud	Cloud
Anonymous	Yes	Yes	No	Yes	Yes
Auditable	No	No	No	Yes	Yes

6 Conclusion

This article proposes an attribute encryption information sharing scheme based on blockchain technology. For many enterprises, storing data is of utmost importance. Both cloud servers and blockchains can store shared data, but cloud servers are more convenient. Smart contracts are all issues to consider and can simplify control. In terms of the level of data sharing, appropriate methods can be selected according to needs. Ensure users' personal information through anonymity. Enhance the system's risk resistance through auditability while providing monitoring capabilities.

The scheme proposed in this article has been explained and compared in the above five aspects. At the same time, this scheme can be used to provide more fine-grained data sharing in practice, and has very good results in data security, integrity, and tamper resistance. In addition, due to the introduction of the encryption process for each column of data in the database. Therefore, there is still some room for improvement in this area.

References

1. Yao, S., Zhang, D.W., Li, Y., Wang, W.: A survey on privacy protection of transaction content in blockchain. J. Cryptol. Res. **9**(4), 596–618 (2022)
2. Tang, Y.L., Xia, F.F., Ye, Q., Wang, Y.J., Zhang, X.H.: Identity-based linkable ring signature on lattice. J. Cryptol. Res. **8**(2), 232–247 (2021)
3. Tang, M.F., Jiang, P., Zhang, Z.X., Li, Y.L., Zhu, L.H.: Blockchain-based PEKS system for reliability-conscious data retrieval. J. Cryptol. Res. **8**(3), 478–497 (2021)
4. Lai, J.Z., Huang, Z.A., Weng, J., Wu, Y.D.: SM2-based multi-recipient public-key encryption. J. Cryptol. Res. **8**(4), 699–709 (2021)
5. Fan, Q., He, D.B., Luo, M., Huang, X.Y., Li, D.W.: Ring signature schemes based on SM2 digital signature algorithm. J. Cryptol. Res. **8**(4), 710–723 (2021)
6. Zhang, M.W., Xia, Y.X., Zhang, Y.D., Chen, Q.X., Yang, B.: An anonymous endorsement scheme based on group signature in permissioned blockchain. J. Cryptol. Res. **9**(3), 496–510 (2022)
7. Zhong, N., Yang, B., Zhang, L.N.: Crowd intelligence network for medical service based on blockchain. J. Cryptol. Res. **9**(3), 511–523 (2022)
8. Zhang, W.B., Wang, T., Yang, B., Li, H.Y., Zhang, W.Z.: Blockchain-based batch digital images right confirmation and fair-trade scheme. J. Cryptol. Res. **9**(5), 851–871 (2022)
9. Yan, G.C., Jiang, S.R., Li, S.L., Zhang, Q.L., Zhou, Y.: Secure and efficient fuzzy search for EHR sharing based on consortium blockchain. J. Cryptol. Res. **9**(5), 805–819 (2022)

10. Chen, L.Q., Zhang, L.Y., Chen, Y.: An attribute-based searchable encryption with verifiable database. J. Cryptol. Res. **9**(5), 910–922 (2022)
11. Wang, Y.J., Cao, C.T., You, L.: A novel personal privacy data protection scheme based on blockchain and attribute-based encryption. J. Cryptol. Res. **8**(1), 14–27 (2021)
12. Wang, Y.T., Su, Z., Xu, Q.C., Li, R.D., Tom, H.L.: Lifesaving with RescueChain: energy-efficient and partition-tolerant blockchain based secure information sharing for UAV-aided disaster rescue. IEEE INFOCOM 2021, IEEE Conference on Computer Communications, pp. 1–10. Vancouver, BC, Canada (2021)
13. Mythili, R., Revathi, V., Raj, T.S.: An attribute-based lightweight cloud data access control using hypergraph structure. J. Supercomput.Supercomput. **76**(5), 6040–6064 (2020)
14. Afnan, A., Bradley, D.T.: Attribute-based access control of data sharing based on hyperledger blockchain. In: ICBCT'20: Proceedings of the 2020 The 2nd International Conference on Blockchain Technology. 1st edn. Association for Computing Machinery, New York, NY, United States (2020)
15. Noh, S., Kim, D., Cai, Z., Rhee, K.H.: A novel user collusion-resistant decentralized multi-authority attribute-based encryption scheme using the deposit on a blockchain. Wirel. Commun. Mob. Comput.. Commun. Mob. Comput. **2021**(1999), 1530–8669 (2021)
16. Guo, L.F., Yang, X.L., Yau, W.C.Y.: TABE-DAC: efficient traceable attribute-based encryption scheme with dynamic access control based on blockchain. IEEE Access **2021**(9), 8479–8490 (2021)
17. Piao, Y., Kim, J., Tariq, U., Hong, M.: Polynomial-based key management for secure intra-group communication. Comput. Math. Appl.. Math. Appl. **65**(9), 1300–1309 (2013)
18. Zhang, S., Li, W.M., Wen, Q.Y., Zhang, H., Jin, Z.P.: A flexible KP-ABE suit for mobile user realizing decryption outsourcing and attribute revocation. Wireless Pers. Commun.Commun. **114**(4), 2783–2800 (2020)
19. Guo, N., Hu, J., Deng, X.Y.: A privacy preserving CP-ABE-based access control on data sharing in VANETs. Int. J. Web Grid Serv. **19**(2), 211–232 (2023)
20. Minhaj, A.K., Khaled, S.: IoT security: review, blockchain solutions, and open challenges. Fut. Gen. Comput. Syst. **82**, 395–411 (2018)

An Epidemiological Control Strategy Model of SVEIMQR

Jingmeng Zhang$^{(\boxtimes)}$, Yulong An, and Shixing Wu

Institute of Systems Security and Control, College of Computer Science and Technology, Xi'an University of Science and Technology, Xi'an 710054, China
21208223073@stu.xust.edu.cn

Abstract. A new epidemiological model SVEIMQR (Susceptible–Vaccinated–Exposed–Infected-Mutant–Quarantined–Recovered) is proposed to explore COVID-19 transmission mechanism. Based on this model, this paper puts forward a hybrid control strategy model by considering "protection strategy" and "blocking strategy" with the purpose to reduce the number of infected. By comparing the proposed hybrid control strategy model with other strategies, the effectiveness of the hybrid control strategy model is verified. In order to reduce the number of infected and the cost as much as possible, we analyze the optimal control of the hybrid control model, prove the existence of the optimal control by using the Pontryagin's minimum principle, and get the optimal control system. The experimental results show that with the help of optimal control theory can minimize the cost of the hybrid control strategy and achieve the optimal control effect.

Keywords: Epidemiological · COVID-19 · SVEIMQR model · Hybrid control · Optimal control

1 Introduction

The outbreak of COVID-19 caused great losses to the world economy and seriously endangered the social order of all countries [1, 2]. By establishing the nonlinear dynamic model, we can deeply understand its mechanism and explore effective prevention and control strategies, which can not only alleviate the crisis, but also improve the ability to deal with new infectious diseases in the future. In the field of infectious diseases, the classic models are SIR [3–5] and SEIR [6–8]. Cooper et al. [9] simulate the trends of COVID-19 in different countries based on the SIR model, predicting the number of infected cases. The results show that the model can provide a reference for implementing interventions in the COVID-19. Denis Efimov and Rosane Ushirobira [10] use the new SEIR model of epidemic mathematics to analyze the epidemic process of COVID-19 in different countries by theoretical proof and numerical simulation.

In order to quickly curb the spread of COVID-19, the government usually formulates some epidemic prevention policies. At present, the epidemic prevention measures commonly adopted include vaccination [11, 12], isolation [13, 14] and treatment [15, 16], and the economic cost of these measures can't be ignored. Therefore, if we want

© The Author(s), under exclusive license to Springer Nature Singapore Pte Ltd. 2024
H. Jin et al. (Eds.): IAIC 2023, CCIS 2058, pp. 389–403, 2024.
https://doi.org/10.1007/978-981-97-1277-9_30

to minimize the economic cost of epidemic prevention and control, it is necessary to study the method of epidemic prevention and control, and explore the best strategy of in COVID-19. Zamir et al. [17] establish an optimal control model for the COVID-19 using non-clinical methods, considering control measures such as isolation, frequent hand washing, mask-wearing, and disinfection. Perkins et al. [18] use non-drug control and preventive intervention measures to study the optimal control analysis of COVID-19 epidemic in the United States, and with good results. Olaniyi et al. [19] propose the optimal control model based on the prevention and management of these two time-varying control measures, and analyze the most cost-effective control measures.

In this paper, a hybrid control strategy model with "protection strategy" and "blocking strategy" is proposed, and the optimal control performance of the hybrid control strategy is discussed. The structure of this paper is arranged as follows. Section 2 proposes a new SVEIMQR COVID-19 propagation model. A hybrid control strategy model is proposed in Sect. 3. In Sect. 4, the optimal control performance of the hybrid control strategy is proved theoretically and simulated experimentally. Section 5 makes a summary of this paper.

2 SVEIMQR COVID-19 Propagation Model

2.1 Model Establishment

The whole population of the SVEIMQR model can be partitioned into seven groups, namely Susceptible, Vaccinated, Exposed, Infected, Mutant, Quarantined and Recovered. At time t, the densities of Susceptible, Vaccinated, Exposed, Infected, Mutant, Quarantined and Recovered are respectively expressed as $S(t)$, $V(t)$, $E(t)$, $I(t)$, $M(t)$, $Q(t)$ and $R(t)$. The state transition diagram of SVEIMQR model is shown in Fig. 1, and all parameters of the model are described in Table 1.

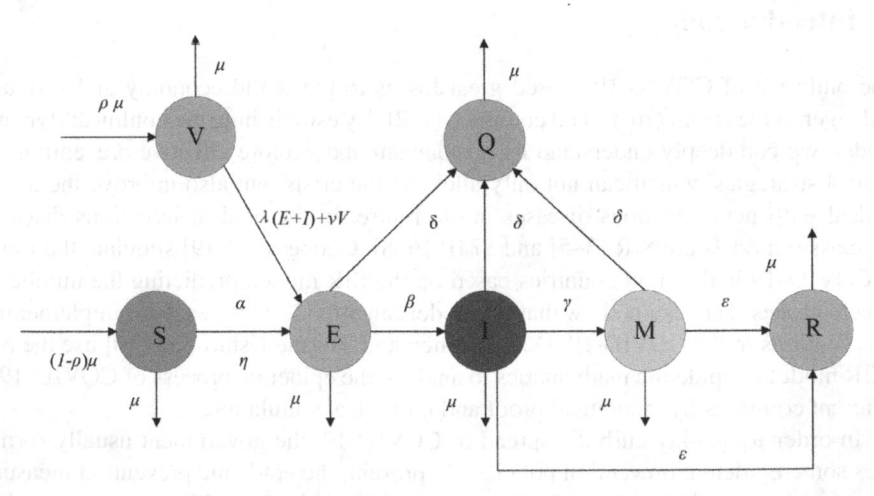

Fig. 1. State transition diagram of SVEIMQR model.

Table 1. Parameters of SVEIMQR model.

Parameter	Physical interpretation	Parameter	Physical interpretation
$S(t)$	Susceptible density	μ	Population entry and exit rate
$V(t)$	Vaccinated density	λ	Infection rate of vaccination in Exposed and Infected
$E(t)$	Exposed density	ν	Infection rate of vaccination in Mutant
$I(t)$	Infected density	α	Infection rate
$M(t)$	Mutant density	β	Exposed explosion rate
$Q(t)$	Quarantined density	γ	Mutated rate
$R(t)$	Recovered density	δ	Isolation rate
ρ	Proportion of vaccination	η	Mutated infection rate

2.2 Dynamics Behavior Analysis

The SVEIMQR (Susceptible–Vaccinated–Exposed–Infected–Mutant–Quarantined–Recovered) model is shown as system (1).

$$
\begin{cases}
\frac{dS}{dt} = (1 - \rho)\mu - \alpha S(I + E) - \eta SM - \mu S \\
\frac{dV}{dt} = \rho\mu - \lambda V(E + I) - \nu VM - \mu V \\
\frac{dE}{dt} = \alpha S(I + E) + \eta SM + \lambda V(E + I) + \nu VM - \beta E - \delta E - \mu E \\
\frac{dI}{dt} = \frac{1}{2}\beta E - \gamma I - \varepsilon I - \delta I - \mu I \\
\frac{dM}{dt} = \frac{1}{2}\beta E + \gamma I - \varepsilon M - \delta M - \mu M \\
\frac{dQ}{dt} = \delta(E + I + M) - \mu Q \\
\frac{dR}{dt} = \varepsilon I + \varepsilon M - \mu R
\end{cases}
\tag{1}
$$

Let all the right-hand terms of the system (1) be equal to 0, and the disease-free equilibrium of the system can be obtained. In addition, there is an endemic equilibrium in the system, but because the model is too complicated, the analytical solution of the disease equilibrium can't be calculated by using related calculation software, so we only analyze the disease-free equilibrium.

The basic reproduction number R_0 [20, 21] is calculated by the method of reproduction matrix [22, 23], assuming that $X = (E, I, M)$ we can know from system (1).

$$
F = \begin{pmatrix} \alpha S(I + E) + \eta SM + \lambda V(E + I) + \nu TM \\ 0 \\ 0 \end{pmatrix}
\tag{2}
$$

$$
V = \begin{pmatrix} \beta E + \mu E + \delta E \\ \delta I + \gamma I + \mu I + \varepsilon I - \frac{1}{2}\beta E \\ \mu M + \varepsilon M + \delta M - \gamma I - \frac{1}{2}\beta E \end{pmatrix}
\tag{3}
$$

$$J_F = \begin{pmatrix} \alpha S + \lambda V & \alpha S + \lambda V & \eta S + \nu V \\ 0 & 0 & 0 \\ 0 & 0 & 0 \end{pmatrix} \tag{4}$$

$$J_V = \begin{pmatrix} \beta + \mu + \delta & 0 & 0 \\ -\frac{1}{2}\beta & \delta + \varepsilon + \gamma + \mu & 0 \\ -\frac{1}{2}\beta & -\gamma & \delta + \varepsilon + \mu \end{pmatrix} \tag{5}$$

$$K = J_F \cdot J_V^{-1} = \begin{pmatrix} \alpha S + \lambda V & \alpha S + \lambda V & \eta S + \nu V \\ 0 & 0 & 0 \\ 0 & 0 & 0 \end{pmatrix}$$
$$\cdot \begin{pmatrix} \beta + \mu + \delta & 0 & 0 \\ -\frac{1}{2}\beta & \delta + \varepsilon + \gamma + \mu & 0 \\ -\frac{1}{2}\beta & -\gamma & \delta + \varepsilon + \mu \end{pmatrix} \tag{6}$$

$$K = J_F \cdot J_V^{-1} = \begin{pmatrix} \alpha S + \lambda V & \alpha S + \lambda V & \eta S + \nu V \\ 0 & 0 & 0 \\ 0 & 0 & 0 \end{pmatrix}$$
$$\cdot \begin{pmatrix} \beta + \mu + \delta & 0 & 0 \\ -\frac{1}{2}\beta & \delta + \varepsilon + \gamma + \mu & 0 \\ -\frac{1}{2}\beta & -\gamma & \delta + \varepsilon + \mu \end{pmatrix} \tag{7}$$

Thus:

$$R_0 = \frac{\begin{matrix} 2(\lambda\rho + (1-\rho)\alpha)(\delta+\varepsilon+\mu)(\delta+\varepsilon+\gamma+\mu) + \beta(\lambda\rho + (1-\rho)\alpha)(\delta+\varepsilon+\mu) \\ + \beta(\nu\rho + (1-\rho)\eta)(\delta+\varepsilon+2\gamma+\mu) \end{matrix}}{2(\beta+\delta+\mu)(\delta+\varepsilon+\gamma+\mu)(\delta+\varepsilon+\mu)} \tag{8}$$

3 Hybrid Control Strategy

3.1 Proposition of SVEIMQR Hybrid Control Strategy

The epidemic prevention in COVID-19 is mainly divided into two parts: one is to make susceptible have certain immunity to the virus through vaccination, which is called "protection strategy"; The other is to block the contact between infected and the outside world through medical isolation, which is called "blocking strategy". At present, the control strategy of COVID-19 model is relatively simple, without considering the combination of multiple strategies, which has certain limitations and is difficult to effectively control the spread of COVID-19. In this paper, a new hybrid control strategy model is proposed by considering "protection strategy" and "blocking strategy". Compared with single protection, single blocking and without control strategy, the effectiveness of the hybrid control strategy model is verified.

The hybrid control of the system is realized with a certain probability, and the protection strategy is realized by introducing the parameter p_1 into the SVEIMQR model, and the blocking strategy is realized by introducing the parameters p_2, p_3 and p_4. We assume that the probability of vaccination of susceptible is p_1, and the probability of isolation of exposed, infected and mutant is p_2, p_3 and p_4 respectively. According to the above analysis, the dynamic system of SVEIMQR hybrid control strategy can be obtained, and as shown in system (9). The state transition diagram of SVEIMQR hybrid control model is shown in Fig. 2.

$$\begin{cases} \frac{dS}{dt} = (1-\rho)\mu - \alpha S(I+E) - \eta SM - \mu S - p_1 S \\ \frac{dV}{dt} = \rho\mu - \lambda V(E+I) - \nu VM - \mu V + p_1 S \\ \frac{dE}{dt} = \alpha S(I+E) + \eta SM + \lambda V(E+I) + \nu VM - \beta E - p_3 E - \mu E \\ \frac{dI}{dt} = \frac{1}{2}\beta E - \gamma I - \varepsilon I - p_2 I - \mu I \\ \frac{dM}{dt} = \frac{1}{2}\beta E + \gamma I - \varepsilon M - p_4 M - \mu M \\ \frac{dQ}{dt} = p_3 E + p_4 M + p_2 I - \mu Q \\ \frac{dR}{dt} = \varepsilon I + \varepsilon M - \mu R \end{cases} \qquad (9)$$

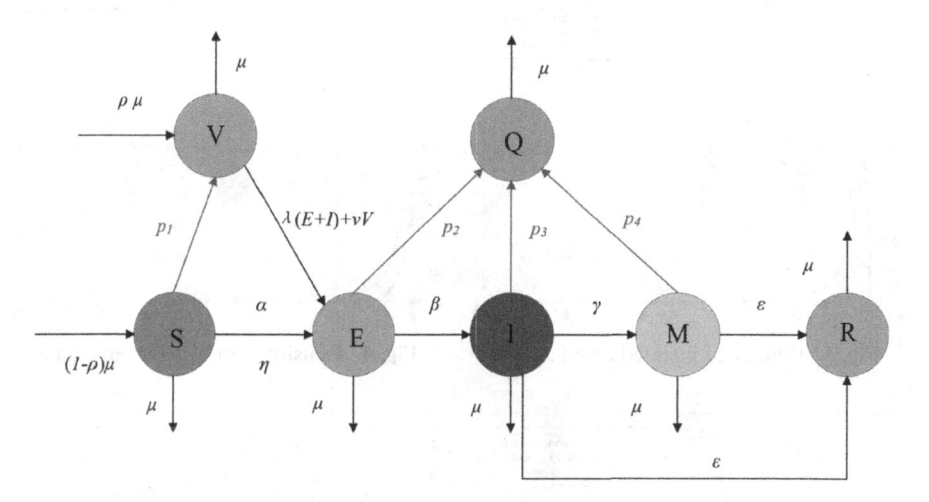

Fig. 2. State transition diagram of SVEIMQR hybrid control model.

3.2 SVEIMQR Hybrid Control Strategy Model Simulation Verification

In order to intuitively judge whether the hybrid control strategy is effective, we compare other strategies with the hybrid control strategy by experimental simulation. The first is the proposed hybrid control strategy model, the second is to add a single blocking strategy, the third is to add a single protection strategy, and the fourth is not to add any strategy. The parameter values of the hybrid control strategy are shown in Table 2, and parameters $\mu \sim \rho$ are common to the four strategies. The difference is that the single blocking strategy contains only parameters p_2, p_3 and p_4, the single protection strategy contains only parameter p_1, and the hybrid control strategy contains parameters p_1, p_2, p_3 and p_4. Using data 1–6 in Table 2, and the curves of infected density under six different data as shown in Figs. 3, 4, 5, 6, 7 and 8.

Table 2. Parameter Values of Hybrid Control Strategy.

Parameter	μ	α	β	γ	ε	η	λ	ν	ρ	p_1	p_2	p_3	p_4
Data 1	0.10	0.60	0.70	0.05	0.15	0.10	0.10	0.10	0.10	0.20	0.20	0.10	0.10
Data 2	0.10	0.60	0.70	0.05	0.15	0.10	0.10	0.10	0.10	0.85	0.50	0.20	0.20
Data 3	0.10	0.20	0.10	0.05	0.10	0.30	0.20	0.10	0.10	0.85	0.50	0.20	0.20
Data 4	0.10	0.20	0.10	0.05	0.10	0.30	0.20	0.10	0.10	0.20	0.20	0.10	0.10
Data 5	0.85	0.50	0.20	0.20	0.85	0.50	0.20	0.20	0.85	0.85	0.50	0.20	0.20
Data 6	0.85	0.50	0.20	0.20	0.85	0.50	0.20	0.20	0.85	0.20	0.20	0.10	0.10

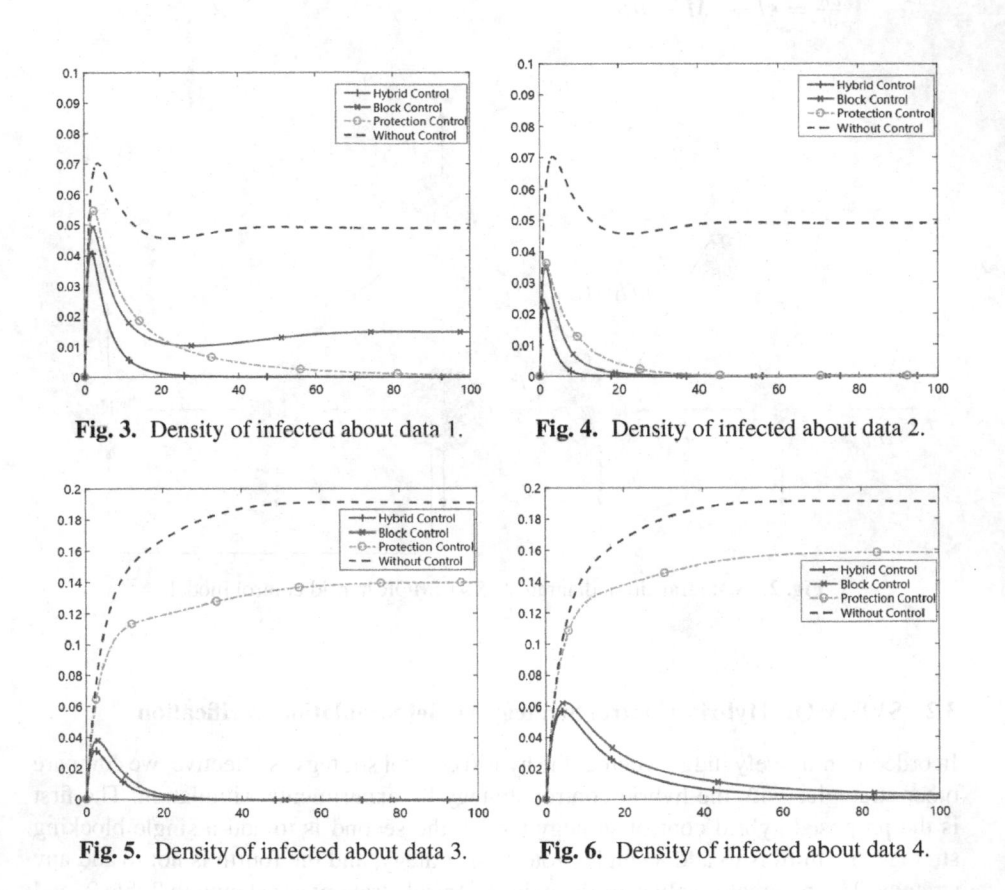

Fig. 3. Density of infected about data 1.　　　**Fig. 4.** Density of infected about data 2.

Fig. 5. Density of infected about data 3.　　　**Fig. 6.** Density of infected about data 4.

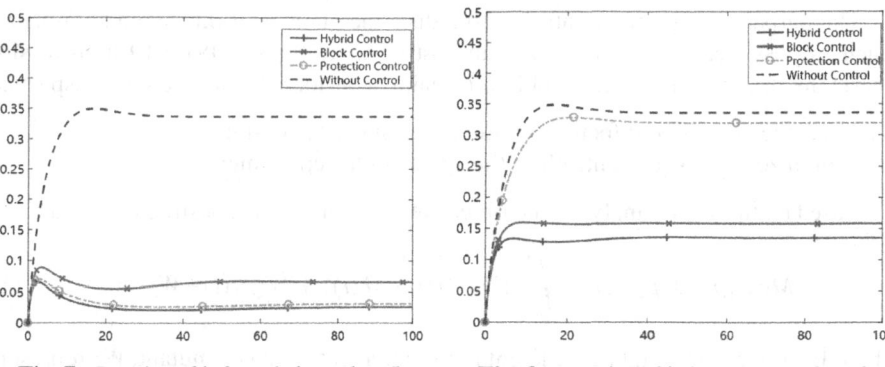

Fig. 7. Density of infected about data 5.   **Fig. 8.** Density of infected about data 6.

Through the above six groups of experimental simulations, it can be seen that the density of infected in COVID-19 is the highest without any control strategy. From Figs. 3, 4, 5 and 6 show that the hybrid control strategy successfully blocked the spread of COVID-19 and effectively controlled the spread of the epidemic. As can be seen from Figs. 7 and 8, the density of infected did not tend to die out but reached a stable state after a period of time, indicating that the hybrid control model failed to control the spread of the virus in the end. While, the simulation results show that the hybrid control strategy is obviously superior to the other two single protection and single blocking strategies in controlling the density of infected.

In a word, applying hybrid control strategy (combination of "protection strategy" and "blocking strategy") is obviously superior to any single control strategy. Therefore, the hybrid control strategy can effectively control the spread speed and infected scale of COVID-19.

4 Optimal Control Strategy

4.1 Theoretical Analysis

The hybrid control strategy model in system (9) is modified and the optimal hybrid control strategy is proposed. In order to simplify the calculation, we choose p_1 as the control parameter of protection strategy and choose p_2 as the control parameter of blocking strategy to realize the optimal hybrid control system.

The optimal hybrid control strategy model is shown in system (10).

$$\begin{cases} \frac{dS}{dt} = (1-\rho)\mu - \alpha S(I+E) - \eta SM - \mu S - p_1(t)S \\ \frac{dV}{dt} = \rho\mu - \lambda V(E+I) - \nu VM - \mu V + p_1(t)S \\ \frac{dE}{dt} = \alpha S(I+E) + \eta SM + \lambda V(E+I) + \nu VM - \beta E - p_3 E - \mu E \\ \frac{dI}{dt} = \frac{1}{2}\beta E - \gamma I - \varepsilon I - p_2(t)I - \mu I \\ \frac{dM}{dt} = \frac{1}{2}\beta E + \gamma I - \varepsilon M - p_4 M - \mu M \\ \frac{dQ}{dt} = p_3 E + p_4 M + p_2(t)I - \mu Q \\ \frac{dR}{dt} = \varepsilon I + \varepsilon M - \mu R \end{cases} \qquad (10)$$

The purpose of optimal control is to reduce the number of infected and mutant as much as possible, and to minimize the cost of controlling COVID-19 transmission. Therefore, the optimal control problem is mainly considered from these two aspects:

(1) Make the density of infected and mutant as small as possible;
(2) Minimize the cost of controlling the spread of the epidemic;

Based on the above analysis, the objective cost function is constructed as Eq. (11).

$$MinJ(p_1(t), p_2(t)) = \int_{t_0}^{t_f} [W_1(I(t) + M(t)) + W_2 p_1^2(t) + W_3 p_2^2(t)] \tag{11}$$

where W_1 is the cost weight coefficient representing infected and mutant, W_2 represents the cost weight coefficient for implementing protection policies, W_3 represents the cost weight coefficient for implementing blocking strategy, $W_2 p_1^2(t) > 0$ and $W_3 p_2^2(t) > 0$ represent the expenses by implementing protection policies and blocking policies, respectively. When $p_1(t) = p_2(t) = 1$, it means that the protection strategy and blocking strategy are the strongest, and the impact on the system is also the greatest.

The optimal control variable is determined by Eq. (12).

$$J(p_1^*(t), p_2^*(t)) = Min\{ J(p_1(t), p_2(t)) | (p_1(t), p_2(t)) \in U\} \tag{12}$$

where $U = \{(p_1(t), p_2(t)) | (p_1(t), p_2(t)) \in L^1[t_0, t_f] p_i \le p_i(t) \le \overline{p_i}, i = 1, 2\}$ is admissible control set, and $p_1(t), p_2(t)$ are optimal control variable.

Theorem 1. Under the restriction of system (10), Eq. (11) has optimal control $p_1*(t)$ 、 $p_2*(t)$, and shown as follows.

$$J(p_1^*(t), p_2^*(t)) = Min\{ J(p_1(t), p_2(t)) | (p_1(t), p_2(t)) \in U\}$$

Proof. According to Fleming and Rishel's theory [24], the Eq. (11) has optimal control and must meet the following conditions:

(1) Existence of $(p_1(t), p_2(t)) \in U$ makes the system (10) have a solution;
(2) The control set U is convex closed set;
(3) The solution of system (10) is bounded by a linear function;
(4) The integrand of the objective cost function (111) is a convex function on the control set U, and exist $c_2, c_1 > 0, k > 1$, make $c_1(|p_1|^k + |p_2|^k) - c_2$ is the lower bound of the integrand of the objective cost function (11).

Because the solution of system (10) is not empty, condition (1) is satisfied. According to the definition, the control set U is a convex closed set, and the condition (2) is satisfied. The solution of system (10) is bounded, so condition (3) is satisfied. The integrand function of the objective cost function (11) $W_1(I(t) + M(t)) + W_2 p_1^2(t) + W_3 p_2^2(t)$ is a convex function on the control set U, make $c_1 = min (W_2, W_3), k = 2$, and because $W_1(I(t) + M(t))$ boundedness, known existence c_2, make $W_1(I(t) + M(t)) + W_2 p_1^2(t) + W_3 p_2^2(t) \ge c_1(|p_1(t)|^k + |p_2(t)|^k) - c_2$.

Thus, the system (10) satisfies the above four conditions, and the system has optimal control, existence $p_1^*(t), p_2^*(t)$ make $(p_1(t), p_2(t)) \in U$.

Next, the necessary conditions of optimal control are discussed by using Pontryagin's maximum principle [25–27]. Taking the system (10) as the constraint condition of the objective cost function, the auxiliary variables, section conditions and control variables of the system (10) are solved by using Pontryagin's maximum principle.

The Hamilton function [28] is shown as Eq. (13).

$$
\begin{aligned}
H =\ & W_1(I(t) + M(t)) + W_2 p_1^2(t) + W_3 p_2^2(t) \\
& + \chi_1(t)[(1 - \rho)\mu - \alpha S(I + E) - \eta SM - \mu S - p_1(t)S] \\
& + \chi_2(t)[\rho\mu - \lambda V(E + I) - \nu VM - \mu T + p_1(t)S] \\
& + \chi_3(t)[\alpha S(I + E) + \eta SM + \lambda V(E + I) + \nu VM - \beta E - p_3 E - \mu E] \\
& + \chi_4(t)[\frac{1}{2}\beta E - \gamma I - \varepsilon I - p_2(t)I - \mu I] \\
& + \chi_5(t)[\frac{1}{2}\beta E + \gamma I - \varepsilon M - p_4 M - \mu M] \\
& + \chi_6(t)[p_2(t)I + p_3 E + p_4 M - \mu Q] \\
& + \chi_7(t)[\varepsilon I + \varepsilon M - \mu R]
\end{aligned}
\tag{13}
$$

Given the controlled variables $p_1^*(t)$, $p_2^*(t)$ and the stable solution of the system, there are auxiliary variables $\chi_i(t)$, $i = 1, 2,\ldots, 7$ satisfying the Eq. (14).

$$
\begin{cases}
\frac{d\chi_1(t)}{dt} = -\frac{dH}{dS(t)} = \chi_1(t)[\alpha(E + I) + \eta M + p_1(t) + \mu] \\
\quad -\chi_2(t)p_1(t) - \chi_3(t)[\alpha(E + I) + \eta M] \\
\frac{d\chi_2(t)}{dt} = -\frac{dH}{dV(t)} = \chi_2(t)[\lambda(E + I) + \nu M + \mu] - \chi_3(t)[\lambda(E + I) + \nu M] \\
\frac{d\chi_3(t)}{dt} = -\frac{dH}{dE(t)} = \chi_1(t)\alpha S + \chi_2(t)\lambda V - \chi_3(t)[\alpha S + \lambda V - \beta - p_3 - \mu] \\
\quad -\frac{1}{2}\chi_4(t)\beta - \frac{1}{2}\chi_5(t)\beta - \chi_6(t)p_3 \\
\frac{d\chi_4(t)}{dt} = -\frac{dH}{dI(t)} = -W_1 + \chi_1(t)\alpha S + \chi_2(t)\lambda V - \chi_3(t)[\alpha S + \lambda V] \\
\quad +\chi_4(t)[\gamma + \varepsilon + p_2(t) + \mu] - \chi_5(t)\gamma - \chi_6(t)p_2(t) - \chi_7(t)\varepsilon \\
\frac{d\chi_5(t)}{dt} = -\frac{dH}{dM(t)} = -W_1 + \chi_1(t)\eta S + \chi_2(t)\nu V - \chi_3(t)[\eta S + \nu V] \\
\quad +\chi_5(t)[\varepsilon + p_4 + \mu] - \chi_6(t)p_4 - \chi_7(t)\varepsilon \\
\frac{d\chi_6(t)}{dt} = -\frac{dH}{dQ(t)} = \chi_6(t)\mu \\
\frac{d\chi_7(t)}{dt} = -\frac{dH}{dQ(t)} = \chi_7(t)\mu
\end{cases}
\tag{14}
$$

Make $\frac{\partial H}{\partial p_1} = 0$, $\frac{\partial H}{\partial p_2} = 0$, we can get Eqs. (15) to (17).

$$
p_1^*(t) = Max\left\{min\left\{\frac{[\chi_1(t) + \chi_2(t)]S}{2W_2}, \overline{p_1}\right\}, \underline{p_1}\right\}
\tag{15}
$$

$$
p_2^*(t) = Max\left\{min\left\{\frac{[\chi_4(t) - \chi_6(t)]I}{2W_3}, \overline{p_2}\right\}, \underline{p_2}\right\}
\tag{16}
$$

4.2 Numerical Simulation

From the experimental results of Figs. 3, 4, 5, 6, 7 and 8, we can know that the hybrid control strategy can control the spread of COVID-19 more effectively than others strategy. The effective control of COVID-19 transmission is achieved by statically adjusting the values of parameters p_1, p_2, p_3 and p_4 in this paper. When the values of p_1, p_2, p_3 and p_4 are all equal to 1, it shows that the hybrid control is the strongest, which will make the disease control effect best, but on the contrary, it will also produce greater costs. Therefore, in order to minimize the number of infected and the cost, we use the optimal control model to dynamically adjust the values of control variables. For the convenience of calculation, parameters p_1 and p_2 are selected to represent the optimal control variables of protection strategy and blocking strategy respectively.

Select the parameter $\mu = 0.10$, $\alpha = 0.10$, $\beta = 0.25$, $\gamma = 0.05$, $\varepsilon = 0.20$, $\eta = 0.12$, $\lambda = 0.25$, $\nu = 0.15$, $\rho = 0.05$, $p_3 = 0.08$, $p_4 = 0.05$, $p_1 = 0.01$, $\overline{p_1} = 1$, $p_2 = 0.01$, $\overline{p_2} = 1$, $W_1 = 0.75$, $W_2 = 0.15$, $W_3 = 0.15$, $t_0 = 0$, $t_f = 140$, $N = 40$. The initial value of the setting state is $(S(0), V(0), E(0), I(0), M(0), Q(0), R(0)) = (0.60, 0.10, 0.05, 0.10, 0.10, 0.05, 0)$. The optimal control variables changing with time is shown in Fig. 9.

$$
\begin{cases}
\frac{dS}{dt} = (1 - \rho)\mu - \alpha S(I + E) - \eta SM - \mu S - p_1(t)S \\
\frac{dT}{dt} = \rho\mu - \lambda V(E + I) - \nu VM - \mu T + p_1(t)S \\
\frac{dE}{dt} = \alpha S(I + E) + \eta SM + \lambda V(E + I) + \nu VM - \beta E - p_3 E - \mu E \\
\frac{dI}{dt} = \frac{1}{2}\beta E - \gamma I - \varepsilon I - p_2(t)I - \mu I \\
\frac{dM}{dt} = \frac{1}{2}\beta E + \gamma I - \varepsilon M - p_4 M - \mu M \\
\frac{dQ}{dt} = p_2(t)I + p_3 E + p_4 M - \mu Q \\
\frac{dR}{dt} = \varepsilon I + \varepsilon M - \mu R \\
\frac{d\chi_1(t)}{dt} = -\frac{dH}{dS(t)} = \chi_1(t)[\alpha(E + I) + \eta M + p_1(t) + \mu] - \chi_2(t)p_1(t) \\
\qquad -\chi_3(t)[\alpha(E + I) + \eta M] \\
\frac{d\chi_2(t)}{dt} = -\frac{dH}{dV(t)} = \chi_2(t)[\lambda(E + I) + \nu M + \mu] - \chi_3(t)[\lambda(E + I) + \nu V] \\
\frac{d\chi_3(t)}{dt} = -\frac{dH}{dE(t)} = \chi_1(t)\alpha S + \chi_2(t)\lambda V - \chi_3(t)[\alpha S + \lambda V - \beta - p_3 - \mu] \\
\qquad -\frac{1}{2}\chi_4(t)\beta - \frac{1}{2}\chi_5(t)\beta - \chi_6(t)p_3 \\
\frac{d\chi_4(t)}{dt} = -\frac{dH}{dI(t)} = -W_1 + \chi_1(t)\alpha S + \chi_2(t)\lambda V \\
\qquad -\chi_3(t)[\alpha S + \lambda V] \\
\qquad +\chi_4(t)[\gamma + \varepsilon + p_2(t) + \mu] - \chi_5(t)\gamma - \chi_6(t)p_2(t) - \chi_7(t)\varepsilon \\
\frac{d\chi_5(t)}{dt} = -\frac{dH}{dM(t)} = -W_1 + \chi_1(t)\eta S + \chi_2(t)\nu V - \chi_3(t)[\eta S + \nu V] \\
\qquad +\chi_5(t)[\varepsilon + p_2(t) + \mu] - \chi_6(t)p_4 - \chi_7(t)\varepsilon \\
\frac{d\chi_6(t)}{dt} = -\frac{dH}{dQ(t)} = \chi_6(t)\mu \\
\frac{d\chi_7(t)}{dt} = -\frac{dH}{dQ(t)} = \chi_7(t)\mu
\end{cases}
\tag{17}
$$

The optimal control of hybrid control strategy is compared with the other three control strategies to verify the optimal control performance. Other three control strategies are hybrid control strategy, single protection strategy and single blocking strategy, respectively. The objective cost function (11) is used to evaluate the cost of implementing these four control strategies. The objective cost functions of optimal control strategy, hybrid control strategy, single protection strategy and single blocking strategy are defined as J, J_h, J_p and J_b, respectively.

Set parameter $p_1 = 0.01$, $\overline{p_1} = 1$, $p_2 = 0.01$, $\overline{p_2} = 1$, $W_1 = 0.75$, $W_2 = 0.15$, $W_3 = 0.15$, $t_0 = 0$, $t_f = 10$, $N = 20$, and set $(S(0), V(0), E(0), I(0), M(0), Q(0), R(0)) = (0.60, 0.10, 0.05, 0.10, 0.10, 0.05, 0)$. The parameter values of different control strategies are shown in Table 3.

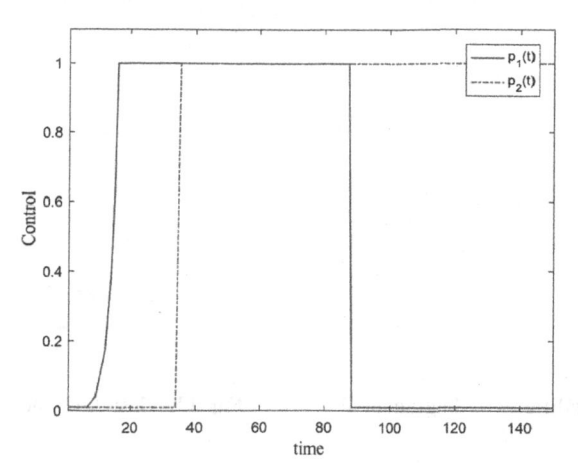

Fig. 9. The graph of optimal control variables with time.

Table 3. Parameter Values of Different Control Strategies.

Parameter	μ	α	β	γ	ε	η	λ	ν	ρ	p_1	p_2	p_3	p_4
Hybrid control	0.10	0.10	0.25	0.05	0.2	0.12	0.25	0.15	0.05	0.75	0.65	0.08	0.05
Single protection	0.10	0.10	0.25	0.05	0.2	0.12	0.25	0.15	0.05	0.75	0	0	0
Single blocking	0.10	0.10	0.25	0.05	0.2	0.12	0.25	0.15	0.05	0	0.65	0.08	0.05
Optimal control	0.10	0.10	0.25	0.05	0.2	0.12	0.25	0.15	0.05	$p_1^*(t)$	$p_2^*(t)$	0.08	0.05

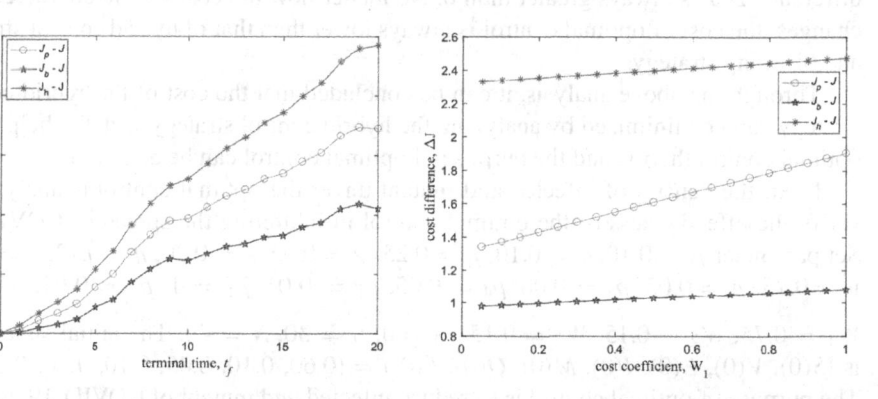

Fig. 10. Cost difference diagram of t_f.   **Fig. 11.** Cost difference diagram of W_1.

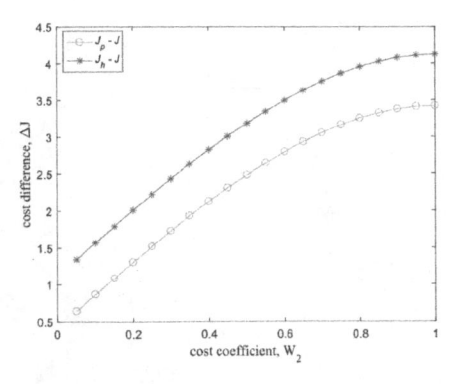

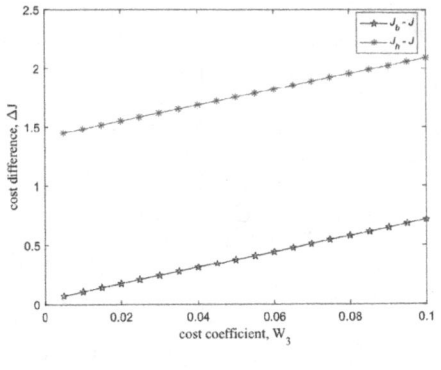

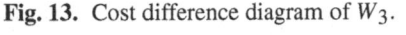

Fig. 12. Cost difference diagram of W_2.　　**Fig. 13.** Cost difference diagram of W_3.

Figure 10 describes the influence of different terminal time t_f on the cost difference of objective cost functions J_h, J_p, J_b and J. It can be seen from Fig. 10 that the cost difference ΔJ is always greater than 0. Experiments show that no matter how the terminal time t_f changes, and the cost of optimal control is always the lowest. Figure 11 describes the influence of different cost weight coefficient W_1 on the cost difference of objective cost functions J_h, J_p, J_b and J. It is known that the cost difference ΔJ is always greater than 0. Results show that the optimal control cost of hybrid control strategy is always the lowest no matter how the cost weight coefficient W_1 changes.

Figure 12 describes the influence of different cost weight coefficients W_2 on the cost difference of objective cost functions J_h, J_p, and J. It can be seen that the cost difference ΔJ is always greater than 0. No matter how the cost weight coefficient W_2 changes, and the cost of optimal control is always lower than that of hybrid control strategy and protection strategy. Figure 13 describes the influence of different cost weight coefficients W_3 on the cost difference of objective cost functions J_b, J_p, and J. It is show that the cost difference ΔJ is always greater than 0. No matter how the cost weight coefficient W_3 changes, the cost of optimal control is always lower than that of hybrid control strategy and blocking strategy.

Through the above analysis, it can be concluded that the cost of the hybrid control strategy can be minimized by analyzing the hybrid control strategy with the help of the optimal control theory, and the purpose of optimal control can be achieved.

Next, the density of infected and mutant under the optimal control is analyzed to verify the effectiveness of the optimal control in inhibiting the spread of COVID-19. Set parameter $\mu = 0.10$, $\alpha = 0.10$, $\beta = 0.25$, $\gamma = 0.05$, $\varepsilon = 0.20$, $\eta = 0.12$, $\lambda = 0.25$, $\nu = 0.15$, $\rho = 0.05$, $p_3 = 0.08$, $p_4 = 0.05$, $p_1 = 0.01$, $\overline{p_1} = 1$, $p_2 = 0.01$, $\overline{p_2} = 1$, $W_1 = 0.75$, $W_2 = 0.15$, $W_3 = 0.15$, $t_0 = 0$, $\overline{t_f} = 30$, $N = 40$. The initial state value is $(S(0), V(0), E(0), I(0), M(0), Q(0), R(0)) = (0.60, 0.10, 0.05, 0.10, 0.10, 0.05, 0)$. The purpose of optimal control is to reduce infected and mutant of COVID-19, and the density curve evolution diagram of infected and mutant under optimal control and hybrid control is shown in Fig. 14. Simulation result show that the optimal control model of hybrid control can effectively reduce the density of infected and mutant, which is helpful for the prevention and control of epidemic situation in COVID-19.

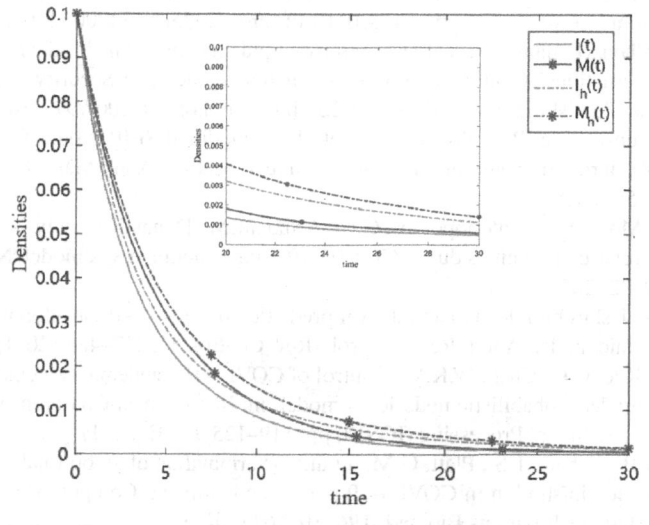

Fig. 14. Evolution diagram of states $I(t)$ and $M(t)$ under optimal control and hybrid control.

5   Conclusion

This paper proposes a new SVEIMQR model to study COVID-19 transmission mechanism, and by next generation matrix method to calculate the basic reproduction number of the proposed model. In order to reduce the number of infected we put forward a hybrid control strategy model by combining "protection strategy" and "blocking strategy" based on SVEIMQR model. Compared with other control strategies, the effectiveness of the hybrid control strategy model is verified. Finally, With the aim of minimize cost expenditure, we propose an optimal control system based on the hybrid control strategy model. The simulation results show that the cost of hybrid control strategy can be minimized and the optimal control effect can be achieved by analyzing the hybrid control strategy model with the help of optimal control theory.

References

1. Huang, C., et al.: Clinical features of patients infected with 2019 novel coronavirus in Wuhan China. Lancet **395**(10223), 497–506 (2020)
2. Li, T., Guo, Y.: Optimal control and cost-effectiveness analysis of a new COVID-19 model for Omicron strain. Physica A Stat. Mech. Appl. **606**, 128134 (2022)
3. Qian, Y.: A Non-autonom SIR model in epidemiology. In: Sun, X., Zhang, X., Xia, Z., Bertino, E. (eds.) ICAIS 2022. LNCS, vol. 13339, pp. 230–238. Springer, Cham (2022). https://doi.org/10.1007/978-3-031-06788-4_20

4. Kudryashov, N.A., Chmykhov, M.A., Vigdorowitsch, M.: Analytical features of the SIR model and their applications to COVID-19. Appl. Math. Model. **90**, 466–473 (2021)
5. Hethcote, H.: The mathematics of infectious diseases. SIAM Rev. **42**(4), 599–653 (2020)
6. Kudryashov, N.A., Chmykhov, M.A., Vigdorowitsch, M.: Comparison of some COVID-19 data with solutions of the SIR-model. AIP Conf. Proc. **2425**(1), 340009 (2022)
7. Qiao, W., Chen, B., Jiang, W., et al.: Research on Epidemic Spreading Model Based on Double Groups. International Conference on Artificial Intelligence and Security. Cham: Springer International Publishing. 1586, 75–85 (2022). https://doi.org/10.1007/978-3-031-06767-9_6
8. Diaz, P., Constantine, P., Kalmbach, K., et al.: A modified SEIR model for the spread of Ebola in Western Africa and metrics for resource allocation. Appl. Math. Comput.Comput. **324**, 141–155 (2018)
9. Cooper, I., Mondal, A., Antonopoulos, C.G., Arindam, M.: Dynamical analysis of the infection status in diverse communities due to COVID-19 using a modified SIR model. Nonlinear Dyn. **109**(1), 19–32 (2022)
10. Efimov, D., Ushirobira, R.: On an interval prediction of COVID-19 development based on a SEIR epidemic model. Annu. Rev. Control.. Rev. Control. **51**, 477–487 (2021)
11. Chen, M., Kuo, C.L., Chan, W.K.V.: Control of COVID-19 Pandemic: vaccination strategies simulation under probabilistic node-level model. In: International Conference on Intelligent Computing and Signal Processing (ICSP), pp. 119–125. IEEE (2021)
12. Libotte, G.B., Lobato, F.S., Platt, G.M., et al.: Determination of an optimal control strategy for vaccine administration in COVID-19 pandemic treatment. Comput. Methods Programs Biomed.. Methods Programs Biomed. **196**, 105664 (2020)
13. Kumar, A., Arora, S., Sambhav, S.: SEIR epidemiology modelling with restricted mobilities in COVID-19. In: IEEE International Conference on Electronics, Computing and Communication Technologies (CONECCT), pp. 1–5. IEEE (2021)
14. Dickens, B.L., Koo, J.R., Lim, J.T., et al.: Modelling lockdown and exit strategies for COVID-19 in Singapore. The Lancet Regional Health–Western Pacific 1 (2020)
15. Nainggolan, J., Harianto, J., Tasman H.: An optimal control of prevention and treatment of COVID-19 spread in Indonesia. Commun. Math. Biol. Neurosci. 2023 (2023). Article ID 3
16. Khoshnaw, S., Mohammed, A.S.: Minimizing the effects of COVID-19 using optimal control strategies (2023). https://doi.org/10.22541/au.168749645.54796660/v1
17. Zamir, M., Abdeljawad, T., Nadeem, F., et al.: An optimal control analysis of a COVID-19 model. Alex. Eng. J. **60**(3), 2875–2884 (2021)
18. Perkins, T.A., España, G.: Optimal control of the COVID-19 pandemic with nonpharmaceutical interventions. Bull. Math. Biol. **82**(9), 1–24 (2020)
19. Olaniyi, S., Obabiyi, O.S., Okosun, K.O., et al.: Mathematical modelling and optimal cost-effective control of COVID-19 transmission dynamics. Eur. Phys. J. Plus. **135**(11), 938 (2020)
20. Alimohamadi, Y., Taghdir, M., Sepandi, M.: Estimate of the basic reproduction number for COVID-19: a systematic review and meta-analysis. J. Prev. Med. Public Health **53**(3), 151 (2020)
21. D'Arienzo, M., Coniglio, A.: Assessment of the SARS-CoV-2 basic reproduction number, R_0, based on the early phase of COVID-19 outbreak in Italy. Biosafety Health. **2**(2), 57–59 (2020)
22. Van, den. Driessche. P., Watmough, J.: Reproduction numbersand sub-threshold endemic equilibria for compartmental models of disease transmission. Math. Biosci. **180**(1–2), 29–48 (2002)
23. Aghdaoui, H., Alaoui, A.L., Nisar, K.S., et al.: On analysis and optimal control of a SEIRI epidemic model with general incidence rate. Results Physis **20**, 103681 (2021)
24. Fleming, W.H., Rishel, R.W.: Deterministic and Stochastic Optimal Control. Springer, New York (2012). https://doi.org/10.1007/978-1-4612-6380-7

25. Chukwu, C.W., Alqahtani, R.T., Alfiniyah, C., et al.: A Pontryagin's maximum principle and optimal control model with cost-effectiveness analysis of the COVID-19 epidemic. Decis. Analyt. J. **8**, 100273 (2023)

26. Madubueze, C.E., Dachollom, S., Onwubuya. I.O.: Controlling the spread of COVID-19: optimal control analysis. Comput. Math. Meth. Med. (2020)

27. Li, C., Lei, H., Hu, Z., et al.: A stochastic model with optimal control strategy of the transmission of Covid-19. In: IEEE International Conference on Emergency Science and Information Technology (ICESIT), pp. 62–66. IEEE (2021)

28. Shen, Z.H., Chu, Y.M., Khan, M.A., et al.: Mathematical modeling and optimal control of the COVID-19 dynamics. Results Phys. **31**, 105028 (2021)

Reducing Overfitting Risk in Small-Sample Learning with ANN: A Case of Predicting Graduate Admission Probability

Mengjie Han[1,2], Daomeng Cai[3,4], Zhilin Huo[4], Zhao Shen[2], Lianghu Tang[1], Shan Yang[6], and Cong Wang[5(✉)]

[1] School of Marxism, University of Electronic Science and Technology of China, Chengdu, China
[2] School of Computer Science and Engineering (School of Cyber Security), University of Electronic Science and Technology of China, Chengdu, China
[3] School of Mechanical Engineering and Automation, Beihang University, Beijing, China
[4] CSCC System Engineering Research Institute, Beijing, China
[5] The Intelligent Policing Key Laboratory of Sichuan Province, Sichuan Police College, Luzhou, China
cong-wang@foxmail.com
[6] Department of Chemistry, Physics, and Atmospheric Sciences, Jackson State University, Jackson, MS, USA

Abstract. AI-assisted personal educational career planning holds immense promise, especially with artificial neural network algorithms demonstrating significant potential in predicting graduate admission probabilities for specific schools. However, issues like graduate admission probability prediction are typical instances of small-sample learning problems. Overcoming neural network overfitting in small-sample learning is a major challenge. To address this, we first introduce a method that cyclically runs shallow neural networks to avoid the traditional use of a high number of epochs called Cyclic-LightNet(CLN). Secondly, in previous research, a validation set is typically employed to promptly detect overfitting. However, this approach further depletes the actual training data in the dataset. Consequently, we employ a dynamic training method during the training process called Dynamic Training Set Augmentation(DTSA), supplementing the training set with the validation set. This method effectively mitigates the risk of overfitting. Finally, we put CLN and DTSA into practice on publicly available small-sample datasets, achieving precise predictions of graduate admission probabilities. Experimental results demonstrate a significant improvement in prediction accuracy and effective resistance against overfitting risk.

This article is a result of three projects: the 2019 National Social Science Fund Major Project "Theory, Methods and Practice of Ideological and Political Education in the Big Data Era "(Project No. 19ZDA007), the University of Electronic Science and Technology of China's key research project on graduate ideological and political education and management, "Precision Research on Graduate Ideological and Political Education in the Big Data Perspective" (YJSSZJJYGL202201) and the Opening Project of Intelligent Policing Key Laboratory of Sichuan Province (No. ZNJW2023KFMS005).

Keywords: Artificial neural networks · Small-Sample Learning · Overfitting risk · Probability of graduate admission

1 Introduction

The application of artificial intelligence in the field of education provides personalized learning support and guidance for students [1], further promoting the progress of education. By analyzing and mining large amounts of student data, AI systems can tailor learning plans to students, providing personalized educational resources and academic advice [2]. As the most common, basic and widely used machine learning algorithm, artificial neural network model, it is used in many predictive engineering. In this study, a neural network model is used to predict the admission probability of graduate students, which is a typical small-sample learning problem [3]. The characteristics of small sample data make it difficult for traditional machine learning methods to accurately predict and model. Therefore, in the face of small sample learning problems, how to effectively prevent the overfitting synthesis of model training is an important challenge [4].

This paper is devoted to solving the overfitting problem in small sample learning. We propose two innovative approaches, **CLN** and **DTSA**, to reduce the risk of overfitting and improve the generalization ability of the model. Through the exploration of this study, we hope to provide new ideas and methods to solve the overfitting problem in small-sample learning, so as to optimize the prediction engineering of postgraduate admission probability, provide more accurate evaluation results for applicants, and contribute to the development of the education sector.

1.1 Motivation

In the context of small-sample learning problems, due to the limited size of the training dataset, the model may excessively focus on the noise and specific examples within the training data, failing to capture the true underlying data characteristics. Common methods like increasing the number of iterations to improve model learning may not be suitable, as higher iteration counts can cause the model to memorize training data details, exacerbating overfitting and reducing its ability to generalize to new data. Furthermore, due to the limited number of samples available for training, neural network models may not receive sufficient training, making them prone to overfitting [5].

Therefore, in small-sample learning problems, two primary challenges exist. The first challenge (**CH1**) is overfitting caused by excessive iterations, and the second challenge (**CH2**) is overfitting due to the scarcity of available training data.

1.2 Our approach and Contributions

To address the aforementioned two challenges, we have introduced an enhanced artificial neural network model that effectively mitigates the risk of overfitting. Our work and main contributions are as follows:

Regarding CH1, we have devised a method of iteratively running a shallow neural network (CLN) as an alternative to training with a high number of epochs. By employing this iterative approach with a shallow neural network, CLN can achieve good training results with fewer iterations, avoiding the overfitting issue caused by an excessive number of iterations. At the same time, it preserves the model's predictive and generalization abilities, effectively reducing the risk of overfitting.

Regarding CH2, we have introduced a dynamic training method (DTSA). Traditionally, a portion of the data is set aside as a validation set to assess the model's performance during training. However, this reduces the size of the training dataset, limiting the model's learning capacity. By using the validation set to supplement the training set during the training process, DTSA can better utilize the data, enhancing the model's fitting and generalization capabilities, thereby effectively reducing the risk of overfitting.

We applied **CLN** and **DTSA** to a publicly available small-sample dataset to predict the probability of students being admitted to specific schools in a "competitive" graduate program. Experimental results indicate that our models significantly improved prediction accuracy while effectively mitigating the risk of overfitting.

The organizational structure of the following chapters is as follows: Sect. 2 introduces the risk of overfitting and relevant research in the graduate admission field. Section 3 elaborates on the two enhancements to the artificial neural network model and provides a detailed introduction to the improvement modules. Section 4 presents the results and analysis of the experiments. The subsequent Section 5 discusses our conclusions and future directions.

2 Related Work

Overfitting is a common issue when using artificial neural network models for small-sample learning [5]. Overfitting occurs when the network structure is overly complex or when there is insufficient training data, causing the model to excessively adapt to the details and noise within the training dataset, ultimately resulting in a decrease in its ability to generalize to new data. In small-sample learning problems, due to the limited number of samples available for training, neural network models are prone to overfitting, leading to inaccurate predictive results. This overfitting phenomenon has a significant impact on predictive engineering.

In past research, regularization methods have been commonly used techniques to prevent overfitting [6, 7]. Among these, L1 regularization and L2 regularization are widely employed to constrain the size of model parameters, reducing the model's tendency to overfit to noise in the training data. Early stopping is a simple yet effective technique to prevent overfitting [8]. By monitoring the model's performance on a validation set, training is halted when there is no further improvement on the validation set, preventing the model from overfitting the training data and improving its generalization ability. Data augmentation is a method to increase the amount of data by reasonably augmenting the training data [9, 10]. By applying transformations like rotation, translation, scaling, and flipping, more training samples can be generated, reducing the risk of overfitting. Dropout is a commonly used regularization technique that reduces overfitting by randomly dropping the output of some neurons in a neural network [11–13]. By setting the output of neurons to zero with a certain probability, it forces different parts of the network to independently learn useful features, enhancing generalization. Model ensembling is a method of combining predictions from multiple different models [14]. By training multiple models and averaging or voting on their predictions, it can reduce the risk of overfitting and improve prediction accuracy.

In addressing overfitting in small-sample learning problems, researchers have proposed various methods and techniques, including regularization, early stopping, data augmentation, dropout, and model ensembling. In our study, while our model shares similarities with most neural network models, we have improved the model training process and provided two new methods to mitigate the risk of overfitting, enhancing the model's generalization and predictive performance with nearly no change in the original algorithm's time and space complexity. This offers a feasible solution to tackle the challenge of overfitting in small-sample learning and provides valuable insights for further research and applications.

3 Methodology

3.1 Cyclic-LightNet

Compared to traditional decision models, such as linear regression or decision trees, neural network models exhibit enhanced learning and expressive capabilities due to their multi-layered architecture, non-linear activation functions, and large-scale parameter optimization. Neural network models can capture intricate relationships and patterns within input data and optimize them end-to-end through backpropagation algorithms. This empowers neural network models with superior performance when dealing with high-dimensional, non-linear, and noise-prone data.

However, in the context of small-sample learning, limited training data often hampers the model from achieving sufficient training. In such cases, it is common to enhance model training by increasing the number of iterations. This approach may introduce potential issues that demand careful consideration. Firstly, an increase in iteration count can lead to overfitting, causing a decline in the model's performance on test data. Furthermore, it substantially raises the training time and computational resource costs, making it unsuitable for resource-constrained scenarios. Moreover, if the label quality of the training data is low, increasing the iteration count may not effectively address label errors or noise; instead, it may exacerbate the model's sensitivity to these issues.

Hence, this paper presents an improved approach to training neural network models, which involves the utilization of cyclic training with shallow neural networks as a substitute for a high number of iterations. This approach mitigates the challenges imposed by limited data. It can be clearly seen in Fig. 1 that after each routine training, the index evaluation function value is calculated once, and then judge whether the maximum cycle number n_{max} is reached. If the maximum cycle number is reached, the cycle is stopped. The following figure shows the specific training steps of the model (Fig. 2):

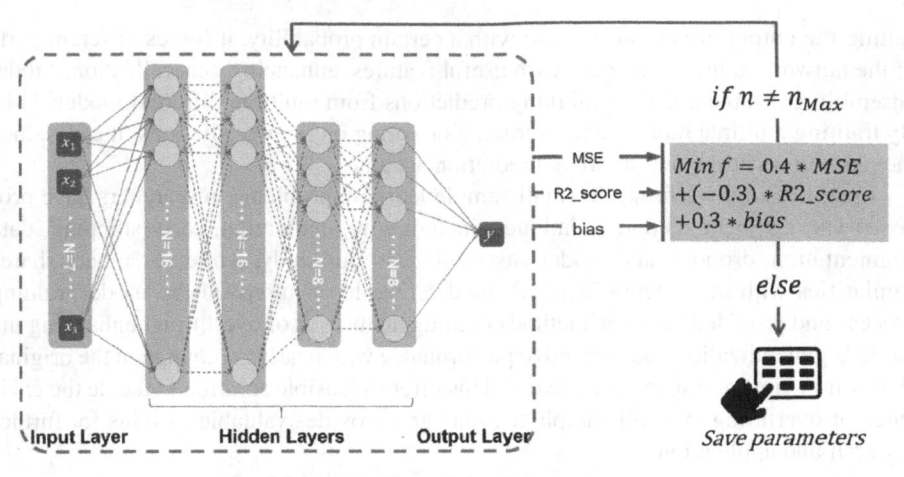

Fig. 1. Cyclic model training process.

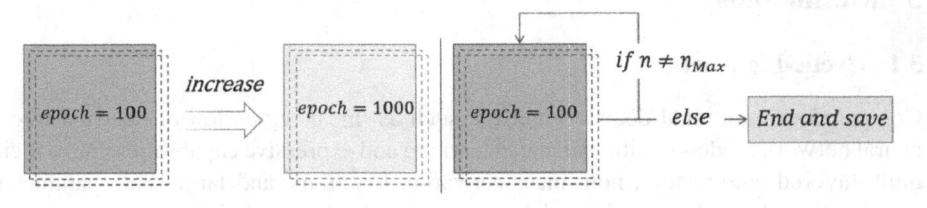

Fig. 2. Comparison of two neural networks.

As is shown in the picture above, when

$$epoch = 100$$

Inside the primary neural network model, the model is iteratively trained 100 times. To get a better parameter, increase the epoch to 1000, i.e.

$$epoch = 1000$$

In this case, the model undergoes 1000 iterations during training. However, experimental results indicate that this operation does not yield globally optimal parameters. On the contrary, after 1000 iterations, the model's predictive performance on the test set does not improve and instead exhibits higher MSE and bias, as well as a lower R2_score. In fact, the R2_score decreases from the initial value of around 0.83 to approximately 0.16, which is clearly an undesirable outcome.

At this point, we can encapsulate the primary neural network model and add an outer layer of iteration, setting the number of iterations to 10. In this way, we obtain a pseudo iteration count:

$$10 * 100 = 1000 \ times$$

In each iteration, we record the model parameters and evaluation metric parameters at the end of the iteration. Upon exiting the loop, we compare the evaluation metric parameters from the 10 iterations to obtain the optimal model parameters. Here, we define a method for comparing the evaluation metric parameters:

$$Min\, f = a1 * MSE + (-a2) * R2\_score + a3 * bias + ... + a_n * other\ criteria$$

s.t.

$$\sum_{i=1}^{n} a_i = 1$$

$$a_i > 0, i = 1, 2, 3, ... \tag{1}$$

In the above formula, $a_1 \sim a_n$ represents the importance weights of each evaluation metric. The parameter before $R2\_score$ is negative, because the higher the $R2\_score$ index, the better the model performance, and this function is set to the minimum value, that is, in order to standardize the objective function, the parameter a_i of $R2\_score$ is artificially changed to $-a_i$.

In practical applications, these weights can be set according to specific needs. It is important to note that before using the above formula, it is necessary to normalize all evaluation metrics to avoid the dominance of large-scale metrics over smaller ones. This normalization step ensures that each evaluation metric contributes proportionally to the overall comparison.

3.2 Dynamic Training Set Augmentation

In Small-Sample Learning, the risk of overfitting is high due to the limited volume of available training data. Typically, a validation set is established to monitor overfitting during the model training process. However, separating a validation set from the overall available data further diminishes the amount of data available for actual training.

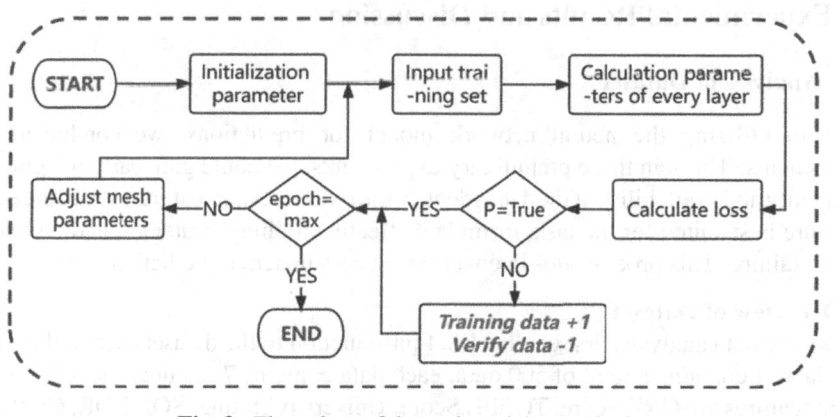

Fig. 3. Flow chart of dynamic training methods.

As the validation set primarily reflects the trend of the loss function without requiring a precise characterization of that function, we have employed a dynamic training approach that supplements the training set with the validation set during the training process. This approach effectively mitigates the risk of overfitting. Figure 3 above shows the dynamic training method of our constructed model.

As can be seen from Fig. 3, with the gradual progress of training, the data volume in the training set "dynamically increases," while the data volume in the validation set "dynamically decreases".

Compared to the conventional neural network model training approach, our proposed dynamic training method involves extracting s data points from the validation set into the training set during each iteration of the model. This study employs random sampling. After the model completes N full forward and backward passes, the transformation of the training and validation sets is as follows:

$$Num\ of\ \textbf{new training data} = Num\ of\ old\ training\ data + N * s$$

$$Num\ of\ \textbf{new verif data} = Num\ of\ old\ verif\ data - N * s$$

So, when does dynamic training come to a halt? We can establish the stopping condition for dynamic training by adjusting the parameter P in Algorithm 2. When the algorithm reaches the validation set quantity threshold we've set, and there are remaining iterations, the subsequent training reverts to the conventional training mode. P represents the current moment's quantity of data in the validation set, given that we initially designed the algorithm model with the presence of a validation set in mind, it is evident that $P \geq 1$.

The minimum value for P is 1, which signifies that at least one data point must be retained in the validation set (in order to maintain the presence of the validation set). Due to the limited data volume, in the subsequent practical experiments, we set P to 1. In other words, when there is only one data point left in the validation set, dynamic training stops, and the training mode reverts to the conventional approach.

4   Experimental Results and Discussion

4.1   Analysis of Dataset

Prior to utilizing the neural network model for predictions, we conducted pre-experiments. Through these preliminary experiments, we could gain early insights into the quality and availability of the data, identify the most relevant features, select the model structure best suited for the task, formulate effective training strategies, and reduce the risk of failure. This process aids in ensuring success in actual prediction tasks.

A. Overview of Dataset
Before the data analysis, first of all, a brief introduction to the dataset used in this study, this dataset contains a total of 500 data, each data contains 7 features and 1 label. The seven features are GRE Score, TOEFL Score, University Rating, SOP, LOR, CGPA and Research, and one label is Chance of Admit. The following table shows the first five pieces of data in the dataset (Table 1).

B. Comprehensive statistical analysis

Table 1. Shows the first five data sets used in this study.

Serial No.	GRE Score	TOEFL Score	University Rating	SOP	LOR	CGPA	Research	Chance of Admit
1	337	118	4	4.5	4.5	9.65	1	0.92
2	324	107	4	4	4.5	8.87	1	0.76
3	316	104	3	3	3.5	8	1	0.72
4	322	110	3	3.5	2.5	8.67	1	0.8
5	314	103	2	2	3	8.21	0	0.65

Table 2. Statistical information on the features of the dataset.

	Max	Min	Mean	Std	Count
GRE Score	340.00	290.00	316.47	11.28	500
TOEFL Score	120.00	92.00	107.19	6.08	500
University Rating	5.00	1.00	3.11	1.14	500
SOP	5.00	1.00	3.37	0.99	500
LOR	5.00	1.00	3.48	0.92	500
CGPA	9.92	6.80	8.58	0.60	500
Research	1.00	0.00	0.56	0.50	500
Chance of Admit	0.97	0.34	0.72	0.14	500

Calculating the maximum (Max), minimum (Min), mean(Mean), standard deviation (Std), and count(Count) values for each of the 7 features provides an overall understanding of the data.

Table 2 above displays the statistical information of the dataset features, including the sample count, mean, standard deviation, minimum, and maximum values for each feature. This table provides essential statistical indicators for understanding the distribution of data and the range of feature variations within the dataset. These statistical insights are crucial for data analysis and model development as they help evaluate the overall characteristics of the data, identify outliers, establish data ranges, and facilitate feature selection and engineering.

C. Correlation analysis

In this section, we conducted an overall correlation analysis of the 8 columns in the dataset, including "Chance of Admit." We have generated the following heatmap to visualize the correlations (Fig. 4).

Upon observing the heatmap, we can identify the three largest correlation coefficients with the admission probability as 0.88, 0.81, and 0.79. These coefficients correspond to the features CGPA, GRE Score, and TOEFL Score, respectively. This finding indicates

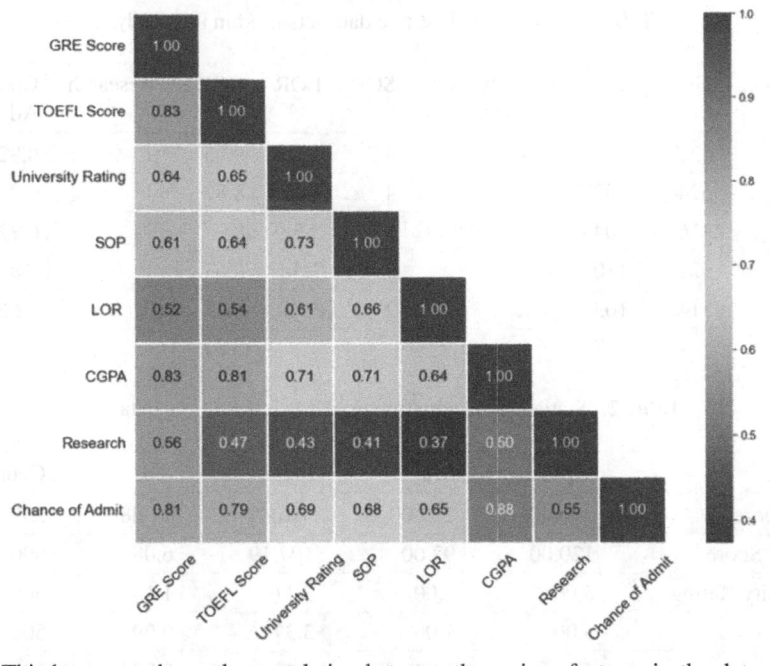

Fig. 4. This heat map shows the correlation between the various features in the data set, with darker colors indicating stronger correlations and lighter colors indicating weaker correlations.

that these three indicators play a significant role in graduate admissions, as they have a substantial impact on the probability of admission.

Furthermore, we conducted correlation analyses for the seven indicators in the dataset, which include GRE Score, TOEFL Score, University Rating, Statement of Purpose (SOP), Letter of Recommendation (LOR), CGPA, Research, and Chance of Admit. These analyses determined whether each indicator exhibited a positive correlation, negative correlation, or no significant correlation with the admission probability. The following composite figures depict the results of the correlation analysis (Fig. 5).

Through the scatter plot above, we can observe that GRE Score, TOEFL Score, and CGPA each exhibit a clear positive correlation with the admission probability. This observation aligns with the analysis results from the heatmap mentioned earlier. In the next section, we will delve into a detailed analysis of the correlations among these three features.

D. Analysis of GRE, TOEFL and CGPA

From the analysis in the previous section, we can see that among the seven characteristics, GRE Score, TOEFL Score, CGPA, and Chance of Admit are most correlated. In this section, we take these four columns of data separately and draw a correlation network diagram to further analyze the three characteristics of GRE Score, TOEFL Score and CGPA.

GRE Score is one of the important indicators for students' graduate school applications as it reflects their academic abilities. TOEFL Score is a crucial measure of students'

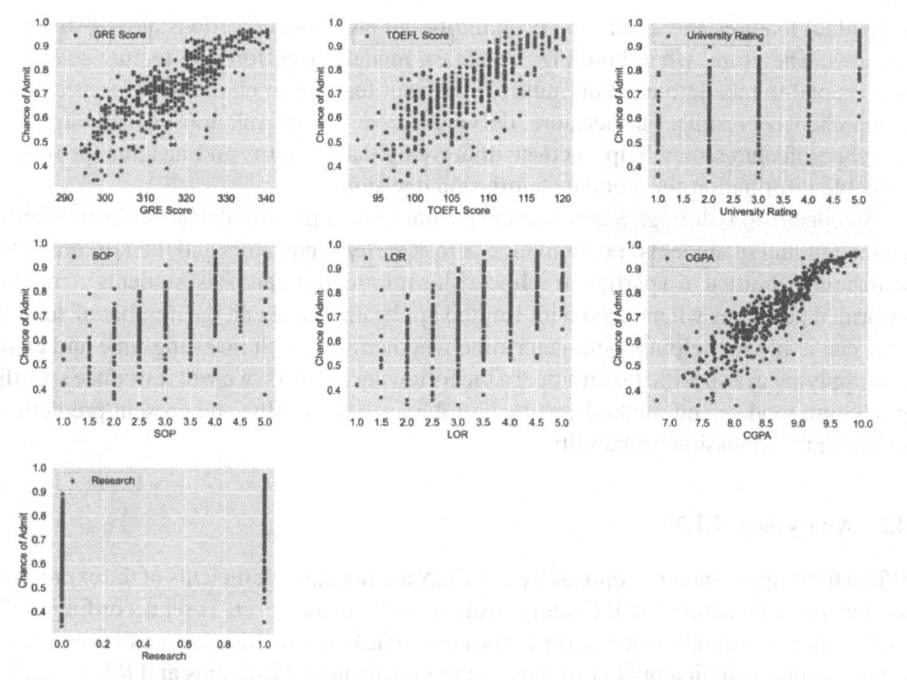

Fig. 5. The graph details the relationship between seven features and the probability of admission.

English language proficiency, particularly for international study and graduate programs. On the other hand, CGPA represents students' academic performance and learning abilities. Observe that in Fig. 6, we can notice a strong correlation between GRE Score, TOEFL Score, CGPA, and Chance of Admit. Moreover, there is a strong correlation among these three features themselves.

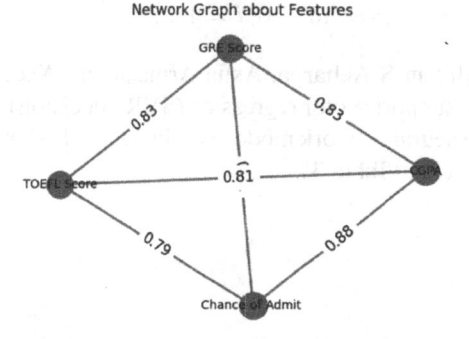

Fig. 6. The graph shows the correlation between GRE Score, TOEFL Score, CGPA, and Chance of Admit.

This insight leads us to the realization that, in the process of predicting graduate admission probabilities, it is essential to consider not only the relationship between individual features and admission probability but also the interactions among different features. Therefore, when building prediction models or performing feature selection, it is crucial to take into account multiple relevant features to obtain more accurate and comprehensive results. Furthermore, delving deeper into the relationships among these correlated features may help uncover underlying data patterns and insights, providing valuable information for graduate admission decisions.

In conclusion, through a comprehensive analysis of the initial data, we gain a better understanding of students' performance across various indicators and their associations with being admitted to a particular school. This information can assist students in making informed decisions when faced with limited application time and the number of schools they can apply to, helping them determine whether it is worth investing time and effort into applying to a specific institution. In the following chapters, we will introduce specific prediction models and methods, showcase the analysis results, and present predictions of students' admission probabilities.

4.2 Analysis of CLN

When training the model proposed by the **CLN** method, the coefficients of the expression (1) must first be determined. Usually, weight coefficient $a_1 \sim a_n$ is set according to the importance of evaluation indicators to the current instance. In this case prediction, we use three evaluation indicators to evaluate the model, namely, MSE, bias and R2_score. We set a_1, a_2, a_3 as 0.4, 0.3 and 0.3 respectively, then the model evaluation index function can be simplified to

$$Minf = 0.4 * MSE + (-0.3) * R2\_score + 0.3 * bias$$

s.t.

$$a_1 + a_2 + a_3 = 1$$

$$a_i > 0, i = 1, 2, 3 \tag{2}$$

Compared with Mohan S Acharya, Asfia Armaan and Aneeta S Antony who use linear regression (LR), support vector regression (SVR), decision tree (DTR) and random forest (RFR) [15], The neural network model used in this study shows obvious advantages in both MSE and R2_score (Table 3).

Table 3. Parameter Values of Different Control Strategies.

Models	MSE	R2_Score	bias
ANN	**0.00202868**	**0.8880500**	**0.14923252**
LR	0.00480149	0.7248631	-
SVR	0.00724206	0.6440130	-
DTR	0.00874299	0.5013442	-
RFR	0.00582112	0.6601701	-

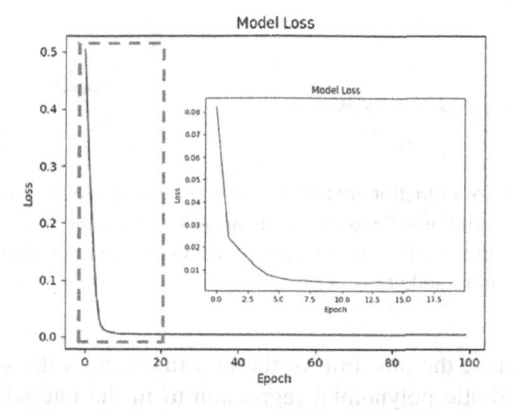

Fig. 7. Loss function graph (global and local).

Moreover, in the model constructed in this study, we observe that the loss function reaches a stationary state when the model is iterated to about the 10th generation, indicating that the model has tended to fit at about the 10th generation. This observation can be verified from Fig. 7 above, where the function line remains almost unchanged after epochs is greater than 10.

4.3 Analysis of DTSA

In this section, we will show how the conventional training mode compares to the dynamic training mode.

First of all, we introduce some important parameter Settings of our experiment, and the volume of training set, verification set and test set used.

Important parameter Settings:

- $P = 1$ (Dynamic training stop condition)
- $s = 1$ (Number of data extractions)
- epochs $= 65$ (Number of model iterations)

Datasets initialization volume:

- Training set: 300 pieces of data
- Verification set: 50 pieces of data
- Test set: 150 pieces of data

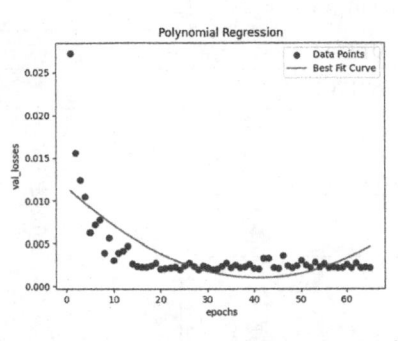 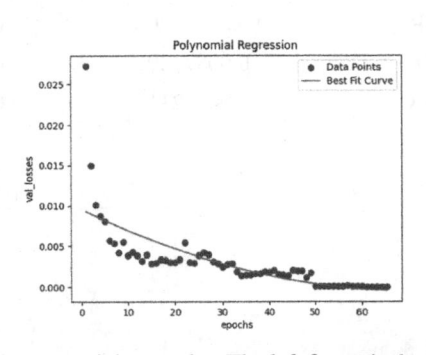

Fig. 8. Comparison of loss function images of the two training modes. The left figure is the loss function image of the training set and the verification set when the conventional training mode is used, and the right figure is the loss function image of the training set and the verification set when the dynamic training mode is used.

In order to describe the direction of the loss function on the verification set more clearly, we use quadratic polynomial regression to fit the interpolation points on the verification set (red curve).

The fitting is carried out by least square method, and the fitting target is:

$$val\_losses = a * epochs^2 + b * epochs + c \tag{3}$$

First, compute the following prefix operator requirements:

$$\sum epochs, \sum epochs^2, \sum epochs^3, \sum epochs^4,$$

$$\sum val\_losses, \sum epochs * val\_losses, \sum epochs^2 * val\_losses$$

Then, construct A third-order square matrix A and a three-dimensional vector B, whose elements are:

$$A = \begin{bmatrix} n & \sum epochs & \sum epochs^2 \\ \sum epochs & \sum epochs^2 & \sum epochs^3 \\ \sum epochs^2 & \sum epochs^3 & \sum epochs^4 \end{bmatrix}$$

In the matrix A, n represents the length of the data brought into the calculation, i.e. the length of the list.

$$B = \begin{bmatrix} \sum val\_losses \\ \sum epochs * val\_losses \\ \sum epochs^2 * val\_losses \end{bmatrix}$$

Finally, the regression coefficient is obtained by solving the equations. According to the regression coefficient, the regression function image is drawn, and the red curve in Fig. 8 is shown.

Comparing the left and right figures in Fig. 8, it is clear that the lost function curve has leveled off after training to 50 generations of verification sets when the dynamic training mode (right figure) is used with the same baseline data set and number of iterations. However, the lost function curve of the verification set of the conventional training mode (left figure) has a tendency of overfitting when training to 40 generations, which can prove that the new method can resist the overfitting risk to a certain extent.

Overall, the experimental results show that the method of dynamically supplementing the training set with the validation set can resist the risk of overfitting. At the same time, considering the extreme case, when the training data is small enough, the overfitting phenomenon becomes a very difficult situation to avoid, then the dynamic training method can also delay the algebra of the overfitting phenomenon to a certain extent, so that the model can be trained to the best effect before the overfitting.

5 Conclusion

In small sample learning problems, we face two major challenges, namely overfitting problem caused by too many iterations (**CH1**) and overfitting problem caused by too little available training data (**CH2**). To address these challenges, we propose two improved methods, namely, the looping shallow Neural Network method (CLN) and the Dynamic Training method (DTSA). We put **CLN** and **DTSA** into practice on a small data set of public samples, and completed the accurate prediction of graduate admission probability. The experimental results show that the artificial neural network model is a more effective prediction tool than the four common decision models (LR, SVM, DT and RF), and the two proposed improvements based on the neural network model effectively reduce the risk of overfitting, while improving the training effect and generalization ability of the model.

Although this study has achieved certain results, there are still some aspects that can be further improved and explored. When training using dynamic training mode, there are three decisions: (1) How many data from the validation set to add to the training set each time? In the process of our experiment, it is found that the ratio of s to the number of training sets and the ratio of s to the number of verification sets have an influence on the training effect of the model. In the later work, we can further explore how to determine the optimal solution. (2) When to stop dynamic training? In our experiment, we take a limit case, that is, when $P = 1$, we stop the dynamic training and return to the normal training. Although this setup is the ideal result of our experiment, we cannot determine whether it is the best effect for the current example. In the following work, we can further explore the optimal stopping strategy of dynamic training. (3) In dynamic training, how to select data from the verification set to join the training set? In our experiment, s data were randomly selected each time and added to the training set. In the following work, we can try to explore whether there is a better extraction strategy to make the model have a better training effect.

Overall, our work provides some useful approaches to small sample learning problems, but there are many areas for further research that could further improve and extend these approaches to address more real-world challenges.

Acknowledgement. We acknowledge the financial supports from the 2019 National Social Science Fund Major Project "Theory, Methods and Practice of Ideological and Political Education in the Big Data Era" (Project No. 19ZDA007),the University of Electronic Science and Technology of China's key research project on graduate ideological and political education and management, "Precision Research on Graduate Ideological and Political Education in the Big Data Perspective" (YJSSZJJYGL202201) and the Opening Project of Intelligent Policing Key Laboratory of Sichuan Province (No. ZNJW2023KFMS005).

References

1. Zhao, W., Wang, Z., Zhu, L.: Research on the application of ai in the field of education big data mining. J. Phys: Conf. Ser. **1992**, 032091 (2021). https://doi.org/10.1088/1742-6596/1992/3/032091
2. Chiu, T.K.F., Xia, Q., Zhou, X., Chai, C., Cheng, M.: Systematic literature review on opportunities, challenges, and future research recommendations of artificial intelligence in education 4 (2023). https://doi.org/10.1016/j.caeai.2022.100118
3. Liu, D., He, Z., Chen, D., Lv, J.: A network framework for small-sample learning. IEEE Trans. Neural Networks Learn. Syst. **31**(10), 4049–4062 (2020). https://doi.org/10.1109/TNNLS.2019.2951803
4. Dong, Y., Li, Y., Zheng, H., Wang, R., Xu, M.: A new dynamic model and transfer learning based intelligent fault diagnosis framework for rolling element bearings race faults: Solving the small sample problem. ISA Trans. **121** (2021). https://doi.org/10.1016/j.isatra.2021.03.042
5. Bejani, M.M., Ghatee, M.: A systematic review on overfitting control in shallow and deep neural networks. Artif. Intell. Rev. **54**, 6391–6438 (2021). https://doi.org/10.1007/s10462-021-09975-1
6. Yang, S., Zhu, X., Zhang, L., Wang, L., Wang, X.: Classification and prediction of Tibetan medical syndrome based on the improved bp neural network. IEEE Access **8**, 31114–31125 (2020). https://doi.org/10.1109/ACCESS.2020.2973304
7. De, S., Doostan, A.: Neural network training using ℓ1-regularization and bi-fidelity data. J. Comput. Phys. **458**, 111010 (2022). https://doi.org/10.1016/j.jcp.2022.111010
8. Xin, Z., Wang, H., Wu, B., Zhou, Q., Hu, Y.: A novel data-driven method based on sample reliability assessment and improved CNN for machinery fault diagnosis with non-ideal data. J. Intell. Manuf. **34**, 1–14 (2022). https://doi.org/10.1007/s10845-022-01944-x
9. Abeysinghe, A., Tohmuang, S., Davy, J., Fard, M.: Data augmentation on convolutional neural networks to classify mechanical noise. Appl. Acoust. **203**, 109209 (2023). https://doi.org/10.1016/j.apacoust.2023.109209
10. Kumar, P., Belchamber, E., Miklavcic, S.: Pre-processing by data augmentation for improved ellipse fitting. PLoS ONE **13**, e0196902 (2018). https://doi.org/10.1371/journal.pone.0196902
11. Ha, N.C., Tran, V.-D., Van, L., Than, K.: Eliminating overfitting of probabilistic topic models on short and noisy text: The role of dropout. International J. Approximate Reasoning **112** (2019). https://doi.org/10.1016/j.ijar.2019.05.010

12. Guo, J., Qi, L., Shi, Y., Gao, Y.: PLACE dropout: a progressive layer-wise and channel-wise dropout for domain generalization. ACM Trans. Multimed. Comput. Commun. Appl. (2023). https://doi.org/10.1145/3624015

13. Li, H., et al.: Adaptive dropout method based on biological principles. IEEE Trans. Neural Networks Learn. Syst. **32**(9), 4267–4276 (2021). https://doi.org/10.1109/TNNLS.2021.307 0895

14. Li, X., Yu, Q., Yang, Y., Tang, C., Wang, J.: An evolutionary ensemble model based on GA for epidemic transmission prediction. J. Intell. Fuzzy Syst. **44**, 1–13 (2023). https://doi.org/10.3233/JIFS-222683

15. Acharya, M.S., Armaan, A., Antony, A.S.: A comparison of regression models for prediction of graduate admissions. In: 2019 International Conference on Computational Intelligence in Data Science (ICCIDS), Chennai, India, 2019, pp. 1–5 (2019). https://doi.org/10.1109/ICC IDS.2019.8862140

16. Tang, X., Zheng, D.*, Kebede, G.S., et al.: An automatic segmentation framework of quasi-periodic time series through graph structure. Appl. Intell. (2023). https://doi.org/10.1007/s10 489-023-04814-y

Comparisonof Chinese College Students'Anxiety During and BeforeCOVID19 Pandemic Based on Clustering Algorithm

Hongjuan Wen[2], Menglu Xu[1], Yang Gong[1], Haiyang Zhang[3], and Xiaoting Zhu[4(⊠)]

[1] School of Marxism, Changchun University of Chinese Medicine, Changchun 130117, China
[2] School of Health Management, Changchun University of Chinese Medicine, Changchun 130117, China
[3] School of Medical Information, Changchun University of Chinese Medicine, Changchun 130117, China
[4] Brain Disease Center, The Third Affiliated Clinical Hospital of Changchun University of Chinese Medicine, Changchun 130118, China
zisedemeng1989@126.com

Abstract. The COVID19 pandemic has had acute and long-term effects on college students' mental health, but there was a lack of research following. This study aims to compared research hot spots of college students' anxiety during and before the epidemic. By analyzing 2,350 pieces of literature from the Chinese National Knowledge Infrastructure (CNKI), researchers utilized LLR clustering algorithms, Price core author algorithm, and Donohue high and low frequency word algorithms to predict future development trends. The results showed that research on college students' anxiety has changed in research hot spots before and after the epidemic, with increase rapidly following the COVID-19 outbreak. Firstly, a core author group was not formed, and most institutions had few collaborations due to geographical restrictions. Secondly the hotspots of research were on the common psychological problems of college students before the epidemic, and after the epidemic they were on the loneliness, information anxiety, sleep, and social support caused by the epidemic, which can be summarized into four aspects of emotional, cognitive, behavioral, and physiological problems. These findings highlight the need for further study of college students' anxiety problems in the post-epidemic era.

Keywords: College Students' Anxiety · COVID-19 Pneumonia · Public Health Emergencies · LLR clustering algorithm · Price's Law · Word frequency analysis algorithm

1 Introduction

The COVID-19 epidemic is a significant international public health emergency that has disrupted the lives of college students and induced psychological stress reactions and post-traumatic stress disorder (PTSD) [2], ultimately affecting their mental health.

The study was supported by the Jilin Provincial Administration of Traditional Chinese Medicine (No.: 2022223) and Changchun University of Chinese Medicine (No.: 2021KC20)

College students exhibit a higher rate of anxiety after COVID-19 than during normal times. Although global prevention and control measures for the COVID-19 epidemic have now relaxed, entering the post-epidemic era will trigger new adaptive challenges [3]. The epidemic's impact is far from over, and new hotspots require further study.

This study employs bibliometric methods to compare anxiety research among college students before and after COVID-19, using the changes in their anxiety research hotspots as a basis. We visually display the changes during and before the epidemic, as well as provide a summary of the responses and proposed future development trends.

2 Data Extraction and Research Methods

2.1 Data Sources

The CNKI database was used to search for the search terms "college students" and "anxiety", published between January 1, 2017 and December 31, 2019, as well as between January 1, 2020 and December 31, 2022, 1091 and 1265 articles were selected for Chinese academic journals, and 2350 articles after deduplication were analyzed. Use Cite Space software to set the time period from JAN 2017 to DEC 2019, JAN 2019 to DEC 2022, and JAN 2017 to DEC 2022, with Years Per Slice set to one year. Select authors, institutions and keywords for the drawing of the cooperative co-emergence map; LLR clustering algorithm is used to realize the drawing of keyword clustering chart; computational core author through Price theorem analysis; high-frequency vocabulary is calculated by the Donohue's formula.

2.2 Methods

LLR Clustering Algorithm. Cluster analysis refers to the analysis process of grouping a collection of physical or abstract objects into multiple classes composed of similar objects based on the similarity of the analyzed objects. Cite Space provides automatic clustering based on spectral clustering algorithm, and in This study, LLR algorithm (Log-likelihood rate) is chosen. Cite Space uses the module value (Q value) and the average profile value (S value) as two indicators to determine the effectiveness of mapping based on the network structure and the clarity of clustering. Q value is generally in the [0,1) interval, when Q > 0.3 then the delineated association structure is significant, when S value > 0.7, the clustering is efficient and convincing, and S > 0.5, the clustering is reasonable[4].

LLR Algorithm. Put M keywords of P_k article with $x^{P_k} = \left(x^{(1,P_k)} x^{(2,P_k)} \cdots x^{(M,P_k)}\right)$ to represent the input set, and then cluster each class of keywords separately. It can be divided into two steps that.

a) Estimating the first i reconstruction coefficient corresponding to the first keyword input α^i:

$$\alpha^i = \left(X^{(i,P_k)}\right) \perp \chi^{(i,P_k)} \tag{1}$$

$X^{i,P_k} = \left(\chi_1^{(i,P_k)} \chi_2^{(i,P_k)} \cdots \chi_N^{(i,P_k)}\right)$ is the P_k article when the set of the i keywords.

b) The corresponding estimated clusters are obtained as:

$$\chi^{(i,P_0)} = X^{(i,P_0)}\alpha^i \tag{2}$$

$X^{(i,P_0)} = \left(\chi_1^{(i,P_0)}\chi_2^{(i,P_0)} \cdots \chi_N^{(i,P_0)}\right)$ is the i keyword corresponding to the cluster set.

Finally, all the keywords are synthesized into a whole set of predicted clusters:

$$\chi^{P_0} = \left(\chi^{(1,P_0)}\chi^{(2,P_0)} \cdots \chi^{(M,P_0)}\right) \tag{3}$$

Network Evaluation Index Q Calculation Algorithm. The modified modularity has been currently extended to text mining considered as a criterion for clustering text data [5].

$$Q = \frac{1}{a_{..}}\sum_{i=1}^{N}\sum_{j=1}^{M}\sum_{k=1}^{C}\left(a_{ij} - \frac{a_{i.}a_{.j}}{a_{..}}\right)z_{ik}w_{jk} \tag{4}$$

$a_{..}$ is the total number of edges. a_i is the degrees of i and j. z_{ik} and w_{jk} is the membership degrees of objects and features.

Silhouette Calculation Algorithm. The silhouettes constructed below are useful when the proximities are on a ratio scale and when one is seeking compact and clearly separated clusters [6].

$$s(i) = \frac{b(i) - a(i)}{\max\{a(i), b(i)\}} \tag{5}$$

a(i) is average dissimilarity of i to all other objects of cluster A. b(i) is minimum average dissimilarity of i to all objects of cluster C which is different from A.

Price's Law. Price's law can reveal the relationship between the number of authors and the journal papers and the quantitative relationship between core authors and other authors [7]. Suppose the number of papers by the author with the highest number of publications is N_{max}, then the total number of publications by all authors is $x(1, N_{max})$, then it can be expressed as

$$\left(\frac{1}{2}\right)x(1, N_{max}) = x(m, N_{max}) = x(1, m) \tag{6}$$

N_{max} is the number of papers by the most prolific authors in the statistical time period, and M is the number assumed by Price, who assumes that the total number of papers by core authors is exactly equal to half of the total number of papers i.e. $x(m, N_{max})$, $m \approx 0.749\left(N_{max}^{1/2}\right)$. Authors whose publications exceed the value of M can be called core authors [8]. The proportion of core authors is $R \approx 0.812 \left/ N_{max}^{1/2} R \approx 0.812 \right/ N_{max}^{1/2}$

Word Frequency Analysis Algorithm. The algorithm allows the calculation of the demarcation of high-frequency words from low-frequency words, and in 1973 Donohue.J.C. based on Zipf's law[9]:

$$I_n/I_1 = 2/n(n+1) \tag{7}$$

I_n is the number of words that appear n times in the text. Donohue proposes a formula for defining high and low frequency words:

$$T = (-1 + \sqrt{1 + 8 \times I_1}) \div 2 \tag{8}$$

I_1 is the number of high-frequency words that appear once, and keywords with a frequency greater than T are high-frequency words, while those less than T are low-frequency words.

3   Results

3.1   Annual Growth of Number of Journal Papers

Statistical analysis of the number of articles published reflects the research trends and development rate of college student anxiety research from 2017 to 2022 [10]. Over this period, the field experienced consistent development, as evidenced by an increasing trend in the number of articles published. Notably, 2020–2021 emerged as a major area of focus, with a high volume of articles published, reaching its peak in 2021. The number of articles declined in 2017–2018, but the subsequent year saw a significant increase(see Fig. 1).

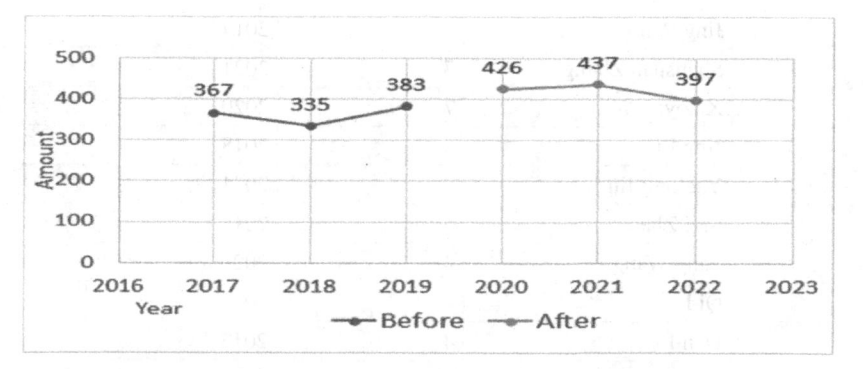

Fig. 1. Number of posts before and after the COVID-19.

3.2   Analysis of Author Collaboration

Mutual cooperation among authors is beneficial for the development of the field. Analysis of author collaboration can determine the core group of authors in the field and the intensity of collaboration between them [11]. Figure 2 has 329 nodes, 301 connections, and a density of 0.0056, forming several significant network structures centered on the core authors(see Fig. 2). According to the calculation of Price's law, M is about 3, and the core authors are 45 about 14% of the total number of authors. The number of publications of core authors is 170, about 7% of the total number of publications(see Table 1).

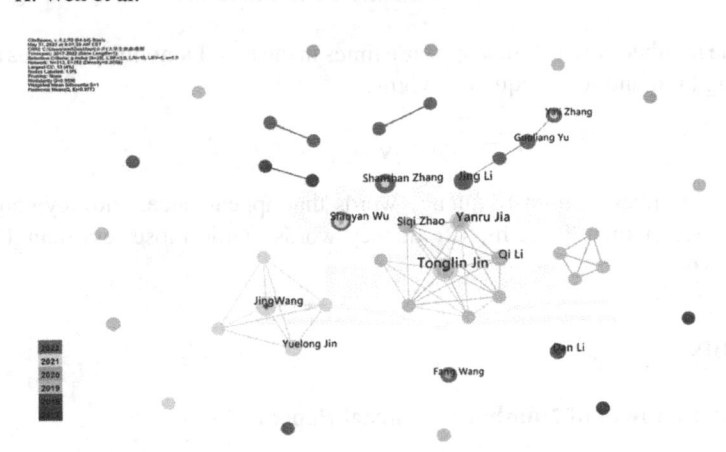

Fig. 2. Author Collaboration Chart.

Table 1. Top 5 authors in terms of number of articles published.

Rank	Author	Amount	First publication year
1	Tonglin Jin	10	2018
2	Yanru Jia	7	2019
2	Jing Wang	7	2017
2	Shanshan Zhang	7	2021
2	Xiaoyan Wu	7	2020
3	Jing Li	6	2018
4	Yuelong Jin	5	2021
4	Siqi Zhao	5	2019
5	Fang Wang	4	2020
5	Qi Li	4	2019
5	Dan Li	4	2017
5	Yali Zhang	4	2021
5	Guoliang Yu	4	2022

3.3 Analysis of Institutional Cooperation

Inter-institutional cooperation across regions can integrate academic resources to make research more focused and promote the development of research in this field. It can be seen that 282 nodes with 72 connected lines appeared, and the density was 0.0018 (see Fig. 3). Inner Mongolia Normal University, the School of Psychology of Guizhou Normal University and the School of Educational Sciences of Harbin Normal University, which have the top number of publications, have formed good cross-regional cooperation. Most of the institutions with the highest number of publications are teacher training colleges, followed by medical colleges, and a few are research institutions (see Table 2).

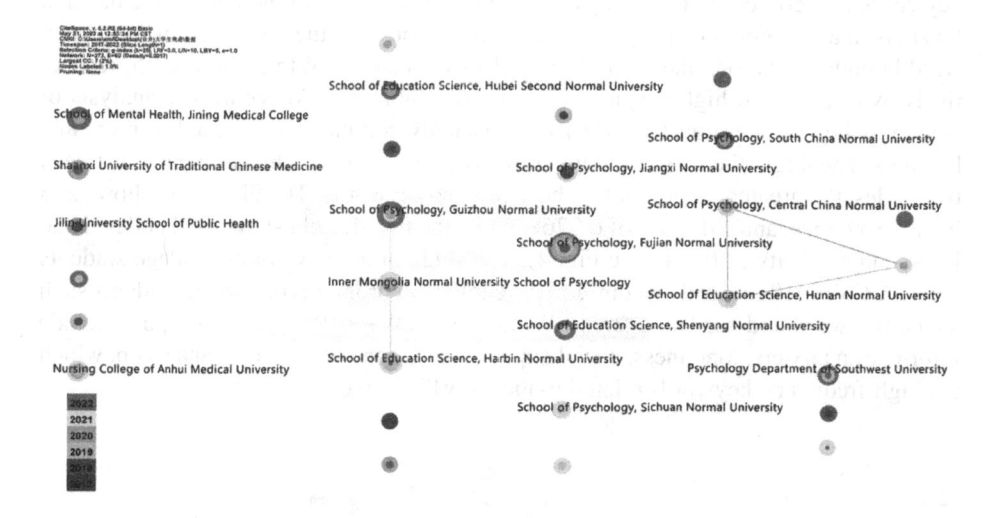

Fig. 3. Mapping of institutional cooperation.

Table 2. Top 5 institutions in terms of number of articles issued.

Rank	Research Institution	Number of institution papers
1	School of Psychology, Inner Mongolia Normal University	9
1	School of Psychology, Fujian Normal University	9
2	School of Educational Science, Harbin Normal University	8
2	School of Psychology, Guizhou Normal University	8
3	School of Nursing, Anhui Medical University	7
4	School of Educational Science, Shenyang Normal University	6
4	School of Mental Health, Jining Medical College	6
5	School of Psychology, Central China Normal University	5
5	School of Psychology, Sichuan Normal University	5
5	School of Education and Science, Hubei Second Normal University	5
5	Shaanxi University of Traditional Chinese Medicine	5

3.4   Research Hotspots

Visual Analysis of High-Frequency and Co-occurrence Keywords. High-frequency keywords and co-occurrence analysis can provide insight into the central themes and hotspots in a research area [12]. Using Donohue's formula, the high and low frequency word boundaries before and after COVID-19 were calculated to be 36 and 25, respectively, with 8 and 16 high frequency words (see Table 3.). Co-occurrence analysis of keywords before and after COVID-19 can visually compare the research centers and hotspots. The larger the node, the higher the frequency, and the thicker the line between two nodes, the greater the co-citation between two keywords. The Fig. 4, left shows 255 nodes, 869 lines, and a density of 0.0268, while the Fig. 4, right shows 340 nodes, 1036 lines, and a density of 0.018 (see Fig. 4). The study of anxiety among college students before COVID-19 focused on common psychological problems of college students such as interpersonal and academic stress, While post COVID-19 research hotspots include information anxiety, loneliness, social support, cell phone dependence and sleep, which are high-frequency keywords related to the COVID-19 (see Table 3).

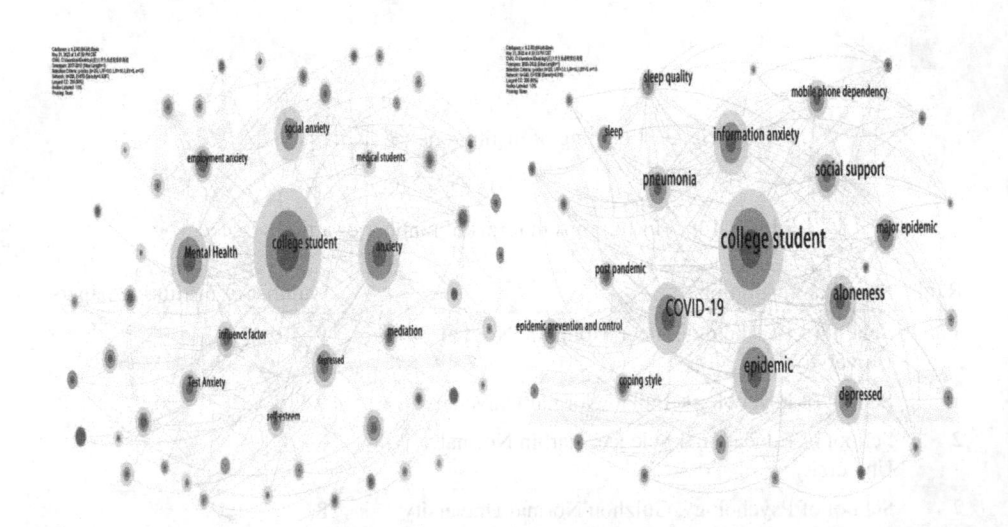

Fig. 4. Pre-epidemic keyword co-occurrence map based on the CiteSpace analysis (left). Post-epidemic keyword co-occurrence map based on the CiteSpace analysis (right).

Keyword Mutation Analysis. The mutation analysis of keywords can further revea the development trend and research frontier of the field [13]. Set γ in the range of 0–1 to 1.0 to obtain top 18 keywords with the strongest citation bursts, were ranked according to their strength, with COVID-19 being the most prominent keyword (see Fig. 5). COVID-19 was discovered in Wuhan, China in December 2019, and the spread is becoming wider as time goes by, causing a certain degree of panic among the population [14]. The emergence of COVID-19 has had a significant impact on anxiety research among college students, with social support, loneliness, information anxiety, and Internet addiction also emerging as hotspots continues to this day and will be a future research trend.

Table 3. High frequency keywords.

Before			After		
Rank	Keywords	Count	Rank	Keywords	Count
1	college student	416	1	college student	599
2	anxiety	154	2	epidemic	265
3	mental health	126	3	COVID-19	219
4	social anxiety	83	4	Information anxiety	182
5	depress	55	5	aloneness	132
6	employment anxiety	37	6	pneumonia	94
7	student	37	7	depress	92
8	examination anxiety	37	8	social support	89
9	influence factor	36	9	major epidemic	70
			10	mobile phone dependence	65
			11	epidemic prevention and control	44
			12	COVID-19	41
			13	sleep quality	41
			14	coping style	35
			15	Post epidemic	34
			16	sleep	33

Cluster Analysis to Detect High-frequency Keywords. Clustering keywords can organize and refine a large number of keywords and reflect the hot research directions in the field [15]. The following conclusions can be found by using the LLR clustering algorithm. Before the epidemic, clustering of keywords formed 10 clusters, and after the epidemic, clustering of keywords formed 8 clusters (see Fig. 6). The clustering of keywords before the epidemic Q value = 0.3734 and S value = 0.7008(Fig. 6, left), after the epidemic Q value = 0.3566 and S value = 0.7398(Fig. 6, right), with a significantly convincing clustering structure. Comparing the cluster analysis before and after the epidemic, it can be found that research on college students' anxiety before the epidemic can be divided into three categories: the first category is about interventions including countermeasures, psychological interventions, and counselors; the second category is about psychological problems including depression, social anxiety, mental health, and validity; the third category is about different populations including college students, students, and female college students. After the epidemic, research on college students' anxiety can be divided into two major categories: the first major category is related to the COVID-19 pandemic including new crown pneumonia and major epidemic; the second major category is emotional, cognitive, physical, and behavioral problems caused by the epidemic, including loneliness, negative emotions, information anxiety, sleep, social anxiety, addiction, and social support.

Keywords	Year	Strength	Begin	End	2017 - 2022
COVID-19	2020	4.93	2020	2022	
behavior	2020	3.26	2020	2022	
interpersonal communication	2017	2.8	2017	2019	
major epidemic	2020	2.79	2020	2022	
habituation	2019	2.33	2020	2022	
psychological counseling	2018	2.21	2018	2020	
loss anxiety	2019	2.11	2020	2022	
consultant	2017	3.26	2017	2018	
group counseling	2018	2.59	2018	2019	
state anxiety	2017	2.52	2017	2018	
health education	2017	2.52	2017	2018	
problem	2017	2.44	2017	2018	
trait anxiety	2017	2.23	2018	2019	
procrastination behavior	2019	2.15	2019	2020	
physical and mental health	2019	2.15	2019	2020	
music therapy	2018	2.07	2018	2019	
consultation process	2017	2.04	2017	2018	
big five personality	2017	2.04	2017	2018	

Fig. 5. Top 18 keywords with the strongest citation bursts.

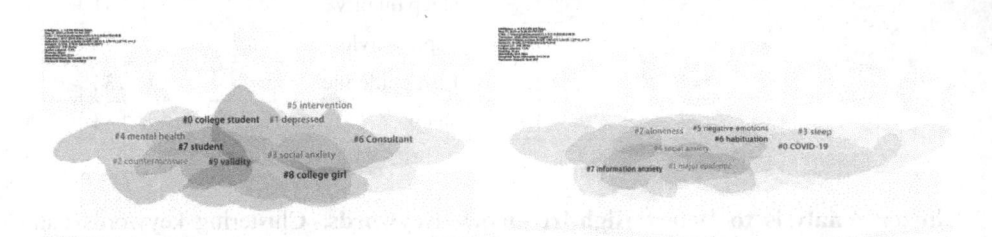

Fig. 6. Keyword clustering mapping before the epidemic (left). Keyword clustering mapping after the epidemic (right).

Evolution of Research Trends. The timeline view of keyword hotspot can visualize the dynamic development trend of the field at different time periods [16]. The hotspots of research attention are becoming more and more diverse, presenting different phenomena and latest responses to the research attention on college student anxiety in different time periods. The research hotspots of college students' anxiety follow the changes of social environment, and many hotspots of research related to COVID-19 Pneumonia have appeared since 2020, and many new hotspots such as loneliness, social support, and information anxiety have appeared after COVID-19 (Fig. 7).

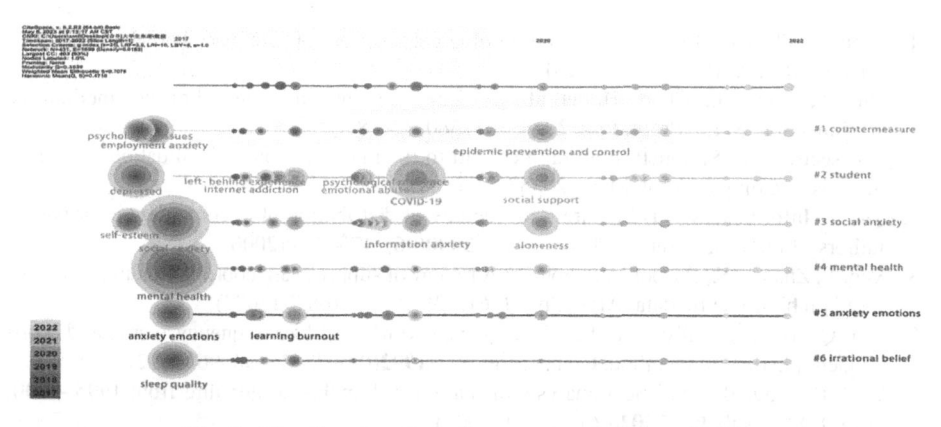

Fig. 7. Timeline view for college student anxiety research.

4 Discussion

The visualization results revealed that research on anxiety among college students developed rapidly after the COVID-19 pandemic. The Price's law, which pertains to the core author algorithm, indicated the absence of a core group of authors in the field. Therefore, core authors should increase their publication volume and enhance their collaboration. Institutions with limited cooperation should strive to achieve a higher number of publications. During the epidemic, the number of publications has significantly increased compared to before. According to the overall trend, it is predicted that the research in this field may become a hot spot in 2023, with the number of publications will reaching a new peak. The high frequency keywords after the epidemic identified through the Word frequency analysis algorithm and the emergent keywords obtained by the Keyword mutation algorithm yielded similar results including information anxiety, loneliness, social support, etc. COVID-19 triggers psychological stress reactions in individuals, especially college students. The LLR clustering algorithm concludes that the COVID-19 leads to issues such as loneliness, information anxiety, cell phone dependence, sleep, social support. These problems can be categorized into four domains: emotion, cognition, behavior, and physiology. These areas are expected to become hot spots for future research.

References

1. The World Health Organization (WHO) has declared COVID-19 an "international public health emergency, https://www.who.int/news-room/detail/30-01-2020-statement-or-the-second-meeting-of-the-international-health-regulations-(2005)-emergency-committee-regarding-the-outbreak-of-novelcoronavirus-(2019-ncov), 2020/01/31
2. Ma, C., Yan, X.: Research progress on psychological stress response and prevention and control strategies of novel coronavirus pneumonia outbreak. J. Jilin Univ. (Med. Ed.) **46**(03), 649–654 (2020)
3. Zhang, Y., et al.: The influence of changes in the Chinese COVID-19 prevention and control policies on mental health of medical staff: a network analysis. J. Affect. Disord. **335**, 10–17 (2023)

4. Chen, Y., Chen, C., Liu, Z., et al.: Methodological functions of CiteSpace knowledge graphs. Scientol. Res. **33**(02), 242–253 (2015)
5. Liu, Y., Chen, J., Chao, H., et al.: A fuzzy co-clustering algorithm via modularity maximization. Math. Probl. Eng. **2018**(3757580) (2018)
6. Rousseeuw, P.J.: Silhouettes: a graphical aid to the interpretation and validation of cluster analysis. J. Comput. Appl. Math. **20**, 53–65 (1987)
7. Qiu, J.: Informetrics (VI) Lecture 6 on the law of distribution of documentary information authors - Lotka's law. Intell. Theory Pract. **2000**(06), 475–478 (2000)
8. Xue, J., Zhang, W., Rasool, Z., Zhou, J.: A review of supply chain coordination management based on bibliometric data. Alex. Eng. J. **61**(12), 10837–10850 (2022)
9. Sun, Q.: Boundary division of high-frequency words and low-frequency words and word frequency estimation method. Chin. J. Lib. Sci. **1992**(02), 78–81+95–96 (1992)
10. Ye, L.P., Ling, B.: Statistical analysis of malignant lymphoma literature from 1995–2000. Chin. J. Med. Lib. Inf. **2002**(06), 58–59 (2002)
11. Hu, Z., Sun, J., Wu, Y.: A review of domestic research on the application of knowledge graphs. Lib. Inf. Work **57**(03), 131–137+84 (2013)
12. Zhao, F., Zhang, Y., Jin, H., et al.: A visual analysis of hotspots and trends of medical students' mental health research at home and abroad based on Citespace and VOSviewer. J. Northwest Univ. Natl. (Nat. Sci. Ed.) **43**(1), 58–65 (2022)
13. Hu, I.X., Liao, J.Y., Dai, N., et al.: Knowledge mapping analysis of Chinese and Western medicine and psychotherapy treatment for post-traumatic stress disorder based on Citespace. J. Zhejiang Univ. Traditional Chin. Med. **46**(7), 816–824 (2022)
14. Jin, Y.F., Cai, L., Cheng, Z.S., et al.: Rapid recommendation guidelines for the diagnosis and treatment of pneumonia in novel coronavirus (2019-nCoV) infection (standard version). PLA Med. J. **45**(1), 1–20 (2020)
15. Li, K., Yu, W., Li, S., et al.: Analysis of hotspots and frontier trends in CiteSpace-based research on social anxiety among college students. Chin. Family Med. **25**(33), 4217–4226 (2022)
16. Hao, W., Han, Y.: Current research status and development trend of cultural competence development in medical education–a visual analysis based on CiteSpace. Med. Educ. Manag. **8**(5), 580–588 (2022)

Risk Perception Visualization of Public Health Emergencies Based on Clustering Algorithms

Zhe Zhao[1], Yiguo Cai[1], Zhen Gong[1], Hao Ni[1], Zhiyong Xu[2], and Li Sun[1(✉)]

[1] School of Health Management, Changchun University of Chinese Medicine,
Changchun 130117, China
775960680@qq.com
[2] School of Medical Information, Changchun University of Chinese Medicine,
Changchun 130117, China
xuzy@ccucm.edu.cn

Abstract. *Objective:* To understand the progress, hotspots, trends and discrepancies of risk perception-related research in the context of public health emergencies, and to provide a reference for further research in the field. METHODS: This paper uses Citespace 6.2 R2 software based on clustering algorithm to search literature on risk perception of public health emergencies among university students in China Knowledge Network (CNKI) and Web of Science core collection database, and visualize and analyze authors, institutions and keywords. Results: In terms of the number of articles published, the proportion of research studies related to risk perception of public health emergencies increased year by year from 2020 to 2023, indicating that risk perception is receiving increasing attention. In terms of keyword clustering effects, both English and Chinese clusters have become tighter after the COVID-19 outbreak, indicating that much research has been conducted around risk perception after the COVID-19 outbreak. In terms of authors and institutions, the number of core authors posting articles is not high, the connection between authors is not strong, and the connection between institutions is low, which should strengthen the cooperation between authors and institutions. Conclusion: A comprehensive analysis shows that there is a growing interest in the study of risk perception in public health events, and although there are some differences in the issues of concern between China and overseas, the overall research trends are similar. Many studies have been conducted on risk perception and people's health, and they have become a hot topic for a while. More research is needed in the future on how to effectively intervene in the adverse effects of risk perception in public health emergencies.

Keywords: Risk perception · Public health emergencies · Clustering algorithms · Citespace

1 Introduction

The outbreak of the COVID-19 pneumonia has had a significant impact not only on individuals physically and psychologically [1], but also on all aspects of society and the country, during which people face a range of risks such as fear of being infected,

H. Jin et al. (Eds.): IAIC 2023, CCIS 2058, pp. 431–441, 2024.
https://doi.org/10.1007/978-981-97-1277-9_33

disruption of their lives and school plans, which may increase their stress levels and lead to social problems [2].

Risk perception refers to an individual's perception and understanding of various objective dangers in the outside world [3], and subjective judgement of the risks being or may be faced. Risk perception can have a large impact on public behaviour in the face of public emergencies [4].With the ever-changing situation of the COVID-19, risk perception has been gradually taken seriously. In view of this, this paper attempts to sort out and summarise the current status of research, research themes and trends in this field from the existing research literature, with the help of bibliometric tools and knowledge graph visualisation, in order to provide valuable references for subsequent research.

2   Data Sources and Clustering Method

2.1   The Tools

Citespace is an information visualisation software developed by Dr Chaomei Chen and his team at Drexel University, USA. Based on co-citation analysis theory and pathfinding network algorithms, the software performs multivariate, time-dependent and dynamic citation analysis on a specific domain literature collection, and finally maps scientific knowledge reflecting the evolution of knowledge and development frontiers of the discipline through visualisation techniques [5].

3   Date Sources

In the data of this study, the period of relevant literature after the outbreak of COVID-19 was limited to 1 January 2020 to 1 May 2023, and the period of relevant literature before the outbreak of COVID-19 was limited to 1 January 2000 to 1 December 2019. Chinese literature was studied using Chinese journal literature in China Knowledge Network (CNKI), and an advanced search was conducted to ensure the comprehensiveness of the included literature by "(Subject = "COVID-19" "infectious disease" "public health event" "influenza") AND (Subject = "risk perception")" in China Knowledge Network After filtering and eliminating literature and news items that were not relevant to the subject matter and duplicates, missing author and year information, and after reading the abstracts, keywords and some of the full text, 336 valid articles remained. The English-language literature was searched in the web of science journals according to "(subject = "risk perception") AND (subject = "Public Health" OR "influenza" OR "SARS " OR "COVID-19"), and 1254 English-language articles were included after screening. 50 Chinese-language articles prior to COVID-19 were searched by "(subject = "risk perception") AND (subject = "epidemic" "SARS" "infectious disease" "public health event" "influenza"), and 50 articles were included after screening; The English-language literature was searched by "(subject = "risk perception") AND (subject = "Public Health" OR "influenza" OR "SARS"), and 529 English-language articles were finally included after screening.

The above Chinese literature was exported in Refworks format and the English literature was exported in txt format, and then processed and imported into Citespace

6.2R2 software in txt format, with parameters set to 2020 -2023 for literature during COVID-19 and 2000 for literature before COVID-19 -2019, the default cut time is 1 year, the node type is keyword, author, and research institution, and the rest of the parameters are set by default for analysis and visualisation. The keywords are used to infer the current research status, hotspots and research trends in risk perception.

3.1 Clustering Method

In this study, the fast clustering function Clustering in Cite Space software was used for fast clustering, and the clustered tag words were extracted with the help of the Log Likelihood Ratio (LLR) algorithm proposed by Dunning in 1993, and the following two optional assumptions were used to explain whether the occurrences of keywords x and y were independent, and 0 and 1 were used to indicate whether keyword x occurred:

Assume H1:

$$p(0|y) = p = p(1|y) \tag{1}$$

Assume H2:

$$p(0|y) = p1 \neq p2 = p(1|y) \tag{2}$$

The maximum likelihood estimation method was used to calculate p, p1 and p2, and c1, c2 and c12 were used to denote the number of occurrences of x, y and xy, with N denoting the total number of occurrences of all keywords.

Assuming H1 is true, we have:

$$p(0|y) = p(1|y) = p = \frac{c2}{N} \tag{3}$$

Similarly, assuming that H2 is true, we have

$$p(0|y) = p1 = \frac{c12}{c1} \tag{4}$$

$$p(1|y) = p2 = \frac{c2 - c12}{N - c1} \tag{5}$$

The likelihood values of hypothesis H1 and hypothesis H2 can be derived using the binomial distribution assumption that

$$L(H1) = b(c12; c1, p)b(c2 - c12; N - c1, p) \tag{6}$$

$$L(H2) = b(c12; c1, p1)b(c2 - c12; N - c1, p2) \tag{7}$$

where the binomial distribution assumes that

$$b(k; n, x) = xk(1 - x)n - k \tag{8}$$

The log-likelihood ratio (log-likelihood ratio) for the likelihood ratio λ is as follows (with a base of 2).

$$LLR = \log\frac{L(H1)}{L(H2)} = \log\frac{b(c12; c1, p)b(c2 - c12; N - c1, p)}{(c12; c1, p1)b(c2 - c12; N - c1, p2)}$$

$$= \log L(c12, c1, p) + \log L(c2 - c12, N - c1, p) - \log L(c12, c1, p1)$$

$$- \log L(c2 - c12, N - c1, p2) \tag{9}$$

of which,

$$L(k, n, x) = xk(1 - x)n - k \tag{10}$$

The likelihood ratio λ gives a clear and intuitive explanation, i.e. if the likelihood ratio is small, it means that it is very likely to meet the hypothesis H2, i.e. that x y does not occur by chance, so the larger the LLR the more representative the word is for this cluster.

4 Results

4.1 Distribution of Posting Time

Keywords are words that express the concept of the topic of the literature. The collection of keywords of research results in a certain field over a period of time can reveal the hotspots of academic research in the field and predict the development pulse and direction of academic research [6]. Citespace 6.2R2 software was used to conduct keyword frequency statistics and cluster analysis on the retrieved literature, and the results are as follows.

Prior to the COVID-19 outbreak, research related to risk perception of public health events was increasing year by year both domestically and internationally, with a general upward trend (Fig. 1); after the COVID-19 public health event in 2020, in terms of the number of literature, the number of literature on risk perception in the English literature increased four-fold, and the Chinese literature increased more than ten-fold compared to the pre-COVID-19 peak, with similar overall publication trends in the Chinese and foreign literature.

4.2 Keyword Contribution Analysis

Citespace is applied to present the keyword knowledge map, nodes represent keywords, the size of the node represents the frequency of the keyword occurrence, the higher the frequency, the larger the node. The Chinese keyword map before and after the COVID-19 outbreak is shown in Fig. 2 Fig. 3, the keywords with higher keyword frequency are shown in Table 1. The top three keywords with the highest frequency before COVID-19 are risk perception (14 times), risk communication (8 times), and avian influenza (5); in the keyword centrality analysis, the top three were risk perception (0.37), avian influenza (0.28), and coping behaviour (0.27). After the COVID-19 outbreak, the top three keyword frequencies were risk perception (101 times), risk perception (47 times),

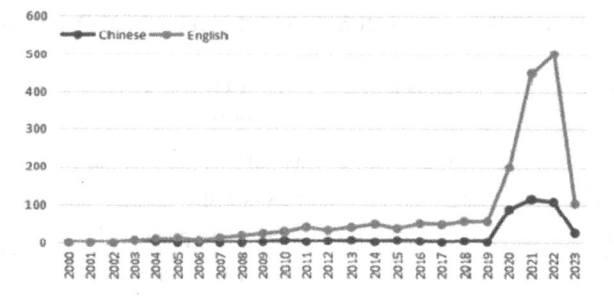

Fig. 1. Number of relevant literature

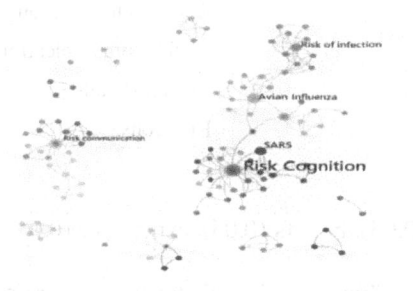

Fig. 2. Chinese keyword mapping before the COVID-19 outbreak

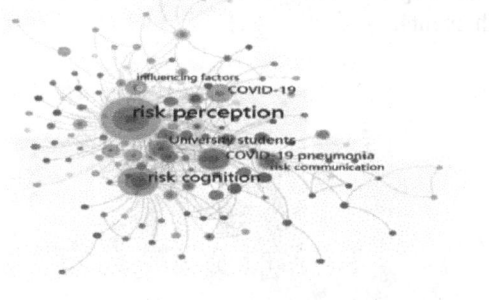

Fig. 3. Chinese keyword mapping after the COVID-19 outbreak

and new coronary pneumonia (27 times); in the keyword centrality analysis, the top three were risk perception (0.65), risk perception (0.27), and risk communication (0.21).

The keyword co-occurrence profile for risk perception before and after COVID-19 in the English literature is shown in Fig. 4 Fig. 5. The top eight keywords in terms of keyword frequency are shown in Table 2. The top three rankings before COVID-19 were risk perception (257 times), knowledge (68 times), and public health (63 times). The top three keywords in the keyword centrality analysis were risk perception (0.28), impact (0.17), and information (0.16). The top three keyword frequencies after COVID-19 were risk perception (582), impact (123), and health (120). Health (120 times), centrality model

Table 1. Frequency of keywords before (left) and after the COVID-19 outbreak (right)

Before COVID-19 Keywords	Number	After COVID-19 keywords	Number
risk cognition	14	risk perception	101
Risk communication	8	risk cognition	41
Avian Influenza	5	COVID-19 pneumonia	27
Risk of infection	3	University students	21
SARS	3	COVID-19	19
		Influencing factors	17
		Risk communication	16
		Health communication	13
		Mental Health	9
		Emotions	9

(0.05), communication (0.05), beliefs (0.05), experience (0.05), uncertainty (0.05), and infection (0.05).

It can be seen that the Chinese and English keywords are more closely linked after the COVID-19 outbreak, and the top ranking of high-frequency keywords shows that the Chinese literature is more concerned with public health events such as avian influenza or COVID-19, while the English literature is more concerned with health and knowledge under public health events.

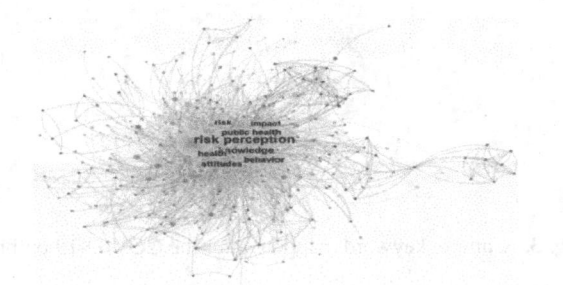

Fig. 4. English keyword mapping before the New Crown outbreak

4.3 Keyword Clustering Analysis

Using the Citespace keyword clustering function, the keywords were clustered using the LLD algorithm. As a reflection of research hotspots and frontiers, the clustered view of keywords reflects the research themes and their development process within a field.

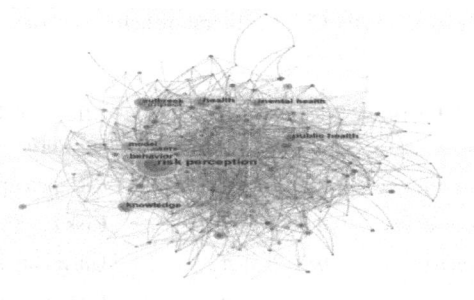

Fig. 5. English keyword mapping after the New Crown outbreak

Table 2. Frequency of keywords in the English literature before (left) and after (right) the COVID-19 outbreak

Before COVID-19 Keywords	Number	After COVID-19 keywords	Number
risk perception	257	risk perception	582
knowledge	68	impact	123
public health	63	health	120
attitudes	58	knowledge	98
behavior	51	behavior	97
health	47	public health	96
risk	36	outbreak	90
impact	36	model	81
communication	31	mental health	78
risk communication	31	SARS	76

The clustering of risk perception keywords before and after the COVID-19 outbreak in the Chinese literature is shown in Fig. 6, Fig. 7, and the clusters and S-values are shown in Table 3.

The English literature COVID-19 Before and After risk perception keyword clusters are shown in Fig. 8, Fig. 9, and the clusters and S-values are shown in Table 4.

A comparison of the keyword clustering shows that the clustering of keywords in both Chinese and English literature before the outbreak was relatively scattered, and after the COVID-19 outbreak, the links between the categories in the Chinese and English clustering plots became relatively close, indicating that research related to risk perception was becoming a hot topic after the COVID-19 outbreak.

4.4 Authors, Institutions and Countries or Regions

From the authors of the publications, the authors are not closely related to each other, among them, the Chinese authors Hao Yanhua, Wu Qunhong and Fan Kaisheng,

Table 3. Clustering of keywords in the Chinese literature before (left) and after (right) the COVID-19 outbreak

Category	Before COVID-19 Keywords	S-value	Category	After COVID-19 keywords	S-value
#0	Risk cognition	1.000	#0	Risk perception	0.830
#2	Infectious diseases	0.948	#1	Risk cognition	0.799
#3	Avian Influenza	0.981	#2	University students	0.909
#4	Consumers	0.957	#3	Risk governance	0.867
			#4	Risk communication	0.953
			#5	Influencing factors	0.900
			#6	Health Behaviour	0.926

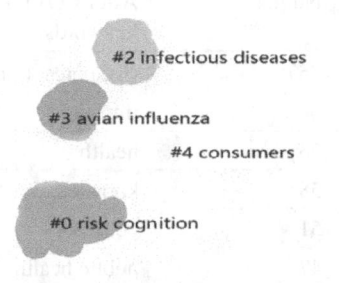

Fig. 6. Chinese keyword clustering before the COVID-19 outbreak

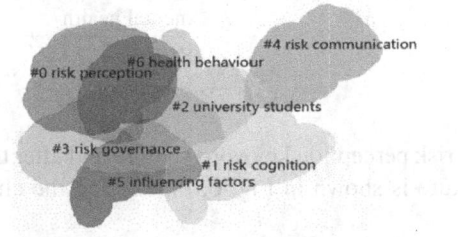

Fig. 7. Chinese keyword clustering after the COVID-19 outbreak

etc., and the English authors Yen.Cheng-Fang, Savoia.Elena, Chang.Yu--Ping and Ferracuti.Stefano have more publications and stronger research strength.

In terms of research institutions, Harbin Medical University, Beijing Normal University, Southwest Jiaotong University and Renmin University of China have published the most articles in Chinese, while in English, RLUK-Research Libraries UK, University of London, Johns Hopkins University, Harvard University, Johns Hopkins Bloomberg School of Public Health, etc. have published 20 or more articles, which have a significant impact on the field.

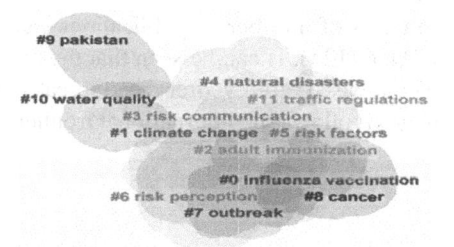

Fig. 8. Clustering of English keywords before COVID-19

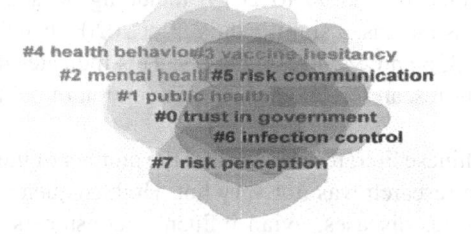

Fig. 9. Clustering of English keywords after COVID-19

Table 4. Clustering of keywords in the Chinese literature before (left) and after (right) the COVID-19 outbreak

Category	Before COVID-19 Keywords	S-value	Category	After COVID-19 keywords	S-value
#0	influenza vaccination	0.632	#0	trust in government	0.611
#1	climate change	0.743	#1	public health	0.609
#2	water quality	0.725	#2	mental health	0.718
#3	traffic regulations	0.741	#3	vaccine hesitancy	0.757
#4	adult immunization	0.854	#4	health behavior	0.693
#5	risk communication	0.710	#5	risk communication	0.693
#6	natural disasters	0.717	#6	infection control	0.708
#7	risk factors	0.814			
#8	risk perception	0.890			
#9	outbreak	0.979			
#10	cancer	0.944			
#11	pakistan	0.982			

In terms of countries with foreign language literature, the top three countries in terms of number of publications before the outbreak of COVID-19 were USA (186), ENGLAND (68), and PEOPLES R CHINA (68), and after the outbreak of COVID-19, the top three countries in terms of number of publications were PEOPLES R CHINA (290), USA (277), and ITALY (125). It can be seen that the number of publications by Chinese scholars on WOS increased rapidly after the outbreak of COVID-19, indicating that China's research direction follows the international frontier and contributes more to frontier research.

5   Discussion

As can be seen from the characteristics of the distribution of literature, the overall trend of fluctuating upward trend in the number of CNKI and WOS research publications, with fewer publications from 2000 to 2019, including 50 and 529 respectively; the number of publications increased significantly after 2020 when the number of included publications reached 336 and 1254 respectively. This indicates that scholars have paid increasing attention to research related to risk perception of public health emergencies during COVID-19.

The number of Chinese literature shows that the number of literature before COVID-19 was small and the research was not very hot. High-frequency words and clustering words such as infectious diseases, avian influenza, consumers, etc., can be seen that prior to the outbreak of COVID-19, domestic research was mainly concerned with the relationship between the governance of public health events and consumer purchasing behaviour [7], etc. After the COVID-19 outbreak, the number of related literature The high frequency words and clustering words such as university students, health communication, risk governance and risk communication show that the focus of the Chinese literature shifted from consumers to students after the COVID-19 outbreak, and at the same time began to focus on the issue of risk governance, which shows that the care for students' physical and mental health in China after COVID-19 as well as the It is evident that after COVID-19, the care of students' physical [8] and mental health and the maintenance of management of the adverse effects caused by public health events [9] have become a hot spot for research.

The English literature began to show a year-on-year increase in the study of risk perception before the COVID-19 outbreak. English literature keywords and clustering terms such as influenza vaccination, climate change, traffic management, natural disasters, risk communication, etc. indicate that the study of risk perception involves many fields and is also relatively scattered, mainly for the prevention of infectious diseases, natural disasters or risk perception of social and public events caused by human factors. After the outbreak of COVID-19, the number of literature has increased significantly compared to the previous year, indicating that foreign countries attach great importance to risk perception of COVID-19. High frequency key words and clusters in the English literature such as health, mental health, trust in government, and vaccine hesitancy show that a large number of studies have begun to focus on public health compared to the pre-COVID-19 outbreak period, with research tending to focus on the relationship between government and the public and public mental health.

Based on the above, and based on the summary of risk perception studies in English and Chinese, and taking into account the current research hotspots in China, the development and application of future research in this field in China can focus on the following aspects. (1) Intervention strategies for risk perception and public health. Previous studies have mostly focused on the identification of health problems and the influencing factors, but future research should focus on exploring intervention strategies for risk perception and public health to fundamentally solve the occurrence of health problems. (2) Risk perception in the post-epidemic era. Under the background of normal epidemic prevention and control, the impact of COVID-19 on people's lives has gradually disappeared, but the psychological trauma caused by COVID-19 may still affect people in their daily lives[10], so we should pay great attention to the current situation of public risk perception in the post-epidemic era and make targeted improvements to further protect their physical and mental health. (3) Drawing on foreign research findings, there are certain differences in the focus of public health events at home and abroad, so we can draw on foreign research findings to fill in the gaps in domestic research or provide research ideas to provide plans or suggestions for future public health emergencies, so as to better prevent the possible harm caused by the risk perception of public health events.

6 Discussion

This study presents a knowledge graph based on a clustering algorithm to analyse and visualise the risk perception of public health emergencies and to explore the frontiers and hotspots of research in this field. Although there are some differences in the issues of risk perception, the overall research trends are similar, and there is a need to explore how to effectively intervene in the adverse effects of risk perception of public health emergencies in the future.

References

1. Teng, D., Li, X.: Exploration of prevention and control strategies in universities during the new coronavirus outbreak. Ind. Technol. Forum **21**(02), 255–256 (2022)
2. Xiaoyu, W., Yifan, L., Boshuang, Y., et al.: A longitudinal follow-up study of changes in life satisfaction and depression among adolescents before and during the New Crown epidemic and their interrelationships. Modern Prevent. Med. **50**(08), 1398–1402 (2023)
3. Entradas, M.: In Science We trust: the effects of information sources on COVID-19 risk perceptions. Health Commun. **37**(14), 1715–1723 (2022)
4. Sun, M.: Research on the influencing factors of public risk coping behavior under emergencies. Harbin Engineering University (2016)
5. Zhao, R., Limin, X.: An exploration of knowledge mapping in the evolution of bibliometrics and research frontiers. Chinese J. Lib. Sci. **36**(05), 60–68 (2010)
6. Chen, Y., Chen, C., Liu, Z., et al.: Methodological functions of CiteSpace knowledge graphs. Sci. Res. **33**(02), 242–253 (2015)
7. Zhang, B., Lin, J.: A study on consumer decision making of chilled chicken under the risk perception scenario of avian influenza–an analysis of the role of cognitive and affective mechanisms based on. Agric. Modernization Res. **38**(05), 772–782 (2017)
8. Jiang, X., Lv, Y.: Research on the mental health counseling of college students under normal epidemic prevention and control. J. Jilin Provincial Coll. Educ. **38**(12), 21–25 (2022)
9. Wang, H.: Public opinion management of public health emergencies: an example of the new coronavirus pneumonia outbreak. J. Hubei Pol. Coll. **34**(01), 103–114 (2021)
10. He, J.: Analysis of the current psychological situation of college students in the post-epidemic era and countermeasures. Int. Publ. Rel. **163**(07), 148–150 (2023)

Visualization of Tumor Treatment Methods Based on Clustering Algorithms

Yiguo Cai[1], Fang Xia[1], Ziying Xu[1], Menglu Xu[1], Siqi Li[2], and Jingshuo Liu[1(⊠)]

[1] School of Health Management, Changchun University of Chinese Medicine,
Changchun 130117, China
13843110147@163.com

[2] School of Medical information, Changchun University of Chinese Medicine,
Changchun 130117, China

Abstract. The purpose of the article is to analyze the research hotspots and trends in tumor treatments using clustering algorithms such as LSI, LLR, Mutual Information and Kleinberg's algorithm. The related articles of CNKI are retrieved with "Tumor treatment methods" as the subject word and keyword for the period 2002–2022. The number of articles published in the field of tumor treatment methods is generally on the rise, and immunotherapy, targeted therapy and photothermal therapy has become a research hotspots and trends. Universities are the main collaborators, followed by research institutes and hospitals, cross-institutional collaboration needs to be strengthened. The main hot keywords are "oncotherapy" "immunotherapy" "targeted therapy".The keyword clusters such as "tumor therapy" "tumor" and "apoptosis" were formed. There are 25 keywords involved in the emergence, with the highest intensity of emergence being immunotherapy, followed by apoptosis, review, etc. The main research in the future will focus on the innovation of tumor treatment methods and drugs to development of anti-tumor with better efficacy and fewer adverse effects.

Keywords: Clustering algorithm · Kleinberg algorithm · Tumors · Immunotherapy · CiteSpace

1 Introduction

Tumor refers to the carcinogenic factors affecting the loss of regulatory ability of cells to mutate, resulting in monoclonal dysplasia and the formation of mass in the local area of the human body [1]. It is the second most deadly disease within the world due to high morbidity and mortality rates that endanger people's lives and health. According to the WHO, the number of people suffering from malignant tumors in China alone reached 4.28 million in 2018, which imposes a huge burden on society and families [2]. More and more tumors are being diagnosed early and cured by surgery, chemotherapy, radiotherapy, bio-targeted therapy and combination therapy [3]. At present, the review articles on tumor treatment methods in China mainly focus on the summary of clinical efficacy and mechanism of action, but there are few systematic reviews and analyses on

© The Author(s), under exclusive license to Springer Nature Singapore Pte Ltd. 2024
H. Jin et al. (Eds.): IAIC 2023, CCIS 2058, pp. 442–453, 2024.
https://doi.org/10.1007/978-981-97-1277-9_34

the research status and trends. The literature on oncology therapies published in China between 2002–2022 visualized and processed using CiteSpace software to quickly grasp current research hotspots and trends for future development.

2 Information and Algorithm

2.1 Literature Search

The literature related to the field of tumor treatment methods was searched in the CNKI database and the search time span from 2002–2022. After manual screening and software elimination of duplicates, 1283 core journals are finally include.

2.2 Clustering Method

CiteSpaceV software developed by Dr. Chen Chaomei is used to draw the knowledge map and keyword co-occurrence map based on the cooperation of authors, institutions, etc., and extract the author and keyword information of higher cited literature for analysis [4]. The algorithms used in this study for keyword co-occurrence and clustering are LSI shallow semantic indexing, LLR logarithmic maximum likelihood rate, mutual information, Kleinberg's algorithm for emergent word detection, and logistic curve model for document quantity prediction.

LSI Shallow Semantic Indexing: LSI is based on a singular value decomposition (SVD) approach to obtain text topics, and the decomposition of SDV can be approximated by writing [5]:

$$LLR(b_i) = ln\frac{P[b_i = 0|y]}{P[b_i = 1|y]} = \frac{1}{\sigma^2}[min_{x:bi=1}\{|y - \beta x|^2\} - min_{x:bi=0}\{||y - \beta x||^2\}] \quad (1)$$

Input m texts with n words in each text. A_{ij} corresponds to the feature value of the jth word in the ith text, commonly based on the preprocessed is normalized TF-IDF value. K is the assumed number of topics, less than the number of texts. After SDV decomposition, U_{il} corresponds to the relevance of the ith text and the lth topic; Σ_{lm} corresponds to the the correlation between the Ith topic and the mth word sense; V_{jm} corresponds to the correlation between the lth word and the mth word.

LLR Logarithmic Maximum Likelihood Ratio: Y is the input symbol to the de-mapper calculation; X is the QAM constellation points; β(BETA)is the constellation energy; $\frac{1}{\sigma^2} = \frac{1}{NV}$ (Noise Variance Inverse-NVI) [6].

$$LLR(b_i) = ln\frac{P[b_i = 0|y]}{P[b_i = 1|y]} = \frac{1}{\sigma^2}[min_{x:bi=1}\{|y - \beta x|^2\} - min_{x:bi=0}\{||y - \beta x||^2\}] \quad (2)$$

Mutual Information: Consider two random variables X and Y whose joint probability density function is p(x,y) and whose marginal probability density functions are p(x) and p(y), respectively. The mutual information I(X;Y) is the relative entropy between the joint distributions p(x,y) and p(x)p(y) [7].

$$KLD = D(p|q) = \sum_x p(x) \log \frac{p(x)}{q(x)} = E_p \log \frac{p(x)}{q(x)} \tag{3}$$

p and q are two probability distributions

$$I = I(X; Y) = \sum_X \sum_Y p(x, y) \log \frac{(x, y)}{p(x)p(y)} = D(p(x, y)\|p(x)p(y)) \tag{4}$$

$$I(X; Y) = H(X) - H(X|Y) = H(Y) - H(Y|X) + H(X) + H(Y) - H(X, Y) \tag{5}$$

$$\text{Conditional mutual information}: I(X; Y|Z) = H(X|Z) - H(X|YZ) \tag{6}$$

$$\text{It is chain rule}: I(X1, X2, LXn : Y) = \Sigma_{i=1}^n I(Xi; Y|Xi - 1, Xi - 2, LX1) \tag{7}$$

$$\text{Symmetric uncertainty}: Su(X; Y) = \frac{2I(X; Y)}{H(X) + H(Y)} \tag{8}$$

Kleinberg Algorithm: There are n sets of data, the t th set of data has a total of d_t papers, of which r_t papers contain emergent words; let $R = \sum_{t=1}^n r_t$, R is the number of all papers related to a topic in the n batches of data; $D = \sum_{t=1}^n d_t$, D is the number of all papers in the domain in the n batches of data. Define the emergence weight index in order to rank the degree of emergence of the manifestation [8].

$$\text{weight} = \sum_{i=t_1}^{t_2} (\sigma(0, r_t, d_t) - \sigma(1, r_t, d_t)) \tag{9}$$

Logic Curve Model: In terms of literature volume prediction, the former Soviet scientists Narimov and Freidutz based on the index model was proposed logical curve model [9], which mathematical formula is:

$$f(t) = \frac{k}{1 + ae^{-bt}} \tag{10}$$

$f(t)$ is t-years of literature accumulation; k is Literature cumulative maximum; a is Parameters; b is continuous growth rate of the literature; t is time.

3   Results

3.1   Distribution of Posting Time

The timing and annual distribution of publications indicator of the hot and the pace of development in the field [4]. Figure 1 shows the changes in the number of publications on tumor treatment methods during 2002–2022. (2023 is the predicted, the data from China Knowledge Network) The publication trend is gradually increasing in 20 years which the number of core journals is 1283 in this field. In detail, we can see that from 2002–2007,the number of relevant research papers has rapid upward, the average number of papers issued is 39; From 2008–2017, the number of relevant research was stable and has a slower growth, with an average annual number of 66. The rapid development stage is 2018–2022, with an average annual number of 98 documents. This indicates that research in this field is developing rapidly in China and will receive more attention in the future.

Fig. 1. Publication trends on Tumor treatment methods (2002–2022)

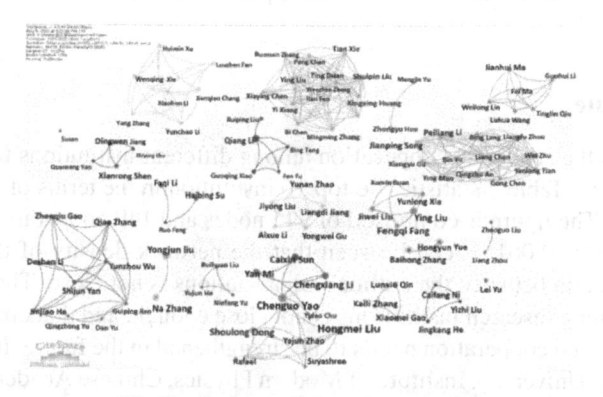

Fig. 2. Author collaboration network

3.2 Co-authorship

The distribution of results and teamwork in this area is determined by analyzing the number of published articles and partnerships [10]. The results from Fig. 2 show that there are 610 nodes, 521 connections, and the network density is 0.0028. It is apparent that many authors tend to collaborate with a relatively stable group of collaborators to generate several major author clusters, and each cluster usually contains two or more core authors. And the most representative author in the field is Chenguo Yao, Qiang Li and Yongjun Liu, etc. The number of authors who published research papers related to

Table 1. Top10 core authors in the field of tumor treatment methods research in China

Rank	Core Authors	Number	Rank	Core Authors	Number
1	Chenguo Yao	12	6	Xinghan Liu	5
2	Qiang Li	8	7	Ying Liu	5
3	Yongjun Liu	6	8	Wenjian Li	5
4	Na Zhang	6	9	Jing Liu	4
5	Caixin Sun	6	10	Chengxiang Li	4

tumor treatment methods in China is 401, and the top 10 authors in terms of the number of core papers published are shown in Table 1.

Chenguo Yao is the most prolific author in his field with 12 publications. According to Price's law [11] we can know the minimum value of core journal authorship is $N = 0.749 \times \sqrt{Npmax}$ (Npmax is the highest yielding authorship), a core team of authors is formed when the core authors publish half the number of articles in the field. The minimum number of articles issued by core authors is two according to the upper limit is rounded. The results from Citespace show that there are 135 core authors with 2 or more publications, and the core authors have published 16.7% of the total literature in the field.

3.3   Co-institute

Figure 3 shows the academic cooperation among different institutions tumor treatment methods research. Table 2 statistic the top 10 institution in the terms of number of articles published. The figure is composed of 545 nodes and 149 cooperation links and the network density is 0.001. It can be seen that the network density of the atlas is low, and the cooperation between the author's organizations is not close. This indicates that cooperation among research institutions is not close enough, and that cross-institutional communication and cooperation needs to be strengthened in the future. In terms of node size, Chongqing University, Institute of Modern Physics, Chinese Academy of Sciences, and Sichuan University, etc. has the most significant node size. The closer collaborations formed in this study are the Institute of Modern Physics of the Chinese Academy of Sciences, the University of Chinese Academy of Sciences, Beijing University of Traditional Chinese Medicine, Guang'anmen Hospital of the Chinese Academy of Traditional Chinese Medicine, and China Pharmaceutical University etc.

Table 2. Top 10 institution in terms of number of articles published

Rank	Institution	Number
1	Chongqing University	28
2	Cancer Hospital, Chinese Academy of Medical Sciences	23
3	Sichuan University	22
4	Institute of Modern Physics, Chinese Academy of Sciences	22
5	China Pharmaceutical University	20
6	Cancer Affiliated Hospital of Tianjin Medical University	19
7	Peking university	19
8	Central South University	18
9	Zhejiang University	16
10	Shanghai Jiao Tong University	15

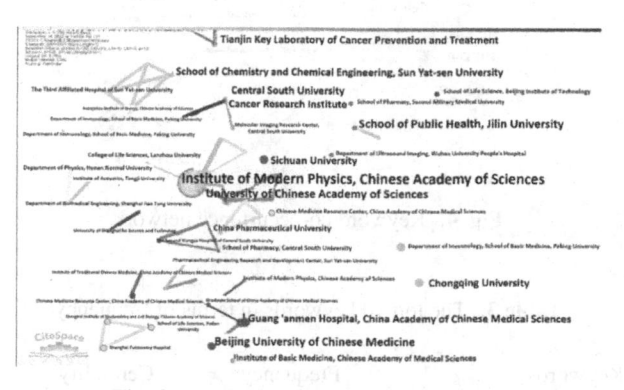

Fig. 3. Institution collaboration network

3.4 Co-occurring Keywords

The main research content and hotspots can be obtained by keyword co-occurrence analysis [12]. The knowledge network of co-occurred keywords is Fig. 4, which consists of 703 nodes and 1506 connections. The top 10 keywords in terms of frequency were displayed (Table 3). The highest co-occurrence of keywords is tumor treatment, followed by various treatment methods and tools including "gene therapy" "photothermal therapy" "inhibitors" "drug delivery". The top 10 most frequently occurring keywords were "oncology" "immunotherapy" "apoptosis" "review" "targeted therapy", "gene therapy" "inhibitors" "radiation therapy" "antitumor" "signalling pathways".

In detail, Table 3 lists the most frequently co-occurred keywords in terms of frequency, centrality, and year of occurrence. The top co-occurred keywords are "tumors" (342 times), "immunotherapy" (68 times), and "apoptosis" (38 times), "overview" (35 times), "targeted therapy" (32 times), "gene therapy" (32 times), "inhibitors" (18 times), "radiation therapy" (17 times), "anti-tumor" (16 times), "signalling pathways" (15 times).

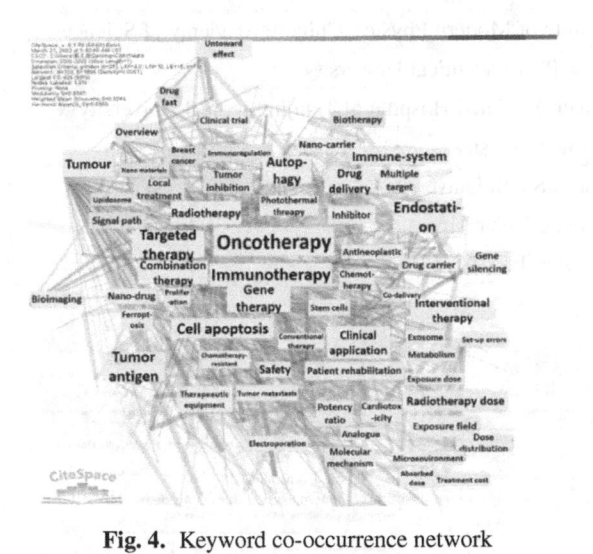

Fig. 4. Keyword co-occurrence network

Table 3. The top 10 keywords in terms of frequency

Rank	Keywords	Frequency	Centrality	Year
1	Tumors	342	0.42	2002
2	Immunotherapy	68	0.05	2002
3	Apoptosis	38	0.02	2002
4	Overview	35	0.04	2012
5	Targeted therapy	32	0.03	2007
6	Gene therapy	32	0.02	2002
7	Inhibitors	18	0.01	2004
8	Radiation therapy	17	0.02	2002
9	Anti-tumor	16	0.02	2006
10	Signalling pathways	15	0.01	2008

3.5 Keyword Clustering

The purpose of cluster analysis is to understand the research hotspots in the field and is based on keyword co-occurrence networks. The results showed that the keywords studied in this field were clustered into 11 categories, which is displayed in Fig. 5, and the sub-categories are #0 oncotherapy, #1 cancer, #2 cell apoptosis, #3 immunotherapy, #4 bioimaging, #5 inhibitor, #6 targeted therapy, #7 drug carrier, #8 breast cancer, #9 antibody and #10 molecular mechanism. Some of the tag words for this cluster are shown in Table 4. Figure 5 and Table 4 it can be seen that the keyword oncology treatment is central. The various treatments form a second level around the central point, with the outermost layer being the detailed treatment and drug administration mechanisms.

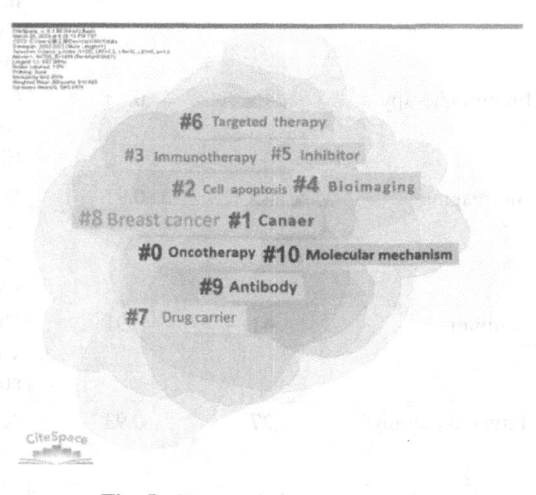

Fig. 5. Keyword cluster network

3.6 Keyword Emergence

The scientific questions or topics explored by a subject of literature based on the knowledge of emergent terms at a given point in time, which can be used as one of the criteria for judging and predicting the frontiers of research [13]. As shown in Fig. 6, this study detects 25 emergent words from 2002 to 2022 with high burst intensity in the tumor treatments methods field. "Immunotherapy" has the highest emergent intensity, followed by "apoptosis" "synthesis" "autophagy" "gene therapy" "nanomaterials" and "drug delivery".

Table 4. Keyword clustering information

Cluster Number	Cluster Name	Frequency	Centrality	Cluster subclusters
#0	Oncotherapy	177	0.97	Oncotherapy, Tumor diagnosis, Thermal stability
#1	Cancer	101	0.83	Famous medical experience, Comb air conditioner, Two wings in one
#2	Cell apoptosis	52	0.90	Apoptosis, Interventional therapy, Application prospects
#3	Immunotherapy	46	0.84	Gene therapy, Nanocarriers, Electrospinning
#4	Bioimaging	45	0.91	Drug delivery, Nanomaterials, Biomedical metal materials
#5	Inhibitor	41	0.91	Clinical trials, lysing viruses, modification strategies
#6	Targeted therapy	27	0.93	Nanomedicines, bionic drug delivery systems, nanoparticles
#7	Drug carrier	26	0.92	Topical treatment, Magnetic nanocarriers, Applied magnetic fields
#8	Breast cancer	23	0.94	Precision therapy, Polycyclic aromatic hydrocarbons
#9	Antibody	20	0.96	Radical therapy, Molecularly targeted therapy, Minimally invasive therapy
#10	Molecular mechanisms	19	0.95	Cell cycle, Cyclin-dependent kinase

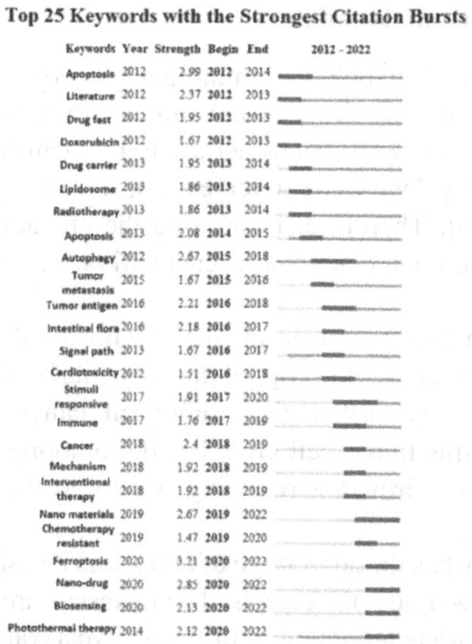

Top 25 Keywords with the Strongest Citation Bursts

Keywords	Year	Strength	Begin	End	2012 - 2022
Apoptosis	2012	2.99	2012	2014	
Literature	2012	2.37	2012	2013	
Drug fast	2012	1.95	2012	2013	
Doxorubicin	2012	1.67	2012	2013	
Drug carrier	2013	1.95	2013	2014	
Lipidosome	2013	1.86	2013	2014	
Radiotherapy	2013	1.86	2013	2014	
Apoptosis	2013	2.08	2014	2015	
Autophagy	2012	2.67	2015	2018	
Tumor metastasis	2015	1.67	2015	2016	
Tumor antigen	2016	2.21	2016	2018	
Intestinal flora	2016	2.18	2016	2017	
Signal path	2013	1.67	2016	2017	
Cardiotoxicity	2012	1.51	2016	2018	
Stimuli responsive	2017	1.91	2017	2020	
Immune	2017	1.76	2017	2019	
Cancer	2018	2.4	2018	2019	
Mechanism	2018	1.92	2018	2019	
Interventional therapy	2018	1.92	2018	2019	
Nano materials	2019	2.67	2019	2022	
Chemotherapy resistant	2019	1.47	2019	2020	
Ferroptosis	2020	3.21	2020	2022	
Nano-drug	2020	2.85	2020	2022	
Biosensing	2020	2.13	2020	2022	
Photothermal therapy	2014	2.12	2020	2022	

Fig. 6. Top 25 keywords with the strongest citation bursts

4 Conclusion

4.1 Literature Characterization

The overall increase in the amount of research literature in this field is on the rise. In terms of author distribution, the number of core authors is low and presents an unstable core group of authors. In the distribution of collaborating institutions, research is still concentrated in higher education institutions, and increased cross-institutional collaboration would improve the quality of cooperation.

The keyword nodes for "oncology", "immunotherapy" and "targeted therapy" are large. Immunotherapy can achieve better therapeutic results which could improve host immunity and induce long-term immune memory effects to inhibit tumor recurrence and metastasis at the time of treatment [14]. The targeted therapy not only ensures the effectiveness of the treatment but also reduces the extent of damage to normal cells and tissues. The "immunotherapy" and "targeted therapies" have reduced the adverse effects of treatment while also improving its efficacy.

Apoptosis is the main focus of research from the key words clustering group and each cluster sub-cluster, and specific treatments are also studied in depth through apoptosis.

The research hotspots have changed from "gene therapy" "apoptosis" and "inhibitors" (which is therapeutic mechanisms) to specific treatments and methods such as "nanomaterials" "targeted therapy" and "photothermal therapy". The reason for this result maybe is that the previous experimental research was focused on exploring and discovering the growth and death mechanisms of tumor cells to find ways to kill them, and now more clinical research is being conducted to improve the effectiveness of the treatment.

4.2 Research Hotspots and Trends

There are hotspots and trends which is immunotherapy, targeted therapy and photothermal therapy. Immunotherapy has shown good clinical efficacy but still faces many challenges. Some patients receiving immunotherapy fail to achieve clinical outcomes [15] and almost all have varying degrees of adverse events in the organ or systems associated with immunotherapy [16]. Therefore, improving the efficacy of immunotherapy and reducing clinical immune-related adverse events will be a research hotspots and trends in this field.

Research in targeted therapies has advanced with the development of molecular biology theories and techniques. Studying of the molecular biology of tumors to identify the malignant genetic phenotypes that affect tumor cells can provide more precise targets for tumor therapy, inhibiting tumor cell growth and promoting apoptosis. Improving the specific anti-tumor effects of drugs and reducing toxicity are the key research components of targeted therapies.

Photothermal therapy has the advantage of being less invasive, highly controllable in spacetime and having fewer adverse effects. The materials are the key that divided into organic and inorganic nanomaterials, both of them have advantages and disadvantages. For example, inorganic nanomaterials have a high photothermal conversion rate, but have some cytotoxicity; organic nanomaterials have a low photothermal conversion rate and poor stability. Reducing the toxicity and increasing the conversion rate of the photothermal materials are the focus of future research.

References

1. Deng, A.: What is a tumor. Diet Health 6(47), 244–245 (2019)
2. Bray, F., Ferlay, J., Soerjomataram, I.: Global cancer statistics 2018: Globocan estimates of incidence and mortality worldwide for 36 cancers in 185 countries. CA Cancer J. Clin. 68(6), 394–424 (2018)
3. Wang, S., Jia, M.: The role of tumor immune microenvironment in the effect of conventional tumor therapy. Chin. J. Tumor Biother. 19(3), 229–238 (2012)
4. Xingen, L., Wei, S.: Hot topics and evolutionary trends of domestic digital governance research–an analysis of knowledge mapping based on CiteSpace. SE Acad. 2, 61–71 (2022)
5. Zhou, Z., Kaihu, H., Hongfa, Y., et al.: Research on an automatic scoring system for subjective questions based on TF-IDF and LSI model. Software 40(2), 158–163 (2019)
6. Chai, X., Shan, S., Chen, X., Gao, W.: Locally linear regression for pose-invariant face recognition. IEEE Trans. Image Proc. 16(7), 1716–1725 (2007). https://doi.org/10.1109/TIP.2007.899195
7. Yuesong, M., Shuxing, Y., Bo, M.: An improved automatic image alignment algorithm based on mutual information. Laser Infrared 37(5), 470–473 (2007)
8. Xu Shuang, X., Dan, H.S., et al.: Application of SemRep and burst monitoring algorithms in bibliometric analysis - an example of disease drug treatment trends. J. Intell. 40(7), 745–755 (2021)
9. Yi, S.: Fitting the growth curve of logical type literature. Intell. Theory Pract. 20(1), 5–8 (1997)
10. Yang, T., Zhang, J.: A visual analysis of the research status and trends of music therapy for insomnia at home and abroad based on citespace. World Sci. Technol. - Modernization Chin. Med.:1–9

11. Xue Yao, C.: Constructing a core author user database for a Sci. J. **29**(1), 64–66 (2017)
12. Zhang, E., Dong, Y., Li, M., et al.: CiteSpace-based visual analysis of Chinese medicine heritage research. World Sci. Technol. Modernization Chin. Med. **21**(12), 2881–2887 (2019)
13. Liang, Z.: Drug TRIPS-Plus protection in U.S. free trade agreements. Comp. Law Res. **2014**(1), 125–140 (2014)
14. Sun, M., Zhang, J.: Research progress and development frontiers of immunomodulatory drug delivery technologies. Adv. Pharm. **46**(09), 641–644 (2022)
15. Emerging immunotherapy strategies: recent advances and future directions. Forward Forum **2023**(03), 29 (2023)
16. Wang, H., Li, H., et al.: Research progress of TCM-assisted cancer immunotherapy. Chin. J. Comp. Med. **33**(3), 130–135 (2023)

Research on the Operational Efficiency of Listed Pharmaceutical Logistics Companies in China Based on DEA-BCC Model and Malmquist Index

Shiyu Gao[1], Yiguo Cai[1], Shan Huo[2], Xin Liu[1], and He Wang[1](✉)

[1] School of Health Management, Changchun University of Traditional Chinese Medicine, Jilin 130117, China
20935443@qq.com
[2] School of Medical Information, Changchun University of Traditional Chinese Medicine, Jilin 130117, China

Abstract. Objective: To analyse the operational efficiency of listed pharmaceutical logistics companies in China and its changes from 2011–2021, and to provide a basis for promoting the efficient operation of pharmaceutical logistics companies as a whole. **Method:** The input-output data of 17 listed pharmaceutical logistics enterprises were selected for the study. The DEA-BCC model of data envelopment analysis and the Malmquist index model were used to measure operational efficiency, and the reasons for the differentiated development of each enterprise were analysed through static and dynamic analysis. **Result:** The changes in the number of effective enterprises in listed pharmaceutical logistics enterprises' technical efficiency (TE = 1) and scale efficiency (SE = 1) are basically similar, showing an upward and then downward trend; from the perspective of the efficiency range, the operational efficiency of Chinese listed pharmaceutical logistics enterprises all range from 0.7 to 1; the Malmquist index and the technological progress index show a wave-like development in parallel, fluctuating above and below 1, with a minimum of 0.887 and a maximum of 1.136. **Conclusion:** During the study period, the overall efficiency of listed pharmaceutical logistics companies showed a fluctuating upward trend; about more than 50% of the companies' operational efficiency was at a moderate or mildly DEA ineffective level; scale efficiency became a key factor limiting operational efficiency; and technological progress was the main reason why the Malmquist Index maintained a positive growth.

Keywords: Data envelopment analysis · Malmquist index method · Listed pharmaceutical logistics company · Operational efficiency

1 Introduction

As an important part of the pharmaceutical industry, the efficiency level of the pharmaceutical logistics industry is an important symbol to evaluate the comprehensive strength of China's pharmaceutical economy [1]. 2021, the total amount of China's pharmaceutical logistics is about 279.51 billion yuan, up 6.9% year-on-year from 2020, and there

H. Jin et al. (Eds.): IAIC 2023, CCIS 2058, pp. 454–463, 2024.
https://doi.org/10.1007/978-981-97-1277-9_35

is still a continuous upward trend in the future. Compared to 2021, the total profit of the national circulation directly reported enterprises is 45.3 billion yuan, the growth rate continues to decrease by 1% on the basis of a 2.7% decline in 2020, showing a trend of slight profitability. Pharmaceutical logistics costs are growing at a much higher rate than enterprise revenues, and profit margins are narrowing, seriously hindering the development of pharmaceutical logistics enterprises. Therefore, how to reduce pharmaceutical distribution costs and achieve efficient operation of the pharmaceutical logistics industry has become a key issue that needs to be addressed for the high-quality development of pharmaceutical logistics.

At present, the number of pharmaceutical logistics enterprises in China is large and unevenly distributed, and the industry as a whole has not yet formed a stable pattern of scale [2]. The implementation of medical reform policies has promoted the development of the pharmaceutical logistics industry in various provinces and cities, and the construction of pharmaceutical logistics in China has achieved initial results [3]. In order to further deepen the reform of the pharmaceutical and health system, enhance the effectiveness of pharmaceutical logistics and improve the operational efficiency of the pharmaceutical logistics industry, in October 2021, the Ministry of Commerce issued the "Guidance on Promoting the High-Quality Development of the Pharmaceutical Distribution Industry during the 14th Five-Year Plan Period" (hereinafter referred to as "Guidance"), pointing out that the pharmaceutical distribution industry is an important part of the national Based on this background, this paper aims to analyze the input-output efficiency of listed pharmaceutical logistics enterprises in China, find out the reasons affecting their differentiated development, and provide reference and reference for the improvement and benign development of the overall pharmaceutical logistics enterprises. This paper aims to analyse the input-output efficiency of listed pharmaceutical logistics enterprises in China, find out the reasons affecting their differentiated development, and provide references and lessons for the improvement and sound development of the overall pharmaceutical logistics enterprises.

2   Sources and Methods

2.1   Source

According to the Pharmaceutical Distribution Classification of the Shen Wan Industry Classification 2021, 25 listed pharmaceutical logistics enterprises were retrieved. In view of the continuity and completeness of the data, the panel data of 17 listed pharmaceutical enterprises from 2011 to 2021 were finally obtained by excluding enterprises with serious data deficiencies. In addition, as there are negative values for individual indicators to measure input-output efficiency, and the DEA model requires that the software input values cannot be negative, the negative values of the indicator data are uniformly adjusted to positive values according to the principle of "linear invariance" in order to To meet the data requirements of the DEA model and ensure that the software outputs the results smoothly.

2.2   Method

Data Envelopment Analysis (DEA). DEA is a commonly used method for measuring input-output efficiency, especially for measuring the efficiency of inputs and outputs with complex production functions, and DEA models can be divided into variable returns to scale (VRS, BCC model) and constant returns to scale (CRS, CCR model), The BCC model was first proposed by Charnes et al. [4] and can be used to compare the relative efficiencies between different decision units of the same type, which decomposes technical efficiency into pure technical efficiency and scale efficiency under the assumption of variable returns to scale.The relationship between the three is: technical efficiency value = pure technical efficiency value x scale efficiency value. The BCC model equation can be expressed as follows.

$$minW_{KD} = \theta$$
$$s.t. \begin{cases} X\lambda + S^- = \theta X_K \\ X\lambda - S^+ = Y_K \\ \sum_{j=1}^n \lambda_j = 1 \\ \lambda \geq 0, S^- \geq 0, S^+ \geq 0, \ \theta \ has \ no \ symbolic \ restrictions \end{cases} \tag{1}$$

X, Y are the input and output vectors of the decision making unit (DUM) respectively. When the technical efficiency value θ of the DUM is equal to 1, it means that the DEA is valid, while less than 1 is relatively invalid and needs to be improved in terms of technology or scale.

Malmquist Index Method. The Malmquist index is an efficiency evaluation model that measures total factor productivity (TFP) in DMUs from a dynamic perspective [5] and is often combined with DEA methods to help researchers better grasp the characteristics and patterns of efficiency evolution, which decomposes TFP into the product of the technical efficiency change index (EC) and the technical change index (TC), whose expression is shown below:

$$tfpch = M(x^{t+1}, y^{t+1}, x^t, y^t) = \left[\frac{D^t(x^{t+1}, y^{t+1})}{D^t(x^t, y^t)} \times \frac{D^{t+1}(x^{t+1}, y^{t+1})}{D^{t+1}(x^t, y^t)} \right]^{\frac{1}{2}} \tag{2}$$

$M(x^{t+1}, y^{t+1}, x^t, y^t)$ is greater than, equal to, or less than 1, respectively, the corresponding total factor productivity rises, remains unchanged, and falls.

Selection of Indicators and Description. The operation of an enterprise is a dynamic process of multiple inputs and multiple outputs, and whether the selection of measurement indicators is reasonable determines the accuracy of the determination of the efficiency of the enterprise. Huang Jin et al. [6] took the number of employees, net fixed assets and total operating costs as input indicators, and net profit and total operating income as output indicators, and concluded that listed logistics companies did not achieve DEA effective and proposed four effective paths based on fsQCA analysis. Fuhua Shu [7] selected 10 core financial indicators such as cost and expense, earnings per share, main operating cost and main operating profit and measured that the profitability of listed pharmaceutical enterprises was fair, but market order and product quality needed further improvement. Combining previous research results and the specific situation of the pharmaceutical logistics industry, this paper uses fixed assets, operating costs, overheads and number of employees as input indicators, and operating income and earnings per share as output indicators, as detailed in Table 1.

Table 1. Evaluation indicators and descriptions

Input/Output	Measurement Indicator	Indicator Description	Unit
Input	fixed asset	Measuring corporate capital investment	Yuan
	operating cost	Measuring the daily operating cost inputs of a business	Yuan
	overhead	Measuring invisible asset inputs into a business	Yuan
	Number of employee	Total number of employees at the end of the year	per
output	operating income	Total corporate revenue	Yuan
	earnings per share	Profit per share of the business	Yuan

3 Analysis of Results

3.1 Static Perspective Analysis Based on the BCC Model

Table 2 and Fig. 1 show the results of the DEA-BCC run, with the number of Chinese listed pharmaceutical logistics companies with effective operational efficiency (TE = 1) and effective scale efficiency (SE = 1) trending slightly upwards from 2011–2019, reaching a maximum in 2019 and then declining. It can be seen that the operating conditions of pharmaceutical logistics enterprises fluctuate and the operational efficiency needs to be further stabilised and improved. The number of purely technically efficient (PTE = 1) effective firms shows an N-shaped trend and the same number of effective firms in 2011 as in 2021.The change in the number of enterprises with effective scale efficiency (SE = 1) is basically similar to the trend of TE = 1, with scale efficiency reaching a maximum of 0.991 in 2019, which fluctuates up and down but is generally stable and has room for increase. Different companies should adjust the scale of resources invested according to their respective stages in order to improve scale efficiency.

As shown in Fig. 1, the overall efficiency of China's listed pharmaceutical logistics enterprises fluctuated from 2011 to 2021, with significant growth in 2011 and 2017. The State Council issued the "13th Five-Year Plan" to deepen the reform of the pharmaceutical and health system in 2017, encouraging the development of green pharmaceutical logistics policy. This shows that since the policy was implemented, enterprises have been paying more attention to pharmaceutical logistics, and the pharmaceutical logistics industry has been rapidly developing and expanding, with economies of scale emerging.

Table 2. Changes in operating efficiency of listed pharmaceutical logistics companies in China.

Type	Means and number of companies	2011	2013	2015	2017	2019	2021
Technical efficiency	Mean	0.920	0.961	0.960	0.965	0.982	0.977
	TE = 1 Number of companies	6	6	8	8	10	7
Pure technical efficiency	Mean	0.957	0.990	0.981	0.990	0.992	0.993
	PTE = 1 Number of companies	11	15	12	11	13	11
Scale efficiency	Mean	0.961	0.971	0.979	0.975	0.991	0.984
	SE = 1 Number of companies	6	7	8	8	11	8

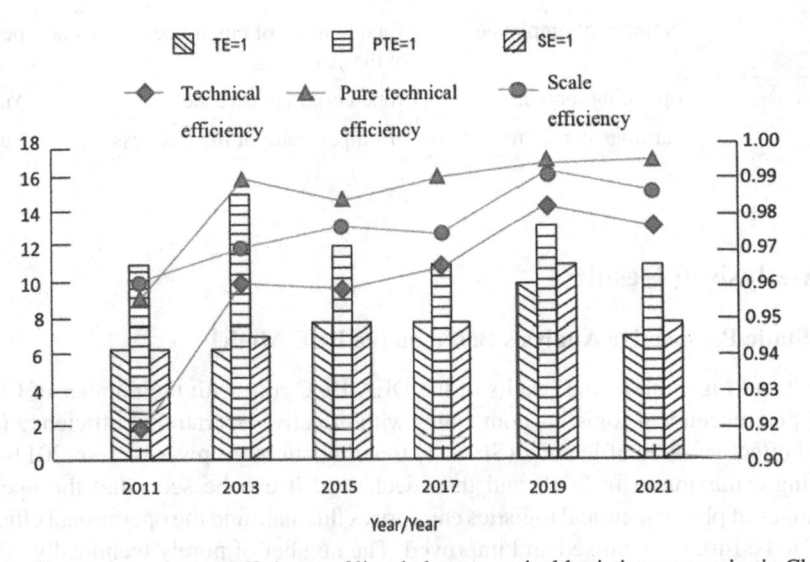

Fig. 1. Trends in operating efficiency of listed pharmaceutical logistics companies in China.

From the perspective of the efficiency range (Table 3), the operational efficiency of listed pharmaceutical logistics companies in China from 2011–2021 all ranged from 0.7 to 1, which was at a moderate ineffective level or above, enterprises with moderately ineffective and mildly ineffective overall efficiency account for more than 50% of the total each year, indicating that more than half of the listed pharmaceutical logistics enterprises have not reached an optimal level of operation, and that each enterprise should make further optimisation and adjustment in the input and use of its resources. From 2011 to 2021, the overall number of companies in the increasing size stage shows a fluctuating downward trend, the overall number of companies in the decreasing size stage shows a fluctuating upward trend, and the number of companies in the constant size stage is highest in 2019.

Table 3. Operational efficiency statistics of listed pharmaceutical logistics companies in China.

Year	Type	Moderately Ineffective (EA)	Moderately Ineffective (EA)	Effective (EA)	IRS (EA)	DRS (EA)	CRS (EA)
2011	TE	5	6	6	5	6	6
	PE	4	2	11			
	SE	1	10	6			
2013	TE	4	7	6	6	4	7
	PE	1	1	15			
	SE	2	8	7			
2015	TE	2	7	8	4	5	8
	PE	1	4	12			
	SE	1	8	8			
2017	TE	1	8	8	4	5	8
	PE	0	6	11			
	SE	1	8	8			
2019	TE	0	7	10	1	5	11
	PE	0	4	13			
	SE	0	6	11			
2021	TE	1	9	7	0	9	8
	PE	0	6	11			
	SE	1	8	8			

Note: IRS is increasing returns to scale, DRS is decreasing returns to scale and CRS is constant returns to scale. 0.7 to 0.9 is moderately ineffective, 0.9 to 1 is mildly ineffective and 1 is effective.

3.2 Dynamic Perspective Analysis Based on the Malmquist Index

The Malmquist index model can reflect the dynamic trend of operational efficiency of pharmaceutical logistics enterprises in different periods [8]. Using DEAP2.1 software, the change of total factor productivity index of listed pharmaceutical logistics enterprises in China from 2011 to 2021 and its decomposition were calculated (Table 4) to further analyse the impact of technical efficiency and technological progress index on total factor.

From the time dimension, the average value of total factor productivity from 2011 to 2021 is greater than 1, indicating that all the operational efficiency has improved

Table 4. Year-on-year results of total factor productivity measurement for listed pharmaceutical logistics companies in China.

Year	Effch	Tech	Pech	Sech	Tfpch
2011–2012	1.021	1.012	1.017	1.004	1.033
2012–2013	1.025	0.865	1.018	1.008	0.887
2013–2014	0.979	1.026	0.977	1.002	1.005
2014–2015	1.020	0.958	1.014	1.005	0.977
2015–2016	0.967	1.051	0.996	0.971	1.017
2016–2017	1.040	1.093	1.014	1.026	1.136
2017–2018	1.006	0.977	1.001	1.005	0.983
2018–2019	1.013	0.959	1.001	1.013	0.972
2019–2020	0.991	1.034	0.998	0.993	1.025
2020–2021	1.003	1.020	1.003	1.000	1.023
Mean	1.006	0.998	1.004	1.002	1.004

during this period, and the overall trend is growing, except for 2012, 2014, 2017 and 2018, which have obvious "short board effect", resulting in efficiency less than 1. The total factor productivity of all years is greater than 1, which shows that the operational efficiency of listed pharmaceutical logistics enterprises in China shows a fluctuating path of "high-low-high" during the study period. From 2012–2013, technical efficiency was greater than 1 and the technical progress index was less than 1. Operational efficiency was affected by the technical progress index in a downward trend. From 2012 onwards when total factor productivity is greater than 1, the technical progress index is also greater than 1 and vice versa, indicating that the technical progress index is the main reason why total factor productivity maintains a positive growth.

From Fig. 2, it can be seen that the Malmquist Index of Chinese pharmaceutical logistics enterprises from 2011 to 2021 shows an overall wavy development, fluctuating around 1, with a minimum of 0.887 and a maximum of 1.136, indicating that there is a certain difference in the degree of productivity progress over the years. The trend of change in the technological progress index keeps pace with the trend of change in total factor productivity, therefore, pharmaceutical logistics enterprises should focus on technological innovation and strengthen the information construction of pharmaceutical logistics in order to improve the operational efficiency of enterprises.

4 Discussion

4.1 Overall Operational Efficiency of Pharmaceutical Logistics Companies is Good, but There is Still Much Room for Improvement

The results of the static analysis show that the number of enterprises with effective DEA (TE = 1) from 2011–2021 tends to rise and then fall with 2019 as the folding point. 2019's unexpected epidemic disrupted the pharmaceutical logistics supply chain

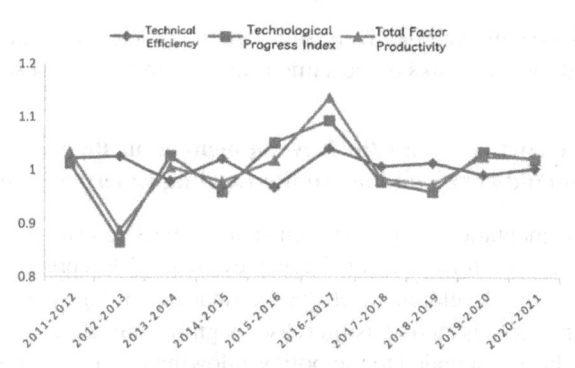

Fig. 2. Year-on-year trends in total factor productivity of listed pharmaceutical logistics companies in China.

[9], causing delays and increasing enterprise costs, indicating that the operational efficiency of pharmaceutical logistics enterprises in China needs to be further stabilized and improved. During the study period, the operational efficiency of listed pharmaceutical logistics enterprises in China were all between 0.7 and 1, with a good development trend, but more than 50% of the enterprises were still in moderate or mild DEA ineffectiveness, mainly influenced by the scale factor. In order to achieve high speed and efficient development, enterprises generally have the problem of blindly expanding their scale. In recent years, China's pharmaceutical logistics industry has vigorously promoted cost reduction and efficiency, and over-emphasis on the growth of scale will inevitably reduce their efficiency level [10]. Companies should understand the competitive landscape of the Chinese pharmaceutical market in line with the needs of market development and scale and the relevant requirements of the state government, and on this basis, propose reasonable targets, integrate their existing resources efficiently and adjust the scale of the enterprise reasonably, thereby improving operational efficiency.

4.2 Pharmaceutical Logistics Companies Are on the Rise, with Technological Advances a Key Factor in Their Advancement

From the Malmquist index, the operational efficiency of listed pharmaceutical logistics enterprises in China from 2011 to 2021 shows a fluctuating path of "high-low-high", and the trend of total factor productivity index and technological progress index is basically similar, technological innovation can not only improve the efficiency of enterprises and reduce costs, but also improve service quality, enhance market competitiveness and promote the transformation and upgrading of the industry [11]. It follows that the technological progress index is the main reason for maintaining positive total factor productivity growth, according to the concept of "intelligent logistics", China's pharmaceutical logistics enterprises should accelerate the transformation, improve the ability of technological innovation, strengthen the application of modern intelligent technology, so that the Internet and logistics entity network integration and innovation, the construction of information network system [12], big data platform for the construction and application of pharmaceutical logistics, especially cold chain logistics information management system is essential [13], only through the establishment of information-based regional logistics system can effectively improve the efficiency of the implementation of pharmaceutical logistics links at the same time as significant cost savings.

4.3 Medical Reform to Promote the Development of the Pharmaceutical Logistics Industry, Should Give Full Play to the Leading Enterprises to Drive the Role

The gradual implementation of government-led medical reform policies such as the "two-ticket system, tax reform and consistency evaluation" has provided a strong impetus to the leapfrogging development of the pharmaceutical logistics industry [13], the combined efficiency in this study has increased significantly after 2011 and 2017, suggesting that firms have responded to the policy, allowing economies of scale to continue to emerge. The Ministry of Commerce put forward relevant opinions and targets on promoting the high-quality development of the pharmaceutical distribution industry during the "14th Five-Year Plan" period, accelerating the development of modern pharmaceutical logistics, strengthening the application of intelligent and automated logistics technology and the upgrading of intelligent equipment [14], and requiring the competent departments of commerce around the world to focus on demonstration and leading, and encouraging through pilot demonstrations and other means backbone enterprises to take the lead first. Give full play to the leading enterprises' role in driving the industry, strengthen experience summing up and promotion, and form more policy toolkits that can be promoted and replicated.

References

1. Xiaolin, W., Xiaoyan, S., Huimin, C.: Demand forecasting for pharmaceutical logistics in Guangxi based on grey forecasting model. Chin. Market (5), 173–174, 192 (2022)
2. Qiutong, Y.: The current development of China's pharmaceutical logistics industry. Res. Prob. Countermeasures (6), 22–24 (2022)
3. Jiao, L., Yanhong, H.: Analysis of the development path of drug distribution enterprises in the context of medical reform. Price Monthly **457**(06), 91–94(2015)
4. Charnes, A., Cooper, W.W., Rhodes, E.: Measuring the efficiency of decision making units. Eur. J. Oper. Res. **2**(6), 429–444 (1978). https://doi.org/10.1016/0377-2217(78)90138-8
5. Yonghe, L., Wei, Z., Bingbing, C.: Coupling analysis of urban input-output and ecological environment in Yangtze river economic zone. J. Huazhong Normal Univ. (Nat. Sci. Edn.) **52**(4), 544–556 (2018)
6. Jin, H., Zhonghua, Y., Yueli, W.: A study on the measurement and improvement path of operational efficiency of listed logistics companies in China - an analysis based on fsQCA. Logistics Technol. (9) (2023)
7. Shu, F.: Research on the evaluation of profitability of listed pharmaceutical companies based on rank sum ratio. China Med. Manage. Sci. **11**(6), 18–23 (2021)
8. Weixia, H., Zhang, Y.: Evaluation of tourism efficiency and analysis of influencing factors in the middle and lower reaches of the Yellow River. Resour. Environ. Arid Reg. **36**(07), 187–193 (2022)
9. Na, L.: A review of pharmaceutical logistics research in China in the post-epidemic context. Logistics Technol. **41**(8), 10–13, 31 (2022)
10. Nannan, Z.: Transformation and development of pharmaceutical logistics in the context of the Internet. Logistics Eng. Manage. **43**(12), 18–19, 27 (2021)

11. Xinyuan, C., Jilin, L.: Construction of a new pharmaceutical logistics system based on big data and artificial intelligence. Logistics Technology **44**(3), 61–63, 81 (2021)
12. Zhang, Y.: Exploring the path of computerized management in hospitals. China Inform. World **357**(03), 84–85 (2023)
13. Cai, J.: Research on the operation of network information technology in Chinese medicine logistics centre. China Logistics Purchasing **623**(19), 86–87 (2021)
14. Lin, L.: Study on logistics integration strategy of pharmaceutical distribution group enterprises in the context of medical reform. China Logistics Purchasing **659**(22), 96–98 (2022)
15. Zhang, L.: Changes and trends in pharmaceutical logistics in the "14th Five-Year Plan" period. Logistics Technol. Appl. **26**(12), 60–61 (2021)

A Study on the Export Trend Forecast of Chinese Pharmaceutical Industry Based on GM (1,1) Model

Xin Liu[1], Fang Xia[1], Lingzhi Kong[2], and Shuo Zhang[1(✉)]

[1] School of Health Management, Changchun University of Chinese Medicine, Changchun 130117, China
zhangshuo@ccucm.edu.cn
[2] School of Pharmaceutical Information, Changchun University of Chinese Medicine, Changchun 130117, China

Abstract. This study aimed to forecast the exports of Chinese pharmaceutical industry by GM (1,1) gray forecasting model. The data of the study are taken from China's pharmaceutical industry exports from 2001 to 2020. The prediction results show that China's pharmaceutical industry exports are still on a steady upward trend and are expected to reach $28,330.5 million by 2027. The prediction accuracy of the model was tested to be superior, the fit between the model and the data was high, and the accuracy of the prediction results was credible, so the gray prediction model can be applied to the research field of pharmaceutical industry export prediction.

Keywords: GM (1,1) model · Trend forecast · pharmaceutical industry exports

1 Introduction

Gray system theory is to take "small sample" and "information-poor" uncertain system, in which some information is known and some is unknown, as the object of research, and extract valuable information through the generation and development of partially known information. By generating and exploiting the partially known information, we can extract valuable information to correctly grasp and describe the operation behavior and evolution law of the system [1]. Among them, the one-variable, first-order differential GM (1, 1) prediction model is an important model for gray system prediction and is more often applied to the case where the long-term trend of the initial data series varies monotonically [2]. The GM (1, 1) method in gray system theory has become a better method in export forecasting due to its simplicity and practicality.

The pharmaceutical industry is one of the industries in which China has achieved stable economic benefits and has made an important contribution to the high-quality development of the Chinese economy [3]. In 2022, the Ministry of Industry and Information Technology, the National Development and Reform Commission and nine other departments jointly issued the "14th Five-Year Plan" for the development of the pharmaceutical industry, emphasizing the comprehensive acceleration of the international

H. Jin et al. (Eds.): IAIC 2023, CCIS 2058, pp. 464–470, 2024.
https://doi.org/10.1007/978-981-97-1277-9_36

development of the pharmaceutical industry and the pharmaceutical export volume keeps growing. At present, China's pharmaceutical exports are still mainly chemical raw materials and Chinese herbal medicines, but in recent years, the export trade volume has not increased much, facing the problems of overcapacity, fierce export competition and low profitability, and the need to study the current situation of the pharmaceutical industry exports and trends [4, 5]. With the new crown epidemic under control, China's pharmaceutical industry exports are facing the opportunities of global economic recovery and digital trade as well as the challenges of changing global foreign trade policy environment and intensifying competition in the international market. Therefore, this study hopes to analyze the current situation and development trend of China's pharmaceutical industry export data in the past 20 years through the gray prediction GM (1,1) model. It provides a forward-looking reference for the development of China's pharmaceutical industry exports.

2 Data and Methods

2.1 Data Sources

The data related to pharmaceutical industry exports in this study are obtained from the China Statistical Yearbook of previous years and the China Pharmaceutical Information and Statistics Network. The model unit is selected as USD billion.

2.2 Research Methodology

2.2.1 Principle of GM (1,1) Model

The GM (1, 1) model is an analytical method based on time series, which transforms irregular raw data to create regular regression equations and apply them to predict the dynamic development of things. Establishing the gray prediction model of first-order variables includes four steps: gray series generation, exponential regularity test, model parameter estimation and model error test [6].

2.2.2 GM (1,1) Modeling Steps

Step 1: Initial data generation. The original data sequences that are equally distant and connected are selected and noted as x^0:

$$x^{(0)} = \left(x^{(0)}(1), x^{(0)}(2), \cdots, x^{(0)}(n) \right) \tag{1}$$

The new sequence is obtained by successive accumulation of the original series super and is denoted as $x^{(1)}$:

$$x^{(1)} = \left(x^{(1)}(1), x^{(1)}(2), \cdots, x^{(1)}(n) \right), x^{(1)}(k) = \sum_{i=1}^{k} x^{(o)}(i), (k = 1, 2, \cdots, n) \tag{2}$$

Step 2: The exponential law test is done on the series. To ensure the feasibility of the gray prediction model, the original data needs to be tested first. Calculation σ^0:

$$\sigma^{(0)} = \frac{x^{(0)}(k-1)}{x^{(0)}(k)} (k \geq 2) \tag{3}$$

If all results fall in the interval $(e^{-\frac{2}{n+1}}, e^{\frac{2}{n+1}})$ within the interval, then $x^{(0)}$ can be used as the modeling sequence.

Step 3: Establish the first-order differential equation GM (1,1) model. For $x^{(1)}$ Establish the model as

$$\frac{dx^{(1)}}{dt} + ax^{(1)} = b \tag{4}$$

where: a is the development coefficient ($a < 2$); b is the amount of gray effect. Denote the parameters as:

$$\hat{\alpha} = a, b^T = (B^T B)^{-1} B^T Y_n \tag{5}$$

Eq:

$$B = \begin{bmatrix} -\frac{x^{(1)}(1)+x^{(1)}(2)}{2} & 1 \\ -\frac{x^{(1)}(2)+x^{(1)}(3)}{2} & 1 \\ \cdots & \cdots \\ -\frac{x^{(1)}(n-1)+x^{(1)}(n)}{2} & 1 \end{bmatrix}, Y_n = \begin{bmatrix} x^{(o)}(2) \\ x^{(o)}(3) \\ \cdots \\ x^{(o)}(n) \end{bmatrix}$$

Least squares calculations are performed on the parameter columns to find a and b.

Step 4: Solve the time-response equation and prediction model. The time response function is given by:

$$\hat{x}^{(1)}(x+1) = \left(x^{(1)}(0) - \frac{b}{a}\right)e^{-at} + \frac{b}{a} \tag{6}$$

The time response equation of the GM (1,1) model is:

$$\hat{x}^{(1)}(k+1) = \left(x^{(1)}(0) - \frac{b}{a}\right)e^{-ak} + \frac{b}{a}, k = 0, 1, 2, \cdots, n \tag{7}$$

where: $x^{(1)}(0) = x^{(0)}(1)$, the prediction model is derived as follows:

$$\hat{x}^{(1)}(k+1) = \left(x^{(0)}(1) - \frac{b}{a}\right)e^{-ak} + \frac{b}{a}, k = 0, 1, 2, \cdots, n \tag{8}$$

Cumulative reduction of predicted values for the prediction model:

$$\hat{x}^{(0)}(k+1) = \hat{x}^{(1)}(k+1) - \hat{x}^{(1)}(k), k = 0, 1, 2, \cdots, n \tag{9}$$

2.2.3 GM (1,1) Model Residual Test

Calculate the residual series of the initial sequence $x^{(0)}(k)$ and the predicted sequence $\hat{x}^{(0)}(k)$:

$$\Delta^{(0)}(k) = x^{(0)}(k) - \hat{x}^{(0)}(k), \Delta^{(0)} = \left(\Delta^{(0)}(k), k = 1, 2, \cdots, n\right) \qquad (10)$$

Relative residual series:

$$\varepsilon(k) = \frac{\Delta^{(0)}(k)}{x^{(0)}(k)} \times 100\% \qquad (11)$$

The average relative residual equation is calculated as:

$$\varepsilon_{avg} = \frac{1}{n} \sum_{k=1}^{n} \backslash \varepsilon(k) \backslash \qquad (12)$$

3 China's Pharmaceutical Industry Export Trends Forecast

3.1 Model Fitting Prediction

In this study, a gray GM (1, 1) forecasting model is established to calculate the forecast data of China's total energy consumption from 2001 to 2020 using the export value of China's pharmaceutical industry from 2001 to 2020 as the original series, and the error test is conducted between the original data and the forecast data, and if the accuracy of the forecasting model meets the standard, then the model can be used to forecast the export value of the pharmaceutical industry. Bring in k = 0, 1, 2...19 The forecast value of China's pharmaceutical industry exports from 2001 to 2020 is obtained by cumulative subtraction $x^{(0)}$. The simulated actual and fitted values are calculated as shown in Table 1. Using the original series $x^{(0)}$ with the predicted series $\hat{x}^{(0)}$ and calculate the residual series $\Delta^{(0)}$ The calculated results are shown in Table 1.

The development coefficient and the gray action quantity can construct a gray prediction model [7, 8]. The development coefficient indicates the development pattern and trend of the series, and the gray action quantity reflects the change relationship of the series. The smaller the posterior difference ratio is, the higher the accuracy of gray prediction is. Generally, the posterior difference ratio C value is less than 0.35 then the model accuracy is high, C value is less than 0.5 means the model accuracy is qualified, C value is less than 0.65 means the model accuracy is basically qualified, if C value is more than 0.65, the model accuracy is not qualified. The value of development coefficient a was calculated to be -0.018, the gray effect degree b was 450.082, the posterior residual ratio c was 0.032, and the average relative error of the fit was 8.806%, and the model fit was ideal. A 7-step prediction was performed on the training sample, and the fitted prediction effect was obtained as shown in Fig. 1.

Table 1. Simulated actual value fitting calculation results.

Year	$x^{(0)}$	$\hat{x}^{(0)}$	$\Delta^{(0)}$
2001	19.79	19.79	0
2002	23.24	20.541	2.699
2003	28.61	28.939	−0.329
2004	32.35	37.49	−5.14
2005	37.78	46.196	−8.416
2006	44.88	55.06	−10.18
2007	60.06	64.086	−4.026
2008	81.04	73.275	7.765
2009	86.35	82.631	3.719
2010	107.05	92.157	14.893
2011	118.33	101.855	16.475
2012	119.35	111.73	7.62
2013	123.2	121.784	1.416
2014	133.78	132.021	1.759
2015	135.06	142.444	−7.384
2016	136.05	153.056	−17.006
2017	150.77	163.86	−13.09
2018	172.67	174.861	−2.191
2019	174.3	186.061	−11.761
2020	220.78	197.465	23.315

3.2 Prediction Results

Substituting the data of China's pharmaceutical industry exports in previous years into the model, the original values fit well with the predicted values, so the GM (1,1) model can be used for prediction. Let K = 10, 11…,16 be brought into the forecasting model to find the forecast value of China's pharmaceutical industry exports by cumulative reduction. As shown in Table 2, China's pharmaceutical industry exports still show an upward trend and are expected to reach 28,330.5 million USD in 2027.

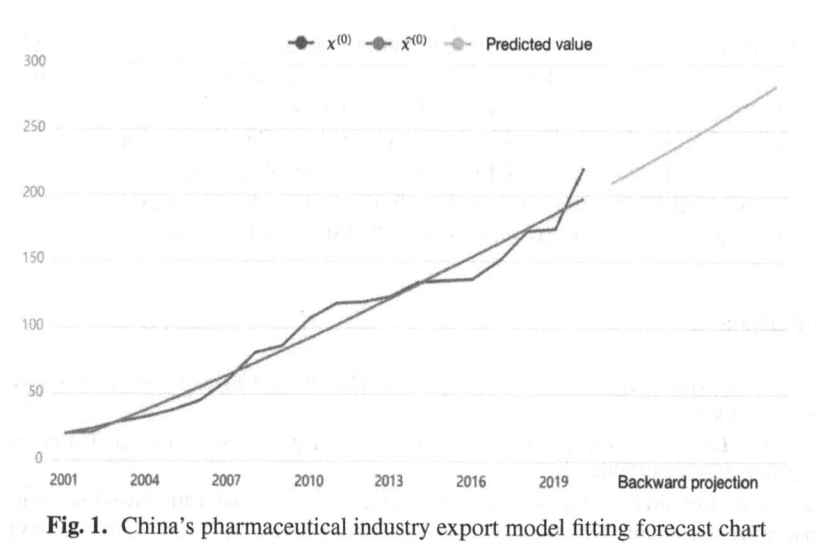

Fig. 1. China's pharmaceutical industry export model fitting forecast chart

Table 2. Forecast value of China's pharmaceutical industry exports / $ billion

Predicted order	Predicted value
2021	209.076
2022	220.898
2023	232.934
2024	245.189
2025	257.666
2026	270.370
2027	283.305

4 Conclusion

With the widespread application of gray forecasting theory, gray forecasting models have been used one after another to forecast the export of pharmaceutical industry. Some studies have concluded that the relative error of GM (1, 1) forecasting model for China's high-tech industry export trade is less than 0.2, and the grade deviation is less than 0.3, which is suitable for medium and long-term economic forecasting with high accuracy [9]. In addition, the model's prediction accuracy is one level, and it can predict the time series trend well, and the prediction accuracy of the model is better [10]. Combining the results of most studies applied to pharmaceutical industry export forecasting, it was found that the prediction results of this study by GM (1, 1) to predict the development of Chinese pharmaceutical industry were credible.

Based on the data of China's pharmaceutical industry exports from 2001 to 2020, a GM (1, 1) gray prediction model was used to forecast the change trend of China's pharmaceutical industry exports. The following conclusions are drawn. China's pharmaceutical industry exports the forecast results fit the actual values to a high degree, indicating that it is feasible to apply it to the forecast of China's pharmaceutical industry exports and make forward-looking data suggestions for the development of China's pharmaceutical industry exports. China's pharmaceutical industry exports still show an upward trend, and if the current export policies and strategies are continued, China's pharmaceutical industry exports will exceed 28,330.5 million USD in 2027.

References

1. Julong, D.: Basic Methods of Gray Systems. Huazhong University of Technology Press, Wuhan (2005)
2. Yue, W.: Discussion on the scope of application of gray system model. J. Beijing Univ. Technol. **1**, 72–75 (1999)
3. Economic Research: China's pharmaceutical economy in the 14th Five-Year Plan. Pharmaceutical Economic News, 2022–03–24 (007). https://doi.org/10.38275/n.cnki.nyyjj.2022. 000323
4. Guangping, W., Shengyi, S.: Research on the optimization path of China's pharmaceutical industry structure adjustment based on export structure. China Pharm. Aff. **32**(05), 575–584 (2018). https://doi.org/10.16153/j.1002-7777.2018.05.001
5. Xiaofang, M., Xiang, Q., Fengxiang, Z.: A comparative study of domestic and foreign pharmaceutical export sales certificate issuance system. China Pharm. Aff. **32**(11), 1549–1557 (2018). https://doi.org/10.16153/j.1002-7777.2018.11.015
6. Zou, F., Wang, L.: Energy consumption forecast for Sichuan province in the 14th five-year plan based on GM (1,1) model. Energy Conserv. **40**(08), 62–64 (2021)
7. Scientific Platform Serving for Statistics Professional 2021. SPSSPRO. (Version 1.0.11) [Online Application Software] (2021). https://www.spsspro.com
8. Julong, D.: Gray Forecasting and Gray Decision Making. Huazhong University of Science and Technology Press, Wuhan (2002)
9. Wei, C., Sumei, W., Jun, Z.: Review of the current situation of export trade of high-tech industries in China–analysis based on gray system model GM (1,1). Sci. Technol. Outlook **26**(34), 236–237 (2016)
10. Huang, Z.E., Chen, S.T., Liu, Y.T., et al.: Research on China's medical device export prospect in the context of the new crown epidemic–based on the gray system GM (1,1) model. Commer. Exhib. Econ. **13**, 153–156 (2022). https://doi.org/10.19995/j.cnki.CN10-1617/F7.2022. 13.153

YOLO-TUF: An Improved YOLOv5 Model for Small Object Detection

Hua Chen, Wenqian Yang, Wei Wang, and Zhicai Liu[✉]

School of Computer and Software Engineering, Xihua university, Chengdu 610039,
People's Republic of China
weiwang@stu.xhu.edu.cn, liuzc.xhu@foxmail.com

Abstract. Traditional target detection algorithms frequently encounter challenges in accurately detecting objects within complex, cluttered environments. This paper presents an optimized YOLOv5-based model to mitigate such limitations. Our contributions are threefold: Firstly, we enhance the upsampling procedure by amalgamating transposed convolution with the CBAM attention mechanism, fortifying the network's fine-grained feature extraction capabilities. Secondly, we introduce an optimized feature-processing module, which enhances feature utilization while maintaining a lightweight architecture. Lastly, we integrate EfficientNet into the backbone architecture to amplify feature extraction performance. We validate our approach using the PASCAL VOC dataset, achieving an mAP0.5 of 84.00% and an mAP0.5:0.95 of 62.10%, while maintaining a modest parameter size of 13.22MB. These results mark an improvement of 4.50% ± 0.12% and 8.20% ± 0.09% over the benchmark, demonstrating an efficient trade-off between computational efficiency and detection accuracy. The proposed model outperforms conventional YOLOv5 algorithms and remains competitive with contemporary state-of-the-art object detection techniques. Code is available at https://github.com/chenxz0906chenxz/YOLO-TUF/.

Keywords: Transposed Convolution · CBAM Attention Mechanism · EfficientNet · YOLOv5

1 Introduction

Small object detection is a hot topic in the field of deep learning, aiming to locate and classify one or multiple targets within images. The current popular object detection algorithms are based on convolutional neural networks [5,11,16,19], which have achieved significant improvements in inference time and accuracy compared to traditional algorithms. In theoretical research, small object detection [10,15,23] has garnered immense interest from researchers due to its unique characteristics of having fewer target pixels and less feature information. In practical applications, there are many scenarios that require the detection of small objects, such as defect detection in industrial products [1], target detection in medical images [12,24], and underwater organism detection [7], to name a few. Small object detection has become the focal point of both theoretical research

© The Author(s), under exclusive license to Springer Nature Singapore Pte Ltd. 2024
H. Jin et al. (Eds.): IAIC 2023, CCIS 2058, pp. 471–484, 2024.
https://doi.org/10.1007/978-981-97-1277-9_37

and practical applications, holding vast potential and a promising future. To enhance the neural network's ability to detect small objects, Lin et al. [13] employed the FPN strategy, merging features from different levels. Hu J et al. [8] introduced the Squeeze and Excite (SE) block, emphasizing relevant features. Zhang Yin et al. [27] proposed a feature enhancement module (FEM), combining features from lower-level maps.

Algorithms for detecting small objects are divided into single-stage and two-stage methods. Classic two-stage algorithms mainly include R2CNN [17], Faster R-CNN [20], and so on. The latest two-stage algorithm, Ganster R-CNN [21], achieved an mAP_0.5 of 80.71% on the PASCAL VOC dataset. These detection algorithms have high accuracy but slower inference speeds. Classic single-stage algorithms mainly include, YOLO9000 [18], YOLOv3 [6], RetinaNet [14] and YOLOv5 [9] among others. YOLOv5 [25] achieved an mAP_0.5 of 78.9% on the PASCAL VOC dataset. These detection algorithms have faster inference speeds but lower accuracy.

Current detection algorithms, while effective, face challenges in enhancing the backbone network's feature extraction and capturing in-depth target region details. Common interpolation methods used for upsampling [2,3,26] can cause distortions, especially during significant resizing, blurring edges and losing details. Specifically, bilinear interpolation, relying on nearby pixels, often fails to restore high-frequency image details. The feature processing network module often needs more layers to obtain richer gradient flow information, resulting in too large parameters. To overcome the limitations of current methods, this article proposes the YOLO-TUF algorithm based on YOLOv5. To our understanding, we are the first to apply standard convolutions, Mobile Inverted Bottleneck convolutions, and transposed convolutions together for object detection. We assessed the performance of all models using the PASCAL VOC datasets.

The main contributions of this paper can be summarized as:

1. A dual-backbone architecture is introduced, integrating EfficientNet B0 with the YOLOv5 backbone. This innovative combination significantly enhances the network's feature extraction capabilities.
2. Transposed Convolution Upsampling (TCU) is introduced as a method to improve upsampled layers. Experimental results confirm that integrating transposed convolution into the upsampling workflow restores the details in low-resolution layers while enriching underlying features. This significantly improves the detection performance for small targets.
3. The introduction of a feature processing module termed CSP Bottleneck with 2 convolutions (C2f) enriches the gradient flow information within the neck network. This ensures that the model remains lightweight while optimizing its feature processing efficiency.

2   Related Work

2.1   YOLOv5

Ultralytics released YOLOv5 [9] in 2020, building upon the strengths of its predecessor, YOLOv4 [4], in both faster processing speed and smaller model size. The YOLOv5 architecture is composed of three primary components: the backbone network, neck, and output. The backbone network extracts features from input images, and the neck fuses these features to enhance overall network performance. The output component outputs the category and location information of the target predicted by the network. The structure of YOLOv5 is shown in Fig. 1. The loss function for the YOLOv5 is formulated with three components: classification loss, bounding box regression loss, and objectness loss. The expression for YOLOv5's loss function is:

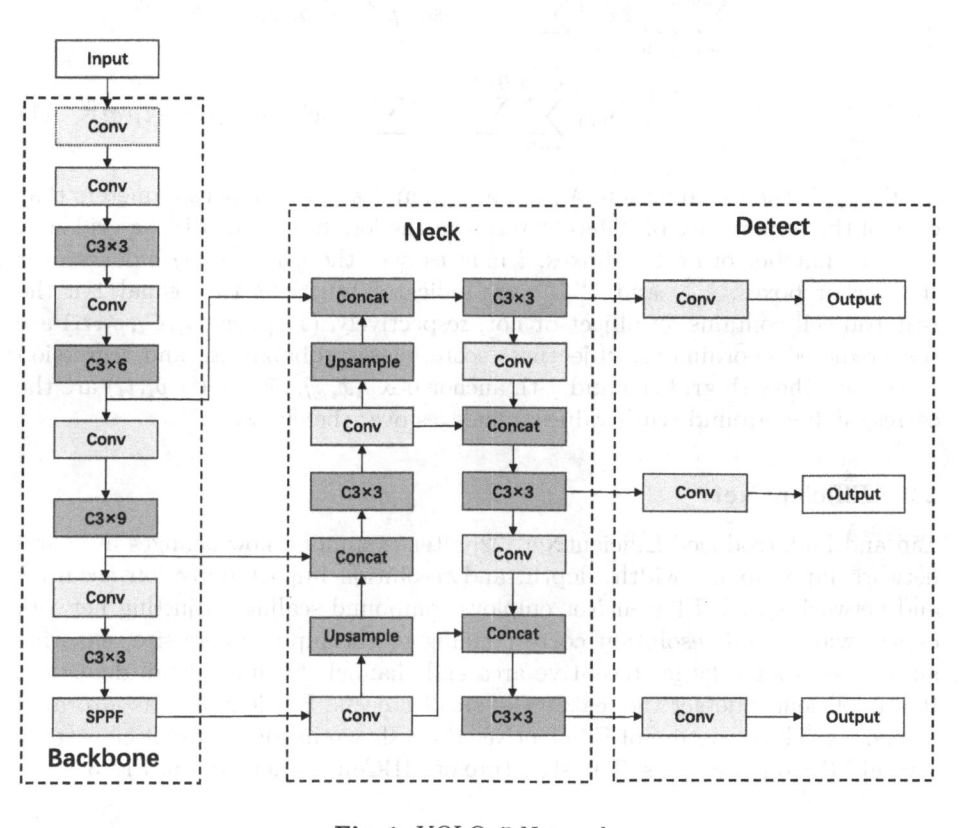

Fig. 1. YOLOv5 Network

$$\mathcal{L}_{\text{object}} = \lambda\text{coord} \sum_{i=0}^{S^2-1} \sum_{j=0}^{B-1} \mathbb{1}_i^{\text{obj}} \left[(x_i - \hat{x}_i)^2 + (y_i - \hat{y}_i)^2 \right] +$$

$$\lambda\text{coord} \sum_{i=0}^{S^2-1} \sum_{j=0}^{B-1} \mathbb{1}_i^{\text{obj}} \left[(\sqrt{w_i} - \sqrt{\hat{w}_i})^2 + (\sqrt{h_i} - \sqrt{\hat{h}_i})^2 \right] +$$

$$\sum_{i=0}^{S^2-1} \sum_{j=0}^{B-1} \mathbb{1}_i^{\text{obj}} \left[C_i - \hat{C}_i \right]^2 +$$

$$\lambda_{\text{noobj}} \sum_{i=0}^{S^2-1} \sum_{j=0}^{B-1} \mathbb{1}_i^{\text{noobj}} \left[C_i - \hat{C}_i \right]^2 +$$

$$\sum_{i=0}^{S^2-1} \sum_{j=0}^{B-1} \mathbb{1}i^{\text{obj}} \sum c \in \text{classes}(p_i(c) - \hat{p}_i(c))^2 +$$

$$\lambda_{\text{obj}} \sum_{i=0}^{S^2-1} \sum_{j=0}^{B-1} \mathbb{1}i^{\text{obj}} \sum c \in \text{classes}(t_i(c) - \hat{t}_i(c))^2 \quad (1)$$

$\mathcal{L}_{\text{object}}$ is the loss function, λ_{coord}, λ_{noobj}, and λ_{obj} are hyperparameters that control the importance of different terms in the loss function, S is the grid size, B is the number of anchor boxes, i indexes over the grid cells, j indexes over the anchor boxes, $\mathbb{1}_i^{\text{obj}}$ and $\mathbb{1}_i^{\text{noobj}}$ are indicator functions that equal 1 if the i-th grid cell contains an object or not, respectively, $(x_i, y_i, w_i, h_i, C_i, p_i, t_i)$ are the predicted coordinates, objectness score, class probabilities, and regression targets for the i-th grid cell and j-th anchor box, $(\hat{x}_i, \hat{y}_i, \hat{w}_i, \hat{h}_i, \hat{C}_i, \hat{p}_i, \hat{t}_i)$ are the corresponding ground truth values, c indexes over the classes.

2.2 EfficientNet

Tan and Le introduced EfficientNet [22] after examining how changes in neural network input image width, depth, and resolution impact detection accuracy and network speed. EfficientNet employs compound scaling, adjusting network depth, width, and resolution corresponding to the input image size, ensuring ample layers for a larger receptive area and channels for finer detail detection. It can efficiently detect the features of small objects. EfficientNet has a total of 8 versions. The structure of EfficientNet B0 is shown in Fig. 2, and it consists of several MBConv modules. The structure of MBConv is depicted in Fig. 3.

3 Model Construction

The YOLO-TUF model proposed in this paper is based on the YOLOv5s algorithm but makes significant improvements from the following three aspects: we first add EfficientNet B0 to the backbone of YOLOv5s to make it a dual-backbone feature extraction network; then use C2f as the feature processing

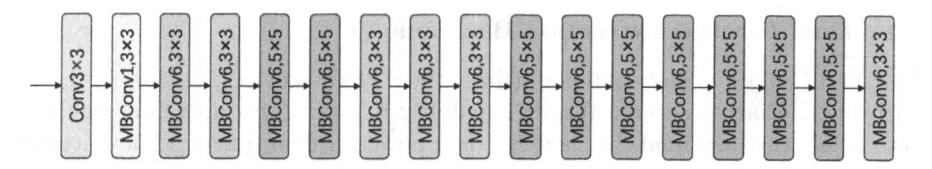

Fig. 2. EfficientNetB0, 3×3 represents the convolution kernel size is 3×3, 5×5 represents the convolution kernel size is 5×5

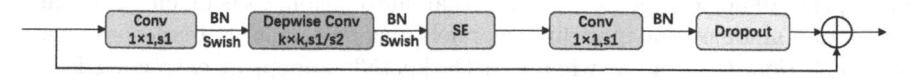

Fig. 3. MBConv is the basic module that makes up EfficientNet B0

module of the neck network to enhance the feature processing capability of the original C3 module; Finally, the original nearest neighbor interpolation upsampling of YOLOv5s is replaced with TCU. TCU integrates the nearest neighbor interpolation method and the transposition mechanism, while also adopting feature fusion concepts and the attention mechanism to enhance the upsampling performance throughout the network. The final architecture is shown in Fig. 4.

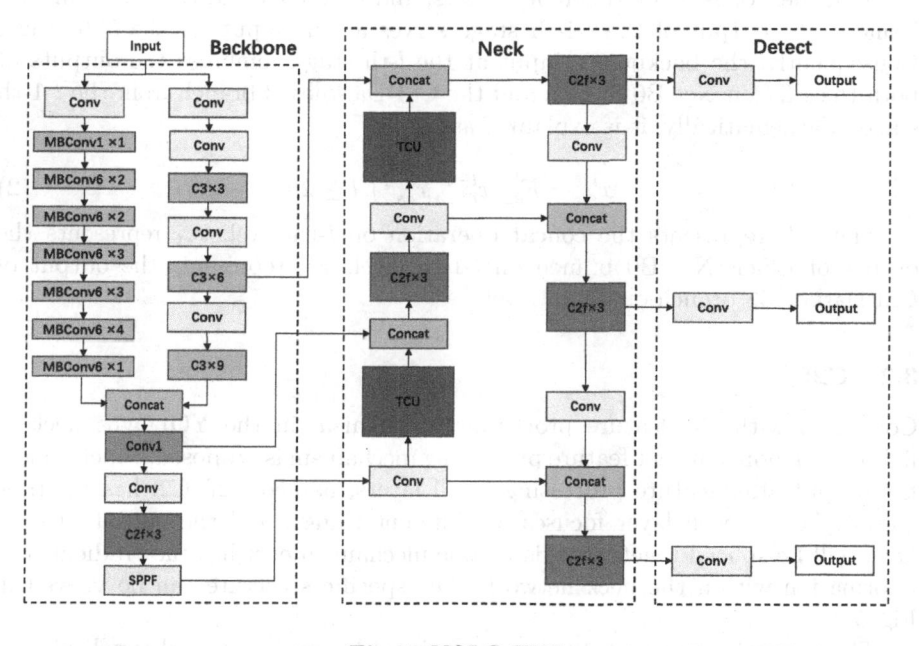

Fig. 4. YOLO-TUF

3.1   Dual Feature Extraction Backbone

The YOLOv5 model's backbone utilizes the CSPDarknet53 network integrated with the CSP architecture. Yet, its capability to extract features from small targets is less than optimal. EfficientNet is a powerful neural network architecture for image classification.

In the field of object detection, the inherent design characteristics of EfficientNet may lead to a tendency to prioritize learning local features, often at the expense of neglecting global contextual information. Concurrently, the lack of correlation amongst the modeling channels may culminate in inadequate feature representation. Contrarily, while CSPDarknet53 is adept at extracting features pertaining to global information and internal channels, it exhibits limitations in its capacity to discern local features from smaller target entities. So, We propose a brand-new backbone network structure, as delineated by the dashed box in Fig. 4. The dual-backbone feature extraction network, which uses two distinct structures (hence 'non-twin'), can capture richer and more comprehensive feature details. It solves the problem of insufficient feature extraction when a single backbone network extracts features of small targets. Simultaneously, it greatly enriches the network's feature information extraction for small target objects. Input a $3 \times 640 \times 640$ picture into the network; after extracting features through EfficientNet B0 and CSPDarknet53 network, get two $256 \times 40 \times 40$ output branches focusing on different aspects, and then concat these two branches. Usually, the output of the l-1-th stage serves as the input for the l-th stage. Consequently, the backbone's input at the l-th stage combines the outputs of both the EfficientNet B0 branch and the CSPDarknet53 branch from the l-1-th stage. Mathematically, it is explained as:

$$x_{ec}^l = F_{ec}^l(x_{et}^{l-1}, x_{cs}^{l-1}), l \geq 2 \tag{2}$$

where F represents the concat operation on l-th level. x_{et} represents the outpot of EfficietNet B0 branch on l-1-th level, x_{cs} represents the output of CSPDarknet53 branch on l-1-th level.

3.2   C2f

Compared with the feature processing mechanism in the YOLOv5s neck, a lighter and more efficient feature processing mechanism is proposed, which splices the output after feature processing of all layers, named C2f. C2f has multiple module layers, each layer focuses on different things, and the output of each layer will be spliced together. this unique mechanism enriches the gradient flow information within the neck network. The specific structure can be viewed in Fig. 5.

The network employs a 1×1 convolutional kernel for inter-channel interactions. The Split layer divides the input into two equal sub-vectors. One of these sub-vectors is sequentially passed through three Bottleneck_a modules, with the output from each Bottleneck_a being retained. These outputs are concatenated and then processed using a 1×1 convolutional kernel to further facilitate

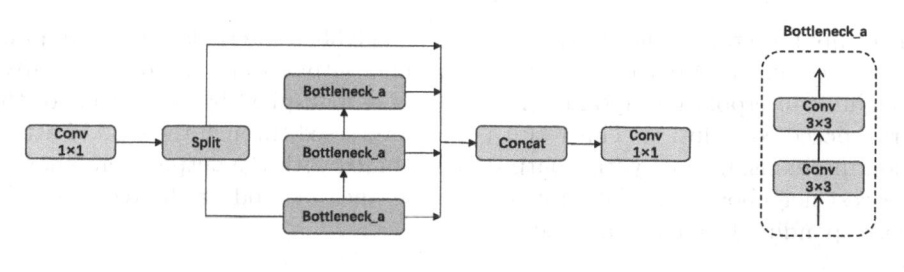

Fig. 5. C2f

interactions among the various channels. This ensures that the model remains lightweight while optimizing its feature processing efficiency.

3.3   TCU Model

Transpose Convolution. Transposed convolution is a learnable upsampling method, where the network automatically learns parameters to achieve the best upsampling information. The formula for transpose convolution (considering single-channel input and single-channel output) is as follows:

This layer receives an image y_i as its input, which has K_0 color channels denoted as $y_1^i, ..., y_{K_0}^i$. Each channel c in the image is represented by a linear combination of K_1 latent feature maps z_k^i when convolved with the filters $f_{k,c}$:

$$y_c^i = \sum_{k=1}^{K_1} z_k^i * f_{k,c} \tag{3}$$

CBAM Attention. CBAM is an attention mechanism module that achieves good results on convolutional neural networks. Its unique Channel Attention and Spatial Attention help the network focus on more important features. It can be added anywhere in the network without introducing excessive overhead. We process the product output of the nearest-neighbor upsampling and transposed convolution upsampling through the CBAM module. This allows the network to efficiently extract spatial features and fine-grained information from the image, enhancing the feature extraction capability for smaller targets.

TCU Model Network Structure. Although the nearest neighbor interpolation is straightforward and preserves essential information, it might lose connections between pixels when enlarged. This can potentially lead to blocky structures and the loss of spatial information for small targets. Transposed convolution autonomously learns upsampling parameters, allowing for a substantial recovery of the image's spatial relationships and fine-grained details. However, transposed convolution might lose some of the salient essential information in the feature map. In light of these issues, this paper introduces the TCU model. The network architecture of the proposed TCU model is illustrated in Fig. 7.

The input vector x undergoes two nearest neighbor interpolation upsampling processes and a transposed convolution. The output from one of the nearest neighbor interpolation upsampling branches is multiplied by the output of the transposed convolution. This result is then processed through the CBAM attention mechanism. Finally, this output is combined with the output from another nearest neighbor interpolation upsampling branch to produce the vector y. The corresponding formula is as follows:

$$y = f_{CBAM}(\sigma(F_{up}(x) * G_{tc}(x))) + F_{up}(x) \tag{4}$$

The $F_{up}(x)$ represents the result of up-sampling the input data x, G_{tc} represents the inverse convolution, σ is the *softmax* function, and the final result y is the output of our proposed up-sampling method. The upsampling method we proposed fully retains the outstanding important information. It effectively restores detailed information and restores the connection between pixels. This approach adeptly addresses the issues previously highlighted.

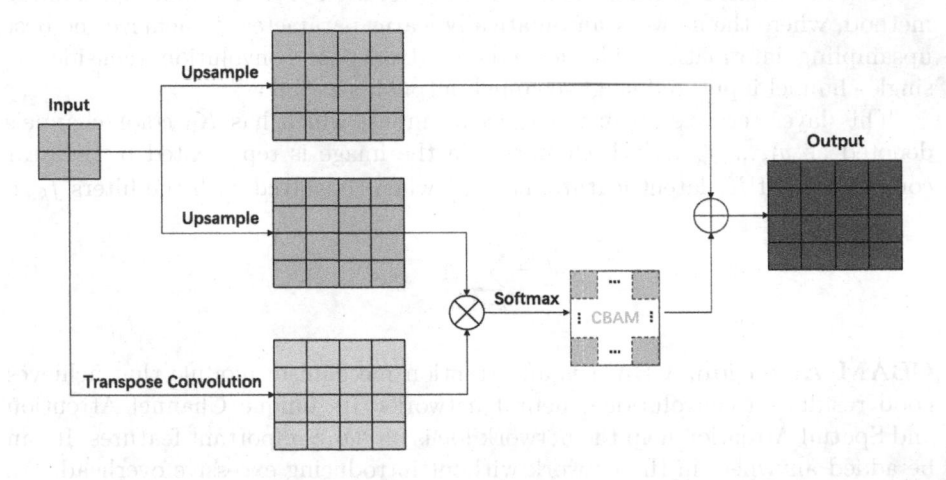

Fig. 6. TCU Model Network Structure

4   Experimental Results and Analysis

We evaluated the YOLO-TUF using the widely recognized PASCAL VOC dataset. Comparative tests with other leading target detection techniques and ablation studies for various components of our approach were conducted on this dataset.

4.1   Network Evaluation

In our research, we assess the effectiveness of the suggested model using metrics such as recall, precision, Average Precision (AP), and Mean Average Precision (mAP). AP indicates the model's average accuracy, determined by the area beneath the Precision-Recall curve. Besides, mAP computes the average of AP scores across all categories, serving to evaluate the model's overall accuracy and robustness. The respective calculations are detailed below:

$$Precision = \frac{TP}{TP + FP} \tag{5}$$

$$Recall = \frac{TP}{TP + FN} \tag{6}$$

wherein TP (True Positives) counts the instances that were correctly identified as positive, FN (False Negatives) captures the positive instances wrongly flagged as negative, and FP (False Positives) tallies the negative instances misclassified as positive. Subsequently, the Average Precision (AP) is derived from the area beneath the Precision-Recall curve, expressed by the integral of Precision (p) over the range of Recall (r) within the interval [0, 1].

$$AP = \int_0^1 p(r)\, dr \tag{7}$$

$$mAP = \frac{\sum AP}{Numclass} \tag{8}$$

The performance of the network is primarily characterized by the aforementioned four evaluation metrics and the number of model parameters.

4.2 Datasets

We train and assess our network using common object detection datasets. The PASCAL VOC dataset offers a benchmark dataset comprising 20 object classes tailored for detection tasks. We employ the VOC2007 and VOC2012 datasets for training, utilizing both their train and val sets (a total of 16,551 images), and test our model on the VOC2007 test set, which contains 4,952 images.

4.3 Training Results and Analysis

To validate the superiority of the improved strategy, we trained both YOLOv5s and YOLO-TUF using the same dataset. All environmental and hyperparameter settings were kept consistent, and neither model utilized pretrained weights or any other conditions that might affect fairness. As depicted in Fig. 7. During the initial 50 epochs, the mAP_0.5 and mAP_0.5:0.95 scores for YOLO-TUF consistently surpass those of YOLOv5s, with the gap progressively widening. Between the 50th and 350th epochs, the growth trajectory of YOLO-TUF's mAP_0.5 decelerated compared to the initial 50 epochs, yet consistently exhibited steady progression. By the culmination of training, YOLO-TUF's mAP_0.5 exceeded that of YOLOv5s by a margin of 4.5%. From the 50th epoch to the 350th epoch, the ascent of YOLO-TUF's mAP_0.5:0.95 consistently outstripped YOLOv5s, maintaining an advantage of over 7.0%. By the end of the training phase, YOLO-TUF achieved an mAP_0.5 that was 8.2% superior to that of YOLOv5s. The experimental results affirm the efficacy of the modifications.

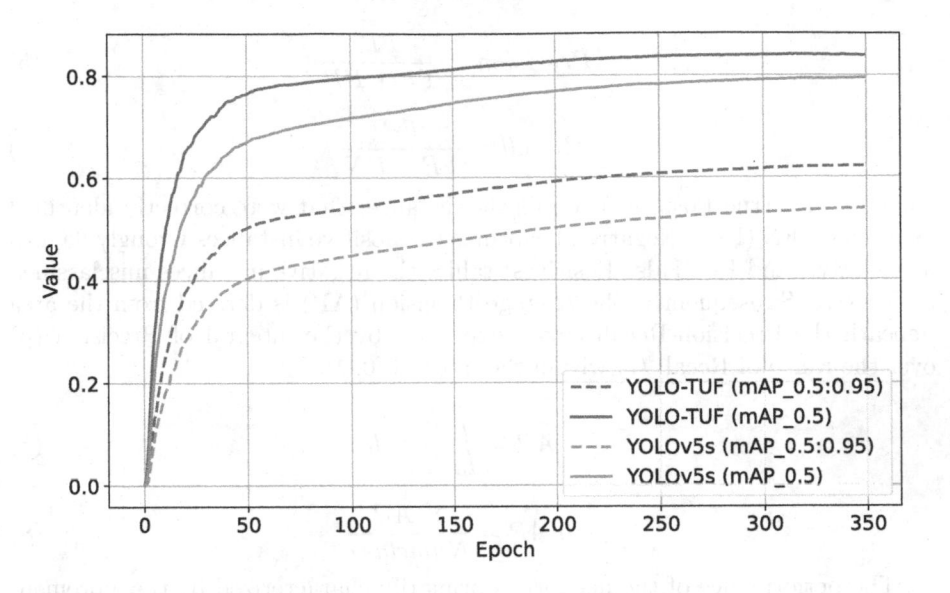

Fig. 7. Comparison of mAP results of YOLO-TUF and YOLOv5s

4.4 Comparison with the State-of-the-Art Modeling

To underscore the merit and superior performance of our proposed method, we conducted experimental comparisons with several state-of-the-art models, namely SSD, SSD300, SSD500, Faster R-CNN, YOLO9000, Improved YOLOv3-Net, EfficientDet-D0, EfficientDet-D1, Faster R-CNN ResNet, Faster R-CNN VGG-16, Improved YOLOv2, YOLOv4-Tiny, YOLOv5s, YOLOv5m, YOLOv8n, YOLOv8s, GL-YOLO, and YOLOD, using the PASCAL VOC dataset.

Compared with the original model, the average accuracy of our YOLO-TUF is improved by 4.4%, which is much higher than the original model. Compared with YOLOv5m, our YOLO-TUF has only half the number of parameters of YOLOv5m, but its average accuracy has increased by 0.9%, which is higher than YOLOv5m. Compared with the latest YOLO algorithm, we have achieved an average detection accuracy increase of 0.8% with a similar number of parameters as YOLOv8s.Table 1 present the comparative results, where YOLO-TUF outperforms the other prominent algorithmic models in detection accuracy. This underscores the robust performance of our refined model, highlighting its advantage over other comparable algorithms.

Table 1. Comparison of Different Object Detection Models

Model Names	VOC Dataset	mAP/%	Parameters/M
SSD	2007+2012	72.3	100.2
SSD300	2007+2012	74.3	–
SSD500	2007+2012	76.8	–
Faster R-CNN	2007+2012	70.5	136.7
Faster R-CNN ResNet	2007+2012	76.4	–
EfficientDet-D0	2007+2012	71.7	3.9
EfficientDet-D1	2007+2012	76.3	6.6
Faster R-CNN VGG-16	2007+2012	73.2	–
YOLO9000	2007+2012	76.8	–
Improved YOLOv2	2007+2012	81.0	–
Improved YOLOv3-Net	2007+2012	77.4	–
YOLOv4-Tiny	2007+2012	70.0	5.9
YOLOv5m	2007+2012	83.0	21.2
YOLOv8n	2007+2012	80.3	3.2
YOLOv8s	2007+2012	83.1	11.12
GL-YOLO	2007+2012	82.5	7.1
YOLOv5s	2007+2012	79.5	7.2
YOLO-TUF	2007+2012	**84.1**	**13.2**

4.5 Ablation Experiment

To ascertain the contribution of each enhancement, we incrementally incorporated the improved modules into the baseline YOLOv5s network, based on our enhancement strategy. Ablation studies were performed on the PASCAL VOC datasets, all hyperparameters are set identically. Post-training, we evaluated the

Table 2. Comparison of Different Object Detection Models

Experiment	DB	C2f	TU	mAP_0.5/%	mAP_0.5:0.95/%	Params/M
1				79.5	53.9	7.24
2	✓			83.1	59.4	11.24
3		✓		81.0	57.0	8.91
4			✓	80.2	55.2	7.57
5	✓	✓		83.9	61.4	12.92
6	✓		✓	83.6	60.8	11.58
7		✓	✓	81.2	58.0	9.25
8	✓	✓	✓	**84.0**	**62.1**	**13.22**

outcomes to confirm the significance of the said improvements. The experimental findings are detailed in Table 2

Utilizing the YOLOv5s as the benchmark model for ablation studies, the checked tick signifies the inclusion of the respective improvement. Experiments on the PASCAL VOC dataset reveal that when combining EfficientNet B0 with CSPDarknet53 features to create a new feature extraction network, mAP_0.5% improved by 3.6%, and mAP_0.5:0.95 increased by 5.5%. Despite a significant increase in the number of parameters compared to YOLOv5s, it achieved higher accuracy compared to YOLOv5m while having only half the parameters of YOLOv5m. This demonstrates that our proposed network, which uses feature fusion, has stronger feature extraction capabilities for images and achieves a maximum balance between parameter count and feature extraction capability, making it more practical. In the NECK section of YOLO V5s, replacing the original C3 module with the C2f module resulted in a 1.5% improvement in mAP_0.5 and a 3.1% improvement in mAP_0.5:0.95.

This shows that the C2f module has better processing capabilities for input features and can more effectively preserve global feature information and prominent detail feature information. When replacing the original upsampling method with the TCU module, mAP_0.5 improved by 0.7%, and mAP_0.5:0.95 improved by 1.3%, demonstrating that the TCU module not only preserves important information but also restores fine details and pixel relationships, enhancing the overall upsampling effect. When we combine these three proposed methods in pairs, they all outperform the accuracy of their respective baseline networks, confirming the generality of our approach. Applying these three improvements to YOLOv5s resulted in a 4.6% improvement in mAP_0.5% and an 8.2% improvement in mAP_0.5:0.95%. Our YOLO-TUF achieves the maximum balance between parameter count and accuracy and is highly competitive compared to all other object detection algorithms.

5 Conclusion

This study tackles pressing challenges in contemporary object detection algorithms, specifically in scenarios characterized by complex environments, dense target distribution, and varied object sizes. Utilizing a refined YOLOv5s framework as the foundation, several key innovations were introduced to overcome these challenges. Firstly, we implemented a dual-backbone feature extraction network that leverages feature fusion to augment the YOLOv5s algorithm's ability to capture salient features from complex scenes. This innovation mitigates the issues stemming from inadequate feature extraction capabilities of the original architecture. Secondly, we developed a learning-based upsampling technique to replace traditional nearest-neighbor upsampling. By combining nearest-neighbor interpolation with transposed convolutions and incorporating the CBAM attention mechanism, this technique not only preserves essential information but also enriches fine-grained details and inter-pixel relationships, thereby elevating the quality of upsampling. In the NECK section of YOLOv5s, we addressed a key inefficiency related to feature processing by incorporating the C2f module. This addition enables more efficient handling and improved utilization of globally available information. While our enhanced algorithm demonstrates substantial improvements in the accurate detection of small objects, there remains a need for optimization in parameter efficiency and inference speed.

References

1. Adibhatla, V.A., Chih, H.C., Hsu, C.C., Cheng, J., Abbod, M.F., Shieh, J.S.: Defect detection in printed circuit boards using you-only-look-once convolutional neural networks. Electronics **9**(9), 1547 (2020)
2. Amanatiadis, A., Andreadis, I.: A survey on evaluation methods for image interpolation. Meas. Sci. Technol. **20**(10), 104015 (2009)
3. Arun, P.V.: A comparative analysis of different dem interpolation methods. Egypt. J. Remote Sens. Space Sci. **16**(2), 133–139 (2013)
4. Bochkovskiy, A., Wang, C.Y., Liao, H.Y.M.: YOLOv4: optimal speed and accuracy of object detection. arXiv preprint arXiv:2004.10934 (2020)
5. Chen, Y., Yang, X., Zhong, B., Pan, S., Chen, D., Zhang, H.: CNNTracker: online discriminative object tracking via deep convolutional neural network. Appl. Soft Comput. **38**, 1088–1098 (2016)
6. Farhadi, A., Redmon, J.: YOLOv3: an incremental improvement. In: Computer Vision and Pattern Recognition, vol. 1804. Springer, Heidelberg (2018)
7. Fayaz, S., Parah, S.A., Qureshi, G., Kumar, V.: Underwater image restoration: a state-of-the-art review. IET Image Proc. **15**(2), 269–285 (2021)
8. Hu, J., Shen, L., Sun, G.: Squeeze-and-excitation networks. In: Proceedings of the IEEE Conference on Computer Vision and Pattern Recognition, pp. 7132–7141 (2018)
9. Jocher, G., et al.: Ultralytics/yolov5: v7. 0-yolov5 sota realtime instance segmentation. Zenodo (2022)
10. Kisantal, M., Wojna, Z., Murawski, J., Naruniec, J., Cho, K.: Augmentation for small object detection. arXiv preprint arXiv:1902.07296 (2019)
11. Krizhevsky, A., Sutskever, I., Hinton, G.E.: ImageNet classification with deep convolutional neural networks. In: Advances in Neural Information Processing Systems, vol. 25 (2012)
12. Li, F., Chen, H., Liu, Z., Zhang, X., Wu, Z.: Fully automated detection of retinal disorders by image-based deep learning. Graefes Arch. Clin. Exp. Ophthalmol. **257**, 495–505 (2019)
13. Lin, T.Y., Dollár, P., Girshick, R., He, K., Hariharan, B., Belongie, S.: Feature pyramid networks for object detection. In: Proceedings of the IEEE Conference on Computer Vision and Pattern Recognition, pp. 2117–2125 (2017)
14. Lin, T.Y., Goyal, P., Girshick, R., He, K., Dollár, P.: Focal loss for dense object detection. In: Proceedings of the IEEE International Conference on Computer Vision, pp. 2980–2988 (2017)
15. Liu, Y., Sun, P., Wergeles, N., Shang, Y.: A survey and performance evaluation of deep learning methods for small object detection. Expert Syst. Appl. **172**, 114602 (2021)
16. Long, J., Shelhamer, E., Darrell, T.: Fully convolutional networks for semantic segmentation. In: Proceedings of the IEEE Conference on Computer Vision and Pattern Recognition, pp. 3431–3440 (2015)

17. Pang, J., Li, C., Shi, J., Xu, Z., Feng, H.: R2CNN: fast tiny object detection in large-scale remote sensing images. IEEE Trans. Geosci. Remote Sens. **57**(8), 5512–5524 (2019)
18. Redmon, J., Farhadi, A.: Yolo9000: better, faster, stronger. In: Proceedings of the IEEE Conference on Computer Vision and Pattern Recognition, pp. 7263–7271 (2017)
19. Redmon, J., Divvala, S., Girshick, R., Farhadi, A.: You only look once: unified, real-time object detection. In: Proceedings of the IEEE Conference on Computer Vision and Pattern Recognition, pp. 779–788 (2016)
20. Ren, S., He, K., Girshick, R., Sun, J.: Faster R-CNN: towards real-time object detection with region proposal networks. In: Advances in Neural Information Processing Systems, vol. 28 (2015)
21. Sun, K., Wen, Q., Zhou, H.: Ganster R-CNN: occluded object detection network based on generative adversarial nets and faster R-CNN. IEEE Access **10**, 105022–105030 (2022)
22. Tan, M., Le, Q.: EfficientNet: rethinking model scaling for convolutional neural networks. In: International Conference on Machine Learning, pp. 6105–6114. PMLR (2019)
23. Tong, K., Wu, Y., Zhou, F.: Recent advances in small object detection based on deep learning: a review. Image Vis. Comput. **97**, 103910 (2020)
24. Wang, S., et al.: Artificial intelligence in lung cancer pathology image analysis. Cancers **11**(11), 1673 (2019)
25. Yang, Y., Zhou, Y., Din, N.U., Li, J., He, Y., Zhang, L.: An improved YOLOv5 model for detecting laser welding defects of lithium battery pole. Appl. Sci. **13**(4), 2402 (2023)
26. Yue, L., Shen, H., Li, J., Yuan, Q., Zhang, H., Zhang, L.: Image super-resolution: the techniques, applications, and future. Signal Process. **128**, 389–408 (2016)
27. Zhang, Q., Zhang, H., Lu, X.: Adaptive feature fusion for small object detection. Appl. Sci. **12**(22), 11854 (2022)

Author Index